Edited by
Roman Gr. Maev

Advances in Acoustic Microscopy and High Resolution Imaging

Related Titles

Marage, J.-P., Mori, Y.

Sonars and Underwater Acoustics

Hardcover
ISBN: 978-1-84821-189-6

Hodges, R. P.

Underwater Acoustics

Analysis, Design and Performance of Sonar

Hardcover
ISBN: 978-0-470-68875-5

Azhari, H.

Basics of Biomedical Ultrasound for Engineers

Hardcover
ISBN: 978-0-470-46547-9

Iniewski, K. (ed.)

Medical Imaging

Principles, Detectors, and Electronics

Hardcover
ISBN: 978-0-470-39164-8

Capelo-Martínez, J.-L. (ed.)

Ultrasound in Chemistry

Analytical Applications

2009
Hardcover
ISBN: 978-3-527-31934-3

Kundu, T. (ed.)

Advanced Ultrasonic Methods for Material and Structure Inspection

2007
Hardcover
ISBN: 978-1-905209-69-9

Kutz, M. (ed.)

Mechanical Engineers' Handbook

Materials and Mechanical Design

2005
Hardcover
ISBN: 978-0-471-71985-4

Oppelt, A. (ed.)

Imaging Systems for Medical Diagnostics

Fundamentals, technical solutions and applications for systems applying ionization radiation, nuclear magnetic resonance and ultrasound

2005
Hardcover
ISBN: 978-3-89578-226-8

Mix, P. E.

Introduction to Nondestructive Testing

A Training Guide

2005
Hardcover
ISBN: 978-0-471-42029-3

Hill, C. R., Bamber, J. C., ter Haar, G. R. (eds.)

Physical Principles of Medical Ultrasonics

Hardcover
ISBN: 978-0-471-97002-6

Edited by Roman Gr. Maev

Advances in Acoustic Microscopy and High Resolution Imaging

From Principles to Applictaions

WILEY-VCH Verlag GmbH & Co. KGaA

The Editor

Prof. Roman Gr. Maev
NSERC Indust. Research Chair
University of Windsor
401, Sunset Avenue
Windsor ON N9B 3P4
Canada

■ All books published by **Wiley-VCH** are carefully produced. Nevertheless, authors, editors, and publisher do not warrant the information contained in these books, including this book, to be free of errors. Readers are advised to keep in mind that statements, data, illustrations, procedural details or other items may inadvertently be inaccurate.

Library of Congress Card No.: applied for

British Library Cataloguing-in-Publication Data
A catalogue record for this book is available from the British Library.

Bibliographic information published by the Deutsche Nationalbibliothek
The Deutsche Nationalbibliothek lists this publication in the Deutsche Nationalbibliografie; detailed bibliographic data are available on the Internet at http://dnb.d-nb.de.

© 2013 Wiley-VCH Verlag & Co. KGaA, Boschstr. 12, 69469 Weinheim, Germany

All rights reserved (including those of translation into other languages). No part of this book may be reproduced in any form – by photoprinting, microfilm, or any other means – nor transmitted or translated into a machine language without written permission from the publishers. Registered names, trademarks, etc. used in this book, even when not specifically marked as such, are not to be considered unprotected by law.

Print ISBN: 978-3-527-41056-9
ePDF ISBN: 978-3-527-65533-5
ePub ISBN: 978-3-527-65532-8
mobi ISBN: 978-3-527-65531-1
oBook ISBN: 978-3-527-65530-4

Cover Design Adam-Design, Weinheim, Germany
Typesetting Toppan Best-set Premedia Limited, Hong Kong
Printing and Binding Markono Print Media Pte Ltd, Singapore

Contents

List of Contributors *XIII*
Introduction *XVII*
Author Biographies *XIX*

Part One Fundamentals *1*

1 **From Multiwave Imaging to Elasticity Imaging** *3*
Mathias Fink and Mickael Tanter
1.1 Introduction *3*
1.2 Regimes of Spatial Resolution *3*
1.3 The Multiwave Approach *4*
1.4 Wave to Wave Generation *5*
1.5 Wave to Wave Tagging *7*
1.6 Wave to Wave Imaging: Mapping Elasticity *8*
1.7 Super-resolution in Supersonic Shear Wave Imaging *14*
1.8 Clinical Applications *16*
1.9 Conclusion *19*
References *21*

2 **Imaging via Speckle Interferometry and Nonlinear Methods** *23*
Jeffrey Sadler and Roman Gr. Maev
2.1 General Introduction *23*
2.2 Part I: Speckle Interferometry *24*
2.2.1 Introduction *24*
2.2.2 Labeyrie's Method *25*
2.2.3 Knox–Thompson Method *29*
2.2.4 Importance of Phase Difference Calculation *32*
2.2.5 Labeyrie and Knox–Thompson in Two Dimensions *33*
2.2.6 Other Improvements to Speckle Interferometry *34*
2.3 Part II: Nonlinear Imaging *34*
2.3.1 Introduction *34*
2.3.2 Deviation (Difference Squared), or Absolute Difference *36*

2.3.3	Fourier Transform-Based Methodology *36*
2.3.4	Fourier Methodology: How to Create an Image *38*
2.3.5	Fourier Transform: Problems with Using *39*
2.3.6	Hilbert Transform-Based Methodology *39*
2.3.7	Hilbert Methodology: How to Create an Image, and 3D Image *42*
2.4	Summary and Closing *44*
	Selected References (By Subject) *45*
	Speckle: Base Methods *45*
	Speckle: More Advanced Methods *45*
	Nonlinear Imaging *45*

Part Two   Novel Developments in Advanced Imaging Techniques and Methods *47*

3 **Fundamentals and Applications of a Quantitative Ultrasonic Microscope for Soft Biological Tissues** *49*
Kazuto Kobayashi and Naohiro Hozumi

3.1	General Introduction: Basic Idea of an Ultrasonic Microscope for Biological Tissues *49*
3.2	Sound Speed Profile *50*
3.2.1	Fundamentals *50*
3.2.2	Specimen to be Observed *50*
3.2.3	Experimental Setup and Acquired Signal *51*
3.2.4	Calculation of Sound Speed *52*
3.2.4.1	Frequency Domain Analysis *52*
3.2.4.2	Time–Frequency Domain Analysis *54*
3.2.5	Two-Dimensional Sound Speed Profiles *56*
3.2.6	Attempts at Better Spatial Resolution *58*
3.3	Acoustic Impedance Profile *60*
3.3.1	Fundamentals *60*
3.3.2	Experimental Setup *61*
3.3.3	Specimen to be Observed *62*
3.3.4	Acquired Signal *63*
3.3.5	Calibration for Characteristic Acoustic Impedance *63*
3.3.6	Observation of Cerebellar Cortex of a Rat *65*
3.3.7	Cell Size Observation *67*
3.3.8	Commercialized Equipment *69*
3.4	Summary *70*
	References *70*

4 **Portable Ultrasonic Imaging Devices** *71*
Sergey A. Titov, Roman Gr. Maev, and Fedar M. Severin
References *91*

5	**High-Frequency Ultrasonic Systems for High-Resolution Ranging and Imaging** *93*	

Michael Vogt and Helmut Ermert

- 5.1 General Introduction *93*
- 5.2 High-Frequency Ultrasonic System Components *94*
- 5.2.1 Ultrasound Echo Systems *94*
- 5.2.2 Transmitter and Receiver Components for High-Frequency Ultrasonic Echo Systems *95*
- 5.2.3 Spectral and Range Resolution Properties *97*
- 5.2.4 Measurement and Optimization of the Pulse Transfer Properties *99*
- 5.2.5 Range Resolution Optimization: Inverse Echo Signal Filtering *101*
- 5.2.6 Measurement of Acoustic Scattering Parameters in Plane Wave Propagation *102*
- 5.3 Engineering Concepts for High-Frequency Ultrasonic Imaging *104*
- 5.3.1 Single-Element Transducer B-Scan Techniques *104*
- 5.3.2 Lateral Resolution Optimization *105*
- 5.3.2.1 B/D-Scan Technique *106*
- 5.3.2.2 Synthetic Aperture Focusing Techniques (SAFT) *106*
- 5.3.3 Limited Angle Spatial Compounding (LASC) *110*
- 5.3.4 Multidirectional Tissue Characterization *112*
- 5.4 High-Frequency Ultrasound Imaging in Biomedical Applications *115*
- 5.4.1 Skin Imaging *115*
- 5.4.2 Imaging of Small Animals *117*
- 5.5 Summary *118*
- References *119*

6 Quantitative Acoustic Microscopy Based on the Array Approach *125*

Sergey Titov and Roman Gr. Maev

- 6.1 General Introduction *125*
- 6.2 Measurement of Velocity and Attenuation of Leaky Waves *126*
- 6.3 Measurement of Bulk Wave Velocities and Thickness of Specimen *141*
- 6.4 Conclusions *150*
- References *150*

Part Three   Advanced Biomedical Applications *153*

7 Study of the Contrast Mechanism in an Acoustic Image for Thickly Sectioned Melanoma Skin Tissues with Acoustic Microscopy *155*

Bernhard R. Tittmann, Chiaki Miyasaka, Elena Maeva, and David Shum

- 7.1 Introduction *155*
- 7.1.1 What Is Melanoma? *155*
- 7.1.2 How Is Melanoma Diagnosed? *156*

7.1.3	Present Problems for Biopsy	*157*
7.1.4	Objective of Present Study	*157*
7.2	Physical and Mathematical Modeling for Five Layer Wave Propagation in an Acoustic Microscope	*158*
7.3	Sample Preparation	*162*
7.4	Digital Imaging – Optical and Ultrasonic	*163*
7.4.1	Optical Image	*163*
7.4.2	Acoustic Imaging Principle (Pulse-Wave Mode)	*164*
7.4.3	Resolution	*168*
7.4.4	Acoustic Images	*169*
7.4.5	Waveform Analysis	*171*
7.5	High Frequency Acoustic Microscopy	*174*
7.5.1	Normal Control Skin Tissue	*174*
7.5.2	Abnormal Skin Tissue	*175*
7.5.3	Acoustic Velocity	*175*
7.5.4	Computer Simulation	*177*
7.5.4.1	Experimental $V(z)$ Curve	*177*
7.5.4.2	Theoretical $V(z)$ Curve (Simulation of $V(z)$ Curve)	*178*
7.6	Conclusions	*181*
	Acknowledgment	*183*
	References	*183*

8 New Concept of Pathology – Mechanical Properties Provided by Acoustic Microscopy *187*
Yoshifumi Saijo

8.1	Introduction	*187*
8.2	Principle of Acoustic Microscopy	*188*
8.3	Application to Cellular Imaging	*189*
8.4	Application to Hard Tissues	*191*
8.5	Application to Soft Tissues	*193*
8.5.1	Gastric Cancer	*193*
8.5.2	Myocardial Infarction	*195*
8.5.3	Kidney	*197*
8.5.4	Atherosclerosis	*197*
8.6	Ultrasound Speed Microscopy (USM)	*200*
8.7	Articular Tissues	*202*
8.8	Summary	*202*
	References	*204*

9 Quantitative Scanning Acoustic Microscopy of Bone *207*
Pascal Laugier, Amena Saïed, Mathilde Granke, and Kay Raum

9.1	Introduction	*207*
9.1.1	Hierarchical Structure of Bone and Properties	*207*
9.1.2	Relevance of Multiscale Elastic Properties	*209*
9.1.3	History of Measurement Principles	*210*

9.2	Quantitative SAM-Based Impedance of Bone	*213*
9.2.1	Theory	*213*
9.2.2	Time-Resolved Measurements	*216*
9.2.3	Measurements with Time-Gated Amplitude Detection	*217*
9.2.3.1	Calibration	*218*
9.3	Tissue Mineralization, Acoustic Impedance, and Stiffness	*219*
9.4	Elastic Anisotropy at the Nanoscale (Lamellar) Level	*222*
9.5	Elastic Anisotropy at the Microscale (Tissue) Level	*223*
9.6	Applications in Musculoskeletal Research	*225*
9.7	Conclusions	*226*
	References	*228*

Part Four Advanced Materials Applications *231*

10 Array Imaging and Defect Characterization Using Post-processing Approaches *233*
Alexander Velichko, Paul D. Wilcox, and Bruce W. Drinkwater

10.1	Introduction	*233*
10.2	Modeling Array Data	*237*
10.2.1	Introduction	*237*
10.2.2	Ray-Based Description of Ultrasonic Array Data	*238*
10.2.2.1	Determining the Ray-Paths	*238*
10.2.2.2	Predicting the Signal Associated with a Ray-Path	*240*
10.2.2.3	Simple Example	*240*
10.2.3	Mathematical Model of Ultrasonic Array Data	*242*
10.3	Imaging with 1D Arrays	*245*
10.3.1	Classical Beam-Forming Imaging Methods in Post-processing	*245*
10.3.2	Total Focusing Method	*246*
10.3.3	Wavenumber Method	*247*
10.3.4	Back-Propagation Method	*249*
10.3.5	Theoretical Comparison of Imaging Methods	*250*
10.3.6	Computational Burden	*251*
10.3.7	Focusing Performance	*252*
10.3.8	Experimental Example	*253*
10.4	Imaging with 2D Arrays	*255*
10.4.1	Optimization of 2D Array Layout	*255*
10.4.1.1	Optimization Criterion	*255*
10.4.1.2	Regular Sampling	*256*
10.4.1.3	Non-uniform Sampling	*257*
10.4.2	Experimental Comparison of 2D Array Layouts	*258*
10.4.2.1	Spherical Inclusion	*259*
10.4.2.2	Aluminum Block with Flat Bottom Holes	*260*
10.4.2.3	Surface-Breaking Fatigue Crack	*260*
10.5	Scattering Matrices and Their Experimental Extraction	*260*

10.5.1	Feature Extraction from Array Data *262*
10.5.1.1	Concept *262*
10.5.1.2	Inverse Imaging *263*
10.5.1.3	Extraction of Scattering Matrix *266*
10.6	Defect Characterization and Sizing *267*
10.6.1	Crack Sizing *267*
10.6.1.1	1D Array *267*
10.6.1.2	2D Array *268*
10.6.2	Experimental Results *269*
10.6.2.1	1D Array *269*
10.6.2.2	2D Array *271*
10.7	Conclusions *272*
	References *273*

11	**Ultrasonic Force and Related Microscopies** *277*
	Andrew Briggs and Oleg V. Kolosov
11.1	Introduction *277*
11.2	Mechanical Diode Detection *279*
11.3	Experimental UFM Implementation *280*
11.4	UFM Contrast Theory *283*
11.5	Quantitative Measurements of Contact Stiffness *287*
11.6	UFM Picture Gallery *289*
11.7	Image Interpretation–Effects of Adhesion and Topography *293*
11.8	Superlubricity *295*
11.9	Defects Below the Surface *297*
11.10	Time-Resolved Nanoscale Phenomena *299*
	Acknowledgments *303*
	References *304*

12	**Ultrasonic Atomic Force Microscopy** *307*
	Kazushi Yamanaka and Toshihiro Tsuji
12.1	Introduction *307*
12.2	Principle *307*
12.2.1	Forced Vibration of Cantilever from the Base *307*
12.2.2	Quantitative Information, Directional Control, and Resonance Frequency Tracking *308*
12.2.3	Effective Enhancement of Cantilever Stiffness *309*
12.2.4	Criterion to Avoid Plastic Deformation *309*
12.3	Theory *311*
12.3.1	Overview *311*
12.3.2	Linear Analysis of Stiffness and the Q Factor *312*
12.3.3	Linear Theory of Subsurface Imaging *314*
12.3.4	Advantage of Appropriate Load *316*
12.3.5	Nonlinear Analysis of Spectra *316*
12.3.6	Duffing Model *318*

12.3.7	Numerical Model with Double Nodes	*319*
12.4	Instrumentation	*320*
12.5	Experiments	*322*
12.5.1	Effort to Avoid Nonlinearity at Tip–Sample Contact	*322*
12.5.2	Relation between UAFM and UFM	*323*
12.5.3	Quantitative Evaluation of Elasticity	*324*
12.6	Observation of Defects in Layered Materials	*325*
12.6.1	Defects in Graphene Sheets	*325*
12.6.2	Dislocation in Molybdenum Disulfide	*328*
12.6.3	Observation of Dislocation Behavior under Different Loads	*329*
12.6.4	Analysis of Dislocation Motion under Varying Applied Load	*331*
12.6.5	Model for the Reversible Long-Range Motion of Dislocation	*333*
12.6.6	Delamination in Microelectronic and Mechanical Devices	*334*
12.7	Conclusion	*335*
	References	*336*
13	**Acoustical Near-Field Imaging**	*339*
	Walter Arnold	
13.1	Principle of Near-Field Imaging	*339*
13.1.1	Early Systems of Acoustical Near-Field Imaging	*339*
13.2	Near-Field Acoustical Imaging and Atomic Force Microscopy	*342*
13.2.1	Force Modulation	*343*
13.2.2	Local Acceleration Microscopy	*344*
13.2.3	Pulsed-Force Microscopy	*345*
13.2.4	Atomic Force Acoustic Microscopy or AFM Contact-Resonance Imaging	*345*
13.2.4.1	Principle of Operation	*345*
13.2.4.2	Flexural Cantilever Resonances	*346*
13.2.4.3	Relationship of Contact Stiffness to Indentation Modulus	*350*
13.2.4.4	Torsional Resonances	*356*
13.2.4.5	Piezo-mode Imaging	*357*
13.2.4.6	Nonlinear Contact Resonances and Related Phenomena	*358*
13.2.4.7	Subsurface Imaging Using Contact Resonances	*359*
	Acknowledgment	*362*
	References	*362*

Index *371*

List of Contributors

Walter Arnold
Saarland University
Department of Material Science and
Technology
Campus D 2.2
66123 Saarbrücken, Germany
and
Göttingen University
1. Phys. Institut
Friedrich-Hund Platz 1
37077 Göttingen, Germany

Andrew Briggs
Oxford University
Department of Materials
16 Parks Road
OX1 3PH Oxford, UK

Bruce W. Drinkwater
University of Bristol
Faculty of Engineering
University Walk
Bristol BS8 1TR, UK

Helmut Ermert
Ruhr-Universität Bochum
Department of Electrical Engineering
and Information Technology
High Frequency Engineering Research
Group
Building ID 03/343
44780 Bochum, Germany

Mathias Fink
Ecole Supérieure de Physique et de
Chimie Industrielles de la Ville de
Paris
CNRS
INSERM
Institut Langevin
10 rue Vauquelin
75005 Paris, France

Mathilde Granke
Université Pierre et Marie Curie
CNRS UMR 7623
Laboratoire d'Imagerie Paramétrique
15 rue de l'ecole de médecine
75006 Paris, France

Naohiro Hozumi
Toyohashi University of Technology
1-1 Hibarigaoka, Tempaku-cho
Toyohashi 441-8580, Japan

Kazuto Kobayashi
Honda Electronic Co., Ltd.
20 Oyamazuka, Oiwa-cho
Toyohashi 980-857, Japan

Oleg V. Kolosov
Lancaster University
Department of Physics
Room A30, Physics Building
Bailrigg, LA1 4YW Lancaster, UK

Pascal Laugier
Université Pierre et Marie Curie
CNRS UMR 7623
Laboratoire d'Imagerie Paramétrique
15 rue de l'ecole de médecine
75006 Paris, France

Roman Gr. Maev
University of Windsor
Institute for Diagnostic Imaging
Research
401 Sunset Avenue
Windsor, ON N9B3P4, Canada

Chiaki Miyasaka
Pennsylvania State University
Department of Engineering Science
and Mechanics
212 Earth-Engineering Sciences
Building
University Park, PA 16802, USA

Kay Raum
Universitätsmedizin Berlin
Julius-Wolff Institut & Berlin-Brandenburg School for Regenerative
Therapies, Charité, Augustenburger
Platz 1
13353 Berlin, Germany

Jeffrey Sadler
University of Windsor
Institute for Diagnostic Imaging
Research
401 Sunset Avenue
Windsor, ON N9B 3P4, Canada

Amena Saïed
Université Pierre et Marie Curie
CNRS UMR 7623
Laboratoire d'Imagerie Paramétrique
15 rue de l'ecole de médecine,
75006 Paris, France

Yoshifumi Saijo
Tohoku University
Biomedical Imaging Laboratory
Graduate School of Biomedical
Engineering
4-1 Seiryomachi
Aoba-ku, Sendai 980-8575, Japan

Fedar M. Severin
University of Windsor
Institute for Diagnostic Imaging
Research
401 Sunset Avenue
Windsor, ON N9B3P4, Canada

David Shum
Hotel Dieu Grace Hospital
Windsor Regional Hospital
Leamington District Memorial
Hospital
1995 Lens Avenue
Windsor, ON N8W 1L9, Canada

Mickael Tanter
Ecole Supérieure de Physique et de
Chimie Industrielles de la Ville de
Paris
CNRS
INSERM
Institut Langevin
10 rue Vauquelin
75005 Paris, France

Sergey A. Titov
Russian Academy of Sciences
N.M. Emanuel Institute of
Biochemical Physics
4 Kosygin st.
Moscow 119344, Russia

Sergey Titov
Russian Academy of Sciences
N.M. Emanuel Institute of
Biochemical Physics
4 Kosygin st.
Moscow 119344, Russia

Bernhard R. Tittmann
Pennsylvania State University
Department of Engineering Science
and Mechanics
212 Earth-Engineering Sciences
Building
University Park, PA 16802, USA

Toshihiro Tsuji
Tohoku University
Department of Materials Processing
Aoba 6-6-02, Aoba-ku
Miyagi, Sendai 980-8579, Japan

Alexander Velichko
University of Bristol
Faculty of Engineering
Room 1.2, Queen's Building,
University Walk
Clifton, Bristol BS8 1TR, UK

Michael Vogt
Ruhr-Universität Bochum
Department of Electrical Engineering
and Information Technology
High Frequency Engineering Research
Group
Building ID 03/340
44780 Bochum, Germany

Paul D. Wilcox
University of Bristol
Faculty of Engineering
Room 2.57, Queen's Building,
University Walk
Clifton, Bristol BS8 1TR, UK

Kazushi Yamanaka
Tohoku University
Department of Materials Processing
Aoba 6-6-02, Aoba-ku
Miyagi, Sendai 980-8579, Japan

Introduction

While a picture is worth a thousand words, in science a single image is often problematic. Imaging technology is largely based on manipulating optical waves, but since optics does not provide all of the information we need, in the twentieth century we turned to other technologies. Acoustic imaging is now an integral and important part of our continuing effort to extend our ability to "see." Although ultrasonic images do not provide the fine details found in magnetic resonance images (MRI) or X-ray methods, the acoustic imaging system provides significant information at one-tenth the cost of MRI, with the added advantage of being completely safe for the patient's health. This form of imaging is particularly useful for obtaining data from inside the human body, for delineating the interfaces between solid and spaces in muscles and soft tissues. Ultrasound renders live images where the operator can dynamically select the most important sections for documenting the changes in structure without long-term side effects in the patient. The introduction of high-resolution acoustic imaging systems in the early 1960s facilitated the examination of the internal microstructure of nontransparent solids and the monitoring of internal stress. In addition to measuring elastic properties, this technique is also used to examine adhesion in multilayered structures and has many other applications. Acoustic microscopy has become not only a new imaging method extensively used in many areas of physics, biology, and technology but also a new efficient tool of quantitative characterization of the microstructure of various species and materials.

The role of high-resolution ultrasonic imaging in academic studies of condensed matter and various applications for microstructural material characterization in physics, biology, and technology is rapidly increasing. The whole spectrum of original physical and methodological approaches to ultrasonic imaging results in a significant improvement in the quality of developed technology. New generations of ultrasonic imaging system devices continue to decrease in size and will soon enter the realm of pocket-sized dimensions. New transducer materials, including advanced composites and recent MEMS applications to novel array solutions, also contribute to substantial changes in the design of ultrasonic imaging systems.

The goal of this book is to provide an overview of recent advances in high-resolution ultrasonic imaging techniques and their applications to biomaterials evaluation and industrial materials. In this book we were lucky to bring together a unique collection of papers presenting novel results and techniques that were developed by leading research groups worldwide.

Novel physical solutions, including new results in the field of adaptive methods and inventive approaches to inverse problems, original concepts based on high harmonic imaging algorithms, and intriguing vibro-acoustic imaging and vibro-modulation technique, have been successfully introduced and verified in numerous studies of industrial materials and biomaterials in the last few years. Together with the above-mentioned traditional academic and practical avenues in ultrasonic imaging research, intriguing scientific discussions have recently surfaced in various fields and will hopefully continue to bear fruit in the future.

This book offers several new results from well-known authors who are engaged in aspects of the development of novel physical principles, new methods, or implementation of modern technological solutions into current imaging devices and new applications of high-resolution imaging systems. I believe that this book will help encourage more research and development in the field to realize the great potential of high-resolution acoustic imaging and its various industrial and biomedical applications. We have also included a biography of every contributor to this book, through which you may be able to trace the progression and future direction of this field. I sincerely hope that you will enjoy reading about these exciting research results.

In closing, I am grateful to all my colleagues, the distinguished contributors to this book, and the co-authors who shared their results and insights, thus lending a unique perspective and voice to this book. I would also like to thank Sabina Baroniciu for her invaluable assistance in the preparation of this book, especially with the arduous task of putting together the collection of all the manuscripts from each respected author. But she did it and did it amazingly well!

Undoubtedly, I would not have been able to work on this book without the support of my family, without the understanding and patience of my wife, Elena Maeva, and my children, Anna and Grigori, who forgave my inattention to them and my preoccupation with the work on this book.

Many thanks to you all

Windsor, Ontario, Canada *Roman Gr. Maev*
2 July 2012

Author Biographies

Walter Arnold received a diploma in physics in 1970 and a PhD in solid state physics in 1974, both from the Technical University Munich, Germany. He then held various positions as a postdoctoral researcher and scientific staff member at the CNRS, Grenoble, France and the Max-Planck-Institute for Solid State Physics, Stuttgart, Germany, the IBM T.J. Watson Research Center, Yorktown Heights, NY, and the Brown Boveri Research Centre, Baden, Switzerland, working on low temperature physics, solid state physics, and applied physics. From 1980 until his retirement in December 2007 he was employed at the Fraunhofer-Institute for Non-Destructive Testing, Saarbrücken, as head of the research department. Parallel to this position, Dr. Arnold was appointed professor of materials technology at the University of Saarbrücken. Since his retirement, he has continued research work with colleagues at the Saarland University and as a guest professor at the 1. Physikalische Institut, Universität Göttingen, Germany.

Dr. Arnold has authored and co-authored about 300 papers including 170 peer-reviewed papers. He has supervised 31 PhD theses and approximately 150 master and diploma students.

Andrew Briggs received his PhD from the Cavendish Laboratory, Cambridge, in 1976. He came to Oxford University as a Research Fellow in 1980 and was appointed a University Lecturer in 1984. He wrote the definitive monograph *Acoustic Microscopy*, which was published by Oxford University Press (OUP) in 1992. The second edition, with a new chapter on acoustically excited probe microscopy written with Oleg Kolosov, was published by OUP in 2010. For his pioneering work in applications of acoustic and scanned probe microscopy he was elected Honorary Fellow of the Royal Microscopical Society in 1999. In 2002 he was appointed professor of nanomaterials at Oxford. From 2002 to 2009 he was Director of the Quantum Information Processing Interdisciplinary Research Collaboration. His current research interests focus on carbon nanomaterials for quantum technologies.

Bruce Drinkwater (PhD, CEng, FIMechE, FInstNDT, DIC) was born in Hexham, England in 1970. He received B.Eng and PhD degrees in mechanical engineering from Imperial College, London, England in 1991 and 1995, respectively. From 1996 to the present he worked as an academic in the Mechanical Engineering Department at the University of Bristol, England. During this time he has published over 80 journal articles on a range of topics connected with ultrasonics and non destructive evaluation.

Between 2000 and 2005 he was an EPSRC Advanced Research Fellow researching ultrasonic wheel probes and the ultrasonic measurement of adhesive joints, thin layers, and interfaces. During this period both his work on array-wheel probes and on bearing condition monitoring was commercialized. He was promoted to professor of ultrasonics in 2007. He currently leads a large collaborative research program that aims to develop ultrasonic array devices for the manipulation of biological particles for applications such as tissue engineering.

Helmut Ermert received a Dipl.-Ing. degree in electrical engineering and a Dr.-Ing. degree from the Technical University (RWTH) Aachen, Germany in 1965 and 1970, respectively. In 1975 he received a Dr.-Ing. habil. degree (Habilitation) from the Engineering Faculty at the University of Erlangen-Nuremberg, Germany.

From 1966 to 1970 he worked on millimeter wave and microwave engineering at the Technical University (RWTH) Aachen. From 1970 to 1975 he was involved in teaching and research in microwave integrated circuits, microwave ferrites, and microwave measurement techniques at the University Erlangen-Nuremberg. From 1978 to 1987 he was a professor of electrical engineering in Erlangen working on microwave and acoustic imaging using various fields and waves (ultrasound, microwaves, thermal waves, and eddy current fields) for diagnostic purposes in medicine and engineering. Since 1987 he has been a professor of electrical engineering and Director of the Institute of High Frequency Engineering at the Ruhr-University in Bochum, Germany.

At present, he is continuing research on measurement techniques, diagnostic imaging, and sensors in the RF and microwave area as well as in the ultrasonic area for applications in medicine, nondestructive testing, and industry.

Mathias Fink received an MS degree in mathematics from Paris University, France, in 1967, and a PhD degree in solid state physics in 1970. He then moved to medical imaging and received the Doctorates-Sciences degree in 1978 from Paris University. His Doctorates-Sciences research was in the area of ultrasonic focusing with transducer arrays for real-time medical imaging.

Dr. Fink is a professor of physics at the Ecole Superieure de Physique et de Chimie Industrielles de la Ville de Paris (ESPCI ParisTech), Paris, France. In 1990 he founded the Laboratory Ondeset Acoustique at ESPCI, which became the Langevin Institute in 2009. In 2002, he was elected to the French Academy of Engineering, in 2003 to the French Academy of Science, and in 2008 at the College de France to the Chair of Technological Innovation.

Dr. Fink's area of research is concerned with the propagation of waves in complex media and the development of numerous instruments based on this basic research. His current research interests include time-reversal in physics, super-resolution, metamaterials, medical ultrasonic imaging, ultrasonic therapy, multi-wave imaging, acoustic smart objects, acoustic tactile screens, underwater acoustics, geophysics, and telecommunications. He has developed different techniques in medical imaging (ultrafast ultrasonic imaging, transient elastography, supersonic shear imaging), wave control, and focusing in complex media with time-reversal mirrors. He holds more than 55 patents and has published more than 350 peer-reviewed papers and book chapters.

Mathilde Granke obtained her Engineer Diploma in computational structural mechanics from the Ecole Centrale Nantes, France in 2007; she then spent six months as a visiting scholar in the Department of Mechanical and Industrial Engineering at the University of Illinois in Chicago. She received a Master's degree in biomechanics from the Ecole des Arts et Métiers Paris in 2008. Since 2008, she has joined the Laboratoire d'Imagerie Paramétrique at University Pierre et Marie Curie, Paris VI, France as a PhD student.

For the past four years, her area of interest has been focused on bone mechanics, with particular applications in biomaterials, bone numerical modeling, and bone pathologies. Her current research is focused on the relationship between the material properties and mechanical characteristics of bone tissue evaluated by scanning acoustic microscopy.

Naohiro Hozumi was born in Kyoto, Japan on April 2, 1957. He received his BS, MS, and PhD degrees in 1981, 1983, and 1990, respectively, from Waseda University. He was employed by the Central Research Institute of Electric Power Industry (CRIEPI) from 1983 to 1999. He was an associate professor of Toyohashi University of Technology from 1999 to 2006, and a professor at the Aichi Institute of Technology from 2006 to 2011. Since 2011, he has been a professor at Toyohashi University of Technology.

He has been engaged in the research of insulating materials and diagnosis for high voltage equipment, as well as the use of acoustic measurement for biological and medical applications.

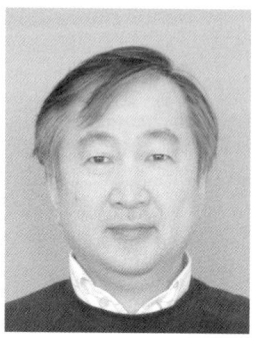

Kazuto Kobayashi was born in Aichi, Japan on June 8, 1952. He received a BS degree in electrical engineering from Shibaura Institute of Technology, Tokyo, Japan in 1976. He is currently a director of the Department of Research and Development at Honda Electronics Co. Ltd. in Toyohashi, Japan. His research activities and interests include medical ultrasound imaging, signal processing, and high frequency ultrasound transducers.

Oleg Kolosov received his PhD from the Moscow Institute of Physics and Technology in 1989, and conducted research at the Russian Academy of Sciences in the group of Roman Maev, authoring his first patents in the field of acoustic microscopy. In 1991 he became a Fellow of the Science and Technology Agency of Japan, working at the Mechanical Engineering Laboratory and Joint Research Centre of Atom Technology, where he filed first patents on ultrasonic force microscopy jointly with Kazushi Yamanaka. He joined the Materials Department at Oxford University as a Research Fellow in 1994, and from 1996 as an Advanced Fellow of Engineering and Physical Sciences Research Council. At Oxford he continued his exploration of ultrasound in scanning probe microscopy, which included the development of time resolved and nanoscale subsurface imaging using ultrasonic scanning force microscopy. In 2000 he was appointed a Group Leader and in 2002 a Director of Innovation of Symyx Technologies—a world pioneer in combinatorial materials discovery, and authored more than 45 patent applications worldwide in this field. In 2006 he was appointed a Reader in Experimental Condensed Matter Physics at Lancaster University.

His research interests span the characterization of microscale and nanoscale properties of materials, combinatorial methods for materials discovery, and nanomechanical quantum sensors.

Pascal Laugier holds a PhD in physical acoustics from the University Denis Diderot in Paris, France. He is currently a Research Director at the French National Scientific Research Center (CNRS) and is head of the Laboratory of Parametric Imaging at University Pierre and Marie Curie, Paris, France. Laugier has 15 years experience in osteoporosis research and more than 25 years experience in ultrasonic biomedical imaging science, developing high frequency imaging and applying tissue characterization techniques to various fields of medicine such as skin, cartilage, eye, and bone.

He was involved in several European research projects and has been an investigator with the European Space Agency on a Microgravity Research Program. He serves as a reviewer of more than 30 scientific journals. He also serves as an Expert Reviewer for major national and international institutions.

He has co-authored over 150 articles in peer-reviewed journals, 160 Conference proceedings papers, and over 750 conference abstracts. He holds 12 patents, all of them in the field of ultrasound imaging for medical applications.

Roman Gr. Maev was born in Moscow, Russia. He received his Master of Science degree in theoretical nuclear physics from the Moscow Physical Engineering Institute followed by a PhD in the "Theory of Semiconductors" from the Physical P.N. Lebedev Institute of the USSR Academy of Sciences. In 1990 he received a Fellowship from Gore-Chernomirdin and as a result successfully attended a course project for the Scientific Business Management Fund for one semester of study at the Harvard Business School (Boston, USA). In 2001 he received a DSc degree from the Russian Academy of Sciences, and in 2005 he received a Full Professor diploma in Physics from the Government of the Russian Federation.

Dr. Roman Gr. Maev is the founding director-general of The Institute for Diagnostic Imaging Research—a multi-disciplinary, collaborative research and innovation consortium with one of its directions in nanotechnology. Dr. Maev is also a full faculty professor in the Department of Physics at the University of

Windsor, Canada, and in 2007 was granted the title of University Professor Distinguished.

The diverse range of disciplines encompassed by Dr. Maev includes theoretical fundamentals of physical acoustics, experimental research in ultrasonic and nonlinear acoustical imaging, nanostructural properties of advanced materials and its analysis. He is the author of four monographs, editor and co-editor of nine books, has published over 350 articles in leading international journals, and holds 23 international patents.

Elena Maeva received bachelor's and master's degrees in 1978 and 1980, respectively, from the Moscow D. Mendeleev Chemical-Technological University. Subsequently, in 1997 she received her PhD in physics and chemistry at the Institute of Chemical Physics within the Russian Academy of Sciences.

Dr. Maeva is an associate professor in physics (cross-appointed with chemistry and bio-chemistry) at the University of Windsor, Windsor, Canada. Her research is mostly related to the area of applied physics and chemistry. She is deeply involved in the application of advanced non-destructive methods of evaluation for the investigation of the structure of different materials and tissues. Her research is focused in three main areas: adhesive and nanocomposite structures, the study of the properties and degradation process of bio and ecology clean material based composites, and investigation of hard and soft biological tissue in biomedical projects.

During her academic career Dr. Maeva has published 26 articles in peer-reviewed journals, 66 articles in peer-reviewed conference proceedings, and has presented invited talks at major national and international conference and symposiums. She also holds seven patents.

Chiaki Miyasaka is a Dr. of Engineering (received from Tokyo Institute of Technology in1996) and adjunct professor in the Department of Engineering Science and Mechanics at Pennsylvania State University.

He is interested in developing sensors (e.g., acoustic lens operating at an ultra high frequency) and applications (e.g., mathematical modeling and novel experimental method) in scanned image microscopy, that is, acoustic microscopy, atomic force microscopy, laser scanning microscopy, scanning electron microscopy, and the like, in the field of biomedical physics. His recent research activities in the biomedical field are focused on medical ultrasonic imaging relating to skin and breast cancers.

Kay Raum graduated from the Martin-Luther-University of Halle-Wittenberg with Diploma and PhD degrees in physics in 1997 and 2002, respectively. From 1995 to 1996 he was with the Bioacoustics Research Laboratory at the University of Illinois at Urbana-Champaign as a Visiting Scholar. From 1997 until 2003 he was a research assistant at the Medical Faculty of the Martin Luther University. In 2004 he received a post-doctoral fellowship from the French National Center of Scientific Research (CNRS) and joined the Laboratoire d'Imagerie Paramétrique at University Pierre et Marie Curie, Paris, France. In 2006 he became the Research Head of the Interdisciplinary Center for Musculoskeletal Diseases and in 2008 he received his Habilitation in "Experimental Orthopedics" at the Medical Faculty of the Martin Luther University. Since 2008 he has been a professor of engineering at the Berlin-Brandenburg Graduate School for Regenerative Therapies, and Head of the Ultrasound Biomicroscopy group of the Julius-Wolff Institute at Charité-Universitätsmedizin Berlin.

He has been working with high frequency ultrasound for more than 15 years, and he has contributed specifically to the establishment and validation of quantitative acoustic microscopy in bone research. His current research is focused on the development of innovative parametric imaging techniques and their application in musculoskeletal research.

Jeffrey Sadler received his Hon. BSc degree in physics from the University of Guelph, and MSc and PhD degrees in physics from the University of Windsor. Recently, he joined the Institute for Diagnostic Imaging Research as a Post Doctoral Fellow. Past research has involved various situations in the area of acoustics, including computer simulations of acoustic waves in various plate structures, calculating the acoustical properties of composite materials, and acoustical imaging though complicated structures.

Amena Saïed received a PhD degree in physics from the University of Science and Technology, Montpellier, France in 1985. For several years she has been involved in the development of VHF acoustic microscopy and specific transducers for nondestructive testing of materials. In 1990, she joined the research department of Schlumberger Industrie (France) where she was in charge of the development of new gas flow-metering techniques using ultrasound. She joined the CNRS (Centre National de la Recherche Scientifique) in 1992 as a research scientist working in the Laboratoire d'Imagerie Paramétrique UMR 7623 (L.I.P) at the University Pierre et Marie Curie in Paris. She was the head of the high frequency ultrasound group. Her research program included technological innovations in the fields of biomedical imaging and the development of new methods of ultrasound signal and image processing for high resolution and quantitative evaluation of tissue composition and pathologies. In particular, she was involved in the development of three-dimensional, high frequency quantitative ultrasonography of eye and articular cartilage.

Dr. Saïed is currently continuing research at L.I.P. on topics including scanning acoustic microscopy of bone tissue and microbubble-mediated sonoporation. Her general research interests include biomedical imaging, tissue characterization, high-frequency transducers, acoustic microscopy, and microbubble-mediated sonoporation for intracellular gene delivery.

Yoshifumi Saijo was born in Yokohama, Japan on July 21, 1962. He received M.D. and PhD degrees in 1988 and 1993, respectively, from Tohoku University. He is currently a professor of the Biomedical Imaging Laboratory at the Graduate School of Biomedical Engineering of Tohoku University. He is concurrently engaged with the Graduate School of Medical Sciences, School of Engineering, Institute of Development, Aging and Cancer of Tohoku University and the Department of Cardiovascular Surgery of Tohoku University Hospital.

His main research interests are high frequency ultrasonic imaging of biological tissues and cells, parametric imaging of intravascular ultrasound, blood flow dynamics imaging by echocardiography and MRI, photoacoustic imaging of biological tissues, and mobile ultrasonic imaging by developing portable ultrasound devices.

In 1997 he received an award for his outstanding research paper in *Ultrasound in Medicine and Biology*. He is a member of The Japan Society of Ultrasonics in Medicine, Japanese Society of Echocardiography, and Japan Circulation Society.

Fedar Seviaryn was born in 1963 in Belarus. He received combined BSc and MSc degrees in physics from the Chair of Acoustics at Moscow State University. After defending his PhD thesis, "Nonlinear acoustical phenomena in layered structures" in 1989, he held the position of researcher at the B. I. Stepanov's Institute of Physics of National Academy of Science of Belarus. Since 1998 Dr. Seviaryn has worked as a research associate in the Department of Physics at the University of Windsor, Windsor, Canada.

As a member of the Centre for Imaging Research and Advanced Materials Characterization he participates in numerous research projects in physical acoustics and the development of ultrasonic nondestructive evaluation applications.

Dr. David Shum

David Shum graduated from the University of Hong Kong, Faculty of Medicine and completed his anatomical pathology residency training at the University of Western Ontario in 1980, becoming a Fellow of the Royal College of Physicians and Surgeons, and a Diplomat of the American Board of Pathology that year. He joined the Department of Pathology at Victoria Hospital and the Faculty of Medicine at the University of Western Ontario and was an associate professor in the Department of Pathology and the Division of Dermatology until 2000. He served as the Head of Surgical Pathology in the London Health Sciences Centre from 1995 to 2000 before becoming the senior dermatopathologist in the Department of Pathology in Vancouver General Hospital. In 2004, he became the Medical Director and Chief Pathologist in the Integrated Hospital Laboratories Service that includes Hotel Dieu Grace Hospital, Leamington District Memorial Hospital, and Windsor Regional Hospital. He is now an adjunct professor in pathology with the Schulich School of Medicine and Dentistry and he is also the Cancer Care Ontario Pathology Lead in Region 1 of the Local Health Integration Network (LHIN).

His medical practice is focused on diagnostic surgical pathology and dermatopathology. His publications include peer-reviewed articles, research papers, book chapters, and an *Atlas of Histopathology of Skin Diseases*. His current research interest is on 3D reconstruction of microscopic images and the use of ultrasound microscope in surgical pathology.

Mickael Tanter is a research professor at the French National Institute for Health and Medical Research (INSERM). For five years, he has headed the team Inserm ERL U979 "Wave Physics for Medicine" at Langevin Institute, ESPCI ParisTech, France. In 1999, he obtained his PhD degree from Paris VII University in physics.

His main activities are centered on the development of new approaches in wave physics for medical imaging and therapy. His current research interests cover a wide range of topics: elastography using shear wave imaging, high intensity focused ultrasound, ultrasonic imaging using ultrafast ultrasound scanners, adaptive beam forming, and the combination of ultrasound with optics and MRI.

Dr. Tanter holds 17patents in the field of ultrasound imaging and is the author of more than 80 technical peer-reviewed papers and book chapters.

Sergey A. Titov was born in Saratov, Russia in 1957. He received a combined BS and MS degree in physics from the Moscow State University in 1980, and a PhD degree in radio physics from the Moscow Institute of Radio Engineering, Electronics and Automation in 1991. Dr. Titov was an assistant professor with the Moscow Institute of Radio Engineering, Electronics and Automation from 1982 to 1992. In 1992, he was appointed associate professor at the same institute. Currently, he is a research associate at the Institute for Diagnostic Imaging Research, University of Windsor, Windsor, Ontario, Canada. In addition, Dr. Titov is a Senior Research Fellow of the Emanuel Institute of Biochemical Physics, Russian Academy of Sciences, Moscow, Russia.

He has published over 100 research papers and tutorials, and holds 11 patents. His research interests include quantitative acoustic microscopy, nondestructive testing, material characterization, electronics, and digital signal processing.

Bernhard R. Tittmann received his PhD from the University of California at Los Angeles in 1965. From 1966 to 1978 he held a post as Member of the Technical Staff at the North American Science Center and from 1979 to 1989 as Manager of the Materials Characterization Department at the Rockwell International Science Center. Then he joined the Engineering Science and Mechanics Department at the Pennsylvania State University at University Park, Pennsylvania as chaired full professor where in 1995 he established a Centre for Engineering Nano Characterization.

Professor Tittmann's research interests focus on the fundamentals of condensed matter, physical acoustics, ultrasonic imaging, and acoustic microscopy. He has contributed to several books, published more than 400 scientific papers, and holds six patents.

Toshihiro Tsuji was born in Shiga, Japan. He was awarded BE, ME, and PhD degrees by the Department of Materials Processing in Tohoku University in 1998, 2000, and 2003, respectively. In 2003, he worked as a JSPS postdoctoral fellow in the Mechanical Engineering Laboratory, Advanced Industrial Science, and Technology, Japan on materials characterization by ultrasonic atomic force microscopy (UAFM). Since 2004, he has worked as a research associate and, since 2006, has been an assistant professor in the Department of Materials Processing, Tohoku University.

His research interests are in materials evaluation, nondestructive testing and sensors, and scanning probe microscopy. One of his present activities is the development and application of UAFM and ball surface acoustic wave (SAW) sensors.

Alexander Velichko was born in Krasnodar, Russia, in 1975. He received a MSc degree in applied mathematics from the Kuban State University, Krasnodar, Russia, in 1998 and a PhD degree from the Rostov State University, Rostov-on-Don, Russia, in 2002. Dr. Velichko has been a researcher in the Department of Mechanical Engineering at the University of Bristol, England since 2003 and was recently appointed to a lectureship.

His current research interests include mathematical modeling of propagation and scattering of elastic waves, ultrasonic imaging using arrays, and guided waves and signal processing.

Michael Vogt was born in Hagen, Germany in 1969. He received a Dipl.-Ing. degree in electrical engineering and a Dr.-Ing. degree from the Ruhr-University Bochum, Germany, in 1995 and 2000, respectively. In 2008, he qualified as a university lecturer at the Ruhr-University Bochum (Habilitation). Since 1995, he has been working on ultrasound imaging and measurement techniques, signal and image processing, and high frequency electronics at the High Frequency Engineering Research Group of the Ruhr-University Bochum. From 2001 to 2006, Dr. Vogt led the interdisciplinary research project "High-Frequency Ultrasound" within the Ruhr Center of Excellence for Medical Engineering (KMR) Bochum, Germany. In 2007, he joined Krohne Messtechnik GmbH, Duisburg, Germany, as an R&D scientist working on ultrasonic flowmeters and electromagnetic level measurement systems.

His research interests include medical imaging systems, high frequency metrology, radar systems, and electromagnetic field simulations. Dr. Vogt is a Senior Member of the Institute of Electrical and Electronics Engineers (IEEE).

Paul D. Wilcox was born in Nottingham, England in 1971. He received an M.Eng. degree in engineering science from the University of Oxford in 1994 and a PhD from Imperial College, London, England in 1998.

From 1998 to 2002 he was a research associate in the Non-Destructive Testing (NDT) research group at Imperial College where he worked on the development of guided wave array transducers for large area inspection. From 2000 to 2002 he also acted as a consultant to Guided Ultrasonics Ltd., a manufacturer of guided wave test equipment. Since 2002 Dr. Wilcox has been at the University of Bristol (Bristol, England) where he is a professor in dynamics and an EPSRC Advanced Research Fellow.

His current research interests include long-range guided wave inspection, structural health monitoring, array transducers, elastodynamic scattering, and signal processing.

**Kazushi Yamanaka** was born in Tokyo, Japan. He was awarded BS and MS degrees by the Department of Applied Physics at the University of Tokyo in 1975 and 1977, respectively.

Since 1978, he has worked in the Mechanical Engineering Laboratory, Ministry of International Trade and Industry, Japan on materials characterization by acoustic microscopy. He obtained his PhD degree from Tohoku University in 1987. From 1987 to 1988, he was a Summit Postdoctoral Researcher at the Industrial Materials Research Institute, Canada. Since 1997, he has been a professor in the Department of Materials Processing, Tohoku University.

His research interests are materials evaluation, nondestructive testing, and sensors, using acoustic microscopy, laser ultrasound and scanning probe microscopy. Some present activities include the development and application of ultrasonic atomic force microscopy (UAFM), subharmonic phased array for crack evaluation (SPACE), and ball surface acoustic wave (SAW) sensors.

Part One
Fundamentals

1
From Multiwave Imaging to Elasticity Imaging

Mathias Fink and Mickael Tanter

1.1
Introduction

Different kinds of waves can be used to provide images of the human body. They propagate in tissues with very different wavelengths ranging from a fraction of micrometer for light, to some tenths of a millimeter for ultrasound, some centimeters for sonic shear waves, to some kilometers for low frequency electromagnetic waves. Each of these waves can provide an image whose contrast and spatial resolution depend on the way the wave interacts with tissues. For example, density and compressibility are the contrasts revealed by ultrasonic waves, while shear waves carry information on the viscoelasticity of tissues (shear modulus and viscosity). For electromagnetic waves, low frequency waves are sensitive to electrical conductivity while optical waves are sensitive to optical absorption coefficient and dielectric permittivity.

1.2
Regimes of Spatial Resolution

The spatial resolution of an image, contrary to common thinking, is not always controlled by the interrogating beam wavelength, as in conventional optical microscopy. There are indeed three different potential regimes describing wave propagation through tissues: coherent, diffusive, and near-field regimes. It is only in the coherent regime that wavelength determines resolution.

Ultrasonic waves, for example, can propagate tens of centimeters without losing their coherence. Since that distance is several orders of magnitude greater than the typical wavelength, spatial resolution in ultrasound depends on the wavelength.

In contrast, light at optical frequencies rapidly loses its coherence when propagating through opaque tissues and scattering off individual heterogeneities. The light transport mean free path l^* – about 1 mm – characterizes the distance after which light loses any memory of its initial direction. For applications like diffuse optical tomography [1] in which the propagation distances are much longer than the mean free path, the spatial resolution is on the order of the observation depth.

Most low frequency electromagnetic imaging methods correspond to the so-called near-field regime, which is characterized by an observation depth much smaller than the wavelength. An example is electrical impedance tomography. With that technique, one generates low frequency alternating currents at multiple electrodes placed on the skin and infers tissue conductivity from potential measurement at the electrodes. In this regime, detectors are able to sense the exponentially decaying evanescent waves radiated by the medium. The spatial resolution is also on the order of the observation distance independent of wavelength.

A simple classification between these three various regimes can be made by comparing the three spatial scales that control any imaging experiment, that is, observation depth z, wavelength λ, and transport mean free path l^*.

The first situation, $\lambda < z < l^*$, corresponds to the *coherent regime* with the possibility of wave focusing and is encountered in the field of ultrasonic imaging and optical computed tomography. The second situation, $\lambda < l^* < z$, corresponds to the *diffusive regime*, where the wave losses its coherence through tissue interactions. The third situation, $z < \lambda < l^*$, corresponds to *near-field imaging* and is encountered in the field of near-field optics, EMG or EEG imaging, and electrical impedance tomography. It is only in the first situation that spatial resolution depends on the wavelength while in the other two regimes the spatial resolution is on the order of the observation distance z.

In all these imaging situations, physicists have striven to reach the optimal limits of the spatial resolution associated with each kind of wave. Today, after having pushed for decades the technological limits of these modalities, physicists are facing the inherent physical limits of the contrast/resolution couple in each modality. For medical imaging and diagnosis, physicians understood rapidly that one way to overcome these limits was to combine different imaging modalities, such as PET/CT, PET/MRI or ultrasound/X-ray mammography. The basic idea of multimodality imaging, such as, for example, the combination PET/CT, is to associate the high-resolution morphological image of a first modality (CT) with an image of the second modality (PET) that is poorly resolved but provides a clinically interesting contrast (i.e., metabolic activity). However, such multimodality imaging remains extremely costly and is limited by the inherent physical limits of each separate modality.

1.3
The Multiwave Approach

A very exciting solution to avoid the use of multimodality imaging is the multiwave imaging concept. It was proposed independently by different groups in the physicist community. It consists of productively combining two very different waves – one to provide contrast, another to provide spatial resolution – to build a new kind of image. Contrary to multimodality imaging that remains the superposition of two images limited by their respective contrast/resolution couples, multiwave imaging [2] overcomes this limitation by providing a unique image of the most interesting contrast with the most interesting resolution.

Multiwave imaging can benefit from three different potential interactions between waves:

- In a first case, *the interaction of the first wave with tissues during its propagation can generate a second kind of wave*. This is the case in thermoacoustic and photoacoustic imaging [3–5] where any kind of electromagnetic wave is absorbed in some region causing a transient change in temperature that radiates an ultrasonic wave *through thermal* expansion.

- In a second case, *a first wave that carries the information about the desired contrast but either completely losses its c*oherence during propagation through tissues or has a large wavelength can be tagged locally by a second kind of wave that remains coherent and well focused. The tagging focal spot can then be steered at different locations to build a complete image. This is the case of acousto-optical imaging (or acousto-optical tomography) [6, 7] where tissue displacements induced by a focused ultrasound beam modulates the optical speckle pattern of photons traveling through tissues. An image of the optical absorption is built with the sub-millimetric resolution of the ultrasonic wave.

- In a third case, *a first wave traveling much faster than the second one can be used to produce a movie of the slow wave propagation*. This is the case of transient *elastography*, where ultrafast ultrasonic scanners can track the motion of tissues scatterers induced by the propagation of low speed shear waves. This last case is relatively unique as it allows us to observe remotely the full movie of the near field of the shear wave around each obstacle (even if these obstacles are located in the far field of the two waves). A local inversion algorithm performed on this near-field movie produces a shear elasticity image relying on a sub-millimetric resolution while the shear wavelength is centimetric.

The concept of multiwave imaging is particularly interesting for the estimation of three physical parameters that remained difficult to map up until recently with a good spatial resolution: shear modulus and shear viscosity that give access to mechanical parameters that doctors feel during palpation, optical absorption, which gives access to tissues color, and finally electrical conductivity that depends on ion concentration and mobility in tissue and on the amount of intra- and extra-cellular fluids. In this chapter we will focus mainly on the third approach, which allows mapping tissue elasticity and viscosity with high precision.

1.4
Wave to Wave Generation

All techniques based on wave/wave generation are related to some dissipative processes that transform one part of a pulsed electromagnetic energy in some transient tissue motion that radiates coherent ultrasonic waves. From the

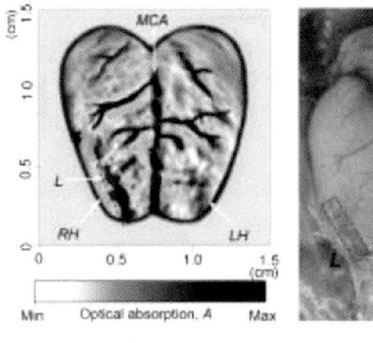

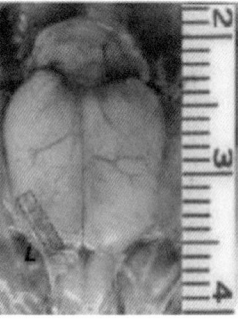

Figure 1.1 (a) Non-invasive photo-acoustic image of a superficial lesion (1 mm × 4 mm) in the right cortex on rat's cerebra acquired through skull. RH is the right cerebral hemisphere, LH the left cerebral hemisphere, and L the lesion. The blood vessels are clearly imaged. (b) Open skull photograph of the rat surface acquired after the photo-acoustic experiment. See Reference [4].

recording of the ultrasonic field on an array of piezoelectric transducers, one can deduce an image of the ultrasonic sources. The fact that the ultrasonic speed is practically uniform in all tissue and has a well-known value greatly simplifies the reconstruction process. An image of the sources is then built with the sub-millimetric resolution of the ultrasonic wave.

In the thermoacoustic approach both microwaves [3, 4] and optical waves can be used. An image of the optical absorption or of the tissue conductivity is built with the sub-millimetric resolution of the radiated ultrasound. Microwave penetration allows deeper exploration and the first conductivity images of breast have been obtained with this modality, while vascular images on small animals have been obtained with the photo-acoustic approach. Figure 1.1 illustrates a spectacular application using as a heating source a laser with 532 nm wavelength and a wideband ultrasonic transducer with a 2.25 MHz central frequency to receive the photo-acoustic wave. Figure 1.1b shows that blood vessels in the cortical surface of small animals can be imaged with the skin and the skull intact. The imaging depth is limited to 1 cm, which is enough to image the entire brain of a small animal.

Others modalities has been proposed to improve electrical impedance tomography, known as magnetoacoustic tomography (MAT), where tissue is displaced by an electric or a magnetic stimulation [5, 6] to produce ultrasound. In the most interesting technique with magnetic induction (MAT-MI) tissues are put both in a strong static magnetic field and in a time-varying magnetic field (MHz range). The time-varying magnetic field induces eddy currents that interact with the static magnetic field to produce a Lorentz force that induces ultrasonic waves, which can also be recorded by ultrasonic transducers. In this approach the acoustic wave amplitude is proportional to the electrical conductivity in the MHz range.

Different approaches can be conducted to map the source terms and the inverse problem image reconstruction is typically based on the fact that in a medium with constant speed the data recorded on the transducers array are spherical integrals of pressure source. A back-propagation algorithm allows recovery of function from integrals over spheres (spherical Radon transform). This step can also be accomplished by time-reversing and back-propagating the acoustic data in a computer model of constant speed.

1.5
Wave to Wave Tagging

Acousto-optic imaging combines, thanks to acousto-optic effects, ultrasound and light in a different way to photoacoustic imaging that is directly related to a dissipative process. A focused ultrasonic beam induces locally an ultrasonic modulation of a light beam traversing a scattering medium. Light transmitted through an organ contains thus different frequency components: the main component (the carrier) is centered at the incident coherent optical beam frequency. It is related to the scattered photons that do not interact with ultrasound. The sideband components are shifted by the ultrasound frequency. The sideband photons that result from the interaction between light and ultrasound are called the "tagged photons." The weight of these tagged photons components depends on the optical absorption in the region of interest. Acousto-optic imaging detects selectively the tagged photons. An image related both to the optical absorption and to optical diffusion is then built up in scanning the focused ultrasonic beam over the whole organ. Marks investigated this tagging technique for the first time in the early 1990s [7]. Since then, many different groups have contributed to this field [8–12]. Two main mechanisms participate in the ultrasonic modulation of light in a scattering media. One is based on the variation of the optical phase in response to ultrasound-induced displacements scatterers. The displacement of scatterers modulates the physical path lengths of light traversing the ultrasonic field. Multiply scattered light accumulates modulated path lengths. Therefore, the intensity of the speckle associated with multiply scattered light fluctuates with the ultrasonic frequency. A second mechanism is based on the variation of the optical phase in response to ultrasonic modulation. As the result of ultrasonic modulation of the index of refraction, the optical phase between scattering events is modulated and the modulated phase causes also the speckle intensity to vary with ultrasound.

Many coherent detection techniques have been proposed to detect the tagged photons. One of the most interesting is the parallel detection scheme that uses a source synchronized lock-in technique in which a CCD camera works as a detector array [10, 11]. An interesting improvement was proposed to increase the axial resolution of the acousto-optic images, which was not as good as the lateral resolution with monochromatic ultrasound. Wang and Ku replaced the ultrasound monochromatic excitation by a frequency-swept (chirped signal) that modulated also the gain of the optical detectors [11].

However, the main difficulty of this technique in living tissues results from both the motion of the scatterers due to the Brownian motion of the scatterers and to the tissue inner motions (blood flow). This speckle decorrelation broadens the carrier and sideband lines. Typically, with 4 cm breast thickness, the speckle decorrelation time is in the ms range and yields a 3 kHz broadening. Detector bandwidth in this range is needed. With a mono-detector (photodiode) there is of course no problem but to achieve a good signal-to-noise one needs a large optical etendue in detector plane (product of the detector area and the detector acceptance angle) to fit the etendue of the tagged photons source, that is, to maximize the scattered light collection. This is why multi-detectors, such as a CCD camera, have been investigated, but they suffer from a low image frequency rate (typically 100 Hz) that is not enough to avoid the broadening effect in living tissues. Faster cameras are not sensitive enough. More recently different groups [12] have proposed very promising tagged-photons detection techniques based on photorefractive crystal based interferometry, which can give both a large etendue and a detection bandwidth in the kHz range (response time of GaAs photorefractive material is of the order of 1 ms).

Interestingly, the tagging concept could be used not only in acousto-optics but in many other fields of medical imaging such as, for example, electric impedance tomography tagged by ultrasonic remote vibrations.

Here we have discussed the concept of wave to wave tagging using two kinds of totally different waves. However, although it is not a multiwave technique, MR imaging can also be interpreted as a tagging technique. It uses only one kind of wave combined with a static field: a radio-frequency electromagnetic wave that causes protons to absorb some of its energy and to release it later at a resonance radio-frequency. The spatial tagging is achieved here through the addition to a static magnetic field of non-uniform magnetic fields whose spatial gradient modifies the local Larmor frequency, allowing during the reception mode a spatial resolution much better than the RF wavelength through a frequency analysis of the received signal.

1.6
Wave to Wave Imaging: Mapping Elasticity

The third approach to multiwave imaging is perhaps the most fascinating. Indeed, the wave interaction is here produced such that the *near field* of the slow wave around each obstacle can be filmed by the faster wave. In this approach, the playground consists of sonic shear waves and ultrasonic waves. These waves interact to produce a quantitative and highly resolved image of the stiffness of deep organs.

Stiffness is characterized by the Young's modulus E (in kPa) and is an important parameter in medicine. Stiffness changes are often linked to pathology [13] and the significant dependence of E on structural changes in the tissue is the basis for the palpatory diagnosis of various diseases, such as detection of cancer nodules in the breast or prostate. Although it is strongly subjective, manual palpation is not only useful for screening and diagnosis but also during interventions to effectively guide the surgeon towards the pathological area. The concept of stiffness imaging was introduced in the early 1990s by J. Ophir et al. and named elastography [14]. Their technique is based on the ultrasonic imaging of tissue deformations induced by a quasi-static compression of organs applied by the operator at the surface of the body. Tissue deformations are obtained by acquiring two *pre-compression* and *post-compression* ultrasonic images of the organ using a conventional ultrasound scanner. Consequently, static elastography is inherently a single wave approach (based on the single use of ultrasonic waves), which implies some important drawbacks. Comparison of the two images enables only the mapping of local tissue strain. This strain image, called an elastogram, is linked to stiffness as soft regions tend to exhibit a higher strain than stiffer areas. However, even for a simple one-dimensional model, the underlying link between local strain ξ and stiffness E (Young's modulus in kPa) is strongly dependent on the local and unknown stress τ via the well-known relation $E = \tau/\xi$. Unfortunately, applying a quasi-static compression at the surface of the body can create a very complex spatial distribution of stress that both prevents the assessment of local stiffness and induces image artifacts.

To understand how to map tissue elasticity without the artifacts observed with static elastography, we will focus on a different approach that uses low frequency shear waves to obtain more precise tissue elasticity information.

To understand the connection between the Young modulus and the shear wave velocity, one can in a first approximation consider soft tissue as an isotropic elastic medium. The mechanical behavior of such a soft solid is characterized by two parameters, K (inverse of the compressibility $\kappa = 1/K$) and μ, which are, respectively the bulk and shear modulus. The relationship between stiffness and these parameters is described by:

$$E = \frac{9K\mu}{3K + \mu} \tag{1.1}$$

The property $K \gg \mu$ is a kind of definition of "soft solids" and human soft tissues belong to this category. It implies straightforwardly a direct link between stiffness E and the shear modulus, $E = 3\mu$. Therefore, an elegant way of accessing stiffness properties consists in using shear waves whose speed c_s depends simply on the shear modulus, $c_s = \sqrt{\mu/\rho}$. Another important consequence of the big discrepancies between K and μ in a soft solid is that the compressional wave speed is much larger than that of the shear waves (from 1540 m s^{-1} for c_p to some m s^{-1} for c_s). This is a unique case where two mechanical waves exhibit totally different wave speeds.

In conventional "single wave" imaging, only the compressional wave (and consequently the contrast of bulk modulus) is used. This is the successful field of medical ultrasound imaging. Now, could we use shear waves to image the shear modulus contrast of tissues? As we have just seen, the MHz frequency range is forbidden due to shear tissue viscosity and shear waves can only propagate on centimetric distances at low sonic frequencies. The typical shear wave frequency ranges between 10 Hz and 1 kHz. For example, to propagate on a 5 cm distance, we are limited to frequencies lower than 100 Hz, corresponding to typical wavelengths of several centimeters. The use of shear waves in a "single wave" imaging approach can only lead to poor results as it will rely on very bad spatial resolution. However, the contrast sensed by shear waves remains very relevant information for the diagnosis.

How can we solve this problem using a "multiwave" approach? We can benefit from the huge discrepancy between shear and compressional wave speeds. The idea is to use the compressional waves at ultrasonic frequencies to observe the tissue motion induced by the propagation of low speed sonic shear waves. During their propagation, shear waves induce local tissue displacements of the order of some tens of microns around their equilibrium position. One approach proposed by Sato and the group of K. Parker [15, 16] consists in observing the shear wave effect with ultrasound Doppler techniques. They use a sinusoidal shear wave excitation to produce a stationary vibrating pattern containing a set of nodes and antinodes. This approach was called sono-elastography. The distance between antinodes is used to deduce the shear wave wavelength and allows an estimate of the shear modulus with a spatial resolution of the order of the shear wavelength (centimeters).

Another approach that allows us to obtain shear modulus estimations with millimetric resolution (instead of centimetric resolution) is to use a short transient shear excitation instead of a sinusoidal shear excitation. This is the field of transient elastography [17, 18]. In this case, one needs to produce ultrafast images of tissues to image the propagation of the transient sonic shear waves. Here, the goal is to obtain a movie of the transient shear wave propagating inside organs with millimetric resolution. As the typical shear wave speed varies between 1 and 10 m s^{-1}, one needs at least to reach 10 000 frames per second to follow the shear wavefront millimeter by millimeter. Using such an ultrafast scanner, it could be possible to estimate these local displacements between successive images.

Our group developed such an ultrafast scanner [18]. This is the first ultrasonic device able to reach more than 10 000 frames per second of deep-seated organs. In this device, whose architecture emerged from the concept of time reversal mirrors [19, 20], one transmits several thousand times per second an ultrasonic beam widely spread in the whole area of interest. This imaging sequence is very different from the one used in conventional ultrasound scanners that insonicate the medium using only a very thin ultrasonic focused beam that needs to be translated step by step to sequentially map the imaged area. Such a conventional echographic image results in more than 128 successive insonications. Taking into account the time of flight of backscattered ultrasound (a 20 cm back and forth propagation requires some 130 µs), a typical frame rate of about 50 images per second can be reached for sequential imaging. Contrarily, in an ultrafast scanner, for each transmit beam, the backscattered echoes coming from a very large region of interest are recorded by an array of some hundreds of piezoelectric transducers and stored in large memories. Then, a fast algorithm transforms several thousand times per second the backscattered echoes into an echographic image. Because the ultrasonic wave speed is known and constant, this operation can be obtained through a numerical time reversal refocusing. To track the local displacements induced by shear wave propagation, successive ultrasonic images are compared. This is possible because the ultrasonic images are dominated by the so-called "speckle" noise that originates from the random distribution of weak scatterers (Rayleigh scatterers much smaller than the wavelength) that exist everywhere in tissues. Note that in soft tissue, contrary to optics, ultrasonic backscattering is dominated by a single scattering process, thus insuring an unambiguous correspondence between the arrival time of the speckle noise and the spatial location of the scatterers distribution. By cross-correlating in the time domain the speckle noise observed from one frame to the other, a motion speckle tracking algorithm enables the reconstruction of a complete movie of the tissue displacements field along the ultrasonic beam direction (Figure 1.2). From this movie, one can locally deduce the shear wave speed and thus the shear modulus μ.

How can we generate shear waves in the human body? Such shear waves already exist naturally in our body. Each heart beat creates transient vibrations that propagate near the cardiac muscles and along arteries. Our vocal cords also produce shear vibrations into nearby organs during speech.

These natural waves usually remain confined close to their source and cannot be used to assess all organs. External vibrators applied at the surface of our body can also produce controlled vibrations that induce shear waves propagating

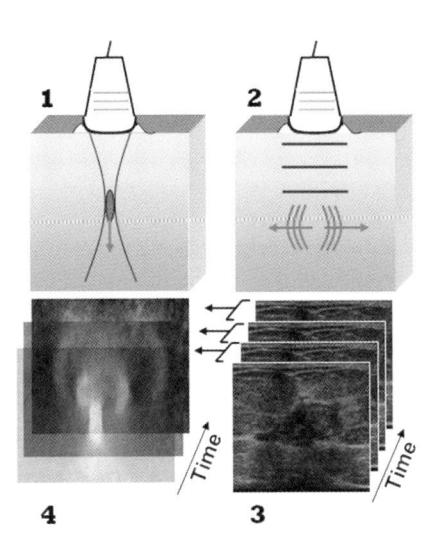

Figure 1.2 Principle of elasticity imaging: In step 1, the ultrasonic probe generates a pushing force in the focal area of the ultrasonic beam. In step 2, this radiation force generates a low frequency shear wave. The ultrasonic array switches into an ultrafast imaging mode. In step 3, the resulting ultrasonic images are built and stored into memories. In step 4, successive ultrasonic images are compared using cross-correlation operations to image tissue displacement induced by the propagation of the shear wave.

from the surface to deeper regions. Finally, instead of using heavy external vibrators, the most elegant way to create a controlled shear wave source consists in using the radiation force induced by ultrasonic focused beams into tissues. Indeed, by focusing an ultrasonic beam at a given location in the organ it is possible to create a volumic radiation force localized in the focal spot and oriented along the beam axis. This force is due to the momentum transfer from the ultrasonic wave to the medium caused by nonlinearities, dissipation, and reflection, and is proportional to the square of the ultrasonic pressure field. Thus, the time profile of the applied force is linked to the beam intensity spatiotemporal profile. The transmission of a 1 ms burst of focused ultrasound with a 5 MHz carrier frequency will lead to an axial force in the kHz range. While remaining below FDA limitations, the radiation force of the ultrasonic beam permit us to remotely create low frequency shear displacements of some tens of microns at depths of several centimeters. Thus, it is possible to generate the shear wave using the same array of piezoelectric transducers that is used in ultrafast imaging. The use of the ultrasonic radiation force as a remote generator of shear waves was proposed by A. Sarvazyan *et al.* in 1998 [21] and is used by several research groups in medical imaging [22–24]. The transducers are used in a first step to transmit a long burst of ultrasound focused at the desired location in the imaged area (Figure 1.1). The beam generates a low frequency pushing force at focus. When the transmission (and consequently the pushing force) ends, tissues displaced in the focal spot come back to their equilibrium position while generating a small localized source of transient shear waves. It is the biomedical analog of a small earthquake created by the shear force of a moving tectonic plate. The resulting shear wave begins to propagate in the organ whereas the ultrasonic probe instantaneously switches into an ultrafast imaging sequence to film this propagation (Figure 1.2). The amplitude of the radiated shear wave decreases quite rapidly due to the natural divergence associated with a small shear source.

To extend the area sensed by the shear wave, we proposed an original solution enabling the generation of weakly diffracting shear waves based on the remote creation of a supersonic source radiating shear waves along a Mach cone [24, 25]. This effect is the analog of the Cerenkov electromagnetic radiation emitted by a beam of high-energy charged particles passing through a transparent medium at a speed greater than the speed of light in that medium [25]. In our configuration, ultrasonic waves are successively focused at different focal depths by changing the electronic delays between the signals transmitted by the transducer elements. By moving the resulting shear source faster than the radiated shear waves, one can induce constructive interference along a Mach cone (Figure 1.3). Such a sonic boom is a very efficient way to create a shear wave of higher displacement that can travel with minimized diffraction in a large area. Compared to the use of external vibrators this technique enables us to optimally polarize the shear displacement field along the ultrasonic beam axis that corresponds to the most sensitive axis of the speckle tracking algorithm.

From the 2D experimental movie of the shear displacements along the ultrasonic beam axis (Oz) $u_z(x,z,t)$ induced by the supersonic push, one can access

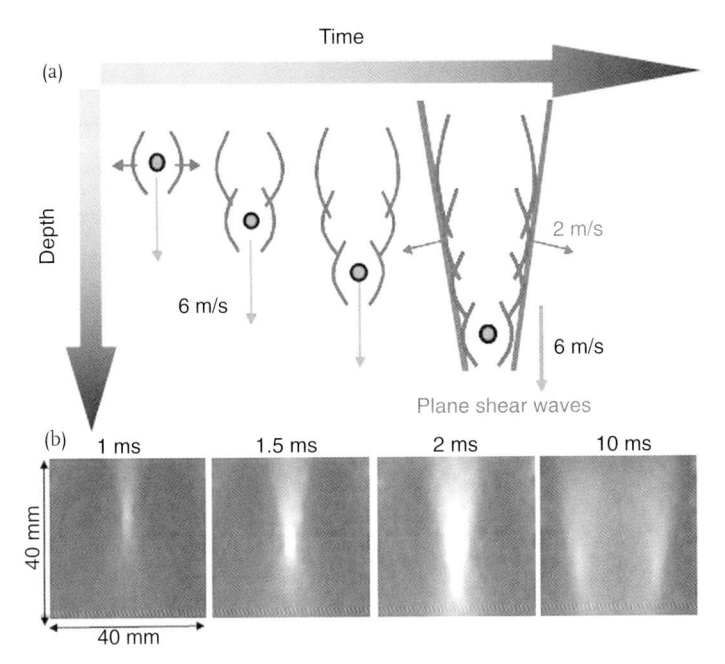

Figure 1.3 Generation of a supersonic shear source: (a) Bursts of ultrasound are focused at successive depths in the organ. Each burst creates a "pushing" radiation force at focus that induces a shear wave. As the "pushing" force is moved faster than the shear waves it generates, a supersonic regime is reached and shear waves accumulate on a Mach cone. (b) Images (40 × 40 mm²) of micrometric displacements induced in a tissue mimicking phantom obtained using the ultrasonic ultrafast imaging mode at different time steps.

the shear modulus map by solving locally the inverse problem of wave propagation. Indeed, the elastodynamic wave equation characterizing the shear wave propagation:

$$\left\{ \Delta u = \frac{\rho}{\mu} \frac{\partial^2}{\partial t^2} u \right\}$$

can be inverted for each pixel (x,z) in the imaging plane and gives access to the local shear modulus $\mu(x,z)$:

$$\mu(x, z) = \frac{\rho \dfrac{\partial^2 u_z(x, z, t)}{\partial t^2}}{\dfrac{\partial^2 u_z(x, z, t)}{\partial x^2} + \dfrac{\partial^2 u_z(x, z, t)}{\partial z^2}} \qquad (1.2)$$

where ρ is the medium density (almost constant in soft tissues). Surprisingly, this very simple inversion approach (based on the calculation of second-order derivatives of the displacement field both in time and space) performs well. This is mainly because the wave field $u_z(x,z,t)$ is experimentally measured everywhere in the region of interest, unlike conventional imaging approaches where the field is known only at boundaries and so requires complex inverse problem approaches. Interestingly, when diffraction in the z-direction remains limited (which occurs in the case of the supersonic shear wave generation), Equation (1.2) can be reduced to a very simple 1D wave equation:

$$\frac{\partial u}{\partial t} - \sqrt{\frac{\mu}{\rho}} \frac{\partial u}{\partial x} = 0 \qquad (1.3)$$

In that case, μ can be estimated using a simple, fast, and robust time-of-flight algorithm [26].

An experiment conducted in a tissue mimicking phantom (made of agar gelatin) illustrates the interest of this technique. A conventional ultrasonic image of this phantom reveals an almost perfectly homogeneous medium in terms of bulk modulus (Figure 1.4a) although the phantom contains a centimetric stiffer inclusion. After generation of the shear Mach cone, high frame rate images of tissue displacements induced in the phantom are recorded during the resulting shear wave propagation. The shear wave is clearly sensitive to the shear modulus contrast as it strongly accelerates while passing through the stiffer region (Figure 1.4b). Then a local estimation of the shear wave speed based on basic time-of-flight estimation enables us to quantitatively map the Young's modulus (Figure 1.4c). The Young's modulus image reveals a highly contrasted spherical inclusion that is twice as hard as the surrounding tissues.

A comparison of the image of the same phantom obtained with static elastography (Figure 1.4d) shows the main attraction of the multiwave approach. Static elastography shows strong artifacts as it is only a strain image and not a true elasticity image. Another elasticity imaging technique called ARFI was proposed in 2001 by Nightingale and Trahey [22]. This technique also uses the ultrasonic radiation force but with the frame rate of conventional ultrasound scanners. The radiation force is induced by a focused ultrasonic beam and tissue displacements are only measured at the location of the focal spot because the limited frame rate of the ultrasound scanner does not allow tracking in real time of the radiated shear wave. To build a complete image, the displacement measurement has to be repeated by successively steering the focal spot at each location in the region of interest. Similarly to static elastography, this strain image is not quantitative as the local stress induced by the radiation force remains unknown. This approach is not a multiwave technique as it does not exploit the radiated shear wave that requires the use of an ultrafast ultrasound scanner.

1.7
Super-resolution in Supersonic Shear Wave Imaging

Interestingly, one can notice in Figure 1.4c that the resolution of the elasticity map is of the order of a millimeter despite the fact that the shear wavelength is of the order of a centimeter (as seen in Figure 1.4a). This super-resolution capability of

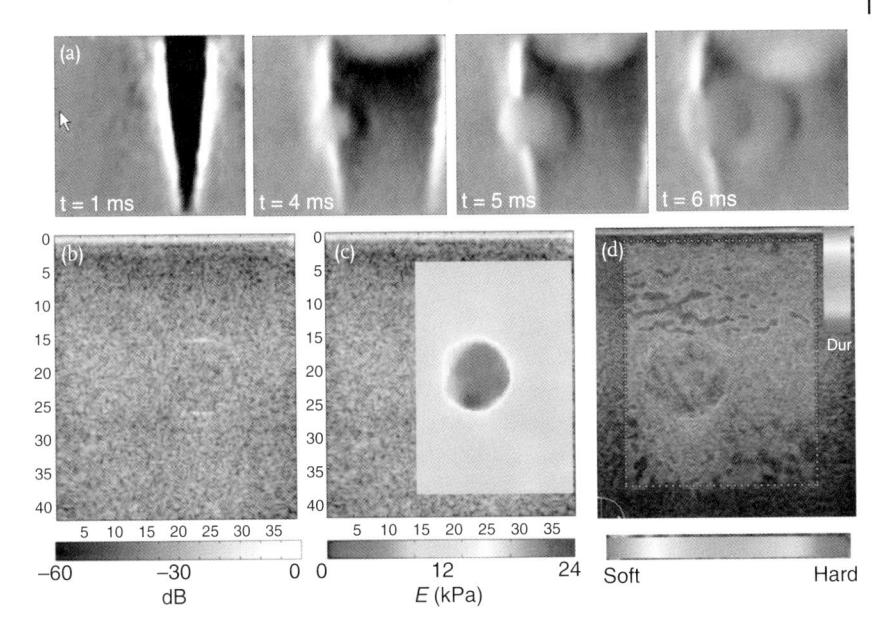

Figure 1.4 Supersonic shear imaging in tissue mimicking phantoms: (a) Images of local tissue displacements (gray scale ranging from -10 μm to +10 μm) at different time steps after the supersonic shear source generation. One clearly sees that the shear wave is sensing the stiffness contrast as it is distorted while passing through a 10 mm stiff inclusion. (b) Conversely, the ultrasonic image of the medium does not reveal the inclusion; (c) quantitative image of the Young's modulus deduced from the shear wave movie (a). (d) Strain image obtained on the same phantom using the static elastography approach of a commercially available imaging device.

SSI can be explained by the fact that the local displacements induced by the wave propagation are not only recorded on the boundaries of the investigated medium (as, for example, in seismology) but at every location deep into the investigated medium. The estimation of local displacement field in the ultrasonic images provides as many virtual motion sensors (accelerometers) deep into tissues – even in the far field of the shear wave source. Such ultrasound-based virtual motion sensors give remote access to the local near-field of the wave-field around each obstacle even if these obstacles are located in the far field of the two waves. The evanescent waves created by the interaction of the incident shear wave with any shear heterogeneities are recorded in the movie with a spatial sampling of some $\lambda/20$ in Figure 1.4. The inverse algorithm based on computation of the spatial and time derivative of the field at each location (Equation 1.3) extracts this information. Therefore, the spatial resolution of the shear modulus image is no longer limited by the classical diffraction limit of shear wavelength but is linked to the much smaller ultrasonic wavelength. We have access to a complete movie of the near field.

1.8
Clinical Applications

A major advantage of this imaging technique, called "supersonic shear imaging" (SSI) or also "shear wave imaging" is that it can be performed using conventional ultrasonic probes and provides an additional imaging modality on a new generation of ultrasound scanners. It gives spectacular *in vivo* results, such as, for example, breast cancer diagnosis [27]. Figure 1.5 shows two interesting cases obtained for breast cancer diagnosis. The 2D map of Young's modulus of a patient breast is superimposed on the conventional ultrasonic image. In Figure 1.5b, two very small and stiff breast lesions (2 mm diameter invasive ductal carcinomas) are visible on the highly contrasted elasticity image but undetected on the conventional ultrasonic image. This clinical case clearly highlights the millimetric resolution of the technique. Figure 1.5a corresponds to a very stiff invasive ductal carcinoma. This example is particularly interesting as it shows that the elasticity information provides new information clearly different from the conventional ultrasonic image. Even more interestingly, the invasive ductal carcinoma presents very stiff boundaries (dark region) and a very soft core (clear region) corresponding

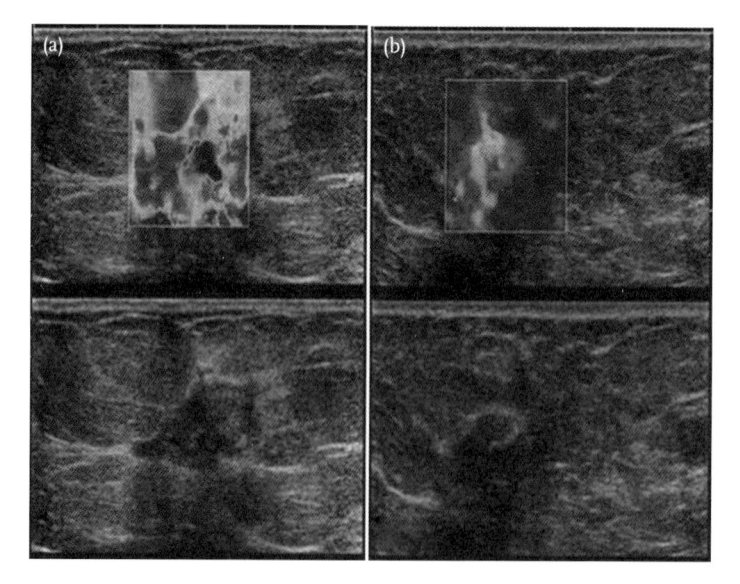

Figure 1.5 *In vivo* clinical application of the supersonic shear imaging (SSI) technique for breast cancer diagnosis. Case (a) corresponds to an infiltrating ductal carcinoma (grade II) and shows the ability of SSI to provide a biomechanical characterization of lesions. The superimposed Young's modulus image shows a lesion much stiffer than surrounding tissue and a very soft core corresponding to a necrosed area. This result was confirmed by histology. Case (b) corresponds to two very small infiltrating ductal carcinomas (grade I &RH+). The millimetric resolution of the Young's modulus image enables us to image these lesions which were not detected by X-ray mammography. Courtesy of Supersonic Imagine, France.

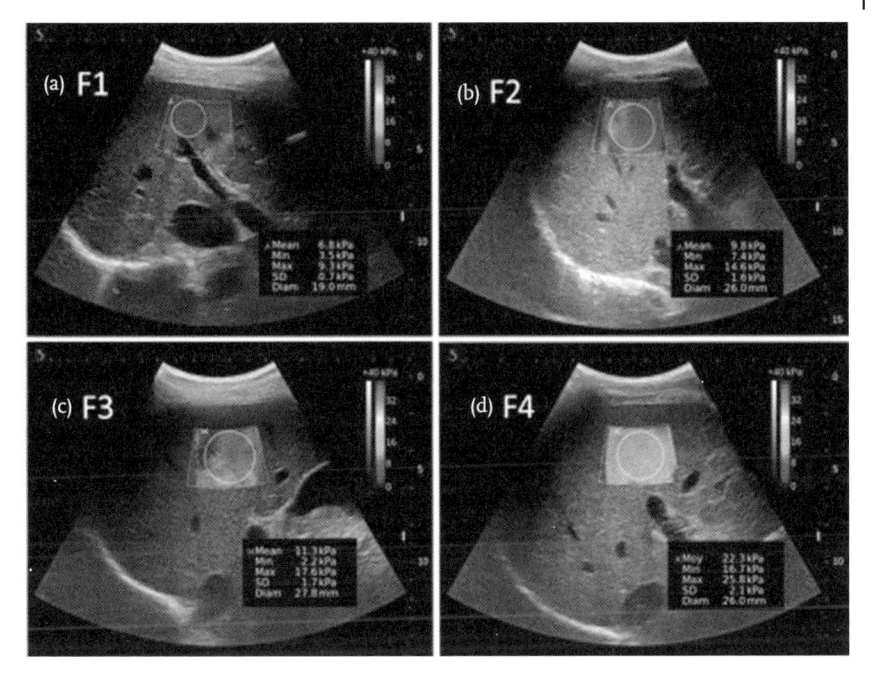

Figure 1.6 *In vivo* clinical application of the supersonic shear imaging (SSI) technique for liver fibrosis staging. Each subplot corresponds to the liver elasticity map of a different patient classified using the conventional fibrosis level classification and ranging from F1 (very low grade fibrosis) to F4 (very high grade fibrosis). Courtesy of Supersonic Imagine, France.

to a necrotic region. These results were confirmed by histology and emphasize the ability of SSI to provide new insights into the radiological characterization of cancer lesions.

Shear wave imaging is currently under investigation for many other clinical applications. Liver fibrosis is one of these applications for which elasticity imaging is envisioned to potentially increase the diagnosis capabilities of ultrasonic scanners. Figure 1.6 shows the elasticity maps acquired *in vivo* on four different patients whose fibrosis level ranges from F1 (very low grade fibrosis) to F4 (very high grade fibrosis). One clearly notices that the global elasticity of the liver increases within the fibrosis classification. Moreover, one can also verify that heterogeneities of liver elasticity increase as expected with the fibrosis level. Current clinical studies tend to evaluate the potential of this technique for accurately staging fibrosis [28].

Shear wave imaging also shows great promises for the imaging and diagnosis of thyroid nodules and liver cancer. Figure 1.7 shows two *in vivo* examples of elasticity for these clinical applications. Figure 1.7a shows both the ultrasonic and elastographic image of an *in vivo* liver cholangitis carcinoma. Whereas the lesion is almost undetectable in the ultrasonic image, it is strongly highlighted in

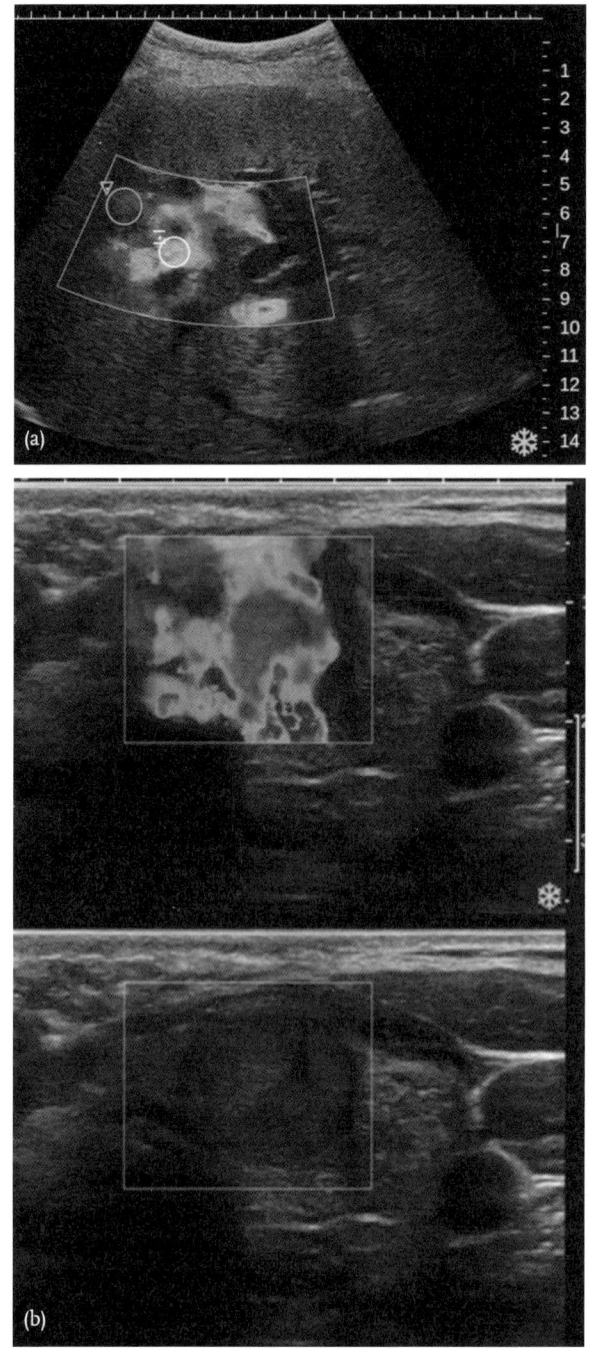

Figure 1.7 *In vivo* clinical application of the supersonic shear imaging (SSI) technique for the imaging of liver cancer and thyroid nodules: (a) *In vivo* imaging of a liver cholangitis carcinoma detected as a much stiffer region (>130 kPa) than surrounding tissues (<40 kPa). (b) Here, thyroid nodules also appear very stiff (>180 kPa) and are clearly visible on the elasticity map. Courtesy of Supersonic Imagine, France.

the stiffness map. The cancer lesion corresponds here to a much stiffer region (>130 kPa) than surrounding tissues (<40 kPa). Again, in Figure 1.7b the elasticity map exhibits a very high contrast between *in vivo* thyroid nodules and surrounding tissues, whereas these nodules are barely visible in the ultrasonic image.

The ability of shear wave imaging to provide quantitative maps of tissue elasticity several times per second enables the study and characterization of organs exhibiting dynamic elasticity changes, such as muscles [29, 30]. Figure 1.8 shows elasticity maps acquired *in vivo* on a healthy subject performing static contractions of his plantar flexors (tibia). In plantar flexors, Young's modulus at rest (Figure 1.8a) was comparable between medial gastrocnemius (area above the mid fascia, 16.5 ± 1.0 kPa) and soleus (area below the mid fascia, 14.5 ± 2.0 kPa) in the knee-extended position. During contraction in the knee-extended position (Figure 1.8b), the increase in Young's modulus was more prominent in medial gastrocnemius (225.4 ± 41.0 kPa), especially about the high-echoed fascia, than soleus (55.0 ± 5.0 kPa). In the knee-flexed position, contraction of plantar flexors (Figure 1.8c) increased Young's modulus in medial gastrocnemius (41.2 ± 2.0 kPa) less than soleus (76.8 ± 7.0 kPa). The results obtained by Shinohara *et al.* [29] show great promise for sport medicine. Coupled with electromyography studies, it could lead to a new tool to help diagnose muscle pathologies. Finally, current work focused on the assessment of these elasticity dynamic changes in the cardiac muscle and common arteries demonstrate in turn that shear wave imaging represents clinical added value in cardiovascular applications.

1.9
Conclusion

Multiwave imaging is a very general concept that can be extended to geophysics as well as non-destructive testing and remote sensing. The future of multiwave imaging for clinical and biomedical applications is bright. The first multiwave imaging application to find clinical success is elasticity imaging. It can be used for breast cancer diagnosis, cardiovascular applications, abdominal or musculoskeletal imaging, and will be soon applied to ophthalmology and dermatology. Two-dimensional ultrasonic arrays will also facilitate new clinical applications such as image guidance for minimally invasive surgery or the monitoring of thermal ablation treatments.

Multiwave imaging also introduces new scientific concepts transcending the conventional limits of wave physics, such as (in the case of supersonic shear wave imaging) providing a way to observe the near field of waves around heterogeneities from far-field detectors. On the technology side, the first clinical ultrafast ultrasonic scanner provides for the first time a way to image in real time many natural transient vibrations occurring (*en permanence*) in our body. This is perhaps the most fascinating kind of multiwave imaging, where one of the waves naturally exists in living organs, such as pulsatility vibrations induced in the heart and along arteries or mechanical vibrations induced in nerves, neurons, and muscle fibers by electric actions potentials.

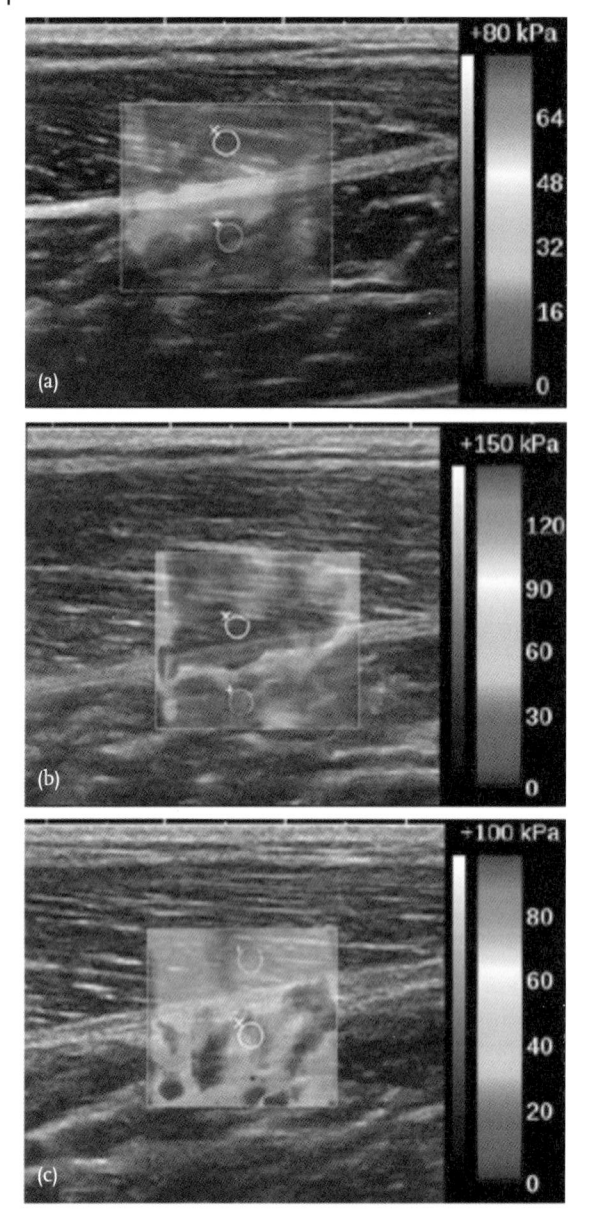

Figure 1.8 Example of the "real time" capability of supersonic shear wave imaging for *in vivo* study of muscle contraction. The subject performed static contraction of his plantar flexors to exert 30% of his maximal voluntary contraction (MVC). (a) At rest, Young's modulus is comparable between medial gastrocnemius (area above the mid fascia, 16.5 ± 1.0 kPa) and soleus (area below the mid fascia, 14.5 ± 2.0 kPa) in the knee-extended position. (b) In the knee-extended position, contraction of the knee-extended position. (c) In the knee-flexed position, contraction of the plantar flexors. Figure adapted from Reference [29].

References

1 Yodh, A. and Chance, B. (1995) *Phys. Today*, **48**, 34–40.
2 Fink, M. and Tanter, M. (2010) *Phys. Today*, **63**, 28–32.
3 Kruger, R.A., Reinecke, D.R., and Kruger, G.A. (1999) *Med. Phys.*, **26** (9), 1932.
4 Wang, X., Pang, Y., Ku, G., Xie, X., Stoica, G., and Wang, L.H.V. (2003) *Nat. Biotechnol.*, **21**, 803.
5 Towe, B.C. and Islam, M.R. (1988) *IEEE Trans. Biomed. Eng.*, **35**, 892–894.
6 Li, X., Xu, Y., and He, B. (2006) *J. Appl. Phys.*, **99**, 066112.
7 Marks, F.A., Tomlinson, H.W., and Brooksby, G.W. (1993) *Photon Migration and Imaging in Random Media and Tissues* (eds B. Chance and R.R. Alfano), Proceedings of SPIE vol. **1888**, 500–510.
8 Wang, L.-H.V., Jacques, S.L., and Zhao, X.-M. (1995) *Opt. Lett.*, **20**, 629–631.
9 Leutz, W. and Maret, G. (1995) *Phys. B*, **204**, 14–19.
10 Leveque, S., Boccara, A.C., Lebec, M., and Saint-Jalmes, H. (1999) *Opt. Lett.*, **24**, 181–183.
11 Wang, L.-H.V. and Ku, G. (1998) *Opt. Lett.*, **23**, 975–977.
12 Gross, M., Lesa, M., Ramaz, F., Delaye, P., Roosen, G., and Boccara, A.C. (2009) *Eur. Phys. J. E*, **28**, 173–186.
13 Sarvazyan, A.P. et al. (1995) *Acoustical Imaging* (ed. J.P. Jones), vol. 21, Plenum Press, New York, pp. 223–240.
14 Ophir, J., Céspedes, I., Ponnekanti, H., Yasdi, Y., and Li, X. (1991) *Ultrason. Imag.*, **13**, 111–134.
15 Yamakoshi, Y., Sato, J., and Sato, T. (1990) *IEEE Trans. Ultrason. Ferroelectr. Freq. Control.*, **37** (2), 45–53.
16 Lee, F., Bronson, J.P., Lerner, R.M., Parker, K.J., Huang, S.R., and Roach, D.J. (1991) *Radiology*, **181**, 237–239.
17 Catheline, S., Thomas, J.-L., Wu, F., and Fink, M. (1999) *IEEE Trans. Ultrason. Ferroelectr. Freq. Control.*, **46** (4), 1013–1019.
18 Sandrin, L., Tanter, M., Catheline, S., and Fink, M. (2002) *IEEE Trans. Ultrason. Ferroelectr. Freq. Control.*, **49** (4), 426–435.
19 Fink, M. (1997) Time-reversed acoustics. *Phys. Today*, **50**, 34–40.
20 Fink, M., Montaldo, G., and Tanter, M. (2003) *Annu. Rev. Biomed. Eng.*, **5**, 465–497.
21 Sarvazyan, A.P., Rudenko, O.V., Swanson, S.D., Fowlkes, J.B., and Emelianov, S.Y. (1998) *Ultrasound Med. Biol.*, **20**, 1419–1436.
22 Nightingale, K.R., Soo, M.S., Nightingale, R.W., and Trahey, G.E. (2002) *Ultrasound Med. Biol.*, **28** (2), 227–235.
23 Fatemi, M. and Greenleaf, J.F. (1998) *Science*, **280**, 82–85.
24 Bercoff, J., Tanter, M., and Fink, M. (2004) *IEEE Trans. Ultrason. Ferroelectr. Freq. Control.*, **51** (4), 374–409.
25 Bercoff, J., Tanter, M., and Fink, M. (2004) *Appl. Phys. Lett.*, **84** (12), 2202–2204.
26 McLaughlin, J. and Renzi, D. (2006) Using level set based inversion of arrival times to recover shear wave speed in transient elastography and supersonic imaging. *Inverse Probl.*, **22**, 707–725.
27 Tanter, M., Bercoff, J., Athanasiou, A., Deffieux, T., Gennisson, J.-L., Montaldo, G., Muller, M., Tardivon, A., and Fink, M. (2008) *Ultrasound Med. Biol.*, **34** (9), 1373–1386.
28 Bavu, E., Gennisson, J.L., Couade, M., Bercoff, J., Mallet, V., Fink, M., Badel, A., Vallet-Pichard, A., Nalpas, B., Tanter, M., and Pol, S. (2011) *Ultrasound Med. Biol.*, **37** (9), 1361–1373.
29 Shinohara, M., Sabra, K., Gennisson, J.-L., Fink, M., and Tanter, M. (2010) *Muscle Nerve*, **42**, 438–431.
30 Gennisson, J.-L., Deffieux, T., Mace, E., Montaldo, G., Fink, M., and Tanter, M. (2010) Viscoelastic and anisotropic mechanical properties of *in vivo* muscle tissue assessed by supersonic shear imaging. *Ultrasound Med. Biol.*, **36** (5), 789–801.

2
Imaging via Speckle Interferometry and Nonlinear Methods

Jeffrey Sadler and Roman Gr. Maev

2.1
General Introduction

This chapter focuses upon a pair of specific imaging techniques: speckle interferometry and nonlinear imaging. Both techniques provide methods to not only vastly improve image quality but also provide for correction of images taken through irregular media. The core of both techniques is their use of multiple sets of flawed images – combining and processing these multiple exposures to create an enhanced final image often near that of the theoretical limits of the imaging machinery. Here both techniques rely upon using advanced averaging methods, in particular using the phase space of the Fourier and Hilbert transforms to achieve their goals. In both cases each of the exposures can be obtained via conventional imaging systems, giving the techniques more strength due to their ease of implementation. While the most common uses of these methods is in optics, using the visible light spectrum, the similarities in the wave-based nature of both optics and acoustics allows the techniques to be useful for imaging over the entire light spectrum (X-ray, infrared, ultraviolet), and also for acoustical imaging.

In its most basic form, most readers have likely already used one of the most simple averaging method to improve data quality; here multiple static images, or more likely multiple sets of experimental data, are averaged together to reduce the noise in the image (or data). Such a method is of course only viable when the subject is static over a set period of time. Both speckle interferometry, and nonlinear imaging go beyond this simple averaging methodology. In fact, both methods are intended to work with dynamic systems, and the motion of the object (or apparent motion) is used to improve the image.

In the case of speckle interferometry, this chapter will examine the version of used in astronomy. Here the position of the object is altered by a secondary factor, which in addition may also blur and distort the image. This distortion can be so severe that only a speckled image is obtained for short exposures, hence the terminology speckle imaging. In contrast, when long exposure images are obtained one obtains a blurry distributed image. By using the Fourier space and combining the multiple short exposures the apparent motion of the object is removed as well as the distortion. Other related speckle techniques also employ these speckle images for different uses. For example, in mechanics laser interferometers can be used to create speckle interference patterns, allowing the distortion of objects to be measured. Other authors have turned to image processing techniques to create speckle tracking algorithms to track the motion of small reflectors or scatterers.

In the second case (nonlinear imaging), the particular case of a flowing media will be discussed. Here the scattering from the flowing media creates dynamic noise in the image. Similar to speckle interferometry it is the random reflections, or speckles, that are used to create the image. Again the Fourier space is used, this time to enhance the small reflections from the flowing media via analysis of the phase of the signal. Comparisons of the differences between pairs of exposures can be used to image the flowing region, while in addition eliminating any static regions of the image. Other forms to analyze flowing media also use these scattered reflections from the flowing media – for example, Doppler imaging, which measures the Doppler shift from the reflections; B-flow imaging, which utilizes specially coded pulses to enhance and detect reflections from blood flow; and the previously mentioned speckle tracking algorithms. Nonlinear imaging also has many other uses, but is generally not specifically related to imaging flowing media. Typically, these are advanced image processing techniques (smoothing, edge detection, noise filtering, feature extraction, etc.) whose physical or mathematical filters and processes contain some form of nonlinear operation.

2.2
Part I: Speckle Interferometry

2.2.1
Introduction

As previously mentioned, with its ability to improve jittery images, one of the initial uses of speckle interferometry was in ground-based astronomy. In this situation fluctuations in the Earth's atmosphere cause the light from stars and planets to be blurred and distorted. This in turn distorts the objects in the image, making it difficult to resolve features such as double stars and other fine astronomy features. Added to this distortion from the Earth's atmosphere are the very weak images caused by the vast distances the light has traveled (in some cases only a small number of photons are detected), and vibrations in the telescope causes additional motion of the image on the telescope's optics. These vast distances also add to the divergence of the optical information, causing the further blending together of any astronomical features. By using speckle interferometry methods this damage to the image can be averaged out over many exposures. In fact, astronomers find the technique is powerful enough to be able to resolve double stars previously unobservable in the original image; and in many cases produce images limited only by the telescope. (i.e., diffraction limited images).

While the specifics of a light-based method will be discussed, the link between light and acoustics due to their common wave-based properties, and similarities in the detection equipment (CCD elements of camera versus 2D elements in acoustic arrays) means this methodology can easily be applied to acoustical applications. Thus, speckle interferometry provides an efficient and powerful tool for image improvement and correction in either field.

2.2.2
Labeyrie's Method

The speckle in "speckle interferometry" refers to the grainy speckled structure of the image obtained from short exposure times (a similar structure is observed when a laser beam is reflected from diffusing surface) [1]. These short exposure speckled images were believed to contain more information on fine astronomical features than long exposure images where the speckle on images is blurred over time. With this basic reasoning as a catalyst, Labeyrie began developing the first speckle interferometry method in 1970, crediting the technique as an extension of Michelson stellar interferometry, and citing as possible applications star diameter measurements (separations) and studies of stellar systems. It was quickly realized this technique also made possible the measurement of separations and magnitude differences of double stars normally masked by the atmosphere [2].

Today Labeyrie's method is well known in astronomy, and the basic algorithm can be described in seven short steps:

Labeyrie's basic algorithm:

1) Take multiple exposures or images using normal techniques:

    $I(x)$

2) Take the Fourier transform (2D) of each image:

    $\tilde{I}(k)$

3) The magnitude, or more properly the square modulus, of the data is taken. This turns the complex valued Fourier data in to a real value (intensity), and loses the phase information:

    $\left|\tilde{I}(k)\right|^2 = \tilde{I}(k)\tilde{I}^*(k)$

4) Average this data across multiple exposures, giving the average intensity of the Fourier transform of the object:

    $\left\langle \left|\tilde{I}(k)\right|^2 \right\rangle$

5) Take a second set of measurements involving a known star, and find the ratio of the two sets of data.

6) As an alternative to step 5, it is possible to measure the scattering of the atmosphere via some other method. Both options, steps 5 and 6, remove the scattering factor, with the first option normalizing the star's magnitude relative to the known star:

$$\langle |\tilde{I}_1(k)|^2 \rangle / \langle |\tilde{I}_2(k)|^2 \rangle$$

7) Take the (inverse) Fourier transform (2D) of this data.

The result is the image.

This image is limited to finding the magnitude of stars, and the separation between two stars.

In the above, $\tilde{I}(k)$ is the Fourier transform of the scattered signal $I(x)$ (x, for special domain, k for spatial frequency or wave number), and the average is over N exposures. With these seven steps Labeyrie's technique relates the original object $\tilde{I}_o(k)$ to the average of multiple exposures $\tilde{I}_1(k)$ (in the Fourier space) via the equation:

$$\frac{\langle |\tilde{I}_1(k)|^2 \rangle}{\langle |\tilde{I}_2(k)|^2 \rangle} = \frac{|\tilde{I}_{o1}(k)|^2 \langle |\tilde{S}(k)|^2 \rangle}{|\tilde{I}_{o2}(k)|^2 \langle |\tilde{S}(k)|^2 \rangle} = \frac{|\tilde{I}_{o1}(k)|^2}{|\tilde{I}_{o2}(k)|^2} \tag{2.1}$$

where $S(k)$, the scattering of the atmosphere, has been eliminated. The full proof of how one gets to this point is well described in Labeyrie's original paper [1], as well as being outlined in many other papers and texts dealing with speckle interferometry [2–4].

As an approximation, or simplification, to Labeyrie's method the details of using a reference start in step 5 are often eliminated. This simplification still adequately describes the basics of the method, and is typically quoted in less scientific situations. It is also useful for cases when one may wish to obtain an image, and is not interested in detailed measurements, but relies upon the scatting of the atmosphere and telescope to be, on average, a constant value over the domain of the Fourier transform (k). What remains is an approximate proportional equation:

$$\langle |\tilde{I}(k)|^2 \rangle \propto |\tilde{I}_{o1}(k)|^2$$

While averaging an image is typically able remove noise from the images, this is not the true mathematical power of the technique. As noted by Knox and Thompson [3] it has been proved that taking only the average does not produce a diffraction limited result, and the image is of little use, being very blurry. The resolution increase comes from a combination of taking both the modulus (step 3) and the average (step 4); these steps together produce the largest benefit to the image, and have been proven to provide the diffraction limited result. This means that the resolution final image is limited only by theoretical limitations of the

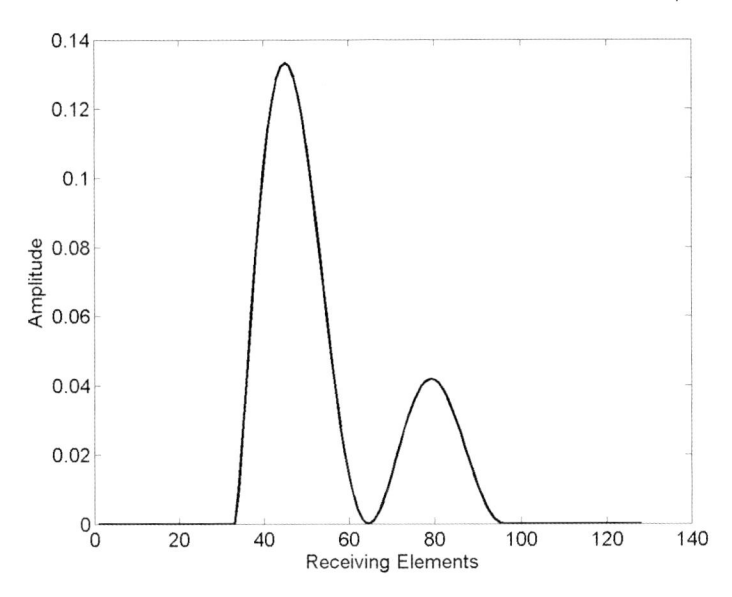

Figure 2.1 Simulated signal of a double star

telescope, and the secondary astronomical problems have been effectively eliminated.

The downside to Labeyrie's method is the loss of the phase information, due to taking the modulus of the Fourier transform. In Labeyrie's initial paper the loss of the phase was said to "make it impossible to reconstruct the object, except if it has a center of symmetry" [1]. This specific symmetry was immediately realized as the symmetry case for double stars, thus giving Labeyrie's methodology its first application. While it was of great interest to use this method for more general objects, or multiple star systems, this restriction to a center of symmetry system caused difficulty in interpreting the results due to the limited information in the resulting image, which described mathematically "only yields the autocorrelation of the object" [3].

For visual reference, sample results using Labeyrie's method can be seen below in Figures 2.1–2.4 and 2.5–2.7. Here two sets of simulated one-dimensional signals are used to help show the process. In the first set (Figures 2.1–2.4) a double star-like (dual source) situation is simulated, via a one-dimensional signal with two nearby peaks of different amplitude (Figure 2.1). This data is randomly scattered, with both the amplitude is altered, and the overall position is shifted left and right (Figure 2.2). This scattering is repeated a number of times to produce the required exposures, which are then processed though Labeyrie's technique, yielding the result shown in Figure 2.3. For astronomical purposes this result is similar to the initial double star signal, with the distance between the center and side peaks being separated by the appropriate distance. Notably, the result does not reproduce the

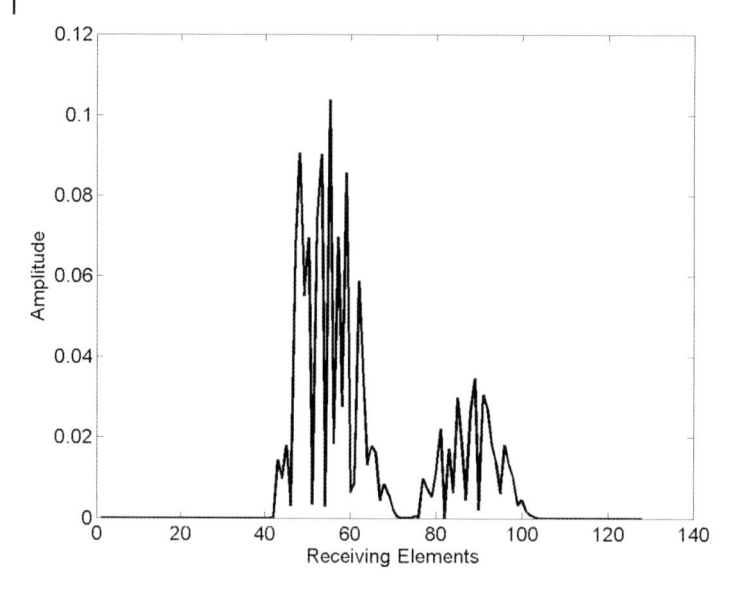

28 | 2 Imaging via Speckle Interferometry and Nonlinear Methods

Figure 2.2 Sample scattered signal from a double star (Figure 2.1).

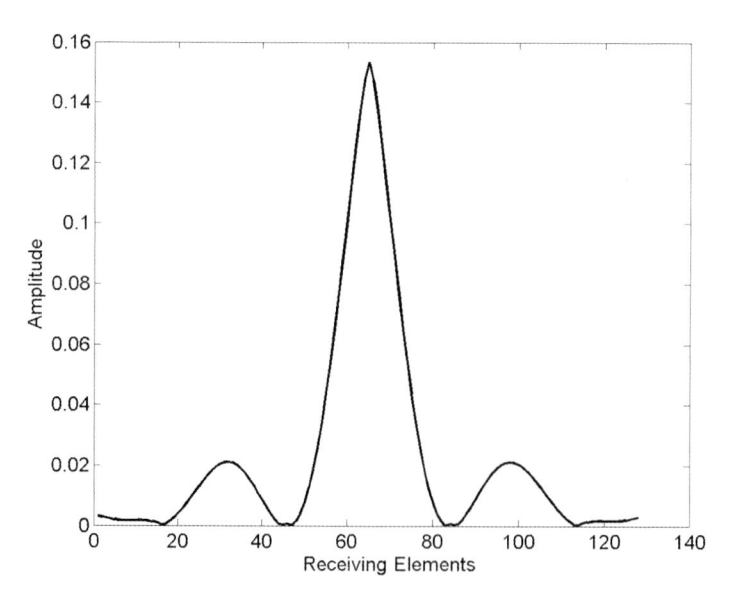

Figure 2.3 Recovered signal using Labeyrie's method for double star (Figure 2.1).

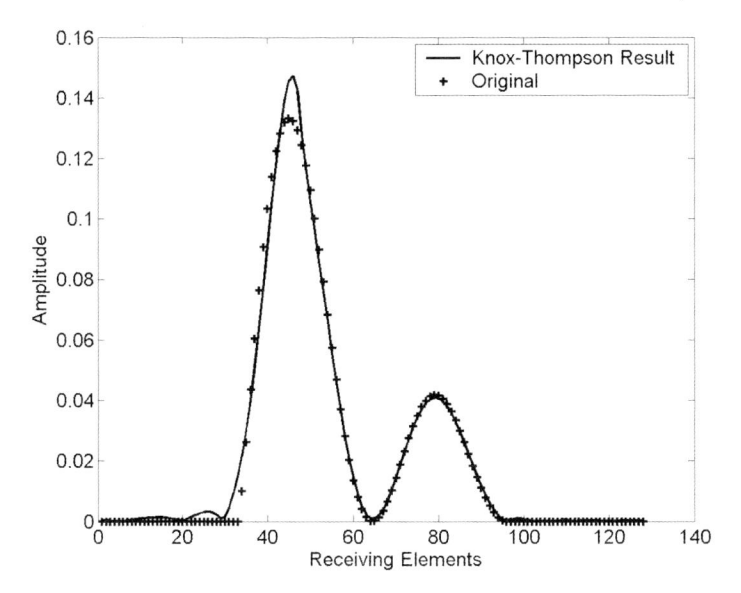

Figure 2.4 Recovered signal using the Knox–Thompson method for double star (Figure 2.1).

exact signal, but instead is a result symmetric about the center. In addition, the overall amplitude of the two peaks is not in the same proportion as the original signal. Here, both observations are a result of the loss of the phase of the signal, and what is pictured does match accurately to reconstructing only the intensity of the Fourier transform, that is:

$$\left|\tilde{I}(k)\right|^2$$

When the process is then repeated for some unknown abstract object, as shown in Figure 2.5, the results are much less useful. While intensity information in the Fourier space is again recovered correctly, the loss of phase information during Labeyrie's process becomes important, and causes the signal to fail to be recovered in a very meaning full manner (Figure 2.6). In comparison the results for the Knox–Thompson method (Figures 2.4 and 2.7), which is able to recover the lost phase information, reproduce the original signal much more accurately. How this is achieved will be discussed next.

2.2.3
Knox–Thompson Method

With Labeyrie's technique presented an initial step forward for image recovery of specific astronomical features, the loss of the phase information during the process did not allow it to be a solution for a more general case image recovery tool. Knox and Thompson [2, 3] would solve this problem, by developing a methodology to determine this phase information, while still being the original idea of statistically averaging many exposures to improve image quality. In fact, the Knox–Thompson

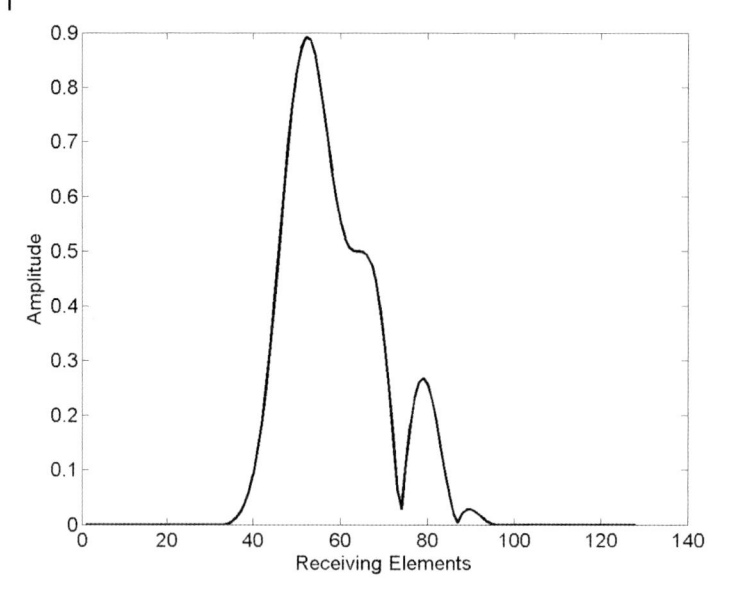

Figure 2.5 Simulated signal of an abstract object.

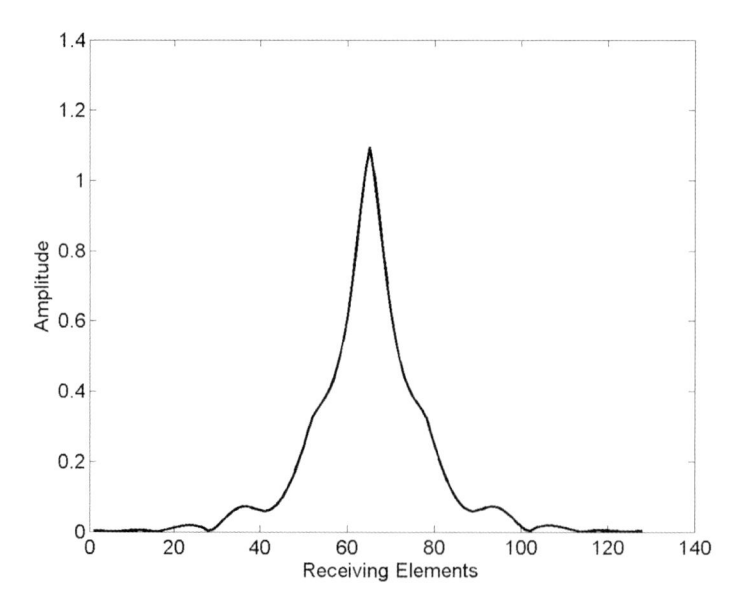

Figure 2.6 Recovered signal using Labeyrie's method for the abstract object (Figure 2.5).

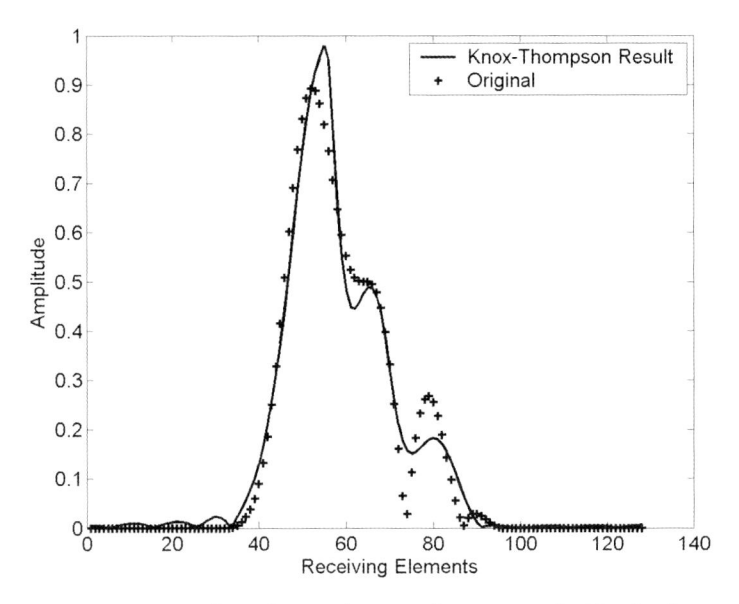

Figure 2.7 Recovered signal using the Knox–Thompson method for the abstract object (Figure 2.5).

method was specifically designed to be an extension of the already proven work by Labeyrie which correctly obtains the intensity distribution By obtaining both the phase and intensity information the combined techniques allowed for not only more complex stellar systems to be analyzed, but creates a system to reconstruct general objects with arbitrary intensity patterns, or arbitrary shapes when examined in two dimensions. Again it is the Fourier space of the image that is used to recover the data, where Knox and Thompson find that the phase information can be retained by examining the product:

$$\tilde{I}(k)\tilde{I}^*(k+\delta) \tag{2.2}$$

where $\tilde{I}(k)$ is the Fourier transform of the intensity $I(x)$ (x, for spatial domain, k for spatial frequency or wavenumber), and δ is a very small shift in the spatial frequency. This small shift δ allows the quantity to remain a complex number, giving it a phase, contrary to the modulus taken during Labeyrie's process:

$$\left|\tilde{I}(k)\right|^2 = \tilde{I}(k)\tilde{I}^*(k) \tag{2.3}$$

which is a real number. Knox and Thompson is able to prove that when the quantity in Equation (2.2) when averaged over multiple exposures the result is an estimate of the phase difference ($\Delta\Phi$), that is:

$$\Delta\Phi = \Phi(k+\delta) - \Phi(k) = \text{Phase}\left[\langle \tilde{I}(k)\tilde{I}^*(k+\delta)\rangle\right] \tag{2.4}$$

Or, in their own words "the phase difference between two nearby points of the object's Fourier transform can be determined by measuring the complex correlation between two points in the Fourier transform of the image photograph" (noted by Knox–Thompson to be a statistical autocorrelation[1]) [2, 3].

As the phase is a relative quantity, an arbitrary point can be defined as zero and the phase values can be obtained by summing the phase differences. As these calculations are being performed in the Fourier space the natural choice is to define zero as the center of the space ($k = 0$), and work outward. Here Knox and Thompson used a shifted Fourier space shifted so that the DC offset ($k = 0$) is centrally located, and the "outward" values refer to the positive and "negative" spatial frequencies. With the phase now defined by the Knox–Thompson method, and the intensity found by Labeyrie's method, the image (or more precisely the Fourier transform of the image) could now be reconstructed:

$$\tilde{I}(k) = |\tilde{I}(k)| \exp[i\Phi(k)] \quad (2.5)$$

As both the phase and magnitude of the Fourier transform is recovered, taking the inverse Fourier transform will now reproduce the original image, with the averaging techniques causing the scattering to be removed. This allows for not only more complex stellar systems to be analyzed but also creates a system to reconstruct general objects scattered in any manner.

An example of once such arbitrarily object has a signal shown in Figure 2.5; as mentioned previously, Labeyrie's method is unable to recover this pattern (Figure 2.6), reproducing only the intensity of the Fourier space. However, as promised, the Knox–Thompson method does adequately recreate the image (Figure 2.7), with only minor discrepancies. The previous results for the simpler dual source case (Figure 2.4) are even more impressive, with the reconstructed pattern being nearly identical to the initial case. Any errors in the recovery process are likely due to faults in recovering the phase information, whose importance will briefly be discussed next. Small faults in the recovery process also occur due to simply using a limited number of exposures in the Knox–Thompson calculations used to create these images.

2.2.4
Importance of Phase Difference Calculation

While not emphasized often in the literature, Knox and Thompson note that this phase difference calculation is in inexact process, as only finite differences can be calculated. This causes the reliability of the phase calculations to slowly degrade as the calculations get farther from the starting point. This small fact actually makes the location of the arbitrary zero starting point very important. By starting at $k = 0$ and working outward, the phase is most accurate at lower frequencies. Typically, this corresponds to the highest intensity region in the Fourier space, thus maximizing the accuracy of the reconstruction in this region. Further from $k = 0$ the phase calculations become less accurate, but the intensities in the Fourier

---

1) Not to be confused with the autocorrelation process from signal processing.

space are most often small and, thus, any errors that have built up do not affect the image to a large degree.

It is entirely possible that there may exist situations that do not fit this pattern. For any such cases the starting point could be adjusted as needed; however, this requires some knowledge of the image to be recovered.

2.2.5
Labeyrie and Knox–Thompson in Two Dimensions

While the above discussion described the process in one dimension, the process is of course two-dimensional, to create an image (as was Labeyrie's techniques). The transition to the Fourier space in two dimensions is no more difficult than in one dimension; here the Fourier transform of the image is taken along one axis, and this result is processed through a second Fourier transform taken along the second axis. The two-dimensional inverse transformation works exactly the same.

Finding the phase information for the Knox–Thompson method can be a slightly more complicated process, as the difference can be calculated on each of the two axes individually:

$$\Delta\Phi_x(kx, ky) = \Phi(kx + \delta, ky) - \Phi(kx, ky) \tag{2.6}$$

and:

$$\Delta\Phi_y(kx, ky) = \Phi(kx, ky + \delta) - \Phi(kx, ky) \tag{2.7}$$

This adds complexity in that the phase can be calculated over more than one path (Figure 2.8); by considering these multiple paths and averaging the results one

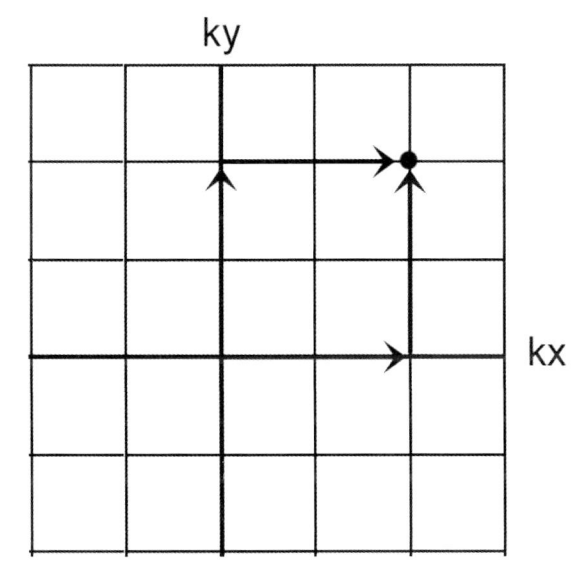

Figure 2.8 Two paths taken to reconstruct the phase.

can improve the accuracy of the resulting phase difference. For example, to obtain the phase $\Phi(kx + \delta, ky + \delta)$ one can calculate:

$$\Phi(kx+\delta, ky+\delta) = \Delta\Phi_y(kx+\delta, ky) - \Delta\Phi_x(kx, ky) \tag{2.8}$$

and:

$$\Phi(kx+\delta, ky+\delta) = \Delta\Phi_x(kx, ky+\delta) - \Delta\Phi_y(kx, ky) \tag{2.9}$$

While mathematically this simplifies to the same result, it is important to remember that each $\Delta\Phi$ term is the results of a single estimated calculation; this estimation makes the values of the two phases slightly different. As it is neglected in much literature, take special note that when averaging many paths one must re-wrap each phase between $(-\pi, +\pi)$ or $(0, 2\pi)$ via a modulo 2π operation [2] before averaging the two results.

2.2.6
Other Improvements to Speckle Interferometry

As the Knox–Thompson phase calculation is an estimate, other authors have tried to improve on this technique. Readers looking for more advanced techniques [5–10] may be interested in processes such as speckle masking and triple-correlation techniques [5, 7] to create higher quality images. Other techniques also exist, such as ones that are geared to specific problems, such as increasing the speed of processing using fast Fourier transform techniques [9], or cases dealing with very low phonon limits [10]. The reference section provides the details on these citations, as well as the papers by Labeyrie [1, 8] and Knox–Thompson [2, 3].

2.3
Part II: Nonlinear Imaging

2.3.1
Introduction

In dealing with the topic of nonlinear imaging (NLI) alongside the areas of acoustical and optical imaging two interpretations come to mind:

1) Imaging that uses the nonlinearity of the medium (i.e., the additional harmonics that are created).

2) The imaging techniques or processes (filters, operations) themselves are nonlinear. This nonlinearity could be due to using polynomial equations, superposition, nonlinear differential equations, or even fuzzy math.

This chapter discusses the second case. Some examples of commonly used nonlinear imaging techniques include a wide variety of options; some of the most

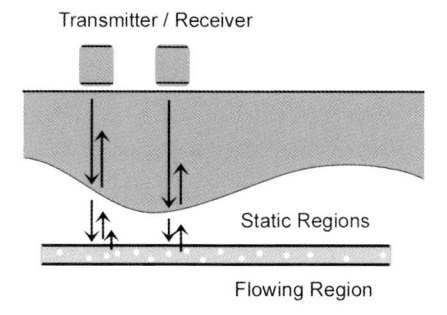

Figure 2.9 A general flow system: the layers have various thickness, and may even be inhomogeneous but are static over time.

common are smoothing, edge detection, noise filtering, feature extraction and so on [11]. Here each technique is often unique, and there is no common framework for nonlinear imaging, but they all do share the link of not being linear; as noted by some texts, though, even this nonlinearity is difficult to define for all techniques.

With no single basis to outline a general nonlinear imaging technique available this chapter will instead choose to examine a specific case, where several of the key processes relate back to those of the speckle interferometry imaging methods from the previous section – in particular, averaging techniques, using the information in the speckles contained in each image, and information from the Fourier space to reconstruct the image. Again a dynamic system will be analyzed, in particular the problem of imaging a flowing media beneath a static structure, or beneath multiple static layers (Figure 2.9). Here it is assumed that the flowing media possesses some scattering property, which creates random reflections, or speckles, in short exposure images, or on each acoustic A-scan. These speckles are superimposed on top of the static response of the system. Nonlinear imaging will be used to analyze the measurable differences between each of the exposures, and image the flowing region. By measuring the differences between the exposures this technique has the additional feature of removing any static sources from the image, leaving only an image of the flowing region.

Such a difference-based analysis is valid due to the principle of superposition. The response from the static regions is roughly the same at each exposure, where this includes locations, phase, and amplitudes of peaks. The random speckles are added to this response, and vary between each exposure. Subtracting two exposures thus will remove any identical static response, leaving the speckles from the flowing image. Such a case of examining differences is noted to work even if the structure is not a simple one (flat, uniform and homogeneous). The flow beneath an irregular, or even inhomogeneous, structure may be imaged by this technique; since even though the response may be complicated it is the same from exposure to exposure.

2.3.2
Deviation (Difference Squared), or Absolute Difference

The most obvious measurement of the difference between two images is the statistical deviation (or simply difference squared), or the sum of the absolute value of the differences. While this is reliable mathematically under very ideal circumstances, it has limited use in the situation presented here due to practical reasons. Here, while the scattering may produce measurable signals in the data, these reflections are most often very small; and even when measured from a zero value the deviation, or differences, will be somewhat minimal in value. To better distinguish the scattering from that of the fluctuations caused by a noisy signal, a more distinguishable measurement for the scattering changes is needed to create a quality image. Much like in speckle interferometry, the information in the Fourier space is useful to analyze the speckles; in this case it is the phase of the signal that is analyzed.

2.3.3
Fourier Transform-Based Methodology

The simplest measure of the phase is via the Fourier transform, where the phase (Φ) is related to the real and imaginary portions of the Fourier transform via the arctangent:

$$\Phi(f) = \arctan\left[\frac{\mathrm{Imag}(f)}{\mathrm{Real}(f)}\right] \tag{2.10}$$

Here the phase can be evaluated at any discrete frequency (f) in the Fourier space, and is typically evaluated with a sign dependent arctangent so that a phase with a full range of 0 to 2π (or $-\pi$ to $+\pi$) is obtained. As an example consider the series of A-scans shown in Figure 2.10, with a reflection from a single scatter flowing between two static walls. In this simple case, as the location of the reflection from

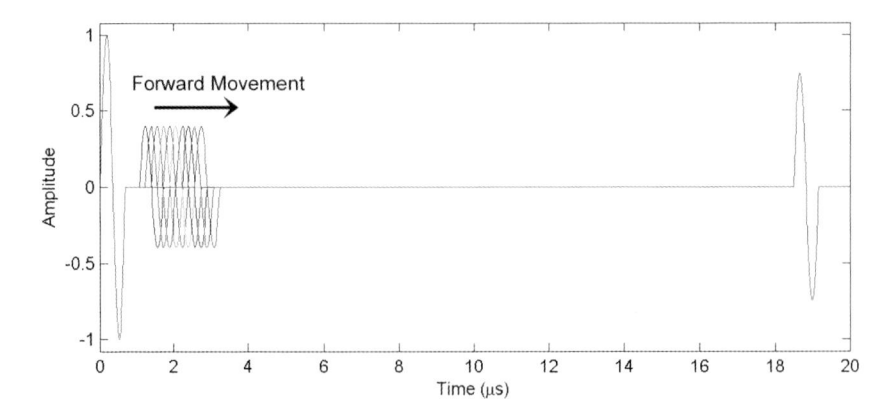

Figure 2.10 Series of simulated A-scans with advancing scatterer location.

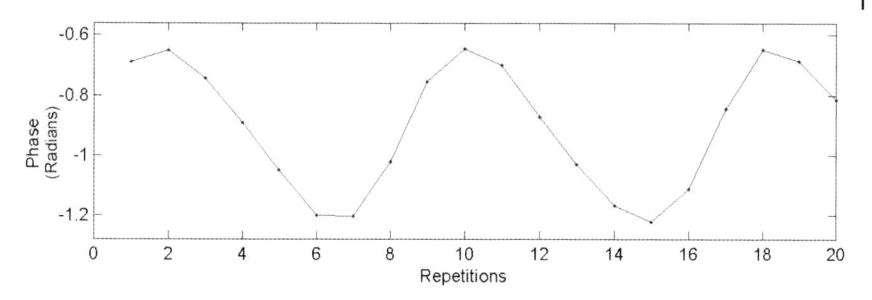

Figure 2.11 Phase of A-scans in each exposure of Figure 2.10.

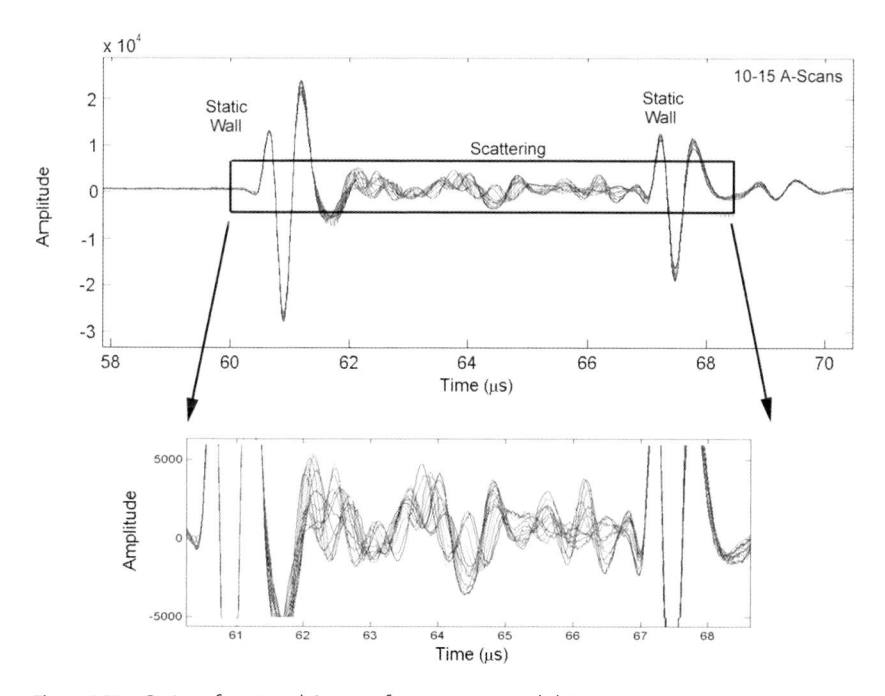

Figure 2.12 Series of scattered A-scans from experimental data.

the single scatter slowly increases in time the phase of the Fourier transform also slowly oscillates, creating a phase change that enables detection of the scatterer (Figure 2.11). While the real-world situation is more chaotic with multiple scatterers and reflections (Figure 2.12), the phase of the Fourier transform still differs from exposure to exposure, thus still allowing the differences between each exposure to be analyzed, and the scatterers to be detected. In the case where there is no flow, each A-scan becomes approximately identical, and the oscillations in the phase are reduced to near zero[2].

2) Any noise present in the experimental system introduces some small phase changes.

In all cases above, the phase is calculated at the frequency corresponding to the reflected pulse, with other phases calculated within the bandwidth limit of this pulse also producing similar behavior. Any phase information from outside the frequency bandwidth of the pulse exhibits a random behavior due to lack of a proper signal at this frequency.

2.3.4
Fourier Methodology: How to Create an Image

With the phase being able to selectively process the distinctness of a flowing region versus a non-flowing region by examining the differences in each exposure it is possible to begin to form an image by examining the quantity:

$$I(x) = \sum_{m=1}^{N-1} |\Phi_m(x) - \Phi_{m+1}(x)|^2 \tag{2.11}$$

where $I(x)$ is a single pixel located at position x, corresponding to the locations where the phases $\Phi(x)$ were obtained, and each phase is calculated at the appropriate frequency for the reflected waves. In its current state these N exposures create $N-1$ differences to create one pixel. However, as long as the system is stable over multiple exposures there is no need to consider only two sets of phase at a time. If N exposures are taken, one can cross-compare the phases of each pair of exposures:

$$I(x) = \sum_{m=1}^{N-1} \sum_{n=m+1}^{N} |\Phi_m(x) - \Phi_n(x)|^2 \tag{2.12}$$

This formulation now produces on the order of $N^2/2$ comparisons, and can dramatically increase the rate at which the image improves with a minimal increase in the number of scans. In this formulation a one-element system is described, with each position corresponding to one pixel, and the result should provide a qualitative measure of flow rate or scatterers per second. The process can then be repeated over multiple pixels to form an image, with a second axis yielding a two-dimensional image. Here each pixel corresponds to its own set of scans, which can be obtained via a motorized system of scans, or an array imaging one element at a time.

When an array is used multiple elements can work together as they are able to send and receive data simultaneously, thus illuminating and receiving data from a large area. If the array has length D the appropriate form of the equation becomes [12–14]:

$$I(p_x) = \sum_m \left| \int_0^D \Delta\Phi_m(x) e^{ikR} dx \right|^2 \tag{2.13}$$

where the sum is over all possible phase differences, and R is the distance from the current pixel (p_x) to the region being imaged at location x, and depth Z:

$$R = \sqrt{Z^2 + (p_x - x)^2} \tag{2.14}$$

Note here that now the location of the pixel (p_x), and the elements of the array (x), are now independent[3], thus allowing for the technique to self-interpolate, and produce an image with pixels located between array elements.

2.3.5
Fourier Transform: Problems with Using

In the example shown in Figure 2.11, one may have noticed that the phase does not have a full 2π range. Such a reduction is due to inclusion of the static walls in the Fourier transform of the data. While a full range of values can obtained if the data internal to the walls is analyzed, this creates problems when there is no flow, and no reflections to detect, leaving only system noise. With lack of a proper signal the phase will oscillate wildly, thus producing very large phase differences when comparing exposures. It is this property that causes trouble in trying to use this technique as a fully formed imaging method. While it can accurately detect flowing regions, and remove static regions, it fails in regions of no response. Any such region is imaged as having a very large result for Equations (2.11) or (2.12) (qualitative scatterers per second measure). Thus the technique relies on knowing in which region the flow is expected, and being able to take the Fourier transform of a region with a detectable static wall to yield reliable results.

Figure 2.13 shows a sample result for the Fourier method. Here to form the image properly the region of the flow was known, and the data external to this was excluded from processing via a simple threshold. While an image of the flowing area was created, and corresponds well to the real world object used, the output of the technique does not separate the flowing regions from the surrounding area by a large threshold in some regions. In addition, the scaling in the flowing region does not seem to be qualitatively rated to the amount of flow (or scatterers per second). Thus, while the methodology creates a very fast way to image the flow in ideal cases, a more suitable methodology should be explored to utilize the idea of analyzing the phase differences for the best results.

2.3.6
Hilbert Transform-Based Methodology

A second way to examine the phase is via the Hilbert transformation, which is able to obtain the phase in a continuous manner by estimating the imaginary portion of the signal (Figure 2.14 shows the Hilbert transform for one exposure). Such an estimation of the imaginary portion of the signal is accomplished by introducing a $\pi/2$ phase shift into the phase information of the signal. While this

---

3) They may be on the same axis, but the 1 pixel does not need to correspond to 1 array element.

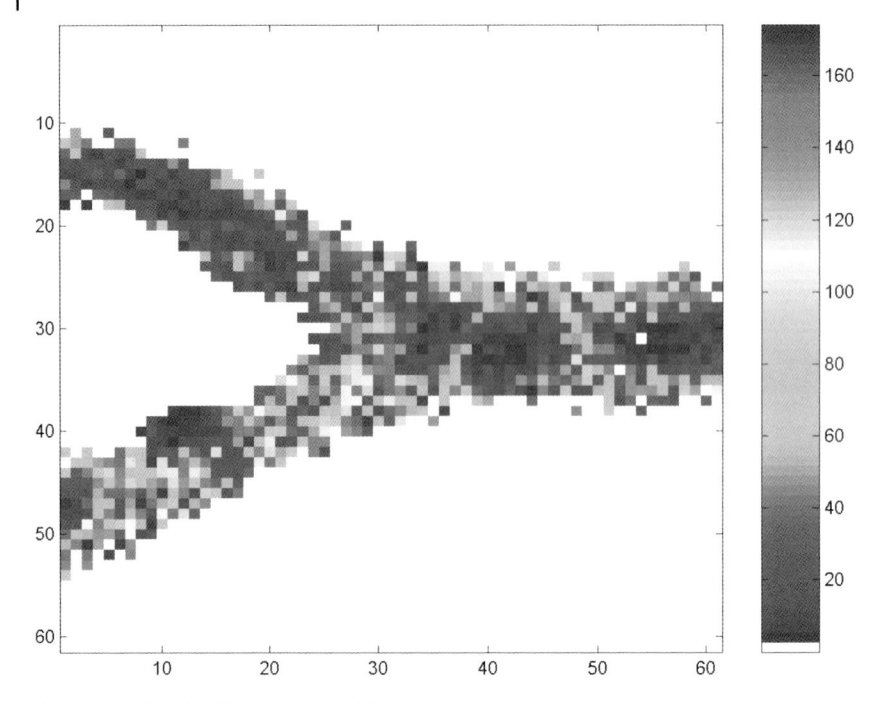

Figure 2.13 Result of Fourier methodology.

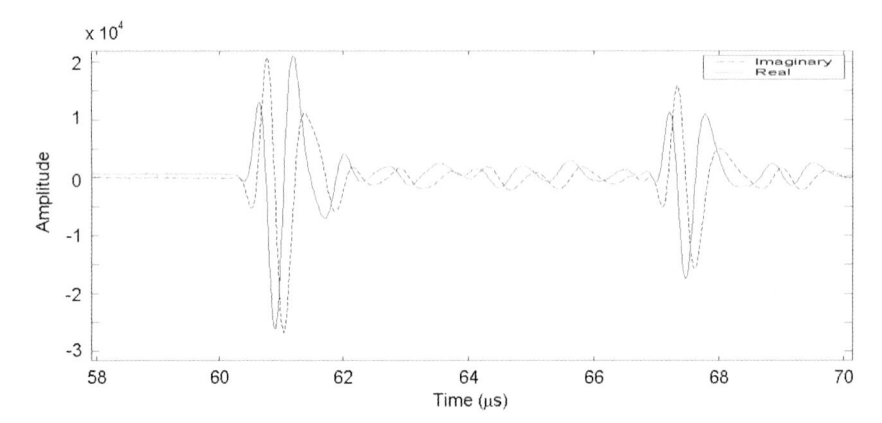

Figure 2.14 Hilbert transform amplitude (1 scatterer).

can be done manually via the Fourier transform, most signal processing software already has this process as a built in function. Once the Hilbert transform is obtained, the phase can be found at each time interval:

$$\Phi(t) = \arctan\left[\frac{\text{Imag}(t)}{\text{Real}(t)}\right] \qquad (2.15)$$

where this continuous phase distribution is noted to no longer require frequency selection, and a sign dependent arctangent should again be used to obtain a full range of phase values.

While the formulation in Equation (2.15) is capable of finding the phase, again there exist problems when dealing with very small signals, particularly for real-world data where the noise in the signal will cause rapid fluctuations around a 0/0 division. (This creates rapid phase changes, and difficulty in imaging using phase differences.) However, the small natural DC offset (d) present in most experiments presents a solution for us in that Equation (2.15) can be expressed more accurately as:

$$\Phi(t) = \arctan\left[\frac{\text{Imag}(t)}{\text{Real}(t)+d}\right] \quad (2.16)$$

With this small finite DC shift Equation (2.16) now becomes a $0/d$ division for small signals, thus producing a more stable phase and, in turn, only minute phase differences in these low amplitude regions. Conversely, when the amplitude is large, either due to the presence of scattering or static reflections, the original phase can approximately be re-obtained:

$$\Phi(t) = \arctan\left[\frac{\text{Imag}(t)}{\text{Real}(t)+d}\right] \approx \arctan\left[\frac{\text{Imag}(t)}{\text{Real}(t)}\right] \text{ for Real} \gg d \quad (2.17)$$

Any discrepancy in miscalculating the phase is lessened further, as one is interested in calculating differences between the phase and not in the exact value of phase. To create the best results for the phase the DC offset needs to be balanced so that it is on the same order as the noise, or slightly larger, but not larger than the scattering signal. With this simple change, the phase differences are now small when needed (static object, areas with no flow, and areas of pure noise and no data) and remain large for scattering regions (flow). This allows the flowing region to be imaged independently of the need to detect said region (Figure 2.15).

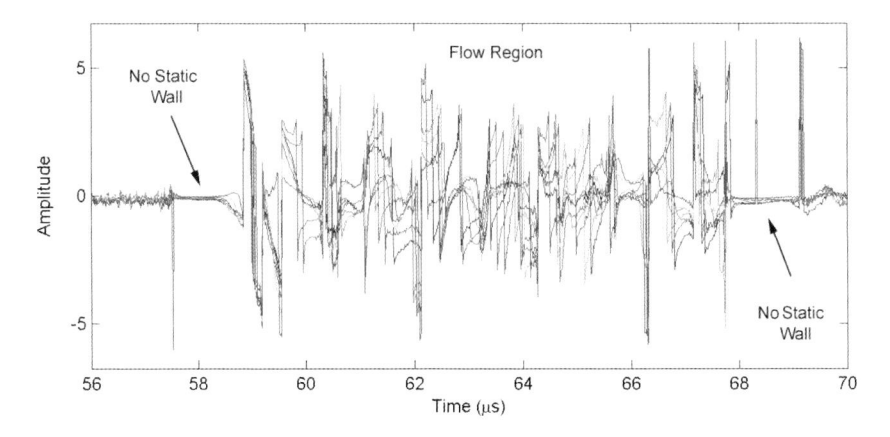

Figure 2.15 Phase differences of Hilbert transform of the multiple scatterers from Figure 2.12: series of scattered A-scans from experimental data.

2.3.7
Hilbert Methodology: How to Create an Image, and 3D Image

Exactly like the Fourier technique the image is formed by summing the absolute value of the phase difference of the scan at position x, and now also along the time axis t [$\Phi(x,t)$]:

$$I(x,t) = \sum_m |\Delta\Phi_m(x,t)|^2 \qquad (2.18)$$

where the sum is over pairs of phase differences [as shown in Equation (2.11) or (2.12)]. Owing to the unpredictable nature of the flow's scattering on the time domain, this process will of course need to be examined over some finite time (t_1 to t_2) to create a quality image in reasonable time constraints; this introduces an additional sum (or integral) into the equation:

$$I(x,t) = \sum_m \sum_{t=t_1}^{t_2} |\Delta\Phi_m(x,t)|^2 \qquad (2.19)$$

When the time summation covers the entire A-scan a flat image (Figure 2.16, with cross section in Figure 2.19 below) similar to the Fourier transform case

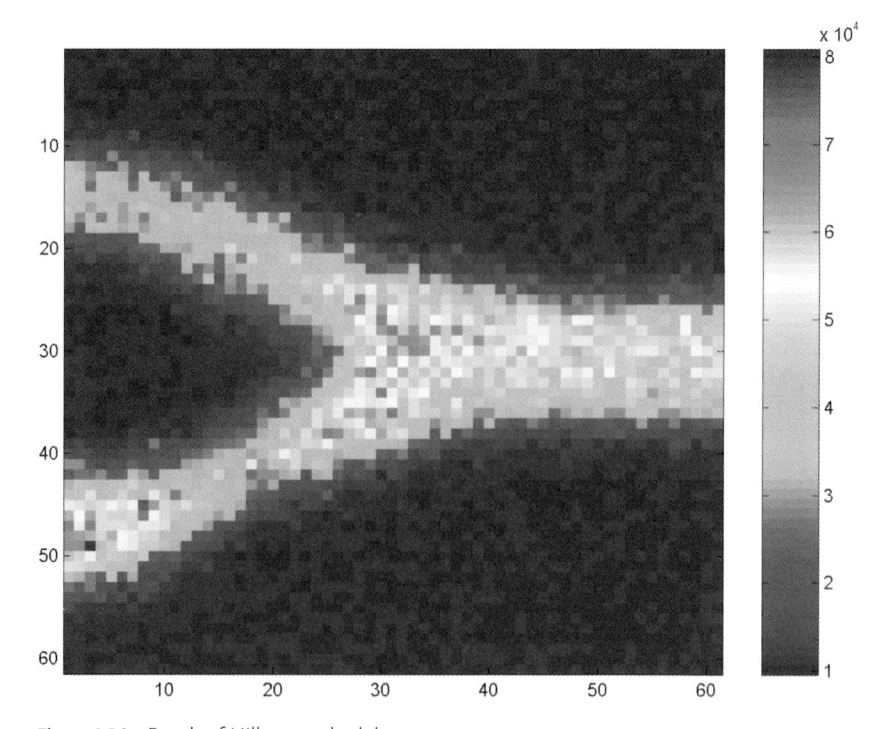

Figure 2.16 Result of Hilbert methodology.

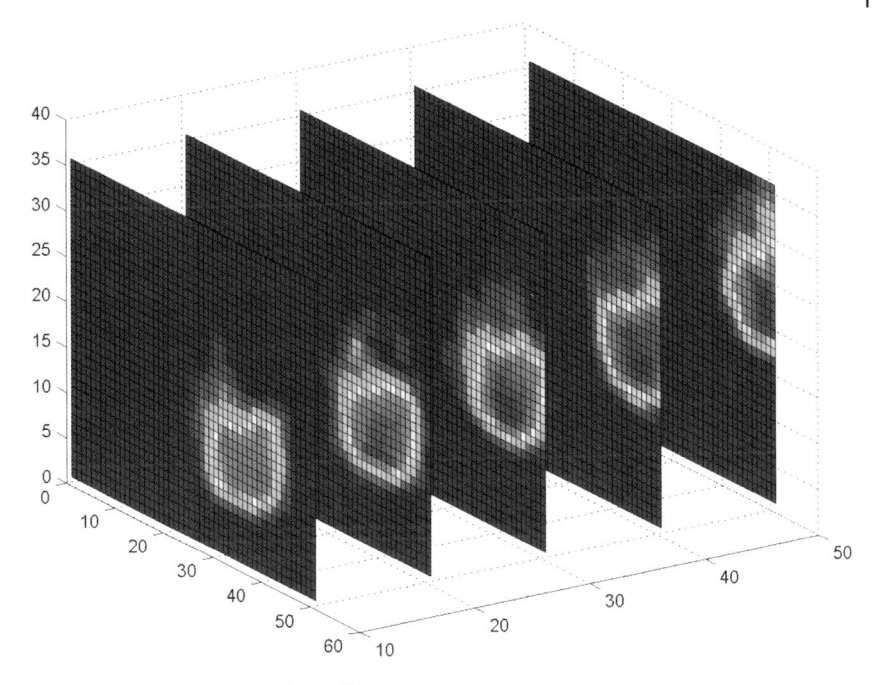

Figure 2.17 Sliced cross sections of the image.

is produced. Here the technique is able to correctly identify the regions of flow, and no flow, without requiring knowledge of where the flow is created, or using thesholding to eliminate these regions from processing as was done for the Fourier transform (Figure 2.15). Here the bifurcation in the flow is evident, as well as a slight narrowing in the lower channel. Examining the scale also gives qualitative information on the number of scatters flowing through each branch.

The time summation in Equation (2.19) could also be processed in several discrete time sections, creating a situation similar to MRI and CAT scan images where one can examine slices at different depths. These various slices can then be processed further to show cross sections of the flow region (Figure 2.17), or reassembled with three-dimensional modeling techniques to show the outer surface of the flow, creating a fully rotatable image (Figure 2.18).

While such 3D images create excellent opportunities to visualize the system, simple one-dimensional images can also be made (or extracted from the previous results), and provide an opportunity to make physical measurements. Figure 2.19 shows an example of measuring the diameter of the flowing region used in the previous results.

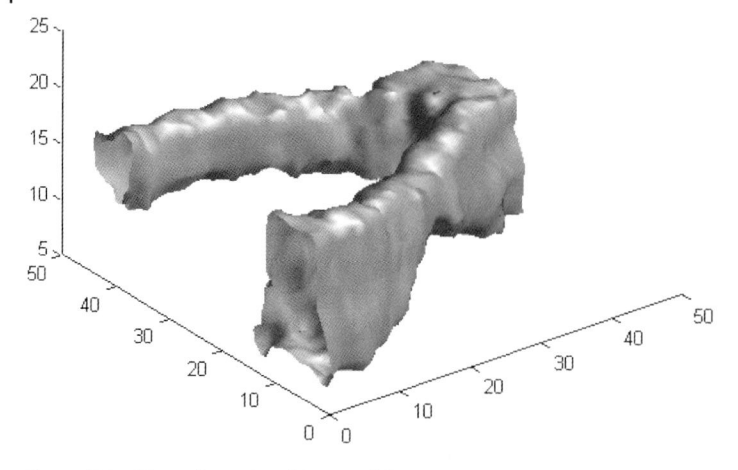

Figure 2.18 Three-dimensional image of flowing region.

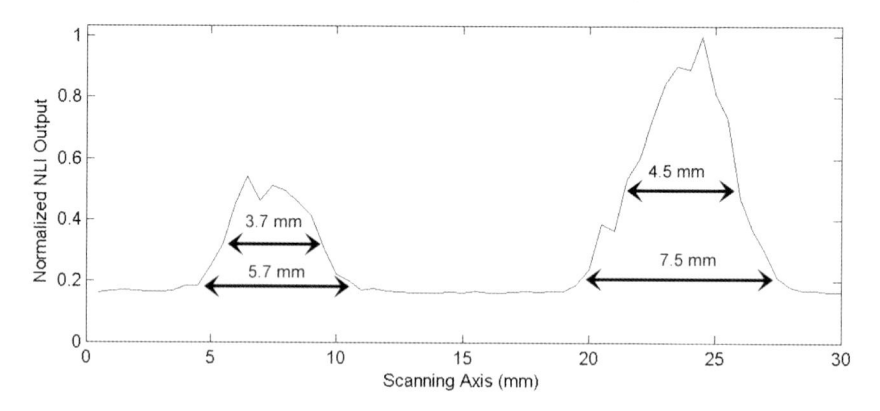

Figure 2.19 Diameter measurement via one-dimensional cross sections.

2.4
Summary and Closing

Both speckle interferometry and the nonlinear imaging flow detection methodologies discussed above have proven to be powerful and reliable techniques to improve imaging results. Each relies upon the idea that the speckles, or seemingly "random noise," in short exposures carry information that when properly processed can create better images than long exposure processes.

Selected References (By Subject)

Speckle: Base Methods

1. Labeyrie, A. (1970) Attainment of diffraction limited resolution in large telescopes by Fourier analysing speckle patterns in star images. *Astron. Astrophys.*, **6**, 85–87.
2. Knox, K.T. (1976) Image retrieval from astronomical speckle pattern. *J. Opt. Soc. Am.*, **66** (11), 1236–1293.
3. Knox, K.T. and Thompson, B.J. (1974) Recovery of images from atmospherically degraded short-exposure photographs. *Astrophys. J.*, **193**, L45–L48.
4. Weigelt, G. (1988) Interferometric imaging in optical astronomy, in *Evolution of Galaxies. Astronomical Evaluations* (ed. I Appenzeller), Lecture Notes in Physics, vol., **333**, Springer, pp. 285–298.

Speckle: More Advanced Methods

5. Aime, C. (2001) Teaching astronomical speckle techniques. *Eur. J. Phys.*, **22**, 169–184.
6. Fontanella, J.C. and Seve, A. (1987) Reconstruction of turbulence-degraded images using the Knox-Thompson algorithm. *J. Opt. Soc. Am.*, **4** (3), 438–448.
7. Lohmann, A.W., Weigelt, G., and Wirnitzer, B. (1983) Speckle masking in astronomy: triple correlation theory and applications. *Appl. Opt.*, **22** (24), 4028–4037.
8. Labeyrie, A., Lipson, S.G., and Nisenson, P. (2006) *An Introduction to Stellar Speckle Interferometry*, Cambridge University Press.
9. Frost, R.L., Rushforth, C.K., and Baxter, B.S. (1979) Fast FFT-based algorithm for phase estimation in speckle imaging. *Appl. Opt.*, **18** (12), 2056–2061.
10. Northcott, M.J., Ayers, G.R., and Dainy, J.C. (1988) Algorithms for image reconstruction from photon-limited data using the triple correlation. *J. Opt. Soc. Am.*, **5** (7), 986–992.

Nonlinear Imaging

11. Mitra, S.K. and Sicuranz, G.L. (2001) *Nonlinear Image Processing*, Academic Press.
12. Zyikova, N.V., Kondrat'eva, T.V., and Svet, V.D. (2001) Visualization of blood flow by ultrasound speckle interferometry. *Acoust. Phys.*, **47** (5), 578–584.
13. Zyikova, N.V., Kondrat'eva, T.V., and Svet, V.D. (2003) Acoustical images of objects moving under an inhomogeneous layer. *Acoust. Phys.*, **49** (2), 148–157.
14. Zyikova, N.V., Svet, V.D., and Shatskov, Y.A. (2006) Determination of the path of a sound source moving in an inhomogeneous medium. *Acoust. Phys.*, **52** (5), 561–570.

Part Two
Novel Developments in Advanced Imaging Techniques and Methods

3
Fundamentals and Applications of a Quantitative Ultrasonic Microscope for Soft Biological Tissues

Kazuto Kobayashi and Naohiro Hozumi

3.1
General Introduction: Basic Idea of an Ultrasonic Microscope for Biological Tissues

This chapter describes microscopic observation of biological soft tissues by ultrasonic measure. In comparison with optical microscopy, ultrasonic microscopy provides quantitative acoustic parameters like sound speed and characteristic acoustic impedance that are relevant to elastic properties. In addition, as it needs no staining process, the observation can be performed rapidly without introducing any chemical or biological damage to the specimen. Two types of ultrasonic microscopes that are being developed by the authors will be described.

If the soft tissue can be treated as fluid-like, we may assume that only pressure waves (longitudinal waves) can propagate through the specimen. The sound speed of a pressure wave is given as:

$$c = \sqrt{K/\rho}$$

where K is the elastic bulk modulus and ρ is the specific gravity. As can be seen, the sound speed strongly reflects its elastic parameter. To measure the local sound speed, a sliced specimen is usually prepared, and the time lag between two reflections from front and rear surfaces of the specimen is measured. In many cases, as the specimen is very thin, two reflections overlap in the time domain. Therefore, some signal processing is needed to separate the reflections.

In some cases, however, it may be required that the observation is performed without slicing the tissue. In such cases, it is not possible to acquire the reflections from both front and rear surfaces, making it difficult to evaluate the sound speed. Conversely, the reflection ratio of the surface (or at the interface between the tissue and a certain medium) is determined as:

$$R = \frac{Z_x - Z_0}{Z_x + Z_0} \quad (3.1)$$

Advances in Acoustic Microscopy and High Resolution Imaging: From Principles to Applictaions, First Edition.
Edited by Roman Gr. Maev.
© 2013 Wiley-VCH Verlag GmbH & Co. KGaA. Published 2013 by Wiley-VCH Verlag GmbH & Co. KGaA.

where Z_x and Z_0 are the characteristic acoustic impedances (CAIs) of the tissue and medium. Equation (3.1) assumes that the incidence of the wave is perpendicular to the interface. The CAI can be obtained by solving the above equation. The CAI reflects the elasticity as:

$$Z = \sqrt{K\rho} \qquad (3.2)$$

Therefore, measurement of the CAI is basically equivalent to that of the sound speed.

3.2
Sound Speed Profile

3.2.1
Fundamentals

The sound speed profile is a conventional means for the quantitative assessment of soft tissues. The sound speed has a close relation with bulk modulus. Much data have been acquired, since this type of microscope has a relatively long history. Recent digital techniques, coupled with a wide band pulse wave, have made it possible to rapidly perform the data acquisition followed by digital signal processing to precisely calculate the sound speed and other parameters. However, the specimen for sound speed measurement needs to be sliced to a thickness of several micrometers before being extended on a substrate.

The sound speed of soft tissue can be assessed by either time domain or frequency domain analysis. Basically it can be determined by comparing the reflections from the front and rear surfaces of a tissue slice dipped into a coupling medium. If the thickness d of the tissue is known, the sound speed c can be calculated as:

$$c = \frac{d}{\Delta t} \qquad (3.3)$$

where Δt is the time lag between the reflections from front and rear surfaces. However, in most cases, it is not easy to precisely measure the thickness at the point where the sound beam is focused. Therefore, often, both thickness and sound speed are simultaneously assessed by referring to the sound speed of the coupling medium.

3.2.2
Specimen to be Observed

Figure 3.1 illustrates the concept of the ultrasonic sound speed microscope for tissue characterization. An acoustic wave is transmitted and received by the same transducer. Distilled water is used as the coupling medium between the specimen

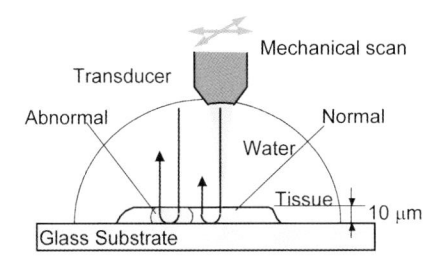

Figure 3.1 Illustration of an ultrasonic sound speed microscope.

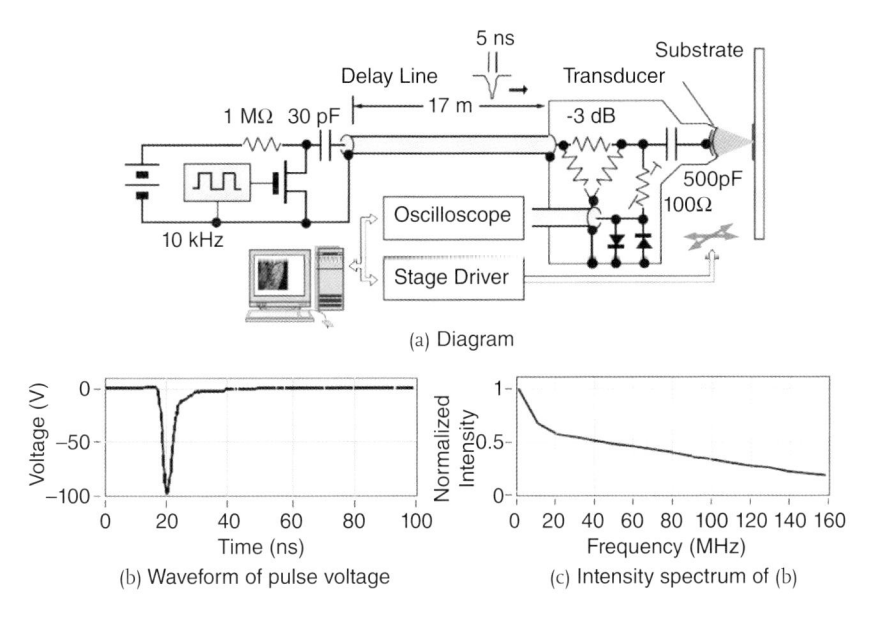

Figure 3.2 Schematic diagram of the ultrasonic sound speed microscope, and waveforms of the electric pulse in the time and frequency domains.

and the transducer. Reflections from both sides of the tissue are compared to determine the sound speed and thickness. Two-dimensional profiles of reflection intensity, thickness, and sound speed can be obtained by mechanically scanning the transducer. A soft tissue is sliced as thin as 10 μm and placed onto a slide glass. In most cases, good adhesion between soft tissue and glass substrate is maintained even after a water droplet, which acts as a coupling medium, covers the specimen.

3.2.3
Experimental Setup and Acquired Signal

Figure 3.2 shows a schematic diagram of a typical measurement system at the authors' laboratory. The 30 pF capacitor was charged at several tens of volts, and discharged through the transistor switch. A pulse voltage about 5 ns in width is generated. The repetition rate of the pulse was 10 kHz. The transducer has an aperture diameter of 1.2 mm and a focal length of 1.5 mm. Its nominal frequency range is 50–105 MHz (−6 dB), with the central frequency being 80 MHz. An acoustic wave with a wide frequency component is generated by applying the voltage pulse, and is then transmitted to the substrate. The reflection was detected by the same transducer, and was introduced into the digital oscilloscope (or a digitizer). The band limit and sampling rate are typically 300 MHz and 2.5 GS s^{-1}, respectively. To reduce random noise, four response times at the same point are averaged in the oscilloscope before being introduced into the computer. The transducer is mounted on an X-Y stage that is driven by the computer through a General Purpose Interface Bus. Considering the focal distance and the sectional area of the transducer, the diameter of the focal spot is estimated to be 20 μm at 80 MHz. Therefore, the distance between the nearest two points is set at 20 μm.

3.2.4
Calculation of Sound Speed

3.2.4.1 Frequency Domain Analysis [1]

Figure 3.3 shows reflected waveforms as examples. A waveform at the glass surface without tissue is shown in Figure 3.3a. This signal is employed as a reference waveform, with the point being defined as the reference point. The decline of the glass surface is compensated by considering the time lags at three different points, including the reference point, without the tissue. Figure 3.3b shows a reflection from where the tissue is placed. The corresponding measuring point will be indicated below in Figure 3.7a. The waveform in Figure 3.3b contains two reflections

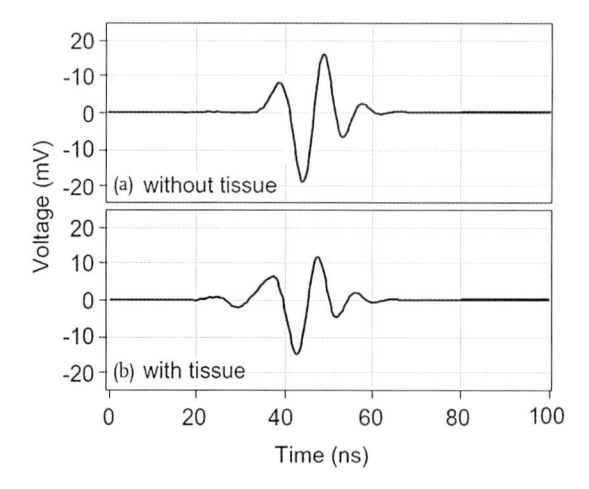

Figure 3.3 Waveforms of the acoustic pulse: (a) from the point where no tissue is placed; (b) from the point where tissue specimen is placed.

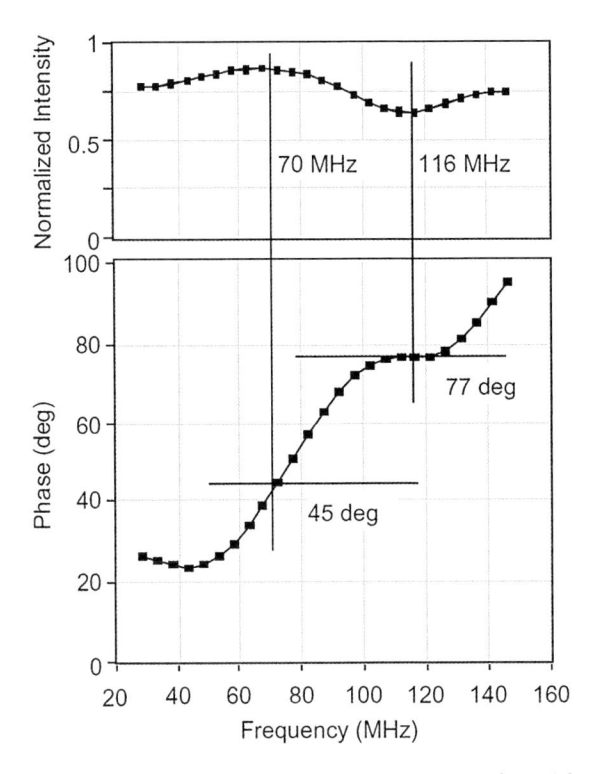

Figure 3.4 Result of a frequency domain analysis performed for the interfered waveform.

at front and rear sides of the tissue; however, it is not easy to separate into two independent signals.

Analysis in the frequency domain is performed by assuming interference of the two reflections. Intensity and phase spectra are calculated by Fourier-transforming the waveform. The result is shown in Figure 3.4. The spectrum is normalized by the reference waveform. Assuming f_m as one of the minimum and maximum points in the intensity spectrum, and φ_m as the corresponding phase angle, the phase difference between the two reflections at the minimum point is $(2n - 1)\pi$, giving:

$$2\pi f_m \times \frac{2d}{c_0} = \varphi_m + (2n + 1)\pi \tag{3.4}$$

where d, c_o, and n are the tissue thickness, sound speed of the water, and a non-negative integer, respectively. The phase difference at the maximum point is $2n\pi$, giving:

$$2\pi f_m \times \frac{2d}{c_0} = \varphi_m + 2n\pi \tag{3.5}$$

3 Fundamentals and Applications of a Quantitative Ultrasonic Microscope

The phase angle φ_m can be expressed by:

$$2\pi f_m \times 2d\left(\frac{1}{c_0} - \frac{1}{c}\right) = \varphi_m \tag{3.6}$$

since φ_m is the phase difference between the wave that has passed through the distance $2d$ with sound speed c and that though the corresponding distance with sound speed c_0. By solving the simultaneous Equations (3.4) and (3.6):

$$d = \frac{c_0}{4\pi f_m}\{\varphi_m + (2n-1)\pi\} \tag{3.7}$$

is obtained for the minimum point. For the maximum point, Equations (3.5) and (3.6) give:

$$d = \frac{c_0}{4\pi f_m}\varphi_m + 2n\pi \tag{3.8}$$

Finally, the sound speed is calculated as:

$$c = \left(\frac{1}{c_0} - \frac{\varphi_m}{4\pi f_m d}\right)^{-1} \tag{3.9}$$

In Figure 3.4, the first maximum in the intensity spectrum appears at 70 MHz, and the phase angle at this frequency is 45°. Substituting them in Equations (3.8) and (3.9), the thickness and sound speed are determined as 11.8 μm and 1665 m s^{-1}, respectively. The second minimum at 116 MHz and its corresponding phase angle give a thickness of 10.9 μm and a sound speed of 1691 m s^{-1}. This result suggests that the sound speed as a function of frequency is not totally flat but tends to increase with increasing frequency.

3.2.4.2 Time–Frequency Domain Analysis [2]

The waveform from where a tissue was attached is subjected to deconvolution processing using the reference waveform from where no tissue is placed. The response obtained by the processing is subsequently subjected to a Gaussian filter to remove high frequency components. Figure 3.5 shows an example of the result. The reflections at the front and rear sides of the tissue are clearly seen as individual peaks.

These two peaks were separated by using proper window functions. The window function was originally a Gaussian function with 1 as its peak value, but the peak was flattened by splitting it at the peak point and inserting 1 with an appropriate length. Intensity and phase spectra of these separated waveforms were then calculated by Fourier transform. The result is shown in Figure 3.6. The spectra of each separated waveform were normalized by those of the reference waveform. The following two expressions derive at any frequency f:

$$2\pi f \times \frac{2d}{c_0} = \varphi_{front} \tag{3.10}$$

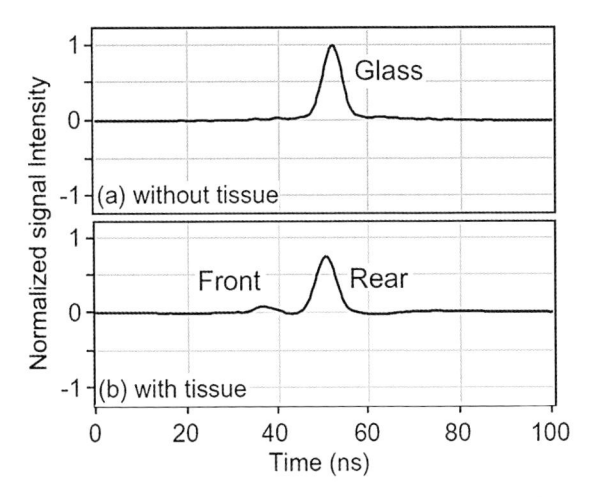

Figure 3.5 Reflected acoustic waves from the specimen after compensation.

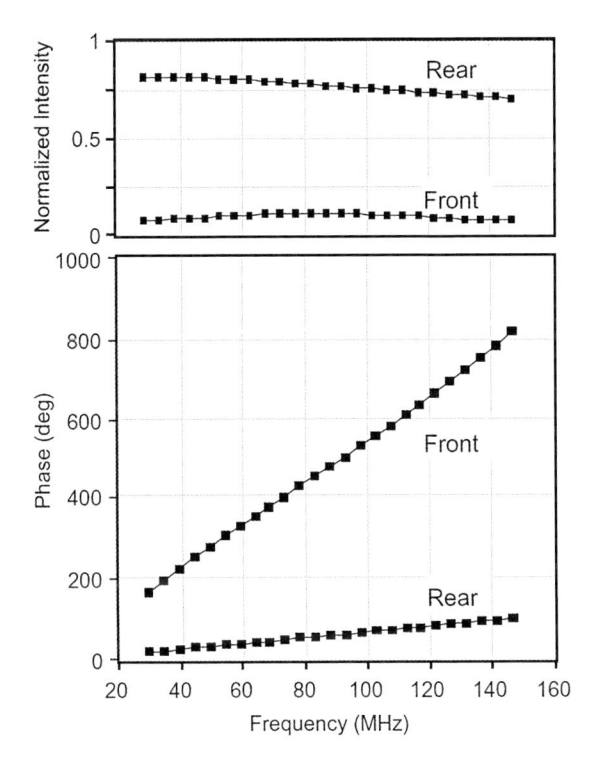

Figure 3.6 Result of a time–frequency domain analysis performed for the compensated waveform.

Table 3.1 Sound speed and thickness at various frequencies calculated by both frequency domain and time–frequency domain analyses.

Parameter	Frequency (MHz)								
	1st max. (70 MHz)	2nd min. (116 MHz)	50	70	90	110	120	130	150
Sound speed c (m s^{-1})	1665	1691	1654	1669	1682	1692	1694	1696	1694
Thickness d (μm)	11.8	10.9	12.0	11.6	11.3	11.1	11.0	11.0	11.2

$$2\pi f \times 2d \left(\frac{1}{c_0} - \frac{1}{c} \right) = \varphi_{\text{rear}} \qquad (3.11)$$

Here, φ_{front} and φ_{rear} are phase angles obtained from the front and rear peaks, respectively. Finally, the thickness d and sound speed c at frequency f are calculated as:

$$d = \frac{c_0}{4\pi f} \varphi_{\text{front}} \qquad (3.12)$$

$$c = \left(\frac{1}{c_0} - \frac{\varphi_{\text{rear}}}{4\pi f d} \right)^{-1} \qquad (3.13)$$

Table 3.1 shows the results calculated in this way at various frequencies within 50–150 MHz, while the result of frequency domain analysis at two different frequencies is shown in Figure 3.4. It is seen that the sound speed is within the range 1650–1690 m s^{-1}, and tends to increase with increasing frequency. Furthermore, the results of the frequency domain analysis at two different frequencies agree with the results of time–frequency analysis at the corresponding frequencies.

3.2.5
Two-Dimensional Sound Speed Profiles

Figure 3.7 shows two-dimensional sound speed profiles at the second minimum in frequency domain analysis and at different frequencies as the results of time–frequency domain analysis. Optical microscopic inspection of the very next slice of the same specimen showed massive hyalinization in the endocardial side (left-hand side of the figure), which was classified as severe allograft rejection. In the case of frequency domain analysis, the sound speed in the hyalinized lesion is 1530–1590 m s^{-1}, which is significantly lower than that of normal myocardium (1600 m s^{-1} or faster). The profiles obtained by time–frequency domain analysis

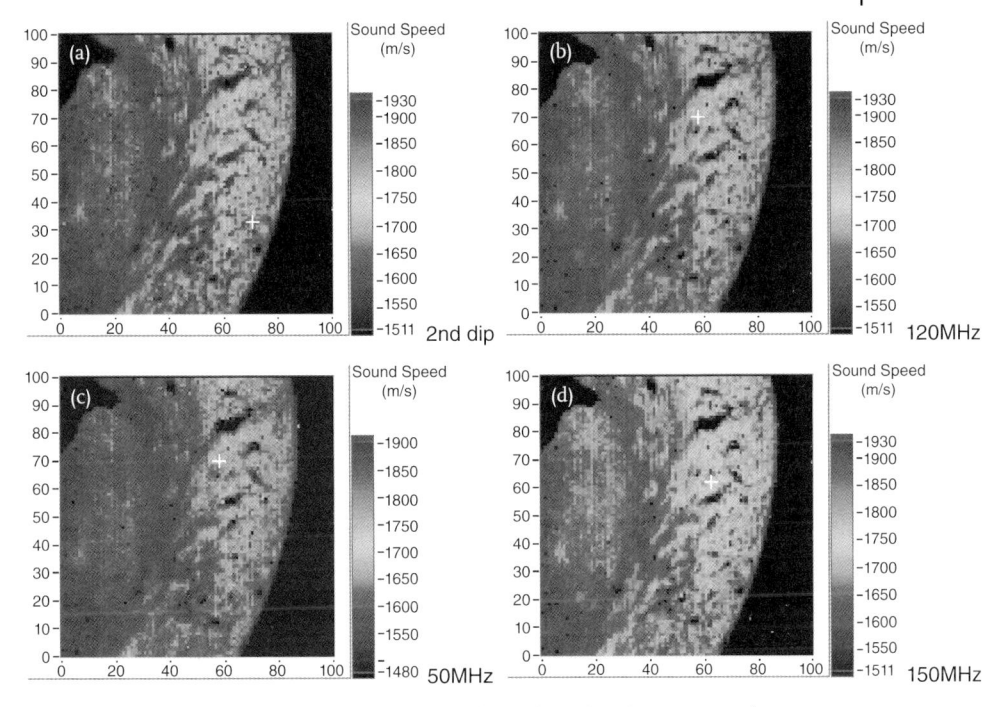

Figure 3.7 Two-dimensional profiles of the specimen with the resolution of 100 × 100 pixels. (a) Frequency domain analysis: 2nd minimum; (b)–(d) time–frequency domain analysis: (b) 50 MHz (note that the scale is different from the other micrographs), (c) 120 MHz, and (d) 150 MHz. Examples of analyses mentioned in the text were performed for the point marked in (a).

look very similar in shape, suggesting its compatibility with the frequency domain analysis. However, profiles by the time–frequency domain analysis suggest that the sound speed tends to be higher with increasing frequency at all parts of the tissue.

As was seen above, sound speed at an arbitrary frequency can be calculated by time–frequency analysis. In addition, the two peaks in the time domain make it easy to understand that the waveform is composed of two reflections at the front and rear sides of the tissue. There may be more than three reflection interfaces depending on the internal structure of the tissue. Such an additional reflection will bring a significant error if the analysis is performed in the frequency domain. In such a case, more than three peaks will appear in the compensated waveform in the time domain. The internal reflection can be eliminated by using an appropriate window function. Similarly, multiple reflections between two surfaces may affect the precision in frequency domain analysis. This can be resolved by time–frequency domain analysis, because multiply reflected signals appear behind two major peaks after the deconvolution process, and will be eliminated using a window function.

In the work reported here, only the sound speed profile was observed. However, as both intensity and phase spectra of two reflections can be obtained, some other parameters such as attenuation and acoustic impedance may be determined and visualized as functions of frequency. The authors believe that the new technique can be a powerful tool for tissue characterization.

3.2.6
Attempts at Better Spatial Resolution

The reflected waveform at one point in the view plane includes information within the focal spot. The obtained sound speed image is therefore blurred depending on the pattern of focal beam. However, as the sound speed at a certain point is determined through a complicated phase analysis along the frequency it is not easy to perform the image processing to restore the blurred sound speed image.

On the other hand, the blurred image of reflection intensity is considered to be approximately the convolution of the spatial distribution of beam intensity and the reflection coefficient (note that this apparent reflection coefficient is composed of those at front and rear surface of the tissue). Therefore, if the beam pattern is known, a deconvolution process may restore the blurred intensity image into a "sharpened" image. Although this restored intensity image is not directly correlated to the acoustic property, this sharpened image would help us to observe minute morphology of the tissue. The system would be improved so that the "fine intensity image for morphological inspection" makes up for the "sound speed image with some restriction in spatial resolution for quantitative inspection" for better characterization of the tissue.

When an image is created by sending and receiving ultrasonic waves with an ultrasonic microscope, the resultant image is less clear than the actual structure of the object (Figure 3.8). The blur depends on the directional properties of the ultrasonic transducer. The blurred image is considered to be the result of applying the blurring function determined by the directional property of the transducer to the true structure of the object. For two-dimensional cases, the relationship between the true distribution of acoustic property and the observed acoustic intensity image is expressed as:

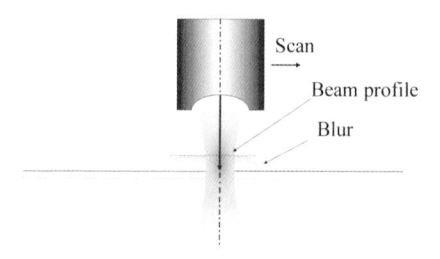

Figure 3.8 Blur under an ultrasonic microscope.

$$y(i,j) = \sum_{n,m} h(n,m)x(i-n, j-m) = h(i,j) \otimes x(i,j) \qquad (3.14)$$

$x(i,j)$ is the true (exact) distribution of the acoustic property,
$y(i,j)$ is the blurred acoustic intensity image,
$h(i,j)$ is the blurring function of the observation system,
$\otimes$ represents convolution.

When the ultrasonic beam is focused at only one point of the view plane, the blurring function becomes a delta function:

$$h(i,j) = \begin{matrix} 1 \{i=0, j=0\} \\ 0 \{ \text{else} \} \end{matrix} \qquad (3.15)$$

The blurring function in the spatial frequency domain (represented by wave numbers) is totally flat. In this case:

$$y(i,j) = x(i,j) \qquad (3.16)$$

the observed acoustic intensity image coincides with the true distribution of acoustic property. On the other hand, when the beam is not focused at one point but spreads on the view plane, the blurring function is no longer a delta function. Consequently, in the frequency domain, the blurring function has a frequency dependence. Its characteristic is such that the signals are more attenuated as the frequency becomes higher. As a result, the observed image becomes blurred and differs from the ideal image.

In the blurred image the high frequency component is less significant. However, the component is not completely lost. By applying the Fourier transform to Equation (3.1), the relationship in the spatial frequency domain is represented as follows:

$$Y(k_i, k_j) = H(k_i, k_j) \cdot X(k_i, k_j) \qquad (3.17)$$

where k represents the wave number. This transforms the convolution relationship in Equation (3.14) into a simple product relationship in Equation (3.17). This way, restoration of the observed image into the true distribution in the frequency domain is expressed by the following equation:

$$X(k_i, k_j) = \frac{Y(k_i, k_j)}{H(k_i, k_j)} \qquad (3.18)$$

The true distribution can be obtained through the inverse Fourier transform. Prior to performing the above process, the blurring function H should be determined. It is realized by observing the reference object for which the true distribution of the acoustic property is known:

$$H(k_i, k_j) = \frac{Y_{\text{ref}}(k_i, k_j)}{X_{\text{ref}}(k_i, k_j)} \qquad (3.19)$$

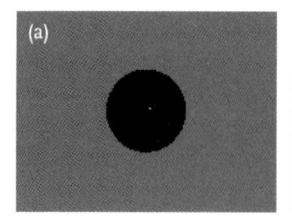

Figure 3.9 (a) Ideal image based on optical microscopic observation and (b) actual ultrasonic image of a pin-hole 200 µm in diameter.

The function Y can be restored into X by applying $1/H$. Hereafter this $1/H$ is defined as the restoring function.

As a reference object for obtaining the blurring function H we used a metal plate sample with a pinhole 200 µm in diameter. First, its fine image was observed by an optical microscope (Figure 3.9a). Figure 3.9b shows the acoustic intensity image observed by the acoustic microscope. The acoustic reflection ratio in the area of the pinhole was assumed to be zero (0) and the other part of the image was filled with the average number of the acoustic intensity at sufficient distance from the edge of the pinhole.

The image processing was applied to the acoustic intensity images of surgically excised tissues such as esophagus, mammary gland, and stomach. In each figure set, the left-hand image is the optical image stained (hematoxylin–eosin staining), the center image is the originally obtained acoustic intensity image with blurring, and the right-hand image is the processed image. Figure 3.10a shows the example of an esophagus. Blur is compensated and the clear layered structure of the esophagus, consisting of mucosal layer, submucosal, and muscularis, is shown. The size of a single gland in the mucosal layer is approximately 40 µm. However, the resolution of the original acoustic image obtained with the central frequency of 100 MHz is approximately 18 µm. Thus, this structure cannot be clearly observed in the original image. In contrast, the structure is clearly observed in the processed image at the same resolution as the optical image. Figure 3.10b is the example of a mammary gland. The muscle fiber orientation is seen in the processed image. Figure 3.10c is the example of a stomach. The layered appearance of the stomach wall is clearly seen in the processed image.

3.3
Acoustic Impedance Profile

3.3.1
Fundamentals

In the previous section, we proposed a pulse driven ultrasonic sound speed microscopy that can obtain a sound speed image. Although a small degree of roughness

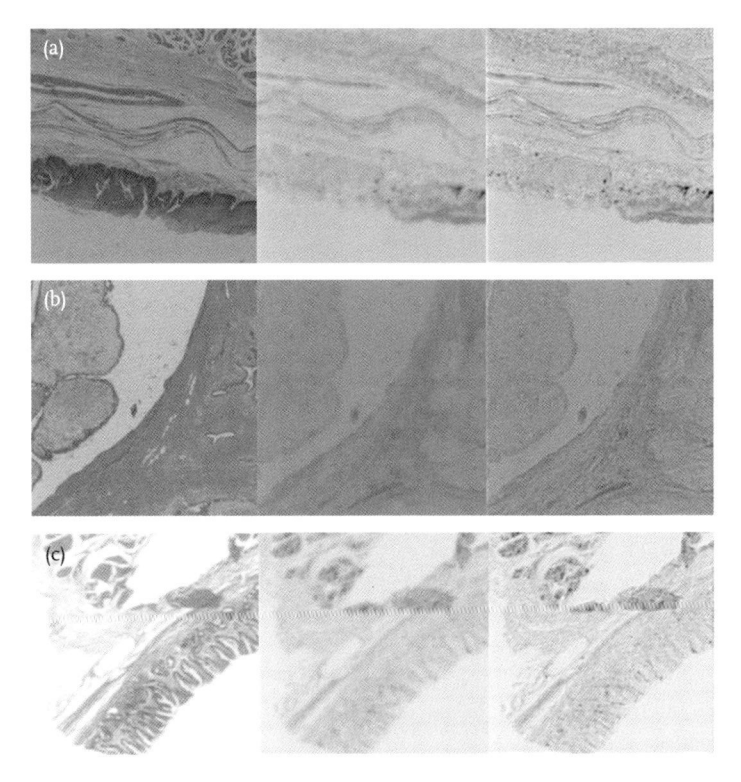

Figure 3.10 Improvement of acoustic image of body tissue: (a) esophagus, (b) mammary gland, and (c) stomach. Optical microscopy, stained (left), original acoustic image (middle) and improved acoustic image (right).

of the specimen may allowed in this type of microscope, slicing the specimen into sections several micrometers thick is still required for the observation. However, in many cases the slicing process needs to be avoided, as slicing may damage some functions of the tissue. In addition, the tissue may also be damaged if it is in contact with a liquid coupling medium such as pure water.

Based on the above, we propose an acoustic impedance microscopy that can image the local distribution of cross sectional acoustic impedance of tissue. As acoustic impedance is given as a product of sound speed and density, it would have a good correlation with sound speed, when the variance in density was not significant. This section describes the methodology for micro-scale imaging of cross sectional acoustic impedance.

3.3.2
Experimental Setup

Figure 3.11 illustrates the outline of the acoustic impedance microscope. Distilled water is used for the coupling medium between "the substrate and transducer."

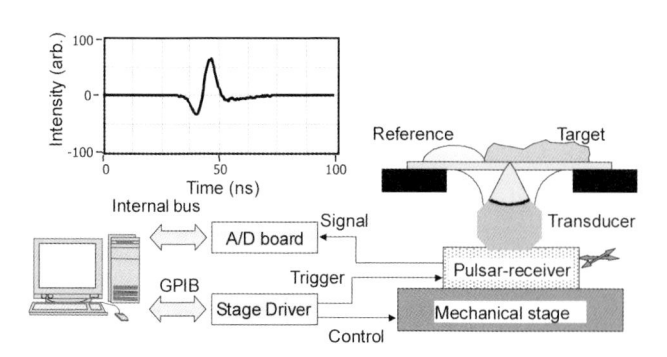

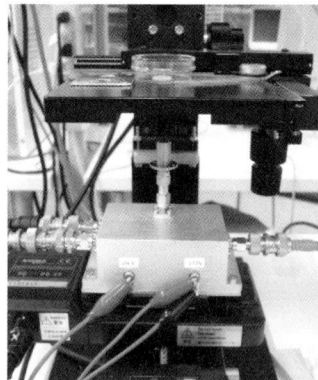

Figure 3.11 Schematic diagram of the measurement system and acoustic waveform reflected from the target.

A step voltage – as high as 40–60 V in peak voltage and as short as 500 ps in rising time – is generated. The maximum repetition rate of the pulse is as high as 10 kHz. The transducer was a PVDF-TrFE type. It is typically about 1.5 mm in aperture diameter, and 3.0 mm in focal length. An acoustic wave with a wide frequency component is generated by applying the voltage pulse. The acoustic wave, being focused on the interface between the substrate and tissue, is transmitted and received by the same transducer.

The reflection is detected and digitized by an oscilloscope or digitizer. Considering the focal distance and the sectional area of the transducer, the diameter of the focal spot is estimated to be about 26 µm at 80 MHz. The distance between the nearest two points is typically set at 10 µm, to retain sufficient lateral resolution. A two-dimensional profile of acoustic impedance is obtained by mechanically scanning the transducer using the stage driver, keeping the focal point on the rear surface of the substrate. A typical field of view of 2 mm × 2 mm is covered with 200 × 200 pixels.

It takes about 1 min for one observation. To save the time for data transfer from the oscilloscope (or digitizer) to the computer, the waveforms through each X-scan are stored in the oscilloscope using its fast-frame mode before being transferred through the interface. To reduce random noise, three or four response times at the same point are averaged.

3.3.3
Specimen to be Observed

The cross section of a soft tissue specimen is in contact with the substrate. The substrate is a flat plastic plate made of poly(methyl methacrylate) (PMMA) or polystyrene that is 0.5–1 mm thick. A reference material, with a known acoustic impedance, is also placed on the same substrate. In many cases, the target tissue is observed together with the reference, by including both of them in the same field of view.

In some cases, the surface of the substrate is treated beforehand with an atmospheric plasma for 3 s, by use of plasma surface treatment equipment, to upgrade its hydrophilicity.

Silicone rubber, distilled water, or agar may be employed as a reference material, with the choice depending on the convenience of the measurement. When using silicone rubber, the observation was performed 24-h later since the rubber had to be hardened, to retain the stability of the material.

Considering the precision of the calibration, the reference material should be stable both physically and chemically, and should be in complete contact with the substrate. It is recommended that the acoustic impedance of the reference is close to that of the target. Furthermore, as for the substrate, most available plastic materials have a higher acoustic impedance than biological tissues. In such cases, the phase of the transmitted signal is reversed at the interface. The acoustic impedance of the substrate should be sufficiently high compared to that of the target, to retain a strong reflection. However, an extremely high acoustic impedance of the substrate may increase the reflection coefficient at the interface between the coupling medium and substrate, and reduce the intensity of transmitted signal to the target. This would obviously reduce the signal-to-noise ratio. Therefore, to obtain a good signal-to-noise ratio, the material should be carefully selected.

3.3.4
Acquired Signal

Figure 3.12 shows the reflected acoustic signals. In this particular case, a water droplet was used as the reference. A part of cerebellum tissue was used as a target. The signal from the target tissue is very similar to that from the reference, suggesting the acoustic impedance of the tissue is close to that of water (1.5×10^6 Ns m^{-3}). An intensity spectrum of the target signal normalized by the reference signal, and a cross power spectrum of the target and reference signals are also shown (Figure 3.12b). The intensity spectrum is almost flat from 15 to 100 MHz, with the intensity being a little less than 1.0. This indicates that the impedance of the target is somehow different from the reference. The calibration method for the acoustic impedance will be described in the following section. The cross-power spectrum of the target and reference signals shows that a wideband acoustic signal had been successfully generated.

3.3.5
Calibration for Characteristic Acoustic Impedance [3]

Figure 3.13 illustrates the calibration of acoustic impedance. The target signal is compared with the reference signal. Hereafter, the signal component at an

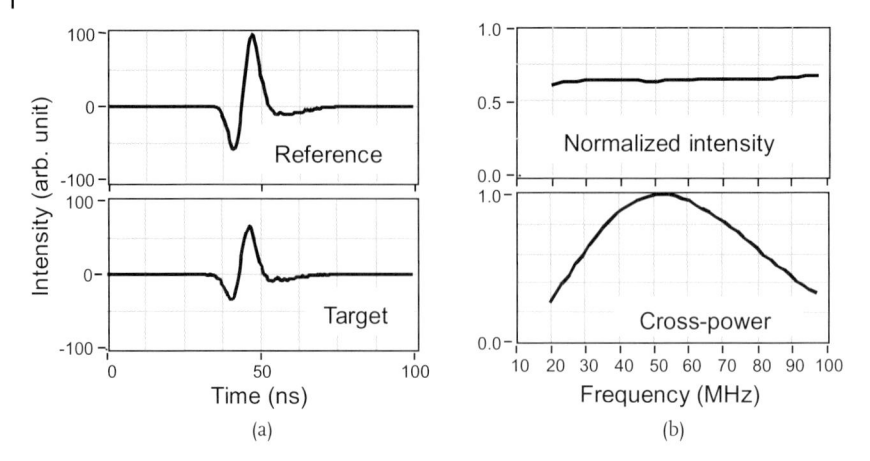

64 | *3 Fundamentals and Applications of a Quantitative Ultrasonic Microscope*

Figure 3.12 Waveforms in (a) time domain and (b) frequency domain.

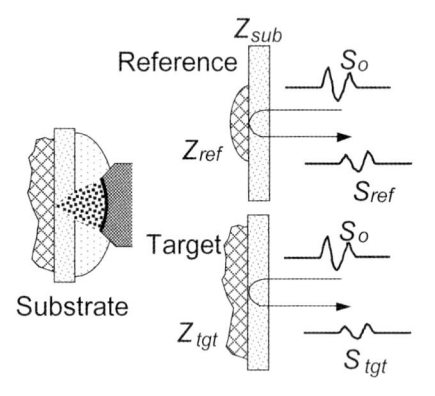

Figure 3.13 Illustration for calibration of the acoustic impedance.

arbitrary frequency will be symbolized by S. Considering the reflection coefficient, the target signal S_{target} can be described as:

$$S_{target} = \frac{Z_{target} - Z_{sub}}{Z_{target} + Z_{sub}} S_0 \qquad (3.20)$$

where S_0 is the transmitted signal and Z_{target} and Z_{sub} are the acoustic impedances of the target and substrate, respectively. On the other hand, the reference signal can be described as:

$$S_{ref} = \frac{Z_{ref} - Z_{sub}}{Z_{ref} + Z_{sub}} S_0 \qquad (3.21)$$

where Z_{ref} is the acoustic impedance of the reference material. We can measure S_{target} and Z_{ref}; however, S_0 cannot be directly measured. The acoustic impedance of the target is subsequently calculated as a solution of the simultaneous equations for Z_{target} and S_0, as:

$$Z_{target} = \frac{1+\dfrac{S_{target}}{S_0}}{1-\dfrac{S_{target}}{S_0}} Z_{sub} = \frac{1-\dfrac{S_{target}}{S_{ref}} \cdot \dfrac{Z_{sub}-Z_{ref}}{Z_{sub}+Z_{ref}}}{1+\dfrac{S_{target}}{S_{ref}} \cdot \dfrac{Z_{sub}-Z_{ref}}{Z_{sub}+Z_{ref}}} Z_{sub} \qquad (3.22)$$

assuming that S_0 is constant throughout the observation process.

When using water as the reference, its acoustic impedance was assumed to be 1.52×10^6 Ns m^{-3}. On the other hand, when using silicon rubber its acoustic impedance was calibrated, using water as the standard reference material. In this case, a figure of 0.98×10^6 Ns m^{-3} was used. The acoustic impedance of agar was calibrated in the same manner, a short time before the observation. It was calculated to be 1.65×10^6 Ns m^{-3}.

As the sound speed of the PMMA substrate at about 50 MHz (25 °C) and its specific gravity (25 °C) were 2.78 km s^{-1} and 1.16 mg mm^{-3}, respectively, the acoustic impedance of the substrate was calculated to be 3.22×10^6 Ns m^{-3}.

As the transducer was designed for usage with water as the coupling medium, the existence of the plastic plate between the transducer and focal point may cause an aberration. This will be significant if the substrate is very thick, and the convergence angle is very large. In this experiment, however, the angle was as small as 14°, suggesting the error caused the aberration would be small. Nevertheless, a quantitative analysis is needed to precisely assess the acoustic impedance, especially when a thick substrate is employed.

3.3.6
Observation of Cerebellar Cortex of a Rat [4]

The cerebellum tissue of a rat was employed as the specimen to be observed. Figure 3.14 illustrates the development of cerebellar cortex observed by the acoustic impedance microscope. Rats were dissected and their whole brains removed. Some of the isolated cerebellum was sliced (200-μm thick) using a rotor slicer (Dohan EM, Kyoto, Japan). The slices were incubated in oxygenated phosphate buffer solution (PBS) on ice for 1 h. They were chemically fixed with 4% formaldehyde fixative, for 20 min. For optical observation, some slices were subjected to

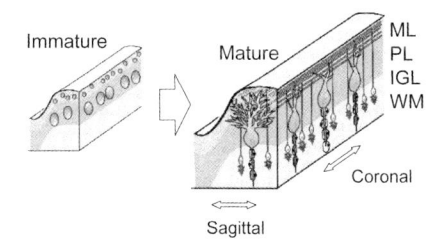

Figure 3.14 Illustration for the development of cerebellar cortex.

immunohistochemical staining against calbindin D-28k. Other specimens, the intact ones, were cut at an appropriate cross section. Both intact and fixed slices were rinsed and observed in the same PBS.

Parallel fibers in a molecular layer are axons of granule cells and play an important role in cerebella neuronal connections. Migrating granule cells elongate them horizontally and form a lot of excitatory synapses to dendrites of Purkinje cells. These are major neuronal circuits of cerebellum, so that parallel fibers are expected to construct a rich molecular layer with development. However, it was hard to evaluate a degree of parallel fiber development over more than a molecular layer. We have insufficient histochemical tools to visualize the developing parallel fibers. Although electron microscopy shows fine structure, it is limited to a very local image. As the field of view of the proposed acoustic technique is as wide as that of an optical microscope, it is expected to be a suitable replacement that can be used without any histochemical tools.

The granule cells migrate away from the ventricular zone, over the top of the developing Purkinje cells to form a secondary zone of neurogenesis, called the external granule layer. After the birth, the cells in this layer continue to actively proliferate, generating an enormous number of granule cell progeny at postnatal 7–10 days.

A short time after their generation, after their final mitotic division, the granule cells change from a very round cell to take on a more horizontally oriented shape as they begin to extend axons tangential to the cortical surface, called parallel fibers. Next, the cell body of granule cells migrate to deep into the cerebellum, and so the cell assumes a T shape. The cell body eventually migrates past the Purkinje cell layer and then begins to sprout dendrites in the granule cell layer until postnatal 20 days.

Purkinje cells, on the other hand, have arranged on the Purkinje cell layer at birth, whereas their dendrites are short and immature. They form synapses to parallel fibers of the granule cells and construct the cerebellar neuronal network. The network layers containing parallel fibers and Purkinje cell dendrites are called the molecular layers. Parallel fibers are thin and unmyelinated neuronal fibers. Because parallel fibers run transverse between the left and right hemispheres, sagittal sections of cerebellar cortex show the cross view of the parallel fibers and coronal sections show the side view.

Figure 3.15 shows the observed images of the cerebellar cortex of a rat at immature (P1; postnatal 1 day), transient (P7), and mature (P20) stages. All the specimens in Figure 3.15 had been chemically fixed.

In the immature cerebellar cortex (P1) (Figure 3.15a), the external granular layer (EGL), the outer layer of the cortex, showed a higher impedance than the inner layer. The area indicated by the rectangle in the acoustic image corresponds morphologically to the immunohistochemical observation, although the scale does not completely corresponded because the tissue was somehow subjected to compression during the acoustic observation. At this stage, as myelin is not yet generated, the existence of white matter (WM) is not clearly observed.

In the transient stage (Figure 3.15b), four different layers become comprehensive: the WM, internal granular layer (IGL), Purkinje layer (PL), and EGL. The

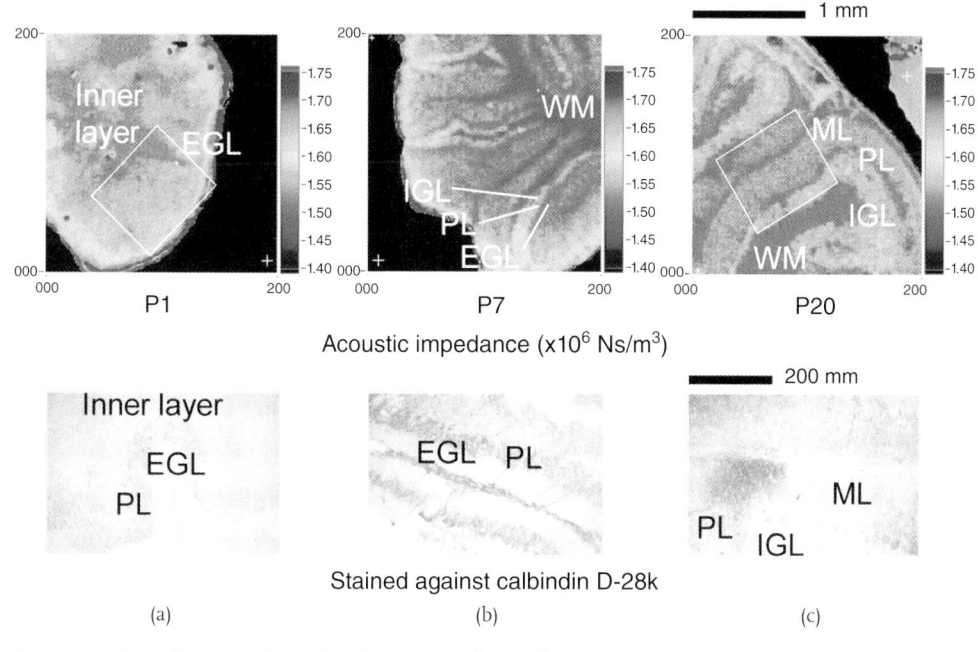

Figure 3.15 Two-dimensional profiles of cerebellar cortex at (a) immature (P1; postnatal 1 day), (b) transient (P7), and (c) mature (P20) stages: acoustic impedance ($\times 10^6$ Ns m^{-3}) (top row) and optical microscopy (bottom row). Specimen: rat, sagittal cross section, chemically fixed. Frequency range: 60–100 MHz.

EGL and IGL showed higher impedance than the PL and WM. Morphological correspondence between acoustic and immunohistological observation is, however, not clear in these images.

In the mature stage (Figure 3.15c), the EGL, which is composed of small neuronal cell bodies, has developed into the molecular layer (ML), which is composed of elongated axons (neurites), called parallel fibers. The four layers – WM, IGL, PL, and ML – are more clearly observed in the acoustic image. The correspondence with immunohistological observation is also clearly seen.

As the WM is rich in fat, its acoustic impedance would be lower than that of the IGL. The ML is composed of axons, which have a lot of actin fibers with high elasticity. It would lead to a high acoustic impedance. The reason why the Purkinje layer has low impedance is, however, not clear. Further pharmacological investigation is required.

3.3.7
Cell Size Observation [5]

For cell size observation, a transducer with a thin ceramic film and a sapphire rod lens was employed. Although the electric pulse was the same as that for tissue

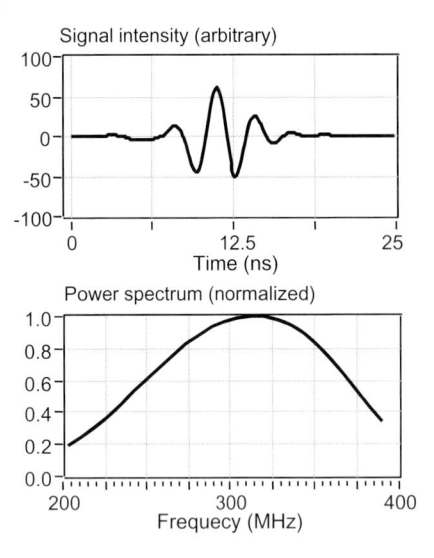

Figure 3.16 Ultrasonic waveform for cell size observation and its frequency spectrum.

observation, the generated acoustic wave had a much higher frequency component (Figure 3.16). Cells were cultured on a specially designed culture dish; its wall was a ca 70 μm-thick polystyrene film.

Figure 3.17 shows the result. The cell size spreads from 30 μm to 100 μm. The morphology observed by acoustic measurement looks similar to the optical microscopy; however, they are not completely the same. The acoustic impedance microscopy visualizes the area that is in contact with the substrate film, whereas the optical microscopy represents the projection of the cells. This leads to some differences in the morphologies of acoustic impedance and optical profiles.

The acoustic impedance profile is composed of four fragments filled with 200 × 200 pixels. The distance between every two pixels is 2 μm. Although precise evaluation has not yet been carried out, the spatial resolution would be about 3–5 μm.

The acoustic impedance profile indicates the nucleus at the center of each cell. The nucleus is round, and its acoustic impedance is as high as 1.6 MNs m^{-3}. The nucleus is surrounded by the portion that has an acoustic impedance as high as 1.65–1.7 MNs m^{-3}. This portion has the highest acoustic impedance in the cell. It is a region close to the nucleus that is filled with fibrous cytoskeleton, microtubules, which may have a high density compared with the other part of the cell. The trace of acoustic impedance along the white arrow in the acoustic profile shows that three peaks are clearly seen, suggesting that the internal cell structure can be assessed quantitatively.

The cumulative probability plot shows the distribution of acoustic impedance in the area surrounded by the white box in the acoustic profile. The acoustic impedance spreads between 1.5 and 1.7.

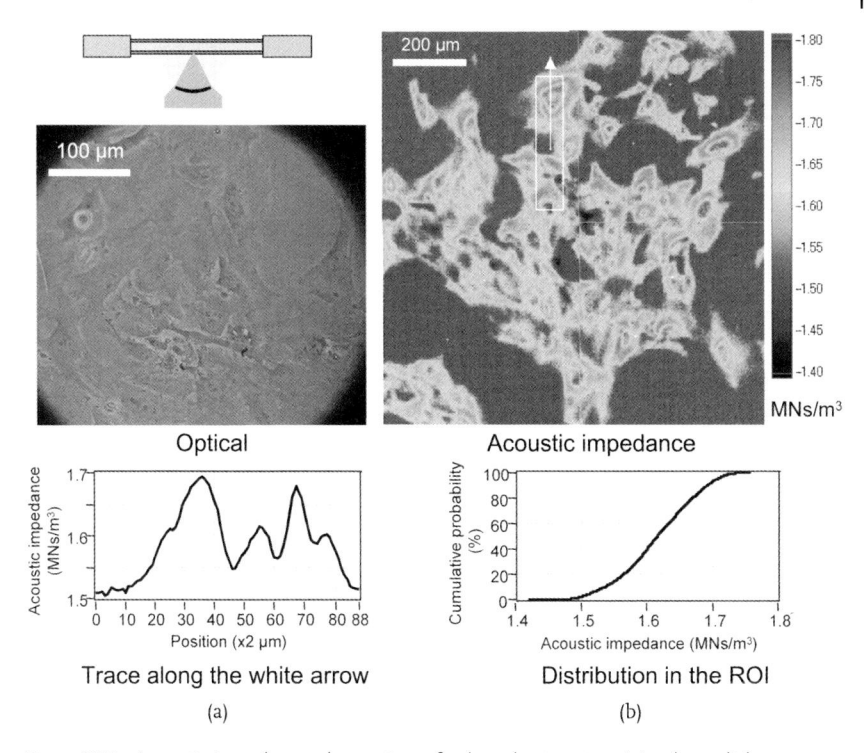

Figure 3.17 Acoustic impedance observation of cultured astrocytes: (a) culture dish specimen; (b) acoustic impedance profile.

3.3.8
Commercialized Equipment

Figure 3.18 shows the equipment commercialized by Honda Electronics Co., Ltd. It has both modes of "sound speed" and "acoustic impedance." In both cases, the focused ultrasound pulse is transmitted from the bottom of the substrate.

For both modes, a transducer with 80 MHz in center frequency is equipped. In addition, optional transducers with 160 and 320 MHz in center frequencies are available depending on the scale of the target. For convenience, it also is equipped with an optical stereoscopic microscope. The XY stage scans with a stroke of 4.8 × 4.8 mm, 2.4 × 2.4 mm, 1.2 × 1.2 mm, or 0.8 × 0.8 mm, with each scan corresponding to 300 × 300 pixels of resolution. The time needed to finish all scanning and calculation is as short as 1 min for the 2.4 × 2.4 mm scan, and even shorter if the stroke is less than that.

The XY stage, on which the transducer and pulsar-receiver is mounted, is driven by linear motors to realize a quick and stable scan.

The control and analyzing program, written in C language, is quite graphical and human friendly, such that bed-side use in a hospital is realistic.

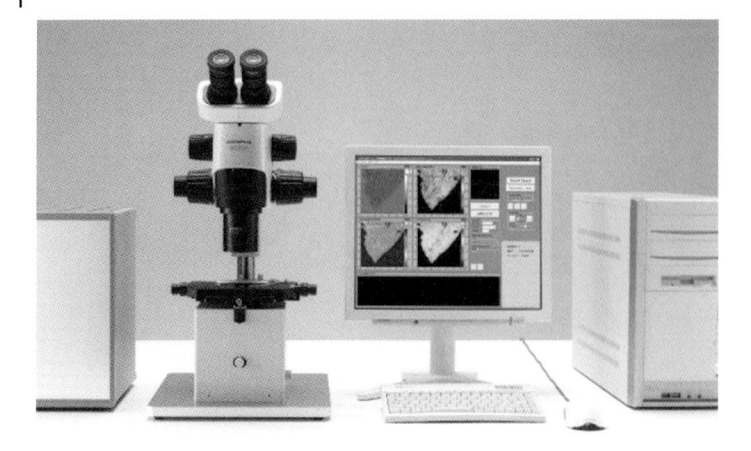

Figure 3.18 Commercially available system (Honda Electronics AMS-50SI).

3.4
Summary

Both sound speed microscopy and acoustic impedance microscopy were proven to be powerful tools for biological tissue observation. The spatial resolution may be higher in sound speed microscopy, because the beam is directly focused onto the object. On the other hand, acoustic impedance microscopy has the advantage that the specimen does not have to be sliced. This may make it possible to develop a probe-type microscope that can observe directly the surface of the tissue. In addition, the target object is completely separated from the substrate. This may prevent contamination of the system, so that *in vivo* observation can be carried out.

References

1 Hozumi, N., Yamashita, R., Lee, C.-K., Nagao, M., Kobayashi, K., Saijo, Y., Tanaka, M., Tanaka, N., and Ohtsuki, S. (2003) Ultrasonic sound speed microscope for biological tissue characterization driven by nanosecond pulse. *Acoustical Science and Technology*, **24** (6), 386–390.

2 Hozumi, N., Yamashita, R., Lee, C.-K., Nagao, M., Kobayashi, K., Saijo, Y., Tanaka, M., Tanaka, N., and Ohtsuki, S. (2003) Time-frequency analysis for pulse driven ultrasonic microscopy for biological tissue characterization. *Ultrasonics*, **42**, 717–722.

3 Hozumi, N., Nakano, A., Terauchi, S., Nagao, M., Yoshida, S., Kobayashi, K., Yamamoto, S., and Saijo, Y. (2008) Development of biological acoustic impedance microscope and its error estimation. *Mod. Phys. Lett. B*, **22** (11), 1129–1134.

4 Hozumi, N., Kimura, A., Terauchi, S., Nagao, M., Yoshida, S., Kobayashi, K., and Saijo, Y. (2005) Acoustic impedance microscopy for biological tissue characterization. *Proceedings 2005 IEEE International Ultrasonics Symposium*, 170–173, September 18–21, 2005, Rotterdam, The Netherlands.

5 Nakano, A., Uemura, T., Hozumi, N., Nagao, M., Yoshida, S., Kobayashi, K., Yamamoto, S., and Saijo, Y. (2008) Non-contact observation of cultured cells by acoustic impedance microscope. *Proceedings 2008 IEEE International Ultrasonics Symposium*, 1893–1896.

4
Portable Ultrasonic Imaging Devices

Sergey A. Titov, Roman Gr. Maev, and Fedar M. Severin

While the principles of scanning acoustic microscopy are still the same as when the first ultrasonic imaging device was introduced by Cal Quate (Stanford University, 1974) [1], the design and characteristics of particular implementations of this instrument and methods vary widely. Different requirements and application areas are reflected in corresponding technical solutions. The scanning system is one of the most essential parts of the acoustic microscope, determining its parameters. Many different types of these systems have been built by research groups and commercial companies – from large stationary machines to compact handheld gadgets. Situated on the edge of this spectrum are portable devices, which require small dimensions, transportability, and robust design to survive rough environmental conditions.

Demand for a portable design was initially generated by the airspace industry, where samples were larger in size and more unique than commonly available scanners were able to accommodate. Only a relatively small area could usually be scanned in a single measurement, with the whole inspection area being covered by a sequence of individual measurements. The long inspection time and related high labor costs were considered acceptable because of the extremely high cost of the samples themselves.

The design of portable scanning devices includes a light but rigid frame with supporting legs that allows for the positioning and adjustment of orientation on the tested surface. The rails, motors, and drives required for transducer motion are mounted on the frame and allow one- or two-dimensional scanning. A focused ultrasonic transducer is attached to the scanner carriage and the acoustical coupling between the transducer and the inspected part is provided by the water layer. In most cases this layer is created by a continuous water stream supplied through an elastic pipe. Another option includes a small volume of water encapsulated in a rubber bubble attached to the transducer. Both variants eliminate the use of a large water bath for sample submersion.

As an example, Figure 4.1 shows one of the recent implementations designed for biomedical applications (detection of small foreign objects in the brain through the skull bone). The frame size (105×105 mm) is small enough to hold the scanner

Advances in Acoustic Microscopy and High Resolution Imaging: From Principles to Applictaions, First Edition.
Edited by Roman Gr. Maev.
© 2013 Wiley-VCH Verlag GmbH & Co. KGaA. Published 2013 by Wiley-VCH Verlag GmbH & Co. KGaA.

Figure 4.1 Small handheld two-dimensional scanner.

with one hand. Raster scanning in an area up to 50×50 mm is provided by two miniature stepper motors. The concurrent electronics, including motor drivers, motion controller, and pulser–receiver circuitry, are connected to the scanner through a flexible cable. The data acquisition and analog-to digital conversion is performed by a standard 14-bit ADC card inserted into the PCI slot of a computer. The sampling rate of ADC (65 MS s^{-1}) determines the upper limit of the frequency range of the whole system (15 MHz). Windows®-based software controls the scanning and data acquisition procedures. At each point of the scanning area an echo signal from the transducer is collected and preprocessed. Obtained data is stored in the form of a 3D cube, allowing further processing based on time-reversal matched filtering methodology [2]. Advanced analysis eliminates wave-front distortion caused by irregular skull structure and recreates the image of the object.

The significant scanning time required for image acquisition may be reduced by using a multi-eyed acoustic unit that consists of several identical focused ultrasonic transducers [3,4]. In an acoustic microscope of this kind, the mechanical motion of the acoustic unit is carried out along with the electronic switching of the transducers.

Figure 4.2 shows the multi-eyed acoustic unit developed for a 100 MHz frequency range. The eight-element unit was designed based on the standard structure for transducers with the fused quartz delay line (1). The piezoelectric lithium niobate plate (2) was attached to the top plane of the delay line on which a thin metal film was deposited as common ground electrode. To form the ultrasonic transducers, eight circular electrodes (3) were deposited on the opposite surface of the piezoelectric plate. With the of a plate thickness of about 30 μm; the central frequency of the transducers was 100 ± 15 MHz, and the bandwidth was about 40%. At the lower face of the delay line the acoustic lenses were fabricated in the form of spherical cavities. The axes of the transducers and lenses were aligned for proper focusing of the ultrasonic waves radiated by the transducers. In this particular assembly, the diameters of the transducers and lenses and the radius of curvature of the lenses were determined to provide the focal distance in water of 6.7 mm.

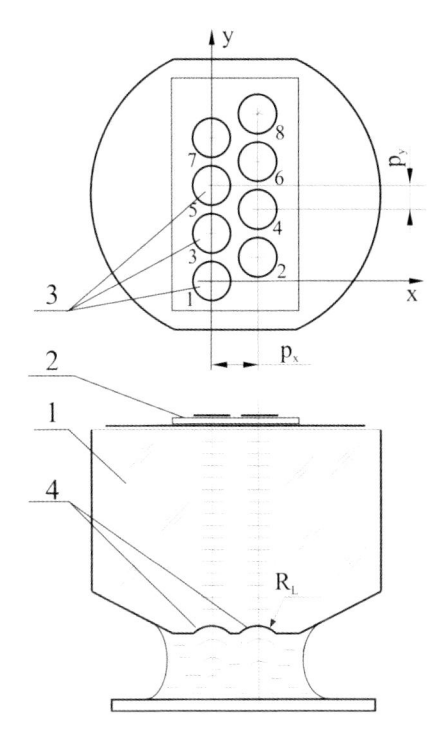

Figure 4.2 Multi-eyed acoustic unit: (1) delay line, (2) piezoelectric plate, (3) electrodes, and (4) acoustic lenses.

Eight transducer–lens pairs were arranged in two lines as shown in Figure 4.2. The transducers with odd and even numbers have offset at a distance where $p_x = 1.8$ mm, and the distance between the transducers along the y axis is $p_y = 1.0$ mm. During data acquisition the acoustic unit is mechanically translated along axis x while the ultrasonic pulse–echo signals are recorded by all transducers with the spatial period Δx. To acquire a complete data set the scans in the x direction are performed N_y times at incremental displacement along the y-axis with step $\Delta y = p_y/N_y$. Thus the number of B-scans in the full data set is $M \times N_y$, where $M = 8$ is the number of transducers in the acoustic unit, and the transverse size of the field of view is $p_y M$. To eliminate image distortion caused by the offset of the transducers with odd and even numbers along the x-axis, it is necessary to neglect the first $N_p = p_x/\Delta x$ samples recorded by the transducers with odd numbers and the last N_p samples recorded by the transducers with even numbers. Thus, the longitudinal size of the corrected field of view is equal to $(N_x - N_p)\Delta x$, where N_x is the total number of samples recorded by one transducer in one pass.

Firing of the transducers and receiving of the echoes were performed sequentially to avoid overlapping of the reflected waves and mutual interferences of the electrical signals. Therefore, the transducers were connected to a single

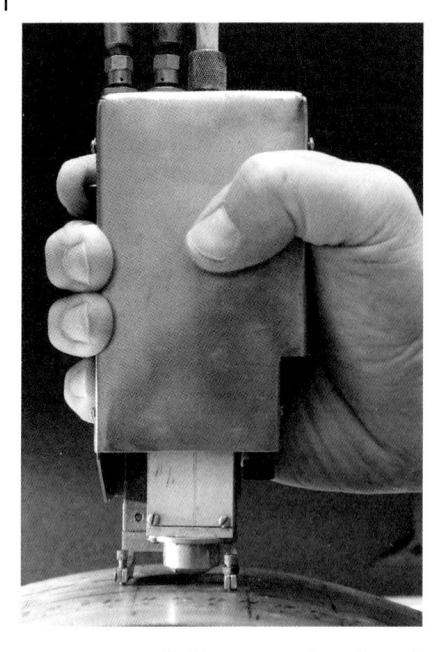

Figure 4.3 Handheld scanner with multi-eyed transducer.

pulser–receiver through a low-noise high-voltage analog multiplexer. The multiplexer, protection circuit, and preamplifier were placed inside an electromagnetically shielded box directly attached to the multi-eyed acoustic unit. This assembly was mounted on the compact handheld two-dimensional scanner (Figure 4.3). The step motor stages provide precise movement of the acoustic unit in the x and y directions at the maximal distances of about 10 and 1.0 mm, respectively. Taking into account that $M = 8$ the corrected field of view is 8 mm × 8 mm. For proper positioning of the head on curved surfaces of specimens the scanner is equipped with three adjustable supports and the manual vertical stage. A drop of water or ultrasonic gel is placed between the head and the specimen to provide an acoustic contact. Since the gap between the lenses and the specimen is small, water is kept there due to the capillary effect.

The described multi-eyed acoustic microscope has been used to visualize the internal structure of spot welds performed on steel and aluminum sheets. Figure 4.4 shows the C-scans of the weld of 1.1 mm aluminum sheets obtained at various number of passes N_y. The spatial sampling interval along the x-axis was constant: $\Delta x = 0.1$ mm, whereas the transverse sampling steps Δy are equal to 1.0, 0.5, and 0.25 mm at $N_y = 1, 2, 4$, respectively. The main part of the weld is the nugget zone where strong metallurgical bonds have been formed between jointed sheets. This nugget zone is transparent for ultrasound; therefore, it looks like dark areas in the C-scans gated around the gap between sheets. Several gas pores developed in the nugget during the welding process appear as light spots in the dark welded region.

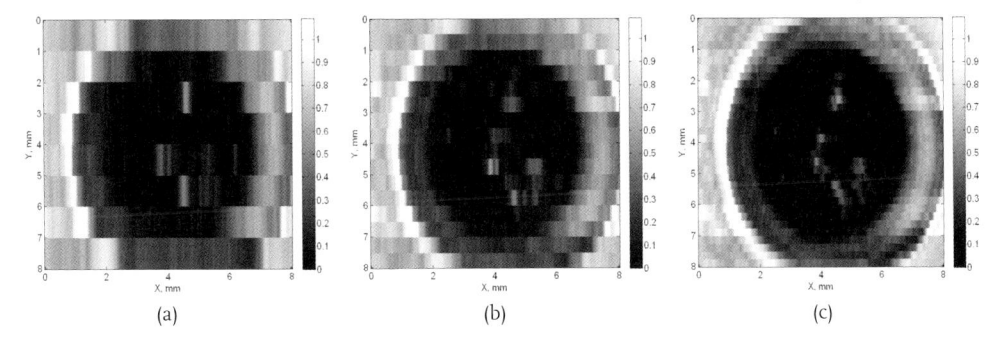

Figure 4.4 C-scans of spot weld of aluminum sheets: (a)–(c) N_y = 1, 2, 4, respectively.

The number and positions of the pores are easy to estimate on the basis of a C-scan taken at $N_y = 4$ (Figure 4.4c). It is still possible to detect the presence of pores using the one-pass scan (Figure 4.4a) but it is difficult to estimate their positions and some of the defects can be missed.

Apparently, the sampling steps Δy represented in Figure 4.4 are too large to achieve the potential resolution of the imaging system with the focused transducer. As a result small defects and tiny details can be missed in the images. However, it is possible to detect the positions of the boundaries of large objects like the weld nugget itself and estimate their sizes with sufficient accuracy even when using a small number of passes N_y.

The scaling up of the system introduces several problems. The number of transducers in the acoustic unit can be increased to achieve a larger scanning area or better resolution at each pass. However, technological capabilities and alignment problems limit the overall size of high-frequency multi-eyed units. The reduced size of each transducer or operation at lower frequency causes smearing of the focal spot and at some point the focusing of individual ultrasonic beams becomes unreasonable. In this extreme case the multi-eyed unit turns into a regular array of plane transducers.

The combination of linear array with perpendicular mechanical displacement has been successfully realized in a number of experimental and commercial ultrasonic imaging systems. Especially effective was the integration of a linear array into a so-called wheel probe [5], which allows the imaging of a wide strip by manual motion of the probe.

The multi-element ultrasonic transducer controlled by a corresponding electronic system allows for the total elimination of moving parts and delicate mechanical components. There are two approaches to realize such a system. First is the phase array system, which was introduced in the early 1980s and was a very efficient electronic scanning device; even today it is widely used in various industrial and medical applications. About 20 years later a new generation of 1–3 piezo-materials allowed researchers to take a completely different approach – the two-dimensional matrix array transducer. The first practical usage of this new

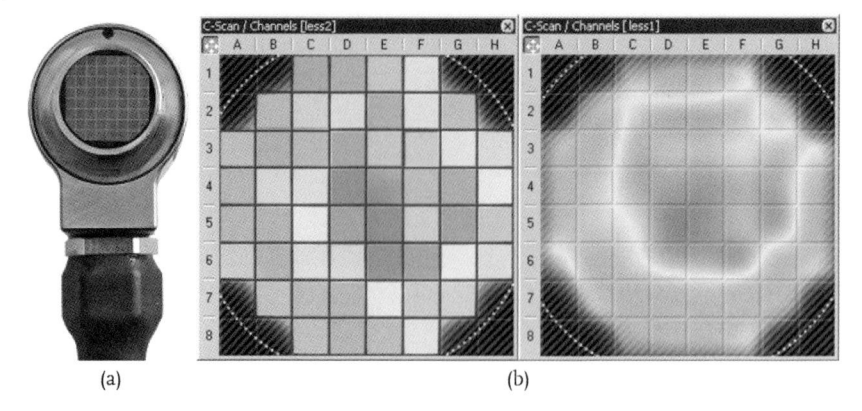

Figure 4.5 (a) Two-dimensional matrix array transducer; (b) the elements response and built on its interpolated image of spot weld nugget.

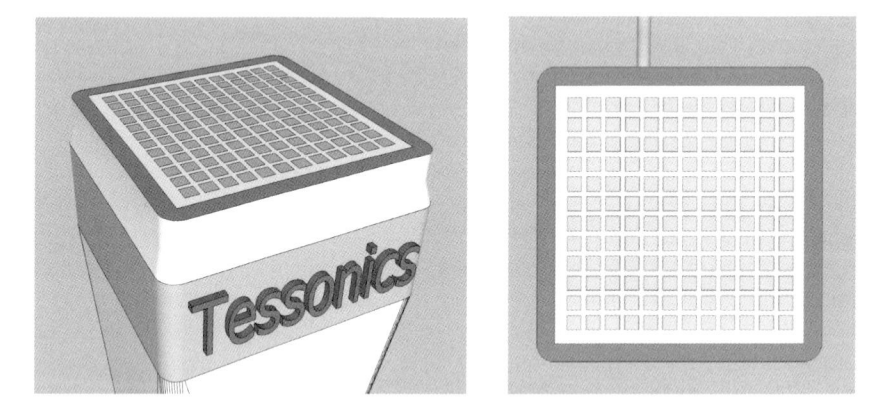

Figure 4.6 Two-dimensional array 12 × 12 with beam-forming ability.

approach was realized in the Resistance Spot Weld Analyzer, produced by Tessonics Inc. [6–10]. It has a 52-element matrix transducer inscribed into a circular outline. Each element of this 2D matrix system, with a size of 1×1 mm and a central frequency 17 MHz, is connected in sequence to the pulser–receiver circuitry. Software controls data acquisition and the analysis of obtained signals, providing a smooth image representing the internal structure of the spot weld (Figure 4.5).

The significant recent progress in technology of transducer fabrication and multichannel electronics allows the combination of this 2D geometry with beam-forming capability. Three-dimensional scanning principles make such a system really unique. Current development of this technology for practical applications (Figure 4.6) allows for the creation of a completely new class of devices with greatly improved scanning speed and ergonomics.

The essential problem associated with two-dimensional arrays is sparse data acquisition. To analyze the problem, let us consider a simple model of the ultrasonic imaging system.

Assuming linearity, and spatial and temporal invariance of the system, the output signal v as a function of the positions of the transmitter $\mathbf{r}_1$ and receiver $\mathbf{r}_2$ and time t can be found as a double convolution over spatial and temporal variables:

$$v(\mathbf{r}_2, \mathbf{r}_1, t) = g(\mathbf{r}_2, \mathbf{r}_1, t) \overset{\mathbf{r}_1, t}{*} h_1(-\mathbf{r}_1, t) \overset{\mathbf{r}_2, t}{*} h_2(\mathbf{r}_2, t) \tag{4.1}$$

where $g(\mathbf{r}_2, \mathbf{r}_1, t)$ is the Green's function of the object, $h_1(\mathbf{r}_1, t)$ is the acoustic field generated by the transmitter, and $h_2(\mathbf{r}_2, t)$ is the impulse response of the receiver.

Suppose that the Green's function obeys the following equation:

$$g(\mathbf{r}_2, \mathbf{r}_1, t) = g(\mathbf{r}_1)\delta(\mathbf{r}_2 - \mathbf{r}_1)\delta(t) \tag{4.2}$$

This means that the acoustic field of the reflected wave at a particular point in the subject plane is entirely determined by the value of the incident field at the same point. For instance, this model is valid if the object can be presented as a set of independent point reflectors.

In this case of a locally reacting object the output signal of the system with the coincident transmitter and receiver $\mathbf{r}_1 = \mathbf{r}_2$ is a convolution of the object function $g(\mathbf{r})$ and the general impulse response of the system $h_0(\mathbf{r}, t)$:

$$v(\mathbf{r}, t) = g(\mathbf{r}) \overset{\mathbf{r}}{*} h_0(\mathbf{r}, t) \tag{4.3}$$

where the general impulse response is defined as follows:

$$h_0(\mathbf{r}, t) = h_1(-\mathbf{r}, t) \overset{t}{*} h_2(\mathbf{r}, t). \tag{4.4}$$

In real imaging systems the output signal is a function over the spatial and temporal variables. The sampling rate of modern electronics can be sufficiently high to neglect the time related discretization effects. Thus the digitized output data can be written in the form [11]:

$$v_d(x, y, t) = \left[g(x, y) \overset{x,y}{*} h_0(x, y, t) \right] \sum_{n=-\infty}^{\infty} \delta(x - n\Delta) \cdot \sum_{m=-\infty}^{\infty} \delta(y - m\Delta) \tag{4.5}$$

where δ is Dirac's delta function and Δ is the spatial sampling period. In the case of mechanical scanning of a focused or unfocused transducer this sampling period can be small enough to satisfy Nyquist's criteria. If the output data set is formed by the individual signals from the independently acting elements of the array the sampling period is equal to the pitch of the array $\Delta = p$. It is reasonable to assume that the element size and the width of the impulse response r_s of the system are approximately equal to the pitch of the array $r_s \approx p$. In this case, according to the uncertainty principle [12] the frequency range of the spatial spectrum of the output signal Δk should satisfy the condition $\Delta k > 1/p$, and the Nyquist frequency in the k-domain should be at least two-times larger. To avoid distortion of the signal due to spectral aliasing the maximal value of the sampling period Δ should not exceed $p/2$. Therefore, the images recorded by the array system with sampling interval p suffer from undersampling.

The spatial resolution of the array system depends on the width of the general impulse response r_s and the sampling interval Δ. Two point reflectors can be recognized as separated objects in the image if there is at least one element with a substantially smaller response between two elements that receive echoes from these reflectors. To ensure separate visibility of these points on the image at any position of the array with respect to the object the minimal distance between them should be $\Delta + r_s \approx 2p$. Since a practical number of elements of a two-dimensional array does not exceed several hundreds the spatial resolution of the system is relatively low. Thus, the portable matrix array system has a limited ability to generate fine ultrasonic images but it can be effective for the detection of small defects and estimation of sizes of large objects.

Let us consider a matrix array used for the detection of a boundary of a flat reflector and estimation of its sizes. Suppose that the boundary of the reflector is smooth in the (x,y) plane and the properties of the reflector are constant inside this boundary. Assume that the boundary can be approximated by a straight line within the impulse response of the element h_0. The spatial step function $u_0(x,y,t)$ can be defined as a response of the system on a half-plane reflector:

$$g(x, y) = \begin{cases} 0, & x < 0 \\ 1, & x \geq 0 \end{cases} \tag{4.6}$$

Using expressions (4.3) and (4.6) the step response can be written in a form of the double integral:

$$u_0(x, t) = \int_0^\infty d\xi \int_{-\infty}^\infty h_0(\xi - x, \eta, t) d\eta \tag{4.7}$$

Usually, the peak value of the envelope of the received echo within a corresponding time gate is used for imaging. Let us consider the normalized step response:

$$u(x) = \max_t \left\{ \sqrt{u_0^2(x, t) + u_G^2(x, t)} \right\} \left[\max_t \left\{ \sqrt{u_0^2(\infty, t) + u_G^2(\infty, t)} \right\} \right]^{-1} \tag{4.8}$$

where $u_G(x,t)$ is the Hilbert transform of $u_0(x,t)$, and the normalization factor is calculated for the array positioned over the reflector far from the edge. This normalization increases the robustness of the method with respect to the variation in sensitivity of the array elements.

Typical behavior of the normalized step function $u(x)$ is shown in Figure 4.7 as a function of spatial coordinate. It is possible to make the following estimation of the boundary coordinate ξ with known $u(x)$ and the system noise level σ. If the normalized amplitude of the echo a_j received by the j-th element of the array is larger than the upper noise threshold $a_j > (1 - \sigma)$ then we should assume that the element is entirely located over the reflector and the boundary cannot be too close

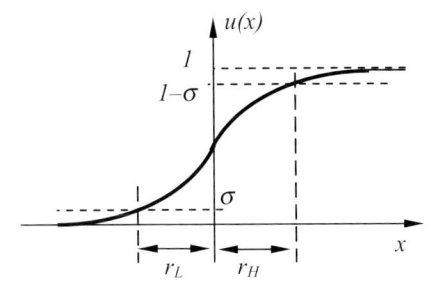

Figure 4.7 Normalized step response $u(x)$.

to the position of the element x_j. Thus $|\xi - x_j| > r_H$, where the minimal distance r_H satisfies the relationship $u(r_H) = 1 - \sigma$. If the amplitude a_j is smaller than the lower noise threshold $a_j < \sigma$, the element is outside the reflector $|\xi - x_j| > r_L$, where $u(-r_L) = \sigma$.

The exact position of the boundary can be determined if the amplitude a_j lies on the slope part of the step response $\sigma \leq a_j \leq (1 - \sigma)$. In this case the coordinate of the boundary can be found from the equation $u(\xi) = a_j$. The error of the obtained boundary coordinate $\delta\xi$ depends on the additive noise value σ and the slope of the step function:

$$\delta\xi = \sigma \left(\frac{du}{dx}\right)^{-1} \tag{4.9}$$

If the width of the impulse response $r_L + r_H$ is larger than the spatial sampling period Δ then several readings may appear in the interval $(-r_L, r_H)$. In this case their average value can be used as the estimation of the boundary position. As follows from (4.9), the accuracy of the measurement increases with increasing slope of the step response and, consequently, with decreasing width $r_L + r_H$. For a narrow impulse response it is possible that all readings are outside the interval $(-r_L, r_H)$. For instance, if $a_j < \sigma$ and $a_{j+1} > (1 - \sigma)$ the boundary is located in the interval $(x_j + r_L, x_j + \Delta - r_H)$. It is reasonable to use the middle of this interval as the estimation of the boundary position:

$$\xi = x_j + \frac{\Delta}{2} + \frac{r_L - r_H}{2} \tag{4.10}$$

The maximal error:

$$\delta\xi = \xi - x_j - r_L = \frac{\Delta}{2} - \frac{r_L + r_H}{2} \tag{4.11}$$

becomes small when the width of the step response approaches the sampling interval Δ.

The presented considerations can be extend for the two-dimensional case. As an example the fragment of the matrix array is shown in Figure 4.8. Let points A,

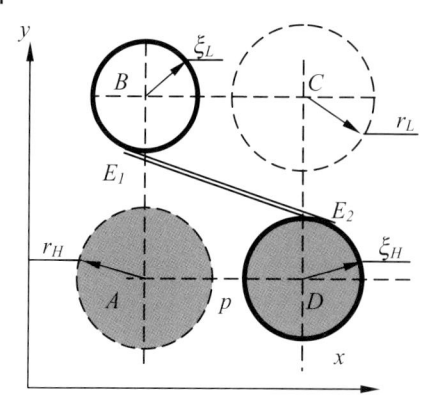

Figure 4.8 Fragment of the matrix array with the boundary construction.

B, C, and D be the centers of the array elements, and a_A, a_B, a_C, and a_D be amplitudes of the echoes received by these elements, respectively.

Suppose that $a_A > (1 - \sigma)$, $\sigma < a_B < u(0)$, $a_C < \sigma$, and $u(0) < a_D < (1 - \sigma)$. Since a_A is larger than the upper threshold and a_C is less than the lower threshold the edge of the reflector passes somewhere between points A and C. Moreover, in the case of axial symmetry of the step response the circle with the center in A and the radius r_H belongs entirely to the reflector area. Similarly, the circle with the center in C and the radius r_L is located outside the reflector. Since $a_B < u(0)$, point B is located outside the reflector as well. Moreover, the distance between B and the boundary ξ_L should satisfy the relationship $u(-\xi_L) = a_B$. Point D is located inside the object, and the distance between D and the boundary can be found from the equation $u(\xi_H) = a_D$. As result, the boundary can be approximated by the tangent line E_1E_2 to the circles ξ_L, ξ_H (Figure 4.8). This procedure should be repeated using rest elements of the array to determine the whole boundary of the reflector.

The proposed method has been verified on several test samples. Figure 4.9 shows the B-scan of a slit in a thin steel plate recorded by one element of the matrix array transducer. The plate was 0.1 mm thick and the width of the slit was $d_s = 2.20 \pm 0.02$ mm. The normalized amplitude of the received echo $a(x)$ calculated using Equation (4.8) is presented in Figure 4.10.

Based on the obtained function $a(x)$ and known width of the slit d_s the experimental step response was estimated through the use of linear approximation. It was assessed that the slope of the response is 1.4 mm^{-1}. Assuming that the noise level of the system is about $\sigma = 0.05$ the parameters of the response r_L and r_H (Figure 4.7) were found to be 0.31 and 0.4 mm, respectively. The total length of the approximated step response is equal to $r_L + r_H = 0.71$ mm. This value is less than the sampling interval $\Delta = p = 1.25$ mm and, therefore, using Equation (4.11) the maximal error can be estimated to be 0.27 mm.

To obtain the experimental error of the method a hole in the steel plate with a special shape was tested. The hole consisted of a slit with width d_s and a round

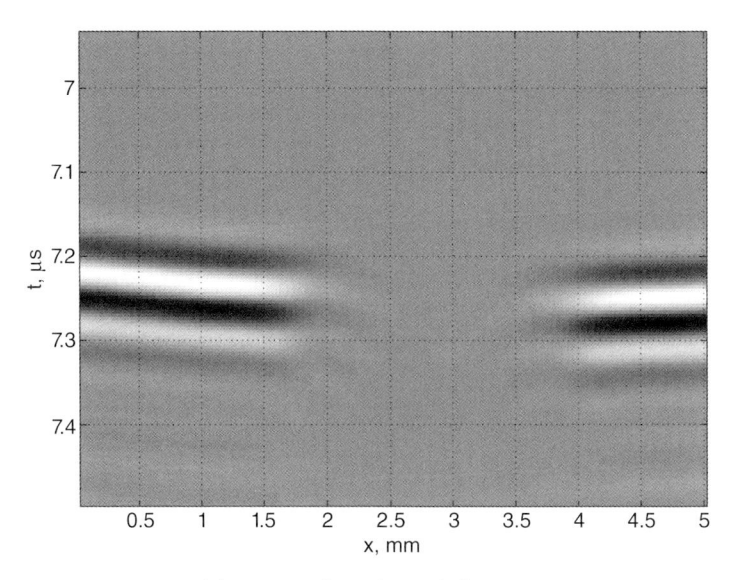

Figure 4.9 B-scan of the 2.2 mm slit in the steel plate.

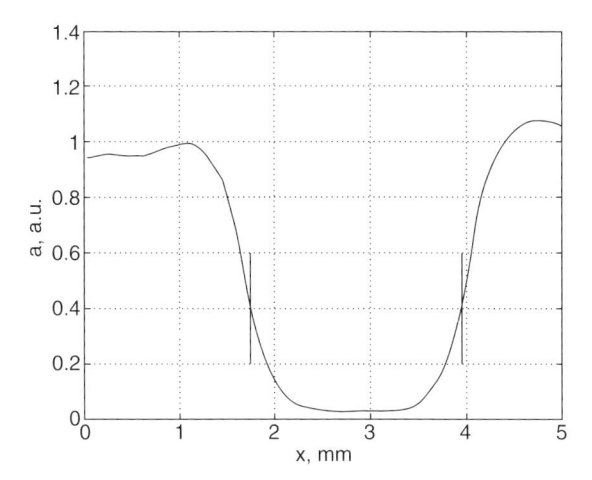

Figure 4.10 Normalized amplitude $a(x)$.

hole with a diameter of $d_0 = 5 \pm 0.02$ mm (Figure 4.11). The normalized amplitudes of the echoes $a(x_j, y_k)$, $1 \leq j, k \leq 8$ received by the elements of the array are presented in Figure 4.12 as a grayscale image. Reconstruction of the boundary of the test hole based on the technique described above is illustrated in Figure 4.13. Each pixel in this image is presented as a circle with a particular color and radius, which depend on the amplitude of the signal at this point $a(x_j, y_k)$. The white and black circles are similar to the circles B and D in Figure 4.8. They correspond to the

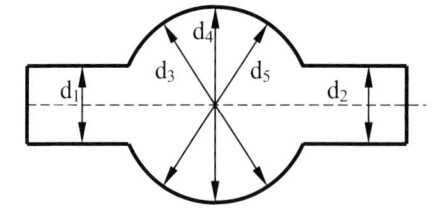

Figure 4.11 Test hole in the steel plate.

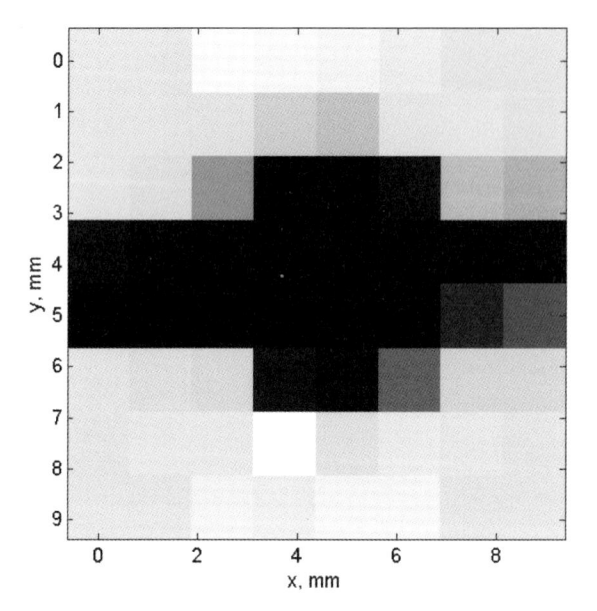

Figure 4.12 Amplitudes $a(x_j,y_k)$ measured for the test hole.

cases $\sigma < a < u(0)$ and $u(0) < a < (1 - \sigma)$, and their radii satisfy the equations $u(-\xi_L) = a$ and $u(\xi_H) = a$, respectively. The segments of the boundary line are tangent to these circles, whereas the white and black circles are located outside and inside the hole. The light- and dark-gray circles represent amplitudes $a > (1 - \sigma)$ and $a < \sigma$, respectively. The radii of these circles are r_H and r_L, and they lie entirely outside and inside of the hole. If a light-gray circle has a neighboring dark-gray circle then the boundary line should be placed in the middle between them. Figure 4.13 shows an example of the boundary reconstruction.

To evaluate the accuracy of this technique the widths of the slit d_1 and d_2 and the diameters of the round hole d_3, d_4, and d_5 (Figure 4.11) were measured at various positions of the array with respect to the sample. Table 4.1 presents the statistically processed data.

Measurements of the slit tilted in the subject plane at 30° (d_{30}) and 45° (d_{45}) were performed to evaluate the stability of the algorithm with respect to the orientation

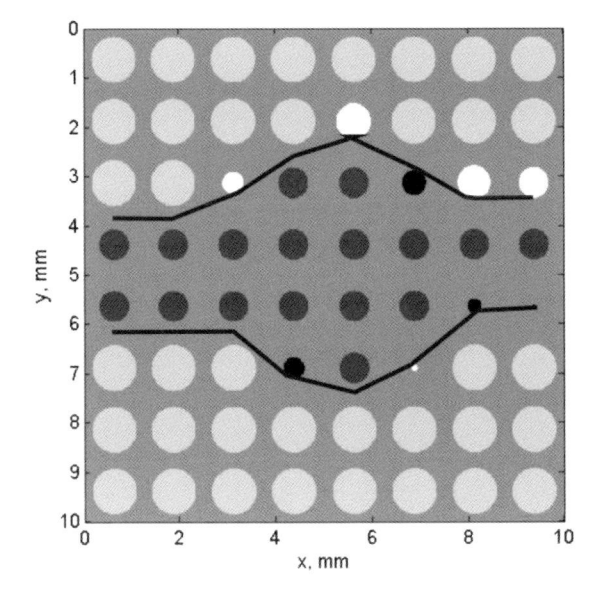

Figure 4.13 Reconstruction of the test hole boundary.

Table 4.1 Measured sizes (mm) of the hole.

	Slit width $d_s = 2.20 \pm 0.02$ mm		Hole diameter $d_0 = 5.00 \pm 0.02$ mm		
	d_1	d_2	d_3	d_4	d_5
Average	2.18	2.27	5.11	5.15	5.07
RMS	0.08	0.09	0.15	0.11	0.14

of the object (Figure 4.14). The obtained average and root-mean-square error (Table 4.2) for the width do not exceed the values measured for the horizontal orientation of the slit.

Summarizing the statistical analysis: the two-sigma confidence intervals can be estimated to be ±0.15–0.20 mm for the slit and ±0.20–0.30 mm for the round hole. These experimental values are in agreement with the error of 0.27 mm that was theoretically predicted above using the relationship (4.11) between the width of the step response and the sampling interval. The accuracy can be increased by reducing the sampling interval. However, the intrinsic sampling interval of the array is equal to its pitch when the one-fold recording of the output signals of the elements is used. To reduce the equivalent sampling interval the electronic scanning can be combined with the mechanical movement of the whole array

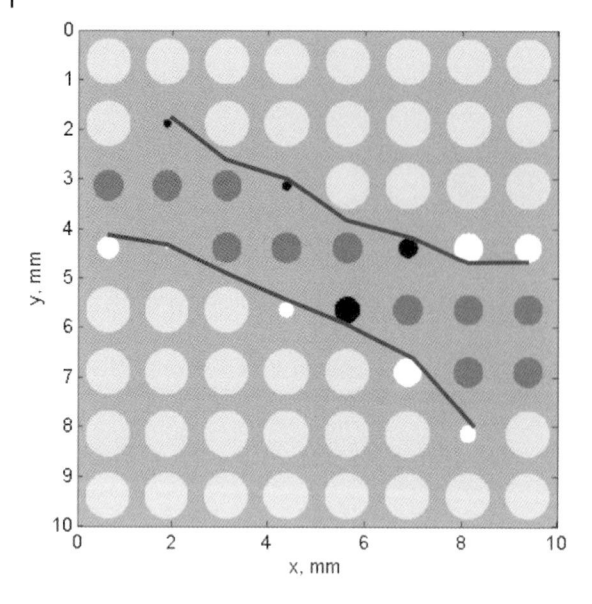

Figure 4.14 Construction of the boundary of the slit tilted at 30°.

Table 4.2 Measured width (mm) of the tilted slit.

	Slit width $d_s = 2.20 \pm 0.02$ mm	
	d_{30}	d_{45}
Average	2.22	2.26
RMS	0.08	0.07

transducer. In this combined oversampling data acquisition mode, to obtain the sampling interval $\Delta_e = p/n_e$ the recording of the array data set should be repeated n_e^2 times at the positions of the array $x_0 + p/n_e j_e$, $y_0 + p/n_e k_e$, where $0 \leq j_e, k_e \leq (n_e - 1)$, and x_0, y_0 are the coordinates of the initial position of the array.

The synthetic image of the test hole consisted of four primary array data sets (Figure 4.15). The spatial oversampling coefficient n_e is equal to 2 in this case, and the shifts of the array transducer over x- and y-axis equal $\Delta_e = p/2 = 0.63$ mm. The resulting data set a_{jk}, $1 \leq j, k \leq 16$ occupies approximately the same area in the object plane but the outline of the test hole appears more clearly in comparison with the one of the primary images presented in Figure 4.12. Estimation of the boundary of the hole is smoother (Figure 4.16) and the errors in the sizes d_1–d_5 (Table 4.3) determined with the oversampled data are significantly lower than the errors listed in Table 4.1.

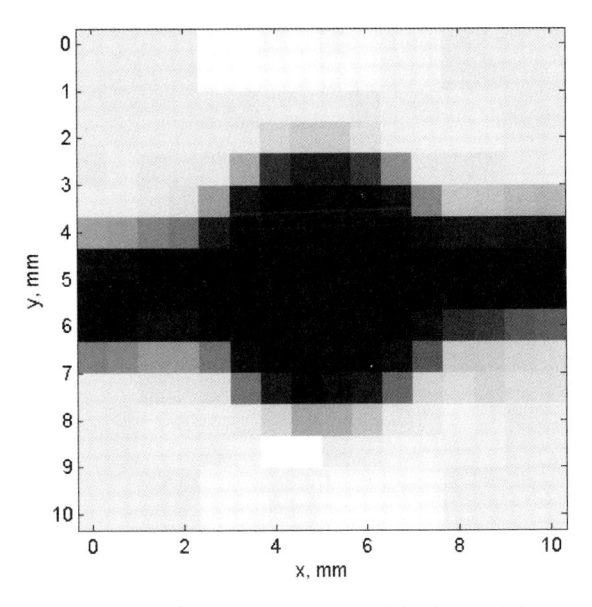

Figure 4.15 Amplitudes $a(x_j, y_k)$ measured for the test hole in the oversampling mode ($n_e = 2$).

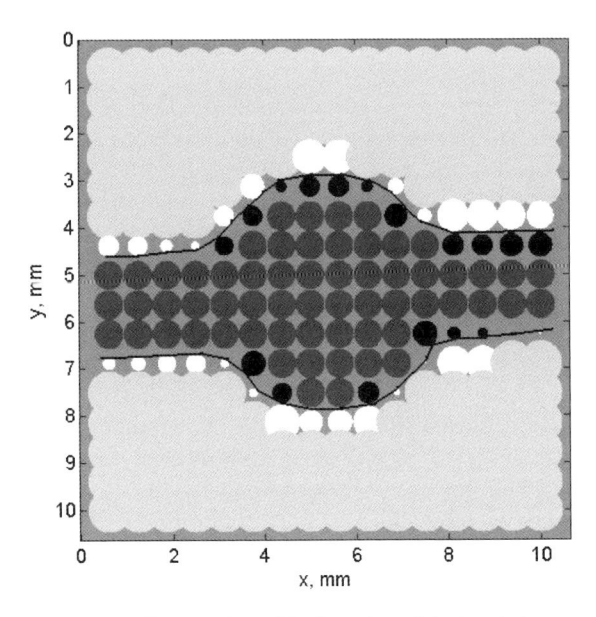

Figure 4.16 Construction of the boundary of the test hole using data presented in Figure 4.10.

Table 4.3 Sizes (mm) of the hole measured using oversampled data.

	Slit width $d_s = 2.20 \pm 0.02$ mm		Hole diameter $d_o = 5.00 \pm 0.02$ mm		
	d_1	d_2	d_3	d_4	d_5
Average	2.19	2.25	5.04	5.02	5.05
RMS	0.04	0.06	0.10	0.07	0.08

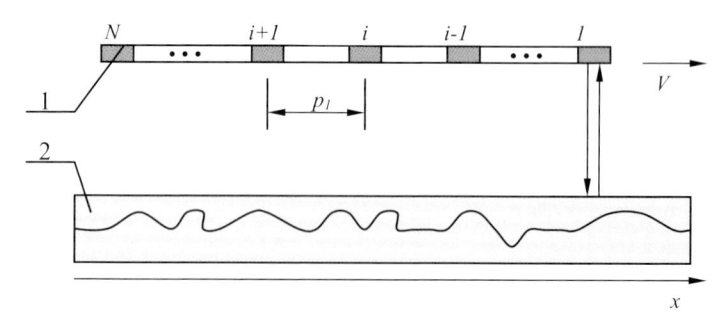

Figure 4.17 Movement of array transducer (1) over the specimen (2).

The described matrix probe technique can be effective for the evaluation of relatively small areas of a specimen. The manual transition of the array probe is often used to test extended regions of the object. However, the manual transition is associated with low accuracy – not sufficient for proper presentation of the spatial data. Precise probe coordinates may be measured with some encoder attached to the probe. However, encoder presence may cause inconvenience for the operator and may restrict the accessibility of testing due to the increased size of the transducer. We can propose the following alternative method: estimation of the probe position using signals of multiple transducers located along the direction of movement.

The basic idea of the method is illustrated in Figure 4.17. The multi-element probe 1 consists of N identical transducers aligned along the axis x and separated by a constant distance p_1. The whole probe is positioned parallel to the surface of the sample 2 and is translated in the same direction x. The transducers, working interdentally in the pulse–echo mode, record the spatiotemporal data $s_i(t,T)$, where i is the element number, $1 \le i \le N$, t is the "fast" time associated with the ultrasonic wave propagation, and T is the "slow" time associated with the probe movement $x(T)$. Since the transducers pass sequentially over the same points on the surface of the sample the signals recorded by the neighboring transducers should be similar but shifted in time T on the delay $\tau(T)$, which depends on the pitch p_1 and velocity of the movement V. Let $s_1(t,T)$ be the data received by the first element, then the rest of the signals can be presented in the form:

$$s_i[t, T + \tau(T)] = s_{i-1}(t, T) + \varepsilon_i(t,T), \quad 2 \le i \le N \tag{4.12}$$

where $\varepsilon_i(t,T)$ are the errors produced by the electrical hardware noise, nonlinearity of the probe movement, non-stability of the acoustical contact between the probe and the sample, and variation of the parameters of the elements. Assuming that the movement of the probe is sufficiently smooth, the velocity averaged over the spatial interval $[x(T), x(T) + p_1]$ can be determined using the simple equation:

$$V(T) = p_1[\tau(T)]^{-1} \tag{4.13}$$

The delay $\tau(T)$ can be estimated using cross correlation over time T between the waveforms recorded by the elements of the probe:

$$B_{i-1}(T, \eta) = \int_{t_1}^{t_2} dt \int_{-T_0}^{T_0} s_{i-1}(t, T + \xi) s_i(t, T + \xi + \eta) d\xi \tag{4.14}$$

where (t_1, t_2) and $(T - T_0, T + T_0)$ are the "fast" and "slow" time windows. The window (t_1, t_2) selects echoes from the sample discontinuities and the window $2T_0$ cuts a portion of the data for the analysis where the velocity of the movement is approximately constant. If the errors are absent $\varepsilon_i = 0$, the signals of the neighboring elements are equal $s_i[t, T + \tau(T)] = s_{i-1}(t,T)$, and the correlation integral (4.14) has a maximum exactly at $\eta = \tau(T)$. Therefore, in the presence of moderate noise, the position of the peak of the correlation function η_0 can be used as an estimation of the delay $\tau(T)$.

The accuracy of this technique depends on the characteristics of the received signals. The accuracy is good when the reflected signals are strong and rapidly varying functions over the time T. For weak and smooth signals the accuracy will be lower. When the signals do not change with the probe movement at all, this method is inapplicable. But this extreme case means that the specimen is laterally uniform and there are no discontinuities for which sizes and positions should be determined.

It is sufficient to have only two elements in the transducer array to implement this method. Usage of a larger number of elements increases the accuracy of the velocity estimation due to averaging of $N - 1$ readings. In addition, the long aperture of the array increases the robustness of the method in situations when at a certain position of the probe the signals of some elements are not suitable for the processing.

This method has been implemented in the ultrasonic imaging system based on the Tessonics 2D matrix probe described above. There are two groups of elements that were selected from the whole matrix. The elements of the first group A1–A11 are arranged in the perpendicular direction with respect to the velocity of the probe movement V (Figure 4.18). The elements of the second group B1–B6 are aligned along the direction of translation. The signals of the A elements are used for the

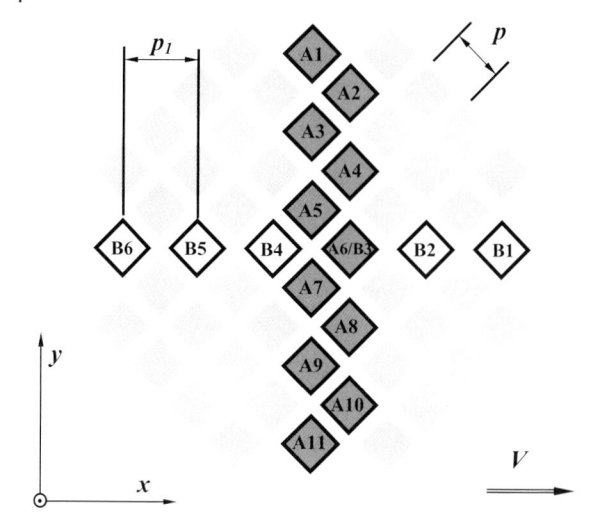

Figure 4.18 Movement of 2D array.

image construction, whereas the elements of the group B are used for the velocity measurement.

A reference sample – a Perspex plate with small notches on the back surface – has been tested to estimate the error of the position measurement. The plate was 1.25 mm thick and the notches were 0.3 mm wide and 0.2 mm deep. Figure 4.19 shows two B-scans, $s_2(t,T)$ and $s_3(t,T)$, recorded by the elements B2 and B3 when the probe was manually translated across the notches. Here, the horizontal strip represents the echo R_0 reflected from the back surface of the plate and its interruptions – responses R_1–R_8 are produced by the notches.

The correlation integrals (4.14) were calculated using five pairs of the data sets $s_{i-1}(t,T)$, $s_i(t,T)$ and the results were averaged to obtain the correlation function $B(T,\eta)$. The duration of the "slow" time window was $2T_0 = 2$ s. The function $B(T,\eta)$ presented in Figure 4.20 for the moment $T = 5$ s has the narrow peak whose time position is equal to the delay of the signals of the neighboring elements $\eta_0 = \tau(T)$. The velocity of the movement $V(T)$ shown in Figure 4.21 was calculated using Equation (4.13) where $p_1 = \sqrt{2}p$, and $p = 1.25$ mm is the pitch of the array.

The position of the probe as a function of time $x(T)$ was obtained by integration of $V(T)$ and the calculated coordinates of the notches were compared with their actual positions. The differences did not exceed ±0.5 mm [9]. Taking into consideration that the size of the individual transducers is 1 mm, it can be acknowledged that the accuracy achieved in this experiment is satisfactory.

The described ultrasonic system was employed for imaging of void-like defects in laminated windshield glass. The upper part of Figure 4.22 shows the optical image of the defect with the path of the matrix probe and the 5 mm grid. The lower ultrasonic image was built using 11 scans recorded by the elements A1–A11 of the probe (Figure 4.18). In the ultrasonic image the data are presented in the

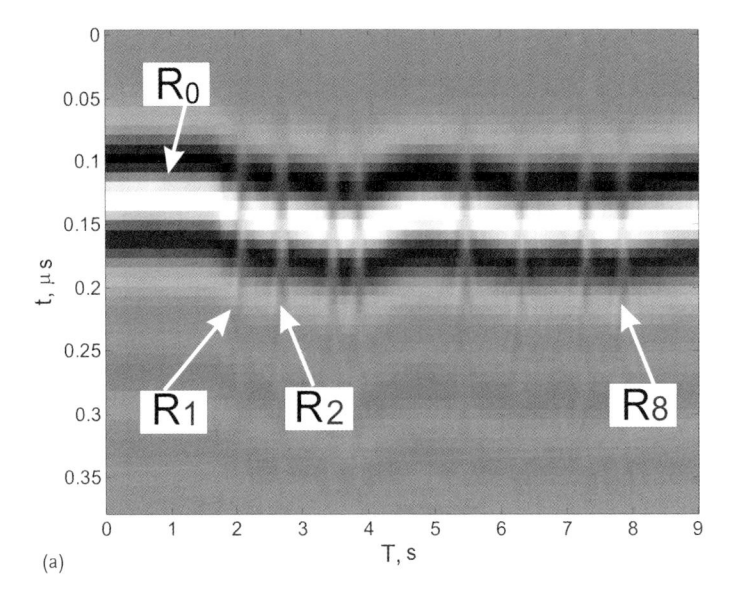

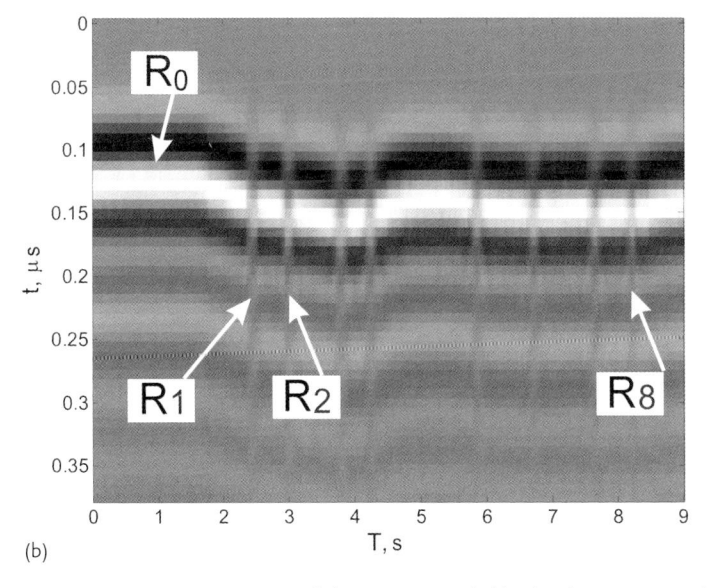

Figure 4.19 B-scans (a) $s_2(t,T)$ and (b) $s_3(t,T)$ recorded by the elements B2 and B3, respectively.

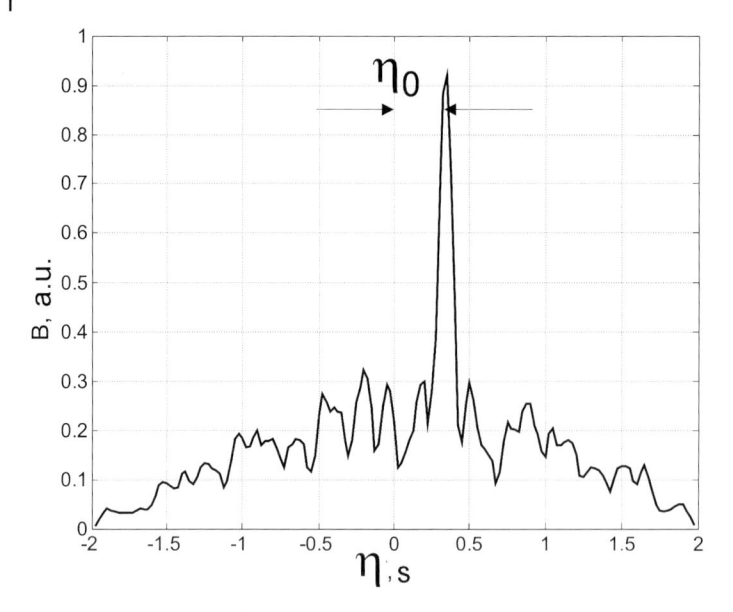

Figure 4.20 Correlation function $B(T,\eta)$, $T = 5$ s.

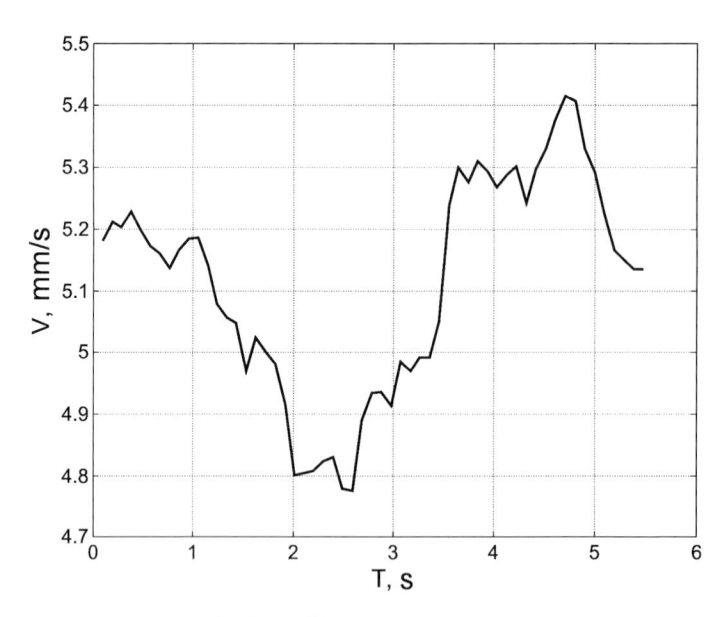

Figure 4.21 Measured velocity of the probe movement $V(T)$.

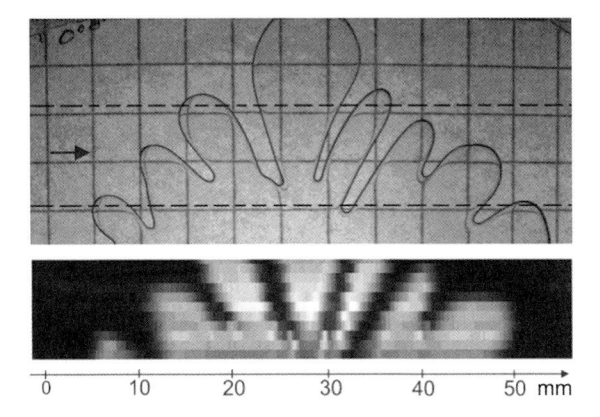

Figure 4.22 Optical and ultrasonic images of void in windshield glass.

spatial domain (x,y). The longitudinal coordinate x was determined using correlation analysis of the signals of elements B1–B6. To compensate for the distortion of the image caused by the zigzag layout of the elements A1–A11 the scans recorded by the elements with even numbers were shifted by the distance $p_1/2$. Although the discrete structure of the image over the transverse coordinate is strong due to the small quantity of the imaging elements, the ultrasonic image correctly represents the shape of the defect within the intrinsic resolution of the system.

In conclusion we would like to underline that huge progress has been made in the development and production of portable ultrasonic imaging systems during last few decades, opening up wide new horizons for their applications. Dramatic miniaturization of complicated systems such as in acoustic microscopy will in the near future give experimentalists, engineers, and doctors a completely new generation of powerful, portable handheld analytical tools for precise quantitative and minute examinations and diagnostics. Our experience and involvement as both witnesses and active participants in the vigorous technological development in various R&D fields, including the creation of new efficient materials, precise nanomachinery, and advanced electronics, give us a strong belief that extraordinary results will follow.

References

1 Maev, R.G. (2008) *Scanning Acoustic Microscopy: Fundamentals and Applications*, Wiley-VCH Verlag GmbH, Weinheim, approx. 450 pp.
2 Sadler, J., Shapoori, K., Malyarenko, E., Seviaryn, F., and Maev, R.G. (2010) Locating an acoustic point source scattered by a skull phantom via time reversal matched filtering. *J. Acoust. Soc. Am.*, **128** (4), 1821–1824.
3 Maev, R.G., Maslov, K., and Titov, S. (2000) Multieyed acoustical microscope

lens system, US Patent Number 6,116,090. Filed: April 30, 1999; issued: September 12, 2000.

4 Titov, S.A., Maev, R.G., and Bogachenkov, A.N. (2009) A small-size multichannel scanning acoustic microscope. *Instrum. Exp. Tech.*, **52** (5), 721–724.

5 Drinkwater, B. and Brotherhood, C. (2008) Coupling element with varying wall thickness for an ultrasound probe, US Patent Number 7,360,427 B2 from April 22, 2008.

6 Maev, R.G., Ptchelintsev, A.A., and Denisov, A.A. (2001) Ultrasonic imaging with 2D matrix transducers. *Ultrasonic Imaging*, **25**, 157–162.

7 Denisov, A.A., Shakarji, G.M., Lawford, B.B., Maev, R.G., and Paille, J.M. (2004) Spot weld analysis with 2D ultrasonic array. *J. Res. Natl. Inst. Stand. Technol.*, **109** (2), 233–244.

8 Maev, R.G., Ewasyshyn, F., Titov, S., Paille, J., Maeva, E.Y., Denisov, A., and Seviaryn, F. (2010) Method and apparatus for assessing the quality of spot welds, US Patent Number 7,789,286. Filed: June 4, 2003; issued: September 7, 2010.

9 Titov, S.A., Maev, R.G., and Bogachenkov, A.N. (2010) Acoustic visualization system employing matrix ultrasonic probe. *J. Sens. Syst.*, **7**, 18–20.

10 Maev, R.G., Titov, S., Bogachenkov, A., Ghaffari, B., Lazarz, K., and Ondrus, D. (2012) Method for assessing of quality of adhesively bonded joints using ultrasonic waves, US Provisional Application No. 61/623838. Filed: April 13, 2012.

11 Goodman, J.W. (1968) *Introduction to Fourier Optics*, McGraw-Hill Publishing House, New York.

12 Papoulis, A. (1968) *Systems and Transforms with Applications in Optics*, McGraw-Hill Publishing House, New York.

5
High-Frequency Ultrasonic Systems for High-Resolution Ranging and Imaging

Michael Vogt and Helmut Ermert

5.1
General Introduction

Ultrasound is widely used in medical diagnostics for soft tissue imaging (sonography), and also for non-destructive evaluation (NDE) of materials and mechanical components. For ranging and imaging, a large center frequency and a large bandwidth of the utilized ultrasound waves are required to obtain a good spatial resolution with respect to the axial direction of sound propagation, and along the lateral and elevational coordinates in the perpendicular directions. This fundamental relationship is valid for all systems that make use of time of flight (TOF) measurements along narrow "beams" with propagating acoustic, ultrasound, or electromagnetic waves. Examples are RADAR ("radio detection and ranging") and LIDAR ("light detection and ranging") systems, utilizing microwaves and light (visible or infrared), and also sonar ("sound navigation and ranging") systems. Applications like level metering (industrial process measurement), ultrasonic parking sensors (automotive), and others do not necessarily require two-dimensional (2D) or three-dimensional (3D) imaging. Instead, typically information about objects only along a *single* one-dimensional (1D) line of sight is of interest and assessed in these *ranging* applications.

The focus of this chapter is on engineering concepts, technical solutions, and new modalities for high-resolution ranging and cross-sectional imaging by utilizing high-frequency ultrasound (HFUS) in the frequency range of 20 MHz and above. Examples of various medical and technical applications are presented. In sonar, NDE, and medical sonography, conventionally ultrasound at relatively low frequencies in the range up to about 10 MHz is utilized. In applications that require a "microscopic" spatial resolution for imaging and ranging, HFUS at higher frequencies has to be utilized, as will be shown in the following. Ultrasonic measurement of the thicknesses of layered structures like lacquer coats and packaging foils requires a resolution of the order of some ten micrometers. In medical diagnostics, HFUS can advantageously be applied to high-resolution imaging of near surface areas (skin) [1–24] and low attenuation biological tissue (eye) [25–31],

Advances in Acoustic Microscopy and High Resolution Imaging: From Principles to Applictaions, First Edition.
Edited by Roman Gr. Maev.
© 2013 Wiley-VCH Verlag GmbH & Co. KGaA. Published 2013 by Wiley-VCH Verlag GmbH & Co. KGaA.

as well as to endosonography (mucosa), intravascular ultrasound (IVUS, blood vessels), and imaging of small animals (rats, mice) for preclinical research [32–39]. From the point of view of HFUS imaging and ranging applications, limitations result from the increasing attenuation of ultrasound waves in material samples and biological tissue with increasing frequency. Viscous attenuation and acoustic impedance mismatching are the predominant causes for attenuation of ultrasound waves in most "technical" objects and NDE. In biological tissue, on the other hand, the ultrasonic attenuation is to a very large extent caused by *scattering* of ultrasound waves. In both kinds of applications, the spatial resolution can only be improved by increasing the center frequency and the bandwidth, but the penetration depth of ultrasound waves into objects and tissue decreases at the same time because of the frequency-dependent attenuation.

This chapter is organized as follows: Section 5.2 gives an overview over of the state of the art of HFUS imaging systems, available ultrasound transducer technology, and basic system design considerations. In addition, system implementations and signal processing techniques for 1D ranging and imaging are discussed. The goal of Section 5.3 is to introduce techniques for 2D cross-sectional imaging. Focusing of ultrasound waves is an important issue in this context, and reconstructive imaging modalities, including spatial compounding and multidirectional tissue characterization, are presented. Section 5.4 gives examples of applications of HFUS for skin imaging and small-animal imaging. Summarizing conclusions are drawn in Section 5.5.

5.2
High-Frequency Ultrasonic System Components

5.2.1
Ultrasound Echo Systems

Figure 5.1a shows a block diagram of a typical ultrasound echo system.

The usual and most relevant approach in ultrasonic imaging and ranging is to perform pulse–echo measurements, that is, exciting an ultrasound transducer with a *pulsed* signal $s_{transmit}(t)$ and emitting ultrasound waves along a narrow "beam." Ultrasound waves – which are backscattered or reflected at acoustical inhomogeneities in the insonified area, and which propagate back to the same transducer (monostatic approach) – are converted into an electrical echo signal $s_{echo}(t)$. In the system in Figure 5.1a, the radio-frequency (RF) echo signal at the transducer is digitized by means of an analog-to-digital converter (ADC), and acquired echo data are transferred into a personal computer (PC) for further processing.

Inside the receiver, echo signals are recorded over TOF, and they are directly assigned to radial distances from the transducer by taking the speed of sound (SOS) inside the propagation medium into account. Consequently, echo signals acquired with a transducer with a fixed sound beam deliver information about backscattering and reflection at acoustic inhomogeneities along a 1D spatial

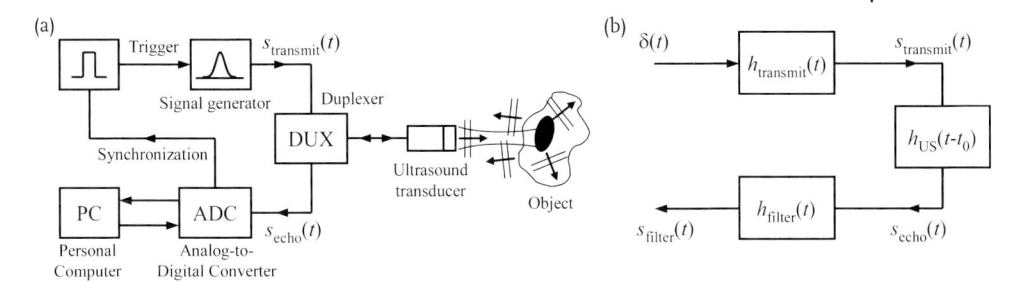

Figure 5.1 Ultrasound echo system: (a) block diagram; (b) system theoretical model.

coordinate. Accordingly, the first implementations of early ultrasound systems were limited to the simple and straightforward approach of visualizing the echo signal amplitude over TOF ("A-scan" technique). In the next section, this technique is extended to 2D imaging concepts by scanning the sound beam along a transverse axis perpendicularly to the axial direction of ultrasound propagation.

Figure 5.1b shows a system model for the ultrasound pulse–echo system in Figure 5.1a. The pulse generator is represented by a linear, time-invariant system with the impulse response $h_{transmit}(t)$ and a Dirac-pulse at its input, driving the ultrasound transducer with the transmit signal $s_{transmit}(t)$. The impulse response $h_{US}(t - t_0)$ describes the electro-acoustic conversion, the propagation of ultrasound waves, and the acousto-electrical conversion with a single, *point-like* scattering object (point-scatterer) inside the transducer's sound beam. Neglecting diffraction effects, a system consisting of the transducer and a point-scatterer can be approximated to be space-invariant, resulting in the impulse response $h_{US}(t - t_0)$, given by an impulse response $h_{US}(t)$ with a delay $t_0 = 2 \cdot r_0/c$ along the distance r_0 between the transducer and the scatterer, and the SOS c. The impulse response $h_{filter}(t)$ in Figure 5.1b describes a linear filter with output signal $s_{filter}(t)$, which is used for processing the echo signal $s_{echo}(t)$.

The transfer characteristics of the ultrasound system, described by the transfer function $H_{US}(j\omega)$, is essentially affected by the ultrasound transducer's bandpass characteristics. Consequently, $h_{US}(t)$ is the impulse response of a bandpass system, whereby the transmit signal's spectrum $S_{transmit}(j\omega)$ should cover the system's passband.

5.2.2
Transmitter and Receiver Components for High-Frequency Ultrasonic Echo Systems

For pulse–echo measurements, so-called "pulser/receiver" units are very common, which consist of a pulse generator in the transmit path, a low-noise amplifier in the receive path, and a "duplexer" for separation of the two paths from each other (Figure 5.1a). Because of the large difference between the amplitudes of the transmitted and the received signal, the task of the duplexer on the one hand is to protect the receiver against the large transmit signal amplitude (typically up to some hundred volts peak amplitude) and, on the other hand, to suppress electronic noise emitted from the pulser electronics during reception. The advantage of acoustic systems compared to electromagnetic systems is the large TOF of ultrasound waves because of the very small SOS (about 1540 m s^{-1} in biological tissue). This makes it very easy to separate transmitted and received pulsed ultrasound waves over TOF.

In many system implementations, the duplexer consists of a nonlinear "expander/limiter"-network, making use of the large difference between the amplitudes of the *large* transmitted signal and the *small* echo signal [40]. Saturations of the amplifier inside the receiver cause no problems if the distance between the transducer and the first interesting object inside the sound beam, and the corresponding TOF, are sufficiently large. Alternatively, an active switch, for example, with pin-diodes, can be utilized to disconnect the receiver from the pulser and transducer during transmission, and to disconnect the pulser from the transducer and receiver during reception, controlled by the ultrasound system [24].

The development of appropriate pulsers for HFUS systems is challenging, because a small pulse width appropriate to cover the bandwidth of the utilized transducer, together with a large pulse amplitude, is required. In addition, the lengths of cables play an important role concerning the system's transfer characteristics in conjunction with electrical mismatching at the interconnection between the transducer and the pulser/receiver [40].

The ultrasound transducer is a crucial component in all ultrasound systems – this is especially true for HFUS systems. PZT (lead zirconium titanate) ceramics, lithium niobate compounds, and PVDF [poly(vinylidene fluoride)] fluoropolymer foils are used as piezoelectric materials in HFUS transducers.

The following approaches for forming a narrow "beam" are common in ultrasound imaging and ranging applications. Piston transducers with a *planar* aperture, from which ultrasound waves emanate, show an almost cylindrical sound beam characteristic in the near field, and the sound beam diverges in the far field with a relatively large aperture angle (Figure 5.2a).

Disadvantageously, the width of the sound beam of piston transducers in the directions perpendicularly to the axial direction of ultrasound propagation is too large for the requirements of most ultrasonic imaging systems. Alternatively, a focused sound beam can be formed by means of a *spherical* aperture. For HFUS transducers, this is preferably achieved by using a curved PVDF foil (Figure 5.2b). As an alternative, a planar piezoceramic element can be equipped with an acoustic lens, in the case of HFUS preferably consisting of quartz glass (Figure 5.2c).

Linear and phased array transducers consist of a large number of piezoelectric elements; they are widely used in conventional, low-frequency ultrasound imaging systems. The challenge in realizing HFUS transducer arrays is that the *spacing* between the array elements has to fulfill strong requirements to generate sound beams with sufficiently suppressed grating lobes. Preferably, the spacing between the array elements should not exceed one wavelength in the case of linear arrays, and half a wavelength in phased arrays. The fabrication of HFUS arrays is challenging and expensive [41–44]. A further problem of HFUS array transducer

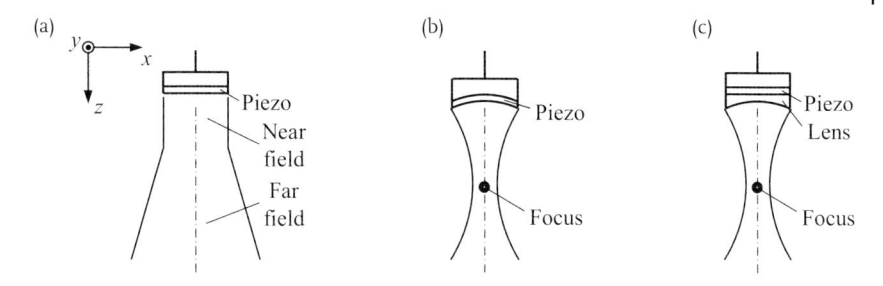

Figure 5.2 Single-element transducers: (a) piston transducer; (b) spherical aperture; (c) acoustic lens.

systems is the need for wideband multiplexers and beamformers, which are used for switching between array elements and for driving the active array elements. For these reasons, single-element transducers are still widely used in HFUS imaging systems [40, 45–52].

In terms of complexity, annular array transducers, which consist of only a few concentric transducer elements, are a good compromise between fixed-focus single-element transducers and linear or phased array transducers [53–59]. Like single-element transducers, annular array transducers produce sound beams rotationally symmetric with respect to the axial direction of ultrasound propagation. Transmit focusing can be performed electronically by driving the array elements with different time delays (or phase shifts in narrowband systems). In the receive mode, dynamic focusing can be performed electronically by applying different time delays (or phase shifts) to the received signals of each array element and coherent summation. As will be discussed below in more detail, synthetic aperture focusing can be applied by performing consecutive echo measurements with each of the array elements, recording the received signals of each element for each measurement, and *numerical* image reconstruction, which allows for full transmit and receive beamforming [60–65]. However, for the acquisition of 2D or 3D echo data with annular arrays, a mechanical scan also has to be performed like in single-element transducers applications.

Annular arrays for HFUS imaging applications have been proposed and developed by several groups [53–59]. Because of the small number of elements, annular array transducers are less complex and much easier to implement than linear and phased arrays [41–44].

5.2.3
Spectral and Range Resolution Properties

The axial resolution of ultrasound systems characterizes the ability to distinguish echo signals from scatterers located at different positions along the axial direction of ultrasound propagation based on the TOF. The lateral and the elevational resolution in the perpendicular directions, on the other hand, depend on the focusing

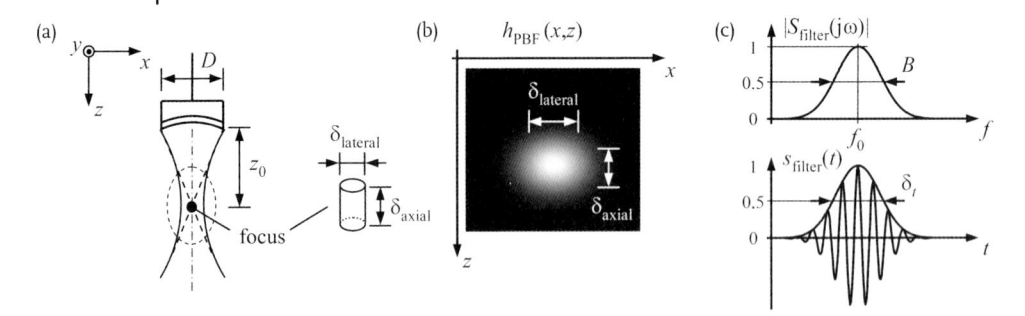

Figure 5.3 (a) Spherically focused single-element transducer; (b) point spread function (PSF); (c) system's response.

of the ultrasound beam. Figure 5.3a shows the example of a spherically focused transducer (aperture diameter D, focus length z_0). This type of transducer is widely used in HFUS imaging systems, and has also been employed in most of the measurements presented below.

In this chapter, the axial and lateral resolution δ_{axial} and $\delta_{lateral}$, respectively, denote the full width at half maxima (FWHM) of the system's point spread function (PSF), which represents the image of a single point-scatterer generated by the ultrasound imaging system [20–23, 40, 48]. Figure 5.3b illustrates the PSF $h_{PBF}(x,z)$, that is, the 2D image of a point-scatterer along the lateral and axial coordinates x and z, respectively. As will be discussed below, imaging can be performed with a mechanical scan of the single-element transducer along the lateral coordinate x by means of a scanning unit with a stepping motor, and by pulse–echo measurements at discrete lateral transducer positions [49–51, 66]. For each measurement, the envelope of acquired echo signals over TOF t is acquired and assigned to image amplitudes along the axial coordinate $z = c/2t$ by taking the SOS c into account.

Figure 5.3c shows the system's response, which is here the filter output signal $s_{filter}(t)$ (Figure 5.1b), the magnitude of the Fourier transform $|S_{filter}(j\omega)|$, and the envelope $|s_{filter+}(t)|$ [analytical signal $s_{filter+}(t)$] for a single point-scatterer. The Gaussian spectrum in Figure 5.3c, associated with a Gaussian shape of the echo signal's envelope, is an appropriate model for typical ultrasound transducers. Under these conditions, the following relationships for the axial resolution δ_{axial} and the lateral resolution $\delta_{lateral}$ can be derived [22, 48]:

$$\delta_{axial} = \frac{c}{2}\delta_t = \frac{2\ln(2)}{\pi}\frac{c}{B} = 0.441\frac{c}{B}, \quad \delta_{lateral} \approx \lambda_0 F, \quad \text{with: } \lambda_0 = \frac{c}{f_0}, \quad F = \frac{z_0}{D}$$

(5.1)

In Equation (5.1), δ_t denotes the pulse width and B the system's bandwidth (FWHM) (Figure 5.3c). For spherically focused transducers, the lateral resolution $\delta_{lateral}$ depends on the wavelength λ_0 at the transducer's center frequency f_0, and the "F-number" F defined as the quotient of the focus length z_0 and the transducer's aperture diameter D, as given in (5.1) [20, 48]. At the transducer's focus, the rotational symmetric ultrasound beam can be described by a cylindrical "resolution cell," which is represented by the axial and the lateral resolution δ_{axial} and $\delta_{lateral}$, respectively (Figure 5.3a).

The relationships in Equation (5.1) show that the bandwidth B and the center frequency f_0 have to be increased to improve the axial and lateral resolution. However, the attenuation of ultrasound waves in biological tissue and technical objects increases significantly with increasing frequency. Accordingly, the maximum penetration depth is limited by the dynamic range of the ultrasound system's electronics. Consequently, a compromise between good spatial resolution and sufficiently large penetration depth has to be found. HFUS systems for superficial tissue imaging are optimized with respect to the spatial resolution and the penetration depth by utilizing ultrasound in the frequency range of 20 MHz and above [48–52, 66].

5.2.4
Measurement and Optimization of the Pulse Transfer Properties

As discussed above, backscattered and reflected ultrasound waves are directly converted into an echo signal. The system's response depends on the transfer characteristics of the ultrasound path ("free space") as well as on the transfer characteristics of the pulser/receiver electronics. This will be discussed now based on the schematic ultrasound echo system shown in Figure 5.4 with a duplexer consisting of a nonlinear "expander/limiter"-network.

In Figure 5.4, the ultrasound transducer is connected through a coaxial cable to the transmit/receive electronics. During transmission, the transducer is driven through the expander network with the *large* transmit signal generated by the pulser, and the receiver is disconnected from the transducer. During reception of

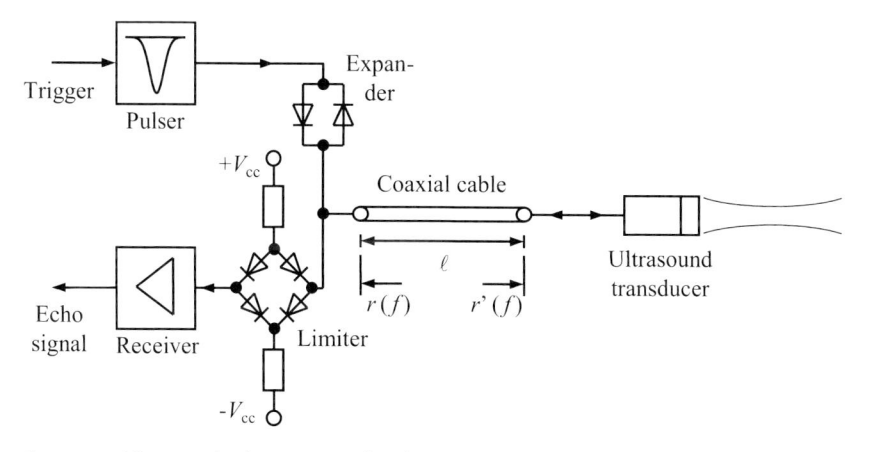

Figure 5.4 Ultrasound echo system with pulser/receiver driving the ultrasound transducer.

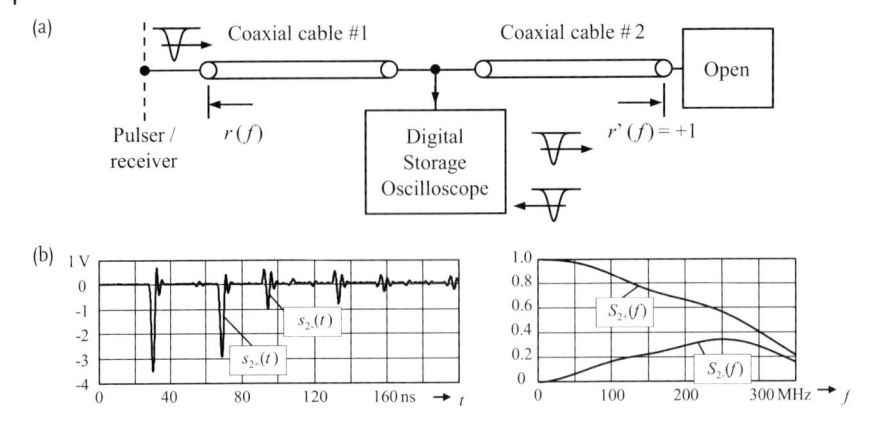

Figure 5.5 (a) TDR measurement setup; (b) pulsed signals and corresponding spectra after time-gating.

the *small* echo signal, the transducer is connected to the receiver and disconnected from the pulser.

Especially in the case of HFUS, the propagation of *broadband* pulsed ultrasound signals along the cable is critical if the length of the cable is not negligibly small. Because the electrical matching of the HFUS transducer is usually relatively poor over the large passband of the transducer [reflection coefficient $r'(f)$ in Figure 5.4], and because the TOF of the electrical signals along the cable is usually not negligibly small against the pulse width, the electrical matching of the pulser/receiver electronics is of great importance [reflection coefficient $r(f)$ in Figure 5.4]. Both the large-signal behavior and the small-signal behavior have to be optimized separately in this regard.

We have analyzed this problem and proposed a time domain reflectometry (TDR) concept for measuring the large-signal reflectance $r(f)$ by using the output signal of the pulser itself [40] (Figure 5.5a). The pulser/receiver is terminated with two coaxial cables, 1 and 2, which are connected in series and terminated with an open loop [reflectance $r'(f) = 1$]. The pulsed signals propagating along the cables are measured by means of a digital storage oscilloscope (DSO), where the lengths of the two cables have to be large enough to be able to separate propagating and reflected pulses over the TOF. Based on the spectra of the time-gated signals, the large-signal reflectance $r(f)$ can be measured (Figure 5.5b). The small-signal reflectance $r(f)$ can easily be measured by means of a network analyzer (NWA).

The proposed measurement concept has been used to optimize a 100 MHz range HFUS imaging system. Figure 5.5b shows the measured time-domain signals and the corresponding spectra for the system's pulser/receiver with 1.8 ns pulse width, 60 V peak amplitude, and 174 MHz receiver bandwidth. Large- and small-signal behavior of the pulser/receiver have been optimized by means of separate matching networks in the transmit and receive paths [40].

5.2.5
Range Resolution Optimization: Inverse Echo Signal Filtering

It was assumed above that the PSF of the HFUS imaging system is space-invariant. Especially in the case of *focused* single-element transducers, this is only a rough approximation for imaging along a large axial depth range z as compared to the transducer's depth of field (DOF) [20, 48]. The PSF is actually significantly depth-variant, which can, at least partly, be compensated by *time-variant* inverse filtering of the echo signal [23]. The filtered echo signal $s_{filter}(t)$ is obtained by convolution of the echo signal $s_{echo}(t)$ with the impulse response $h_{filter}(t,t')$ of the time-variant filter (compare Figure 5.1b):

$$s_{filter}(t) = s_{echo}(t) * h_{filter}(t, t'), \quad \text{with: } H_{filter}(j\omega, t') = \frac{S_{echo}(j\omega, t'_0)}{S_{echo}(j\omega, t')}, \quad t' = \frac{2}{c}z', \quad t'_0 = \frac{2}{c}z_0$$

(5.2)

In Equation (5.2), z denotes the axial depth, z_0 the focus length, and $H_{filter}(j\omega,t')$ the transfer function of the inverse filter. The latter is calculated by the inverse of the echo signal spectrum $S_{echo}(j\omega,t')$ for a point-scatterer at a depth z' with the echo signal spectrum $S_{echo}(j\omega, t'_0)$ from the focus as the *desired* optimum response of the system.

The potential of inverse filtering to optimize the system's imaging properties has also been evaluated with the above-mentioned HFUS imaging system working in the 100 MHz range [23]. In a first step, the system's PSF is assessed by means of echo measurements on a glass plate surface positioned at different axial depths z and calculating the corresponding echo signal spectra $S_{echo}(j\omega,t')$ (Figure 5.6a).

As expected, the echo signal spectrum $S_{echo}(j\omega, t'_0)$ with the glass plate surface at the transducer's focus ($z = 4.3$ mm focus length) shows the maximum amplitude. This spectrum is used as the desired optimum response of the system to calculate the transfer function $H_{filter}(j\omega,t')$ of the inverse filter according to Equation (5.2) (Figure 5.6b). For a further quantitative analysis of the performance of

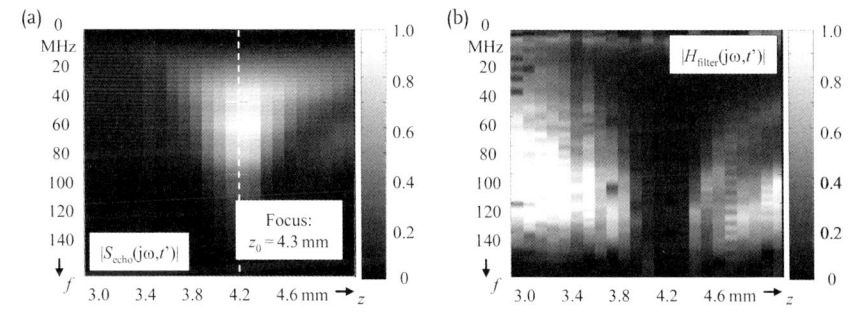

Figure 5.6 (a) Measured echo signal spectra (echoes from glass plate surface); (b) inverse filter transfer function.

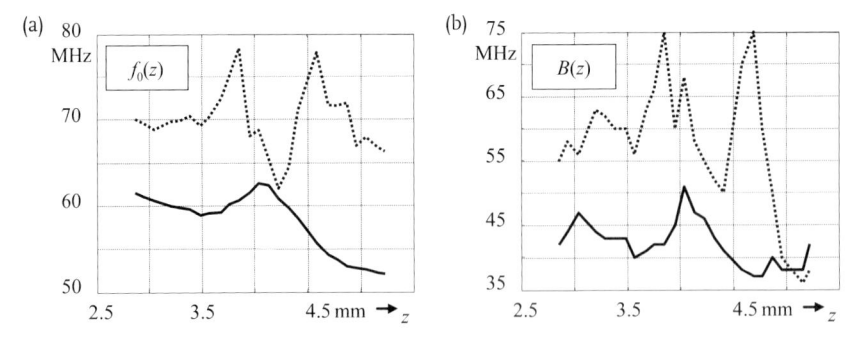

Figure 5.7 Echo signal characteristics; unfiltered echo signals (solid lines), inversely filtered echo signals (dotted lines): (a) center frequency; (b) bandwidth.

the inverse filter, the center frequency f_0 and the bandwidth (−6 dB) of the echo signals without and with the inverse filter have been analyzed (Figure 5.7a and b).

Both the center frequency and the bandwidth are increased by using the inverse filter, and the lateral and the axial resolution are improved, accordingly. It has also been shown by means of measurements on a speckle phantom and by *in vivo* images of skin that the proposed time-variant inverse filter enables us to improve the system's imaging properties. Electronic noise in the images, on the other hand, is slightly increased, because of the inverse filtering operation in Equation (5.2) [23].

5.2.6
Measurement of Acoustic Scattering Parameters in Plane Wave Propagation

Above, concepts and techniques for *echo* measurements along a 1D spatial coordinate with a narrow sound beam have been discussed, which are designed for *imaging* backscattering and reflecting objects. The discussed concepts of assigning echo signal amplitudes over TOF to morphological structures of objects along the axial direction of ultrasound propagation by assuming a constant SOS is a rather qualitative approach. The following presents a concept for combining both *echo* and *transmission* measurements, and for *quantitative* measurement of acoustic parameters in an 1D configuration by means of HFUS. As will be shown, the proposed technique can be applied to the analysis of layered plane-parallel objects, which are of interest in many NDE applications.

Figure 5.8a shows the proposed two-port "ultrasound network analyzer," consisting of two HFUS transducers opposite each other, with the device under test (DUT) positioned in-between.

The DUT consists of multiple homogeneous plane layers with different acoustic properties (SOS c_n, acoustic impedances Z_n) and different thicknesses D_n, being parallel to each other, and water is used as coupling medium between the transducers and the DUT. Propagating ultrasound waves can be approximated by *planar* waves, and the reflection and transmission of ultrasound waves at the planar

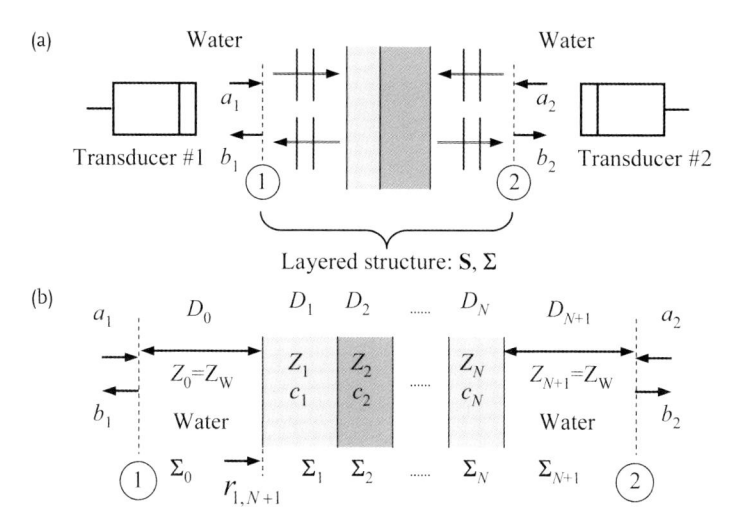

Figure 5.8 Ultrasonic analysis of layered structure: (a) measurement setup; (b) 1D network model.

boundaries between neighboring layers can be described based on a 1D propagation model [67]. For a first measurement, transducer #1 is excited with a pulsed signal, and the reflected and the transmitted signals $m_{11}(t)$ and $m_{21}(t)$, respectively, at the two transducers #1 and #2 are recorded. In another measurement, transducer #2 is excited, and again the reflected and transmitted signals $m_{22}(t)$ and $m_{12}(t)$, respectively, at the transducers #2 and #1 are recorded for further processing.

In the frequency-domain, the network between the two ports 1 and 2 in Figure 5.8a and b can be described by the scattering and transmission matrices S and Σ, respectively:

$$\begin{pmatrix} b_1 \\ b_2 \end{pmatrix} = S \begin{pmatrix} a_1 \\ a_2 \end{pmatrix}, \quad \begin{pmatrix} b_1 \\ a_1 \end{pmatrix} = \Sigma \begin{pmatrix} a_2 \\ b_2 \end{pmatrix}, \quad S = \begin{pmatrix} S_{11} & S_{12} \\ S_{21} & S_{22} \end{pmatrix}, \quad m_{ij} = k_{ij} S_{ij}, \quad i, j \in \{1, 2\}$$

(5.3)

As given in Equation (5.3), the spectra m_{ij} of the acquired time-domain signals $m_{ij}(t)$ are linear functions of the scattering parameters S_{ij}. Based on the network model in Equation (5.3) and on the assumption of a *lossless* propagation of ultrasound waves inside the layers, a method for the reconstruction of the acoustic parameters and thicknesses of the individual layers by analyzing the measured scattering parameters has been developed [67].

Measurements in a setup according to Figure 5.8a have been performed using *planar* HFUS piston transducers with 25 MHz center frequency, 13 MHz bandwidth (−6 dB), and 6.3 mm aperture diameter. Figure 5.9a shows an example of acquired time-domain echo and transmission signals for a two-layer DUT (layer 1: aluminum, $D_1 = 1$ mm; layer 2: brass, $D_2 = 0.5$ mm).

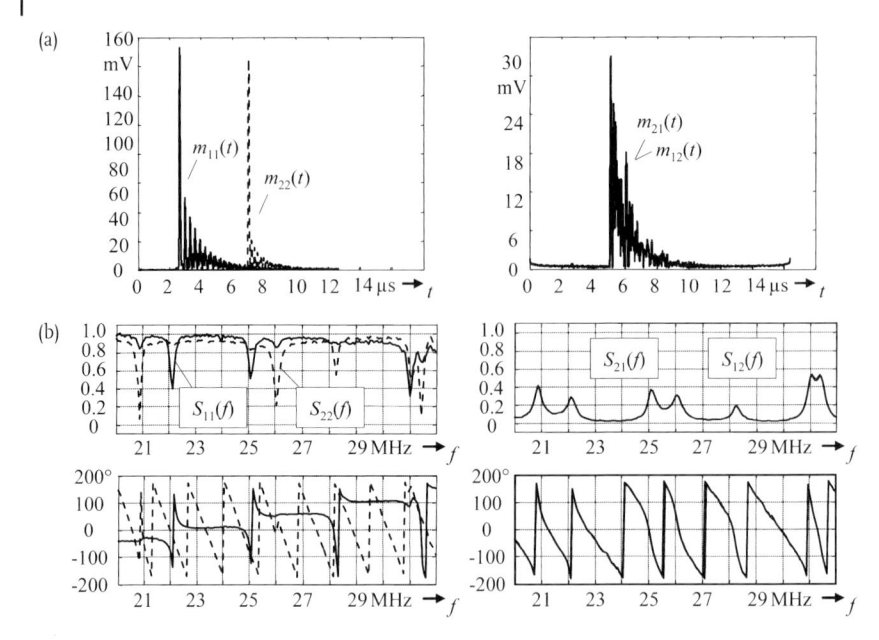

Figure 5.9 Ultrasonic analysis of layered structure: (a) time-domain responses; (b) scattering parameters (frequency-domain).

In Figure 5.9b, magnitude and phase of the scattering parameters of the DUT are represented as a function of frequency. As expected, measured scattering parameters $S_{21}(f)$ and $S_{12}(f)$ turn out to be equal because of the reciprocity of the network. Furthermore, "resonances" at discrete frequencies are found in all scattering parameters $S_{ij}(f)$, which result from the propagation and interference of ultrasound waves inside the layers with different SOS and thicknesses, and from the reflection of ultrasound waves at the boundaries between layers with different acoustic impedance. Acoustic parameters and thicknesses of the layers have been reconstructed using the measured scattering parameters [67].

5.3
Engineering Concepts for High-Frequency Ultrasonic Imaging

5.3.1
Single-Element Transducer B-Scan Techniques

Above, the "A-scan" technique (A: "amplitude"), the relatively simple approach of acquiring echo signals along a 1D spatial coordinate using an ultrasound transducer with a *spatially fixed* sound beam, has been discussed (Figure 5.1). As a result, only information about backscattering and reflection from inhomogeneities of the acoustic impedance along a 1D spatial coordinate is obtained. Although the very first implementations of ultrasound systems were based on this A-scan

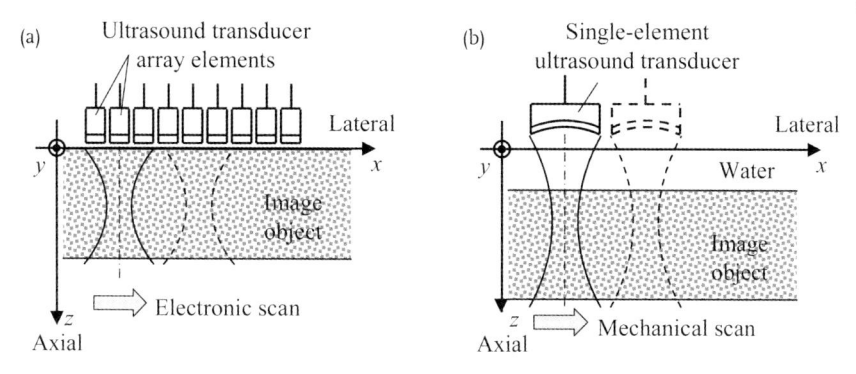

Figure 5.10 B-mode imaging approach: (a) electronic scan with ultrasound transducer array; (b) mechanical scan with single-element ultrasound transducer.

concept, more sophisticated scanning schemes are required for cross-sectional, tomographic ultrasound imaging.

For 2D sonography, scanning along the *lateral* coordinate perpendicularly to the *axial* direction of ultrasound propagation has to be performed. In the "B-mode" imaging approach (B: "brightness"), the amplitude of echo signals in the resulting 2D image plane along the axial and the lateral coordinate are displayed using a gray-scale map. While the axial scanning of the imaged object is achieved by echo measurements with broadband ultrasound waves, scanning along the lateral coordinate x is usually performed by *electronically* switching single elements or multiplexing groups of elements of an ultrasound transducer *array*, which consists of a large number of transducer elements (Figure 5.10a).

In HFUS imaging systems, however, mechanical scanners with focused single-element transducers are still widely applied, because the implementation of HFUS transducer arrays is still technically challenging and difficult [41–44]. In the scenario shown in Figure 5.10b, a mechanical scan using a spherically focused single-element transducer along the lateral direction x is performed for echo measurements at adjacent lateral transducer positions. A water path is used as sound propagation medium between the ultrasound transducer and the imaged object (biological tissue, technical object, etc.). For skin imaging and for small-animal imaging, preferably a water tank containing the ultrasound transducer, equipped with a hole in the bottom of the tank and the bottom placed upon the object surface to guarantee a good acoustic coupling to the biological tissue, is used [48]. Mechanical scanners are also used for human eye imaging with some preference for a sector scanning geometry. A real-time operation mode of HFUS systems can be realized by using sufficiently fast scanning mechanics.

5.3.2
Lateral Resolution Optimization

The motivation behind using spherically *focused* single-element transducers in HFUS imaging systems is to optimize the lateral and the elevational resolution,

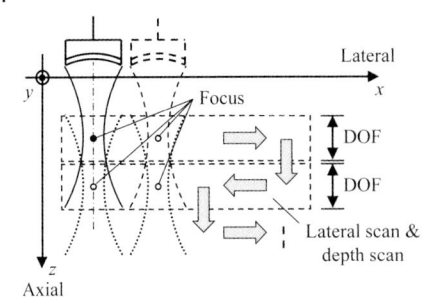

Figure 5.11 B/D-scan technique: composition of axially "short" B-mode images acquired from within the transducer's DOF with the focus positioned at different axial depths.

which has already been discussed above in the context of Figure 5.3a. However, the DOF, that is, the axial dimension of the focus zone (−6 dB), of ultrasound transducers with a fixed focus is a limiting factor concerning B-mode imaging. The DOF decreases with increasing focusing, and the axial field of view (FOV) is limited, accordingly [48]. This section presents and discusses more sophisticated scanning and image reconstruction techniques that can cope with the aforementioned problem.

5.3.2.1 B/D-Scan Technique

Providing it is feasible regarding the minimum acceptable frame rate in the application, a reasonable approach, which has been referred to as the "B/D-scan" technique ("brightness/depth") [24], is to acquire multiple *short-range* B-mode images from within the DOF of the focused transducer with the focus located at different depths (Figure 5.11).

The ultrasound transducer and, consequently, its focus is stepwise positioned at different equidistant axial depths by means of a motor unit, and, as above, a mechanical scan in lateral direction is performed at each depth. A B-mode image covering a *large* axial FOV is composed by adding subsequently acquired short B-mode images. The proposed B/D-scan technique has been implemented in a 100 MHz range HFUS imaging system for dermatologic and ophthalmologic applications, and it has been very successfully applied for *in vivo* imaging of human skin and *in vitro* imaging of pig eyes [24].

5.3.2.2 Synthetic Aperture Focusing Techniques (SAFT)

Synthetic aperture focusing is another approach for depth-invariant imaging by performing echo measurements at different transducer positions and numerically *reconstructing* B-mode images from recorded echo signals. In the field of sonography, this concept is referred to as synthetic aperture focusing technique (SAFT), which is the equivalent of synthetic aperture RADAR (SAR) applied in Radar systems for remote sensing and imaging with electromagnetic waves in the microwaves region.

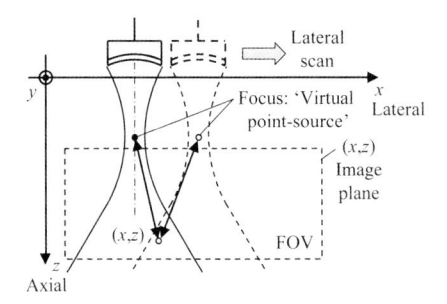

Figure 5.12 SAFT reconstruction: focus of spherically focused single-element transducer assumed as virtual point-source of spherical ultrasound waves.

Especially for HFUS imaging, we have proposed a SAFT concept based on approximating the focus of a spherically focused single-element transducer as a *virtual point-source* for spherical ultrasound waves, which propagate in the far field inside the transducer's opening angle [24, 61, 63]. Similar to the scanning scheme discussed above, echo signals are acquired at equidistant lateral transducer positions by means of a 1D scan along the lateral coordinate x (Figure 5.12). Thereafter, numerical SAFT reconstruction is applied to the recorded echo data in order to extend the FOV beyond the transducer's DOF into the far field. Under the assumption of spherical waves emanating from the virtual point-source, "delay and sum" (DAS) reconstruction can, preferably, be applied to achieve a focusing with respect to the lateral direction, and to reconstruct 2D B-mode images in the axial/lateral image plane. The two-path TOFs of ultrasound waves traveling from the focus of the transducer to each point (x,z) in the image plane to be reconstructed (Figure 5.12) is calculated from the known distances by taking a constant SOS into account. For each individual point (x,z) of the image plane, focusing is achieved by applying delays equal to the calculated TOFs to the recorded echo signals at the different lateral transducer positions and summing up the delayed signals [24, 61, 63].

A problem of this approach is that with only the 1D scan along the lateral coordinate x focusing is also only performed with respect to the lateral direction x. When using a spherically focused ultrasound transducer, the resolution along the *elevational* direction y (Figure 5.12) remains poor. As a solution to this problem, a 2D mechanical scan along *both* the lateral *and* the elevational coordinate can be carried out. Based on the recorded echo signals, focusing can be performed inside a single 2D (x,z) image plane at the elevational coordinate $y = 0$. In addition, *multiple* (x,z) image planes at different elevational positions y can be reconstructed, which is of interest for 3D sonography. Preferably, DAS can be applied for image reconstruction by applying delays to the recorded echo signals from the transducer positions in the 2D (x,y) scanning plane, and by summing up the delayed signals.

SAFT reconstruction with echo data from 1D and 2D scans has also been evaluated using the 100 MHz range HFUS imaging system mentioned above. It has been shown that the proposed SAFT reconstruction allows us to significantly extend the FOV into the transducer's far field and to increase the maximum depth range of the system. A significant improvement of B-mode image quality concerning the lateral resolution and the signal-to-noise ratio (SNR) has been achieved with the 1D scan approach, as compared to unfocused B-mode images. However, to achieve an image quality comparable to the B/D-scan technique, SAFT reconstruction with echo data from a 2D scan is required [24].

More recently, we have proposed a modification of SAFT, referred to as "PSF-SAFT," which can advantageously be applied using *strongly focused* single-element transducers [60, 62, 65]. The goal of this approach is not only to use the transducer's focus region for imaging, but also to account for *diffraction effects* and to extend the DOF towards the far field *and* the near field.

Numerical reconstruction in SAFT can generally be understood as a cross-correlation operation and a matched filtering of the acquired echo signals with the system's PSF, where the latter in particular depends on the characteristics of the utilized ultrasound transducer [65]. A quite simple *analytical model* can be used to calculate the PSF if *spherical waves* emanating from a point-source can be assumed. As described above, this leads to the relatively simple DAS approach for SAFT reconstruction, but because of the assumption of spherical waves the approach is only applicable in the transducer's far field.

The PSF in the *focus region* and in the extended focus region is much more difficult to describe, and in the case of spherically focused transducers, analytical solutions can only be derived with many limitations and based on approximations. For this reason, we have proposed to numerically *simulate* the PSF, which can carried out with sufficiently good precision by discretizing the transducer's aperture and calculating the corresponding radiation fields. As an input for SAFT reconstruction, the PSF is simulated for a point-scatterer at varying axial depth, and the results are stored. SAFT reconstruction is then performed by 2D linear depth-variant filtering of the acquired and recorded echo signals with a 2D filter kernel, given by the depth-dependent simulated PSF [65]. In addition, coherence factor (CF) weighting of the processed echo signals has been proposed for further suppression of undesired side-lobes in the SAFT-reconstructed B-mode images. This concept has also been implemented and evaluated in HFUS imaging applications with a 20 MHz range focused single-element transducer [65].

Figure 5.13a shows a 20 MHz range B-mode post mortem image of the abdominal region of a rat pup, which has been acquired from within the DOF of the transducer *without* synthetic aperture focusing.

The electronic noise level in the unprocessed B-mode image increases with increasing axial depth z, and, accordingly, the SNR decreases. The B-mode image in Figure 5.13b was obtained after SAFT reconstruction with the above-discussed PSF-SAFT concept. It can be seen that the image contrast is significantly improved as opposed to the unfocused B-mode image. In addition, the lateral resolution at large depths is improved. In conclusion, SAFT has been very successfully applied to improve the quality in HFUS B-mode imaging with a spherically focused single-element transducer [60, 62, 65].

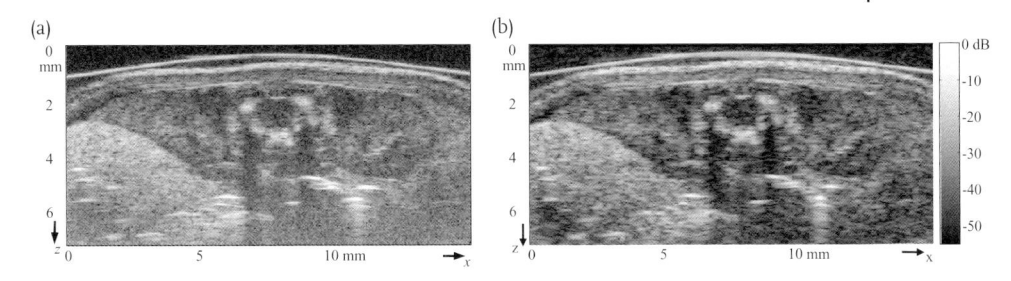

5.3 Engineering Concepts for High-Frequency Ultrasonic Imaging

Figure 5.13 Abdominal region of rat pup cadaver: (a) B-mode image *without* synthetic aperture focusing; (b) PSF-SAFT reconstructed B-mode image.

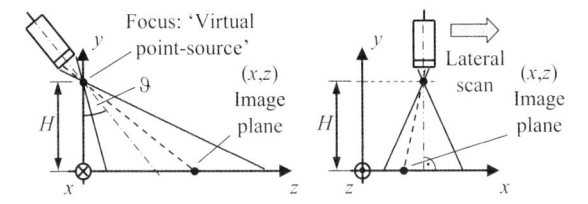

Figure 5.14 Imaging of planar objects: 1D mechanical scan with spherically focused single-element transducer in a side-looking configuration, point (x,z) in image plane.

The aim of the concepts discussed above is the reconstruction of 2D cross-sectional, tomographic ultrasound images of biological tissue inside 3D objects. The reconstruction and imaging of *surfaces* and 2D *planar* objects, on the other hand, is a completely different problem. Ultrasound imaging of planar material samples is of interest in many NDE applications, for example, for imaging superficial cracks. In scanning acoustic microscopy (SAM), the surface of a material sample is usually aligned along the lateral and elevational (x,y) plane at a constant axial depth z. In this configuration, echo measurements are performed with a 2D scan of the transducer along the lateral and the elevational coordinate, and with the ultrasound transducer's beam oriented *perpendicularly* to the surface of the sample.

As an alternative, we have proposed an adaption of side looking airborne radar (SLAR) systems for imaging planar objects with HFUS by using a spherically focused single-element transducer in the "side-looking" configuration with the viewing angle ϑ shown in Figure 5.14 [61].

Advantageously in this approach, only a 1D scan along the lateral direction x with the transducer's focus at a constant height $y = H$ has to be performed for SAFT reconstruction of the 2D planar material sample located in the (x,z) image plane at $y = 0$ (Figure 5.14).

This concept has been implemented in a 20 MHz range HFUS ultrasound system, and DAS has been applied for image reconstruction. It was again assumed that spherical waves emanate from the transducer's focus as a virtual point-source [61]. Evaluations have been performed using the circuit board shown in Figure

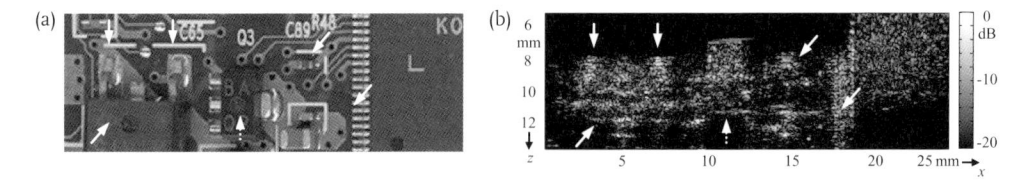

Figure 5.15 PCB with SMD components: (a) photograph; (b) SAFT reconstructed ultrasound image.

5.15a as a test object, consisting of the planar printed circuit board (PCB) with surface-mounted devices (SMD) soldered onto the PCB.

Scans with the 20 MHz ultrasound transducer have been performed at a height $H = 8.6$ mm and with a viewing angle $\vartheta = 35°$. Figure 5.15b shows the B-mode image after SAFT reconstruction. The SMD components mounted onto the PCB and their pins can be identified in the ultrasound image by comparison with the photograph (see arrows in Figure 5.15a and b). The surfaces of some SMD components are shown at *smaller* axial depths z along the axial direction of ultrasound propagation than expected, because of the so-called "lay-over effect," which is well-known in SAR. Echoes from reflecting and backscattering structures at a significantly large height (as compared to the transducer's height H) *above* the reconstructed image plane (at $y = 0$) actually have a significantly *smaller* TOF than a scatterer at the same point (x,z) but at a height $y = 0$. For this reason, the corresponding echoes are virtually shifted to smaller depths z within the SAFT reconstruction.

5.3.3
Limited Angle Spatial Compounding (LASC)

As discussed above, the conventional approach in B-mode imaging is to perform lateral and elevational scans with a *fixed* orientation of the transducer's beam, usually perpendicularly to the scan directions. Ultrasound spatial compounding is an approach for *multidirectional* echo measurements and *incoherent* superposition of B-mode images from different directions. Advantageously in this approach, the image contrast is improved, speckle and electronic noise are suppressed, and imaging artifacts are reduced as opposed to conventional B-mode images [68–70]. Figure 5.16 presents a mechanical scanning system for HFUS multidirectional imaging with a single-element transducer.

The transducer is tilted by an angle α in the (x, z) image plane by means of a motor unit, and mechanical scans along the lateral direction x are performed for a series of different tilting angles. Ultrasound compound images are reconstructed by separately calculating a B-mode image for each tilting angle α and scan-converting these B-mode images into the (x,z) image plane. The necessary transformation of coordinates is performed by taking the geometry shown in Figure 5.16 into account. The compound B-mode image is obtained by the superposition

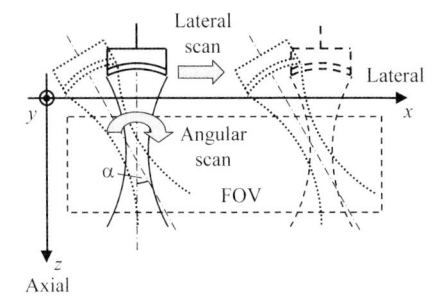

Figure 5.16 Multidirectional echo measurements: lateral scans with different tilting angles α in the (x, z) image plane.

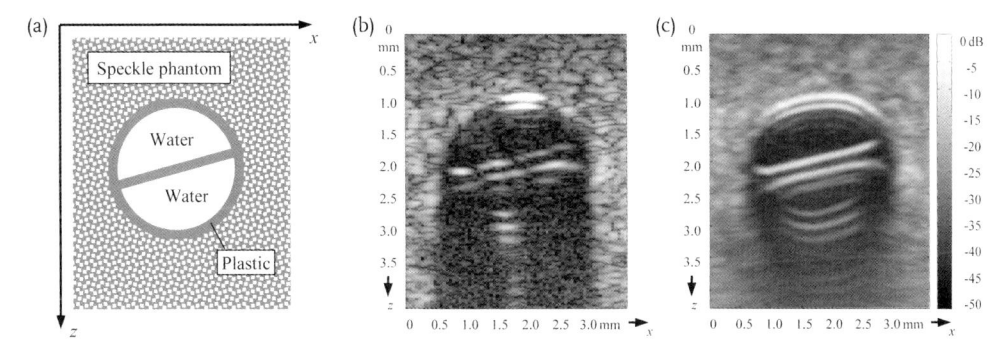

Figure 5.17 Phantom measurements: (a) schematic of water-filled two-compartment phantom surrounded by speckle phantom; (b) conventional B-mode image; (c) spatial compound image.

of scan-converted B-mode images, which is equivalent to an incoherent superposition of echo signals, because the summation is applied to the *envelope* signals. We have equipped a 20 MHz range HFUS single-element transducer system with separate motor units for both the scanning in the lateral direction and the angular scanning [68]. For a precise assignment of acquired echo signal data to the system's center of rotation, a sophisticated calibration scheme based on measurements on a wire phantom has been developed.

The implemented system has been evaluated by means of phantom and *in vivo* measurements. Figure 5.17a shows a schematic of the cross-section of the utilized phantom (plastic straw with two water-filled compartments, surrounded by a speckle phantom). The phantom was designed with the aim of evaluating spatial compound imaging of specularly reflecting surfaces with different spatial orientations, and also of objects with randomly distributed point-scatterers, which cause speckle.

Figure 5.17b and c show a conventional B-mode image ($\alpha = 0°$) and a spatial compound image composed of echo signal data from 27 different directions (tilting angles $\alpha = -32.5°, -30°, \ldots, +30°$).

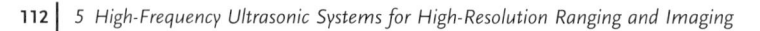

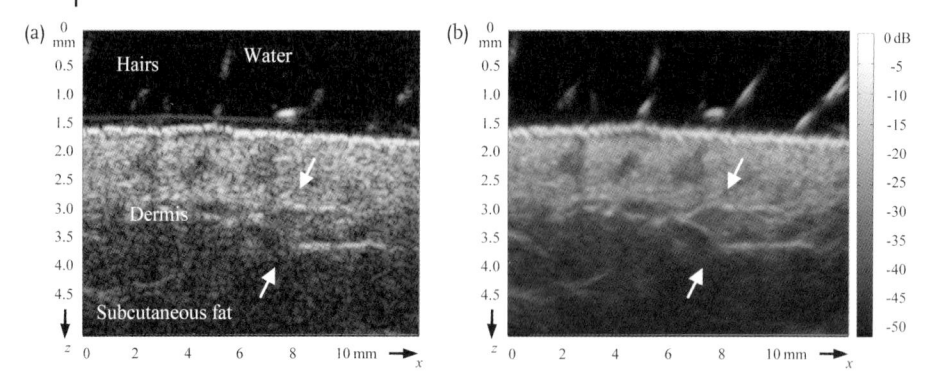

Figure 5.18 *In vivo* measurements of skin at forearm: (a) conventional B-mode image; (b) spatial compound image.

In the conventional B-mode image, the typical speckle pattern caused by backscattering from randomly distributed scatterers in the background of the phantom is visible. Behind the plastic phantom, a strong acoustic shadowing is visible, and only the upper and lower parts of the specularly reflecting circumference of the plastic straw are visible. In the spatial compound image, on the other hand, speckle effects are largely suppressed and the contours of the straw are more completely imaged as opposed to the conventional unidirectional B-mode image [68].

As another example, Figure 5.18a and b shows *in vivo* HFUS images, acquired with the same configuration of the system, of human skin at the forearm of a proband.

The skin surface and the layered skin consisting of the hyperechoic dermis and the hypoechoic subcutaneous fat are visible in both images, and hair follicles appear as hypoechoic structures inside the dermis. Again, speckle is largely suppressed in the spatial compound image compared to the conventional B-mode image, and the lamellar structures at the boundary between the dermis and the subcutaneous fat (see arrows in Figure 5.18a and b) are much better and more completely imaged.

The results of both phantom and *in vivo* measurements show that the visibility of surfaces in the spatial compound images is largely independent of the spatial orientation, whereas surfaces perpendicularly to the axial direction of sound propagation are predominant in the conventional B-mode ultrasound images. Echogeneity, morphology, and thickness of the skin can be much better assessed with the implemented HFUS spatial compound system.

5.3.4
Multidirectional Tissue Characterization

Above, multidirectional echo measurements have been used for limited angle spatial compounding. The same multidirectional measurement approach can also

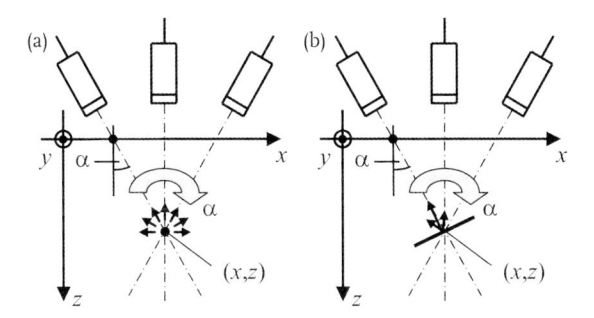

5.3 Engineering Concepts for High-Frequency Ultrasonic Imaging

Figure 5.19 Multidirectional echo measurements: (a) diffuse backscattering; (b) specular reflection.

be applied to multidirectional *tissue characterization*. Ultrasound B-mode imaging is a rather *qualitative* approach with the aim of assessing the *morphology* of biological tissue. In the field of ultrasonic tissue characterization, manifold work on *quantitative* analysis of echo signals with the goal of assessing *histological* tissue information has been carried out by many research groups. We have proposed a *multidirectional* approach for HFUS tissue characterization with the aim of differentiating between specular reflections and diffuse backscattering from biological tissue [71].

Figure 5.19a and b shows the configurations of multidirectional echo measurements for a point-like, diffuse scattering object, and for a planar, specularly reflecting acoustic inhomogeneity.

In the case of *diffuse backscattering* from point-like acoustic inhomogeneities, as in Figure 5.19a, the echo signal's amplitude is to a large extent invariant with respect to the insonation angle α. For *specular reflections* at spatially extended planar boundaries, on the other hand, the echo signal amplitude shows a maximum at 90° insonation angle relative to the boundary, and decreases with increasing angular deviation (Figure 5.19b).

For a quantitative analysis the echo signal amplitude at each point inside the (x,z) image plane is analyzed as a function of the insonation angle α (Figure 5.20a and b).

First-order statistical parameters (MN: mean, SD: standard deviation, MX: maximum value, CT: centroid) are directly calculated from the amplitude of the multidirectional echo signals over the insonation angle. For a second group of parameters, a parametric model function is fitted to the measured amplitudes, and quantitative parameters (PK: peak value, OF: offset, PP: peak position) are derived from the fitted analytical model function (Figure 5.20a and b).

Quotients SD/MN and MX/MN are parameters that do not depend on linear amplitude scalings and, consequently, they are invariant with respect to changes of the *absolute* echo signal amplitude. Both parameters are *small* in the case of diffuse backscattering and *large* in the case of specular reflections. The parameter CT is a measure of the surface orientation of specular reflectors. In addition, the

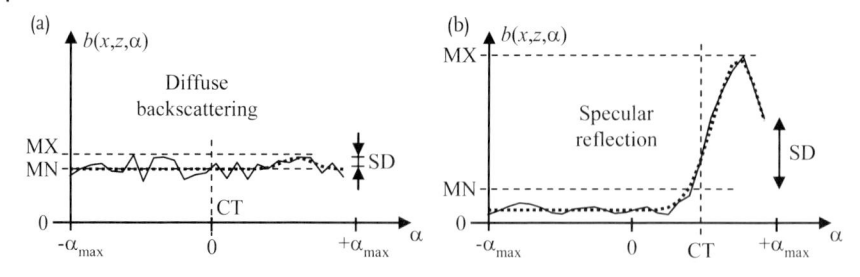

Figure 5.20 Echo signal amplitude as a function of the insonation angle: (a) diffuse backscattering; (b) specular reflection.

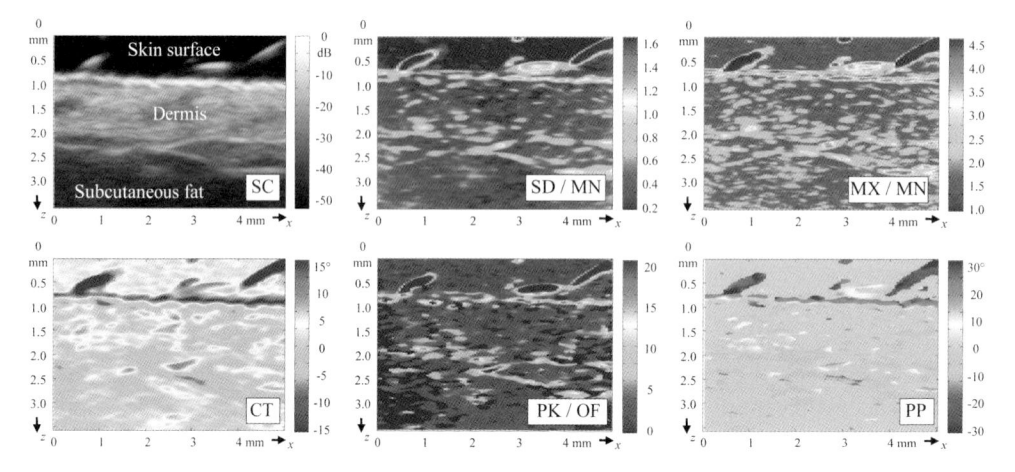

Figure 5.21 *In vivo* skin imaging, normal healthy skin: spatial compound image (SC) and parametric images (SD: standard deviation, MN: mean, MX: maximum value, CT: centroid, PK: peak value, OF: offset, PP: peak position).

amplitude-invariant parameter PK/OF is likewise small in the case of diffuse backscattering and large for specular reflections. Similar to CT, the parameter PP delivers the surface orientation of specularly reflecting structures [71]. The proposed parametric imaging concept has been evaluated with phantom and *in vivo* measurements.

Figure 5.21 shows a spatial compound image and parametric images of normal healthy skin, which have been obtained from echo data acquired with the above-discussed 20 MHz HFUS multidirectional imaging system.

In the spatial compound image, a strong echo from the skin surface appears, and the layered skin as well as the subcutaneous fat can be identified. The dermis shows strong backscattering caused by the network of collagen fibers, and also hypoechoic hair follicles can be identified inside the dermis. Below, lamellar structures at the boundary between the dermis and the subcutaneous fat are visible. In the parametric images SD/MN and MX/MN, specular reflections appear at the skin surface and from the lamellar structures, and mainly diffuse backscattering from the dermis. The two parameters are strongly correlated with each other, and the surface orientation of the lamellar structures is represented by the parameter CT as expected. The parameter images PK/OF and PP show very similar results.

In conclusion, the example shows that the proposed concept for multidirectional tissue characterization enables us to distinguish between specular reflections and diffuse backscattering from tissue.

5.4
High-Frequency Ultrasound Imaging in Biomedical Applications

In this chapter, skin imaging and small-animal imaging, which are two important examples of HFUS-based sonography of biological tissue, are discussed in more detail. Technical developments of imaging systems are discussed, and some of our contributions to both fields of application are presented.

5.4.1
Skin Imaging

In dermatology, *visual* inspection of the skin is the most common method to investigate suspicious skin lesions and to evaluate the success of therapies. However, the information obtained with *optical* techniques is limited to the skin surface and the uppermost skin layers. On the other hand, non-invasive imaging of skin lesions over a large depth range is of interest to assess the thickness of tumors and inflammatory processes, and also for preoperative planning to determine the safety distance before the excision of tumors.

HFUS in the 20 MHz range has found its way into skin imaging applications relatively early, because of the potential to visualize skin morphology non-invasively over depth with a spatial resolution on a microscopic scale. Many clinical studies on normal and pathological skin have been performed with 20 MHz range ultrasound [12, 13]. Skin echogeneity, burn scars, wound healing, skin aging, and skin tumors have been extensively studied.

Skin imaging with HFUS began in the late 1970s using A-scan techniques. Later, significant improvements were achieved by the design and implementation of B-mode imaging systems for 2D cross-sectional imaging using single-element transducers with mechanical scanners. With the progress of B-mode ultrasound systems, the experience with HFUS imaging in dermatology has grown continuously, and 20 MHz range ultrasound is nowadays widely and routinely used for diagnostics [12, 13].

With the goal of visualizing the uppermost skin structures with a better resolution, HFUS imaging systems at higher frequencies have been designed and implemented by several research groups. Spatial resolutions in the range 20–50 µm have been obtained by using HFUS in the 50 MHz range, because of the higher center

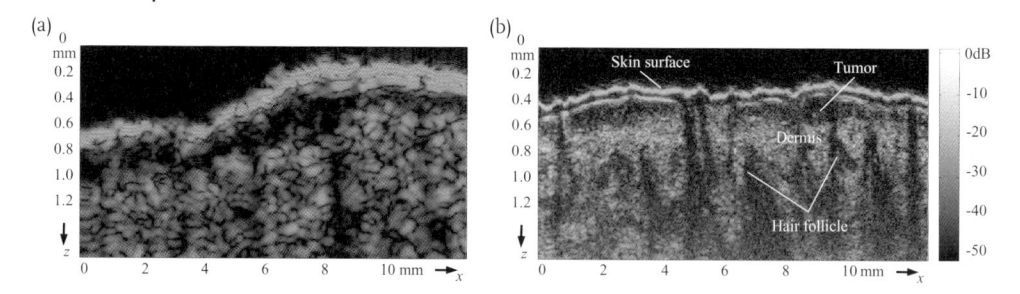

Figure 5.22 In vivo skin imaging: (a) 20 MHz range ultrasound; (b) 100 MHz range ultrasound.

frequency and bandwidth. Later, further system developments enabled utilization of HFUS in the 100 MHz frequency range for very high-resolution skin imaging, with a resolution down to 10 μm. In clinical diagnostics, HFUS delivers valuable information about the size and the depth of skin tumors and inflammatory processes. In the case of the malignant melanoma, HFUS supports the preoperative planning of skin tumor excision. Clinical studies have also shown that high-resolution sonography is an advantageous diagnostic tool for follow-up studies of inflammatory diseases [12].

Our latest technical developments in the field of HFUS sonography of skin have been aimed at improving the quality of B-mode images and at the implementation of new imaging modalities. Based on experience with an earlier prototype system, we have developed a clinical HFUS scanner utilizing both the 20 MHz and 100 MHz frequency range, for high resolution *in vivo* skin imaging. The B-mode images of skin at the forearm in Figure 5.22a and b have been acquired with HFUS in the 20 MHz range and in the 100 MHz range, respectively.

In both images, strong echoes appear at the skin surface, because of the large difference between the acoustic impedances of the skin and of water. The water is used as the sound propagation medium between the transducer and the skin. Below, the hyperechoic dermis can be seen, with hair follicles appearing as hypoechoic structures.

The spatial resolution of the implemented HFUS imaging system has been assessed by means of phantom measurements. Axial and lateral resolutions $\delta_{axial} = 76\,\mu m$ and $\delta_{lateral} = 170\,\mu m$, respectively, have been measured using the 20 MHz range transducer. With the 100 MHz range transducer, on the other hand, much better axial and lateral resolutions $\delta_{axial} = 10\,\mu m$ and $\delta_{lateral} = 55\,\mu m$, respectively, are obtained. The improved spatial resolution with the increased frequency range can also clearly be seen in the two B-mode images. Speckle patterns in the 100 MHz range image are much smaller than those in the 20 MHz range image, and the skin tumor at the upper dermis can be much better delineated. The hypoechoic band directly below the skin surface, which is only visible due to the better resolution of the 100 MHz range system, has been identified as the stratum corneum [19].

As another imaging modality, we have developed a HFUS system for *in vivo strain imaging* of skin [72]. In this approach, suction is applied to the skin surface, and frames of echo signals from inside a 2D axial/lateral image plane are acquired and recorded at stepwise decreased pressure levels by using a 20 MHz range single-element transducer. *Elastic* properties of skin are assessed by estimating the resulting displacements in the recorded frames of echo signals. Results obtained from *in vivo* measurements show that the strain inside the dermis is small and large inside the subcutaneous fat in the case of healthy skin. Based on these findings, it was concluded that the resistance to the suction applied at the skin surface was essentially due to the dermis rather than the subcutaneous fat. In further measurements on burned skin, though, relatively large strains were found as compared to the healthy skin, which resulted from the disarrangement of elastic and collagen fibers [72].

5.4.2
Imaging of Small Animals

Small animals imaging is another field of application of HFUS, for example, for non-invasive imaging in cancer research on mouse models. Requirements for small-animal imaging techniques include a high frame rate and high-resolution imaging capabilities. HFUS shows a large potential in this respect, and research in this field has continuously achieved improvements in the design of appropriate systems, although it is challenging to fulfill all requirements.

A 50 MHz range HFUS system for non-invasive *in utero* imaging of mouse embryos and for monitoring of the progression of melanomas in mice with axial and lateral resolutions of 30 and 60 µm, respectively, with a maximum frame rate of 10 frames per second have been presented in References [32, 33]. Further studies were aimed at the cardiac development of mice using HFUS in the 40 MHz range [33]. The very high heart rate of 120–200 beats per minute of mice causes severe problems with regard to the required frame rate of HFUS imaging systems, as artifacts arise if the heart significantly moves within one frame. The problem of manufacturing suitable HFUS transducer arrays has not yet been completely solved, and the frame rate with single-element transducers is of course limited by the speed of the mechanical scanner. Latest developments allow imaging with a maximum frame rate of 240 frames per second over a limited FOV by using a mechanical sector scanner (Vevo 770, VisualSonics Inc., Toronto, Ontario, Canada). However, the frame rate is still a limiting factor of systems with single-element transducers and mechanical scanners. *Retrospective imaging* is another approach by which to cope with this problem by *stopping* the transducer at consecutive discrete lateral positions and acquiring echo signals with a high pulse repetition frequency (PRF) of typically 10 kHz. At each transducer position, echo signals synchronous to the animal's heart beat are acquired by triggering with an electrocardiogram (ECG). Consequently, a *periodic* ECG-signal is required for reliable measurements, and arrhythmia limits the applicability of this method [38].

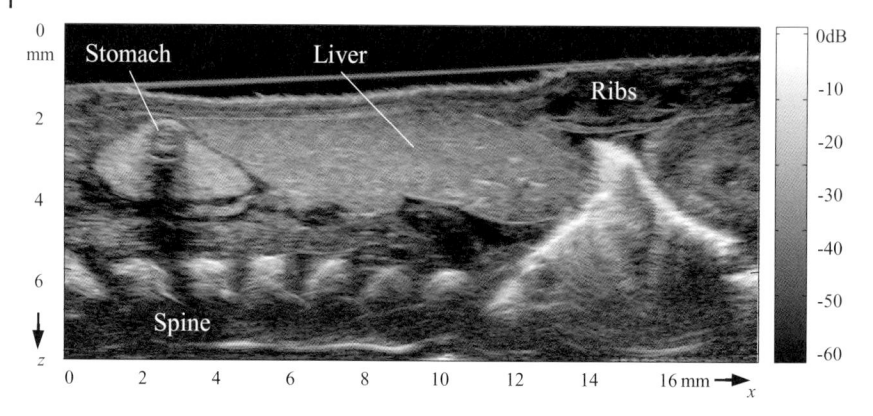

Figure 5.23 Imaging of small animals: abdominal region of rat pup cadaver – 20 MHz range ultrasound utilizing PSF-SAFT and LASC.

For applications that do not require real-time imaging, we have utilized both the above-discussed *multidirectional* limited angle spatial compound imaging approach together with the PSF-SAFT reconstruction technique for small-animal imaging with 20 MHz range HFUS [65]. Figure 5.23 shows a reconstructed post-mortem B-mode image of the abdominal region of a rat pup with a body length of about 50 mm.

The B-mode image was obtained by performing lateral scans with a HFUS single-element transducer with different tilting angles in the (x,z) image plane (25 image frames, insonation angles $\alpha = -30°, -27.5°, \ldots 30°$), applying PSF-SAFT separately to each frame of recorded echo signals for each tilting angle, and by incoherent superposition of the multidirectional echo signal amplitude data. Stomach, liver, and spine of the animal can clearly be seen in the B-mode image in Figure 5.23, and also the contours of the ribs can be identified. As expected, speckle, electronic noise, and shadowing effects are largely suppressed, organs and other structures are very completely imaged, and the spatial resolution varies only slightly with the axial depth.

5.5
Summary

This chapter has presented a detailed overview of engineering concepts, technical solutions, and new modalities for high-resolution imaging of biological tissue and technical objects with HFUS in the 20 MHz and 100 MHz range.

The concept and the architecture of ultrasound echo systems and HFUS transducers have been discussed in detail. The center frequency and the bandwidth are the parameters that determine the spatial resolution of ultrasound imaging systems. Because of the frequency-dependent attenuation of ultrasound waves in tissue, a compromise between the opposing demands for a large penetration depth and good spatial resolutions has to be found. It has also been discussed that the driving electronics have to be designed appropriately in the case of HFUS systems to cover the passband of HFUS transducers and to avoid electrical mismatching. Time-variant inverse echo signal filtering has been shown to allow for at least some compensation of center frequency and bandwidth variations over depth, and for an optimization of the resolution in the axial direction of ultrasound propagation. Furthermore, a concept for combining both echo and transmission measurements, and for quantitative measurements of acoustic parameters of layered structures by means of HFUS, has been presented.

In the context of engineering concepts for cross-sectional imaging with HFUS, B-scan techniques with single-element transducers have been presented. Furthermore, the B/D-scan technique and synthetic aperture focusing have been discussed as approaches that allow an optimization of the lateral resolution. It has been shown that synthetic aperture focusing can be applied in the far field of single-element transducers based on the assumption of spherical waves emanating from the transducer's focus. The presented PSF-SAFT has been proposed as a new reconstruction technique for HFUS imaging with strongly focused transducers taking simulated responses of the system into account. HFUS limited angle spatial compounding and multidirectional tissue characterization has been implemented by acquiring echo signal data from different insonation angles with a single-element transducer in a mechanical scanner. As has been shown, the former leads to strongly improved B-mode imaging in terms of suppression of speckle and shadowing artifacts, an improved and more complete visualization of objects details like surfaces and internal structures, and an improved image contrast. The proposed latter technique enables quantitative distinction between diffuse backscattering and specular reflections.

Finally, the application of HFUS for high-resolution skin imaging and small-animal imaging has been discussed, and examples of B-mode images have been presented in both fields of application.

References

1 Alexander, H.D. and Miller, L. (1979) Determining skin thickness with pulsed ultrasound. *J. Invest. Dermatol.*, **72**, 17–19.
2 Rukinava, B. and Mohar, N. (1979) An approach of ultrasound diagnostic techniques of the skin and subcutaneous tissue. *Dermatologica*, **158**, 81–92.
3 Payne, P.A. (1985) Application of ultrasound in dermatology. *Bioeng. Skin*, **1**, 293–320.
4 Querleux, B., Leveque, J.L., and de-Rigal, J. (1988) In vivo cross-sectional ultrasonic imaging of human skin. *Dermatologica*, **177**, 332–337.
5 Breitbart, E.W., Müller, C.E., Hicks, R., and Vieluf, D. (1989) New developments in ultrasound diagnostic in dermatology. *Akt. Dermatol.*, **15**, 57–61.
6 Serup, J. (1984) Localized scleroderma (morphoea): thickness of sclerotic plaques as measured by 15 MHz pulsed ultrasound. *Acta Derm. Venerol.*, **64** (3), 214–219.
7 Serup, J. (1984) Non-invasive quantification of psoriasis

plaques-measurement of skin thickness with 15 MHz pulsed ultrasound. *Clin. Exp. Dermatol.*, **9** (5), 502–508.

8 Dines, K.A., Sheets, P.W., Brink, J.A., Hanke, C.W., Condra, K.A., Clendenon, J.L., Goss, S.A., Smith, D.J., and Franklin, T.D. (1984) High frequency ultrasonic imaging of skin: experimental results. *Ultrason. Imaging*, **6** (4), 408–434.

9 Yano, T., Fukukita, H., Ueno, S., and Fukumoto, A. (1987) 40 MHz ultrasound diagnostic system for dermatologic examination. *Proc. IEEE Ultrason. Symp.*, 875–878.

10 Serup, J. and Staberg, B. (1987) Ultrasound for assessment of allergic and irritant patch test reactions. *Contact Dermatitis*, **17** (5), 80–84.

11 Murakami, S. and Miki, Y. (1989) Human skin histology using high-resolution echography. *J. Clin. Ultrasound*, **17** (2), 77–82.

12 Altmeyer, P., el Gammal, S., and Hoffmann, K. (1992) *Ultrasound in Dermatology*, Springer, Berlin.

13 Fornage, B.D., McGavran, M.H., Duciv, M., and Waldron, C.A. (1993) Imaging of the skin with 20-MHz US. *Radiology*, **189**, 69–76.

14 Hoess, A., Ermert, H., el Gammal, S., and Altmeyer, P. (1989) A 50 MHz ultrasonic imaging system for dermatologic application. *Proc. IEEE Ultrason. Symp.*, 849–852.

15 Höß, A., Ermert, H., el Gammal, S., and Altmeyer, P. (1992) Signal processing in high-frequency broadband imaging systems for dermatologic application. *Acoust. Imag.*, **19**, 243–249.

16 Turnbull, D.H., Starkoski, B.G., Harasiewicz, K.A., Semple, J.L., From, L., Gupta, A.K., Sauder, D.N., and Foster, F.S. (1995) A 40–100 MHz ultrasound backscatter microscope for skin imaging. *Ultrasound Med. Biol.*, **21** (1), 79–88.

17 Ermert, H., Vogt, M., Paßmann, C., el Gammal, S., Kaspar, K., Hoffmann, K., and Altmeyer, P. (1997) High frequency ultrasound (50–150 MHz) in dermatology, in *Skin Cancer and UV Radiation* (eds P. Altmeyer, K. Hoffmann, and M. Stücker), Springer, Berlin, pp. 1023–1051.

18 el Gammal, S., Pieck, C., Auer, T., Kaspar, K., Hoffmann, K., Altmeyer, P., Vogt, M., and Ermert, H. (1998) 100 MHz sonography of psoriasis vulgaris plaques. *Ultraschall Medizin*, **19** (6), 270–274.

19 Gammal, S.E., El Gammal, C., Kaspar, K., Pieck, C., Altmeyer, P., Vogt, M., and Ermert, H. (1999) Sonography of the skin at 100 MHz enables in vivo visualization of stratum corneum and viable epidermis in palmar skin and psoriatic plaques. *J. Invest. Dermatol.*, **5**, 821–829.

20 Vogt, M., Kaspar, K., Altmeyer, P., Hoffmann, K., and El Gammal, S. (2001) High frequency ultrasound for high resolution skin imaging. *Frequenz*, **55**, 12–20.

21 Vogt, M. and Ermert, H. (2006) High-resolution ultrasound, in *Bio-Engineering of the Skin: Skin Imaging and Analysis*, 2nd edn (eds K.P. Wilhelm, P. Elsner, E. Berardesca, and H.I. Maibach), CRC Press, Boca Raton, FL, pp. 83–97.

22 Vogt, M., Scharenberg, R., Moussa, G., Sand, M., Hoffmann, K., Altmeyer, P., and Ermert, H. (2007) A new high frequency ultrasound skin imaging system: imaging properties and clinical in vivo results. *Acoust. Imag.*, **28**, 137–144.

23 Vogt, M. and Ermert, H. (2008) In vivo ultrasound biomicroscopy of skin: spectral characteristics analysis and inverse filtering optimization. *IEEE Trans. Ultrason. Ferroelectr. Freq. Control*, **54** (8), 1551–1559.

24 Paßmann, C. and Ermert, H. (1996) A 100 MHz ultrasound imaging system for dermatologic and ophthalmologic diagnostics. *IEEE Trans. Ultrason. Ferroelectr. Freq. Control*, **43** (4), 545–552.

25 Goldberg, R.E. and Sarin, K. (1967) *Ultrasonics in Ophthalmology. Diagnostic and Therapeutic Applications*, W.B. Saunders, Philadelphia and London.

26 Pavlin, C.J. and Foster, F.S. (1995) *Ultrasound Biomicroscopy of the Eye*, Springer, New York.

27 Fledelius, H.C. (1997) Ultrasound in ophthalmology. *Ultrasound Med. Biol.*, **23** (3), 365–375.

28 Foster, F.S., Pavlin, C.J., Starkoski, B., and Harasiewicz, K. (1990) Ultrasound

backscatter microscopy of the eye in vivo. *Proc. IEEE Ultrason. Symp.*, 1481–1484.

29 Silverman, R.H., Reinstein, D.Z., Raevsky, T., and Coleman, D.J. (1997) Improved system for sonographic imaging and biometry of the cornea. *J. Ultrasound Med.*, **16** (2), 117–124.

30 Lizzi, F.L., Feleppa, E.J., and Kalisz, A. (2000) High-resolution, three-dimensional visualization and morphology assays of the in-vivo ciliary body. *Proc. IEEE Ultrason. Symp.*, 1421–1423.

31 Kim, H.H., Chang, J.H., Cannata, J.M., and Shung, K.K. (2007) Design of 20 MHz convex array transducers for high frequency ophthalmic imaging. *Proc. IEEE Ultrason. Symp.*, 88–91.

32 Turnbull, D.H., Bloomfield, T.S., Baldwin, H.S., Foster, F.S., and Joyner, A.L. (1995) Ultrasound backscatter microscope analysis of early mouse embryonic brain development. *Proc. Natl. Acad. Sci. USA*, **92**, 2239–2243.

33 Turnbull, D.H., Ramsay, J.A., Shivji, G.S., Bloomfield, T.S., From, L., Sauder, D.N., and Foster, F.S. (1996) Ultrasound backscatter microscope analysis of mouse melanoma progression. *Ultrasound Med. Biol.*, **22** (7), 845–853.

34 Srinivasan, S., Baldwin, H.S., Aristizabal, O., Kwee, L., Labow, M., Artman, M., and Turnbull, D.H. (1998) Noninvasive, in utero imaging of mouse embryonic heart development with 40-MHz echocardiography. *Circulation*, **98**, 912–918.

35 Foster, F.S., Liu, G., Mehi, J., Starkoski, B.D., Adamson, L., Zhou, Y., Harasiecwicz, A., and Zan, L. (2000) High frequency ultrasound imaging: from man to mouse. *Proc. IEEE Ultrason. Symp.*, 1633–1638.

36 Foster, F.S., Zhang, M.Y., Zhou, Q., Liu, G., Mehi, J., Cherin, E., Harasiewicz, K.A., Starkoski, B.G., Zan, L., Knapik, D.A., and Adamson, S.L. (2002) A new ultrasound instrument for in vivo microimaging of mice. *Ultrasound Med. Biol.*, **28** (9), 1165–1172.

37 Sun, L., Richard, D., Cannata, J.M., Feng, C.C., Johnson, J.A., Yen, J.T., and Shung, K.K. (2007) A high-frame rate high-frequency ultrasonic system for cardiac imaging in mice. *IEEE Trans. Ultrason. Ferroelectr. Freq. Control*, **54** (8), 1648–1655.

38 Cherin, E., Williams, R., Needles, A., Liu, G., White, C., Brown, A.S., Zhou, Y.Q., and Foster, F.S. (2006) Ultrahigh frame rate retrospective ultrasound microimaging and blood flow visualization in mice in vivo. *Ultrasound Med. Biol.*, **32** (5), 683–691.

39 Liu, J.H., Jeng, G.S., Wu, T.K., and Li, P.C. (2006) ECG triggering and gating for ultrasonic small animal imaging. *IEEE Trans. Ultrason. Ferroelectr. Freq. Control*, **53** (9), 1590–1596.

40 Vogt, M. and Ermert, H. (2005) High frequency ultrasonic imaging: system design and performance optimization. *Frequenz*, **59**, 150–153.

41 Xu, X.C., Hu, C.H., Sun, L., Yen, J., and Shung, K.K. (2005) High-frequency high frame rate ultrasound imaging system for small animal imaging with linear arrays. *Proc. IEEE Ultrason. Symp.*, 1431–1434.

42 Cannata, J.M., Williams, J.A., Zhou, Q., Ritter, T.A., and Shung, K.K. (2006) Development of a 35-MHz piezo-composite ultrasound array for medical imaging. *IEEE Trans. Ultrason. Ferroelectr. Freq. Control*, **53** (1), 224–236.

43 Lukacs, M., Yin, J., Pang, G., Garcia, R.C., Cherin, E., Ross, W., Mehi, J., and Foster, F.S. (2006) Performance and characterization of new micromachined high-frequency linear arrays. *IEEE Trans. Ultrason. Ferroelectr. Freq. Control*, **53** (10), 1719–1729.

44 Brown, J.A., Foster, F.S., Needles, A., Cherin, E., and Lockwood, G.R. (2007) Fabrication and performance of a 40-MHz linear array based on a 1–3 composite with geometric elevation focusing. *IEEE Trans. Ultrason. Ferroelectr. Freq. Control*, **54** (9), 1888–1894.

45 Cannata, J.M., Ritter, T.A., Chen, W.H., Silverman, R.H., and Shung, K.K. (2003) Design of efficient, broadband single-element (20–80 MHz) ultrasonic transducers for medical imaging applications. *IEEE Trans. Ultrason.*

46 Foster, F.S., Pavlin, C.J., Lockwood, G.R., Ryan, L.K., Harasiewicz, K.A., Berube, L., and Rauth, A.M. (1993) Principles and applications of ultrasound backscatter microscopy. *IEEE Trans. Ultrason. Ferroelectr. Freq. Control*, **40** (5), 608–617.

47 Lockwood, G.R., Turnbull, D.H., Christopher, D.A., and Foster, F.S. (1996) Beyond 30 MHz. *IEEE Eng. Med. Biol.*, **15** (6), 60–71.

48 Vogt, M., Opretzka, J., Perrey, C., and Ermert, H. (2010) Ultrasonic microscanning. *Proc. Inst. Mech. Eng. H*, **224** (2), 225–240.

49 Knapik, D.A., Starkoski, B., Pavlin, C.J., and Foster, F.S. (1997) A realtime 200 MHz ultrasound B-scan imager. *Proc. IEEE Ultrason. Symp.*, 1457–1460.

50 Foster, F.S., Pavlin, C.J., Harasiewicz, K.A., Christopher, D.A., and Turnbull, D.H. (2000) Advances in ultrasound biomicroscopy. *Ultrasound Med. Biol.*, **26** (1), 1–27.

51 Knapik, D.A., Starkoski, B., Pavlin, C.J., and Foster, F.S. (2000) A 100–200 MHz ultrasound biomicroscope. *IEEE Trans. Ultrason. Ferroelectr. Freq. Control*, **47** (6), 1540–1549.

52 Chen, W.H., Gottlieb, E.J., Cannata, J.M., Chen, Y.F., and Shung, K.K. (2000) Development of sector scanning ultrasonic backscatter microscope. *Proc. IEEE Ultrason. Symp.*, 1681–1684.

53 Brown, J.A., Démoré, C.E.M., and Lockwood, G.R. (2004) Design and fabrication of annular arrays for high-frequency ultrasound. *IEEE Trans. Ultrason. Ferroelectr. Freq. Control*, **51** (8), 1010–1017.

54 Ketterling, J.A., Aristizábal, O., Turnbull, D.H., and Lizzi, F.L. (2005) Design and fabrication of a 40-MHz annular array transducer. *IEEE Trans. Ultrason. Ferroelectr. Freq. Control*, **52** (4), 672–681.

55 Ketterling, J.A., Ramachandran, S., and Aristizábal, O. (2005) Operational verification of a 40-MHz annular array transducer. *IEEE Trans. Ultrason. Ferroelectr. Freq. Control*, **53** (3), 623–630.

56 Ketterling, J.A. and Silverman, R.H. (2006) 20-MHz annular arrays for ophthalmic imaging. *Proc. IEEE Ultrason. Symp.*, 252–255.

57 Snook, K.A., Hu, C.H., Shrout, T.T., and Shung, K.K. (2006) High-frequency ultrasound annular-array imaging. Part I: array design and fabrication. *IEEE Trans. Ultrason. Ferroelectr. Freq. Control*, **53** (2), 300–308.

58 Hu, C.H., Snook, K.A., Cao, P.J., and Shung, K.K. (2006) High-frequency ultrasound annular-array imaging. Part II: digital beamformer design and imaging. *IEEE Trans. Ultrason. Ferroelectr. Freq. Control*, **53** (2), 309–316.

59 Aristizábal, O., Turnbull, D.H., and Ketterling, J.A. (2006) Analysis of 40 MHz annular array imaging performance in mouse embryos. *Proc. IEEE Ultrason. Symp.*, 872–875.

60 Opretzka, J., Vogt, M., and Ermert, H. (2009) A model-based synthetic aperture image reconstruction technique for high-frequency ultrasound. *Proc. IEEE Ultrason. Symp.*, 377–380.

61 Vogt, M., Opretzka, J., and Ermert, H. (2009) Synthetic aperture focusing technique for high-resolution imaging of surface structures with high-frequency ultrasound. *Proc. IEEE Ultrason. Symp.*, 1514–1517.

62 Opretzka, J., Vogt, M., and Ermert, H. (2010) A synthetic aperture focusing technique with optimized beamforming for high-frequency ultrasound. *Proc. IEEE Ultrason. Symp.*, 2303–2306.

63 Vogt, M. (2010) Synthetic aperture focusing techniques in radar and ultrasound imaging. Proc. 8th European Conference on Synthetic Aperture Radar (EUSAR 2010). CD ROM, pp. 426–429.

64 Opretzka, J., Vogt, M., and Ermert, H. (2010) A model-based synthetic aperture focusing technique for high-frequency ultrasound imaging. Proc. 8th European Conference on Synthetic Aperture Radar (EUSAR 2010). CD ROM, pp. 430–433.

65 Opretzka, J., Vogt, M., and Ermert, H. (2011) A high-frequency ultrasound imaging system combining limited-angle spatial compounding and model-based synthetic aperture focusing. *IEEE Trans. Ultrason. Ferroelectr. Freq. Control*, **58** (7), 1355–1365.

66 Erickson, S., Kruse, D., and Ferrara, K. (2001) A hand-held, high frequency ultrasound scanner. *Proc. IEEE Ultrason. Symp.*, 1465–1468.

67 Vogt, M. and Ermert, H. (2009) Ultrasonic two-port measurements for the reconstruction of layered media properties. *Frequenz*, **63**, 147–151.

68 Vogt, M. and Ermert, H. (2008) Limited-angle spatial compound imaging of skin with high-frequency ultrasound (20 MHz). *IEEE Trans. Ultrason. Ferroelectr. Freq. Control*, **55** (9), 1975–1983.

69 Opretzka, J., Vogt, M., and Ermert, H. (2008) 3D small animal imaging with high-frequency ultrasound (20 MHz) using limited-angle spatial compounding. *Proc. IEEE Ultrason. Symp.*, 130–133.

70 Opretzka, J., Vogt, M., and Ermert, H. (2008) A correction scheme for refraction and time-of-flight artifacts in limited-angle spatial compound imaging with high-frequency ultrasound. *Proc. IEEE Ultrason. Symp.*, 1378–1381.

71 Vogt, M., Opretzka, J., and Ermert, H. (2008) Parametric imaging of specular reflections and diffuse scattering of tissue from multi-directional ultrasound echo signal data. *Proc. IEEE Ultrason. Symp.*, 1963–1966.

72 Vogt, M. and Ermert, H. (2005) Development and evaluation of a high-frequency ultrasound-based system for in vivo strain imaging of the skin. *IEEE Trans. Ultrason. Ferroelectr. Freq. Control*, **52** (3), 375–385.

6
Quantitative Acoustic Microscopy Based on the Array Approach
Sergey Titov and Roman Gr. Maev

6.1
General Introduction

Several quantitative acoustic microscopy methods have been developed for the measurement of parameters of surface and bulk ultrasonic waves in a local area of a flat specimen immersed in a liquid. One of the most popular techniques is based on recording the output voltage $V(z)$ of the focused transducer during the mechanical translation of the probe along a perpendicular to the surface of the specimen [1–9]. The phase velocity and propagation attenuation of leaky surface waves such as leaky Rayleigh, Lamb, skimming longitudinal waves as well as a reflectance function for the specimen–water interface can be obtained from the recorded $V(z)$ data. Many modifications of the $V(z)$ technique employing various types of focused transducers, processing algorithms of the analog and digital signals, and electronic equipment have been proposed.

Recently, several acoustic microscopy techniques employing two transducers have also been developed for quantitative material characterization. In an ultrasonic microspectrometer, spherical-planar-pair lenses were used to measure the reflection coefficient over a wide frequency range [10]. An angular spectrum of the reflected wave is detected in this system by tilting the spherical–planar–pair lenses as a unit. An ultrasonic system employing separated transmitting and receiving transducers has been developed to study the anisotropic propagation of leaky surface waves. In these methods, the focus of the transmitting transducer was located on the liquid–solid interface and the scattered acoustic field was recorded by rotation of the specimen [11, 12] or by two-dimensional scanning of the receiving transducer [13, 14]. In addition, several ultrasonic material characterization systems based on lateral scanning of the receiving transducer along the x axis have been proposed and developed [15–17]. In comparison with the $V(z)$ system, the $V(x)$ scheme potentially possesses better angular resolution and temperature stability.

All these quantitative methods are based on the mechanical scanning of the receiving transducer. In this case, the acquisition of a sufficient spatiotemporal

Advances in Acoustic Microscopy and High Resolution Imaging: From Principles to Applictaions, First Edition.
Edited by Roman Gr. Maev.
© 2013 Wiley-VCH Verlag GmbH & Co. KGaA. Published 2013 by Wiley-VCH Verlag GmbH & Co. KGaA.

data set is rather time-consuming and the accuracy of the measurements depends on the precision of the mechanical movement. To overcome these drawbacks, ultrasonic arrays can be used to record the spatiotemporal field distribution of the scattered wave. Recently, ultrasonic arrays have been widely used in medical diagnostic and nondestructive testing for imaging of the internal structure of specimens. On the other hand, only a few ultrasonic array applications for quantitative characterization of local areas of a specimen are described in the literature (see, for example, References [18, 19]. The velocities of the bulk waves in geological materials have been measured with the arrays directly attached to contiguous sides of prism samples [18]. The use of an array to measure the velocity of the lateral wave propagating along the soft tissue–cortical bone interface has been proposed [19]. This chapter considers the application of ultrasonic arrays for measurement of the velocity and attenuation of the leaky surface wave and for simultaneous measurement of the bulk wave velocity and the thickness of the plate specimen.

6.2
Measurement of Velocity and Attenuation of Leaky Waves

Figure 6.1 shows a scheme of the setup for measurement of the leaky wave parameters. A single line-focused transducer is used for the generation of leaky surface acoustic wave (LSAW). The axis of the transducer is tilted at the angle γ_0 and the line focus is located at the interface of the specimen. The incident cylindrical wave can generate a leaky wave, if the critical angle of the LSAW θ_R is within the angular range $\gamma_0 - \gamma_m < \theta_R < \gamma_0 + \gamma_m$, where γ_m is a half-aperture angle of the transducer. The leaky wave propagates along the interface reradiating back in the immersion liquid and the reflected wave is received by the linear ultrasonic array tilted at the angle θ_0.

The linear array consists of n equal rectangular elements separated by the distance p. The long sides of the elements are parallel to the line focus of the transmitting transducer and to the surface of the specimen. The width of the elements

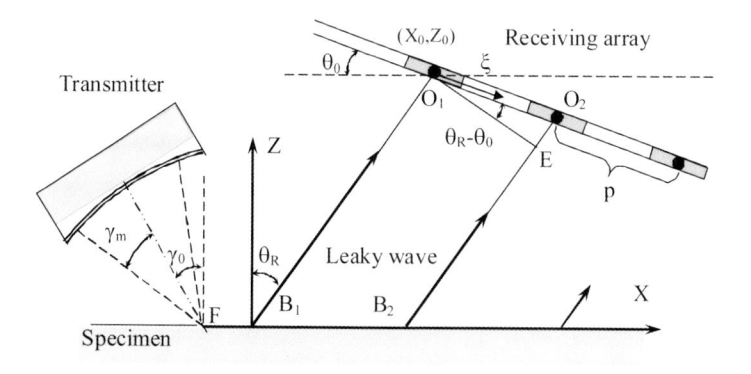

Figure 6.1 Scheme of the LSAW measurement.

6.2 Measurement of Velocity and Attenuation of Leaky Waves

is relatively small to have sufficiently wide angular sensitivity for the LSAW receiving.

In an immersion liquid the acoustic field of a harmonic leaky Rayleigh wave can be written as follows [20]:

$$g(x, z) = \exp(-\alpha_x x - \alpha_z z)\exp\{ik(x\sin\theta_R + z\cos\theta_R)\}\exp(-i\omega t) \quad (6.1)$$

where ω is the frequency, k is wave number, and α_x and α_z are attenuation coefficients of the wave along the x and z axes, respectively. The leaky wave is an inhomogeneous plane wave; therefore, the wave number k satisfies the relation:

$$k^2 = k_0^2 + \alpha_x^2 + \alpha_z^2 \quad (6.2)$$

where $k_0 = \omega/C$ is the wave number of the homogeneous plane wave, and C is the velocity of sound in the liquid. The amplitude of the LSAW decreases along the x-axis due to energy radiation into the liquid ($\alpha_x > 0$); however, the amplitude increases in the z direction:

$$\alpha_z = -\alpha_x \cdot tg\theta_R < 0 \quad (6.3)$$

The attenuation factor is small in comparison with the wave number, $\alpha \ll k_0$; thus $k \approx k_0$ and it is possible to neglect the deviation of the phase velocity of the LSAW from the velocity of homogeneous plane wave C.

Omitting the time factor $\exp(-i\omega t)$, the LSAW field can be rewritten:

$$g(x, z) = \exp[-\alpha_x(x - ztg\theta_R)]\exp\{ik_0(x\sin\theta_R + z\cos\theta_R)\} \quad (6.4)$$

This wave is recorded by the elements of the array. Suppose that the first element is located in the point (x_0, z_0) and ξ is the coordinate that originates from the center of the j-th element ($j = 0, \ldots n - 1$), and it is directed along the aperture of the array (Figure 6.1); in this case, the coordinates of points in the aperture of the element are:

$$x - x_0 + (jp + \xi)\cos\theta_0, \; z = z_0 - (jp + \xi)\sin\theta_0 \quad (6.5)$$

and the wave field distribution in this aperture can be presented as a function of ξ:

$$g_j(\xi) = u_0 \exp\left[-\alpha_x(jp + \xi)\frac{\cos(\theta_R - \theta_0)}{\cos\theta_R}\right]\exp\{i(jp + \xi)k_0\sin(\theta_R - \theta_0)\} \quad (6.6)$$

where:

$$u_0 = \exp[-\alpha_x(x_0 - z_0 tg\theta_R)]\exp\{ik_0(x_0\sin\theta_R + z_0\cos\theta_R)\} \quad (6.7)$$

is the value in the point (x_0, z_0).

Assume that all elements of the array are identical and their spatial response at a given frequency is $h(\xi)$. In this case the output signal of the j-th element can be expressed in the convolution form:

$$s_j = \int_{-\infty}^{\infty} g_j(\xi)h(\xi)d\xi \quad (6.8)$$

Using Equations (6.6) and (6.7), Equation (6.8) can be rewritten:

$$s_j = u_1 \exp\left[-\alpha_x jp \frac{\cos(\theta_R - \theta_0)}{\cos\theta_R}\right] \exp\{ijp \cdot k_0 \sin(\theta_R - \theta_0)\} \quad (6.9)$$

where the coefficient u_1 does not depend on the element number j:

$$u_1 = u_0 \int_{-\infty}^{\infty} h(\xi) \exp\left[-\alpha_x \xi \frac{\cos(\theta_R - \theta_0)}{\cos\theta_R}\right] \exp\{i\xi \cdot k_0 \sin(\theta_R - \theta_0)\} d\xi \quad (6.10)$$

Since $|\xi| < p/2$, and the pitch of the array is small, the attenuation exponential factor can be neglected. In this case the integral (6.10) is a Fourier transform of the spatial response of the element $h(\xi)$:

$$u_1 \approx u_0 H[k_0 \sin(\theta_R - \theta_0)] \quad (6.11)$$

where H is the directivity function of the element.

It follows from Equation (6.9) that phases of the received signals are proportional to j and depend on the difference between the critical angle θ_R and the angle of the array tilt θ_0:

$$\varphi_j = \varphi_0 + jp \cdot k_0 \sin(\theta_R - \theta_0) \quad (6.12)$$

where φ_0 is the phase of the coefficient u_1.

Let us consider nondispersive leaky waves. In this case the phase factor $pk_0 \sin(\theta_R - \theta_0)$ is proportional to the frequency ω. In addition, the attenuation coefficient is proportional to the absolute value of the frequency: $\alpha_x = \gamma|\omega|C^{-1}$, where the normalized attenuation coefficient γ depends on acoustic parameters of the media. The waveforms received by the elements of the array can be obtained using the inverse Fourier transform of Equation (6.9):

$$r_j(t) = F^{-1}[s_j] = r_0(t) * F^{-1}[\exp(-aj)] * \delta(t - jb) \quad (6.13)$$

Here $r_0(t) = F^{-1}(u_1)$ is the signal received by the first element ($j = 0$), the symbol $*$ denotes the convolution operation, $\delta(\ldots)$ is the Dirac delta function, b is the time delay between signals of two neighboring elements:

$$b = pk_0 \sin(\theta_R - \theta_0)/\omega = pC^{-1}\sin(\theta_R - \theta_0) \quad (6.14)$$

and a is the logarithm of the ratio of their amplitudes at the particular frequency:

$$a = \gamma|\omega|p\frac{\cos(\theta_R - \theta_0)}{C\cos\theta_R} \quad (6.15)$$

Taking into account that:

$$C_R = C(\sin\theta_R)^{-1} \quad (6.16)$$

the velocity of the LSAW C_R can be calculated if the time delay b is known:

$$C_R = C\left[\sin\left\{\arcsin\left(\frac{bC}{p}\right) + \theta_0\right\}\right]^{-1} \quad (6.17)$$

Since the time shift b between any two neighboring elements is constant, it can be found by solving the linear regression equation:

$$t_j = t_0 + bj \qquad (6.18)$$

where t_j is the time of flight of the corresponding ultrasonic pulse. If the attenuation of the leaky wave is small, the waveforms $r_j(t)$ have various delays and amplitudes, but their shapes are almost identical for all receiving channels. In this case the delays t_j can be correctly determined using, for instance, the zero crossing method [21]. In the case of large LSAW attenuation, the exponential factor $\exp(-aj)$ in Equation (6.9) may have a significant effect on received waveforms. However, as this factor is a real function of the frequency, the phase spectra of the waveforms $r_j(t)$ do not depend on the attenuation coefficient, and they contain the linear components ωt_j. Thus, the delays t_j can be found by comparing of the phase slopes of the calculated spectra [21].

As follows from Equation (6.9) and Equation (6.15), taking the logarithm of the amplitude spectra $\mu_j = \log|s_j|$ gives the linear regression equation:

$$\mu_j = \mu_0 - aj \qquad (6.19)$$

where $\mu_0 = \log|u_1|$ is a constant. Solving these equations for the appropriate frequency band it is possible to determine the frequency dependence of the attenuation coefficient $\alpha_x(\omega)$ and the normalized coefficient γ.

This method can be illustrated based on the ray model (Figure 6.1). Let the linear segment O_1E be parallel to the wave front of the leaky wave, then the rays B_1O_1 and B_2O_2 arrive simultaneously at points O_1 and E, respectively. Therefore, the time delay between signals of these two elements O_1 and O_2 is equal to $b = O_2E/C$. Since $O_2E = p\sin(\theta_R - \theta_0)$, the ray approach also leads to Equation (6.14). If the attenuation of ultrasound in the immersion liquid is negligible, the amplitude of the leaky wave is constant along the rays B_1O_1 and B_2O_2; therefore, the amplitude ratio of the signals at the points O_2 and O_1 equals $\exp(-\alpha_x B_1 B_2)$. Based on the geometry of the problem, the distance B_1B_2 can be obtained as follows:

$$B_1B_2 = \frac{O_1E}{\cos\theta_R} = \frac{p\cos(\theta_R - \theta_0)}{\cos\theta_R} \qquad (6.20)$$

Thus the value of the attenuation parameter $a = \alpha_x B_1 B_2$ derived from the ray model is in agreement with Equation (6.15).

Measurement of the velocity C_R and attenuation coefficient α_x of the leaky wave is indirect. To calculate the velocity C_R using Equation (6.17) it is necessary to obtain the parameter b and know the pitch of the array p, the tilt angle θ_0, and the velocity of sound in liquid C. The error v in the velocity C_R can be estimated based on the partial differentiation approach:

$$v = \sum_k v_k = \frac{1}{C_R} \sum_k \left|\frac{\partial C_R}{\partial \gamma_k}\right| \delta\gamma_k \qquad (6.21)$$

where $\delta\gamma_k$ are errors in the parameters $\gamma_k = (b, p, \theta_0, C)$. The attenuation coefficient γ depends on a, p, θ_0, and the measured value of C_R. The error in the attenuation coefficient α_x can be written in a similar way:

$$\eta = \sum_k \eta_k = \frac{1}{\alpha_x} \sum_k \left|\frac{\partial \alpha_x}{\partial \gamma_k}\right| \cdot \delta\gamma_k \tag{6.22}$$

here $\delta\gamma_k$ are errors in the parameters $\gamma_k = (a, p, \theta_0, C_R)$.

To obtain the parameters b and a the linear regression Equations (6.18) and (6.19) should be solved using, for instance, the least-squares method [22]. Let us estimate the error δb. Taking into account measurement error, Equation (6.18) can be rewritten:

$$t_j = t_0 + bj + \tau_j + \varepsilon_j \tag{6.23}$$

where τ_j and ε_j are random independent variables of two different kinds. Suppose τ_j is generated by the hardware jitter, variations of the phase impulse responses of the receiving elements, fluctuations of the elastic properties of the immersion liquid, and mechanical vibrations of the setup. In which case, the variance of the noise of this kind $\sigma_\tau^2 = \langle \tau_j^2 \rangle$ does not depend on the signals $r_j(t)$. Another time delay uncertainty ε_j is generated by the additive electromagnetic noise and fluctuations of the ultrasonic absorption in the immersion liquid and other origins. Let σ_0^2 be the total variance produced by all these sources. The time error ε_j should be inversely proportional to the amplitude of the signal and the frequency, and the corresponding time variance σ_j^2 can be estimated as follows:

$$\sigma_j^2 = \langle \varepsilon_j^2 \rangle = \left(\frac{\sigma_0}{|s_j|\omega_0}\right)^2 = \left(\frac{\sigma_0}{|u_1|\exp[-a(\omega_0)j]\omega_0}\right)^2 \tag{6.24}$$

The time delay error of this kind depends on the signal-to-noise ratio and increases with increasing element number j.

The coefficient b can be calculated using the method of least squares:

$$b = \frac{n\sum jt_j - \sum j \sum t_j}{n\sum j^2 - \left(\sum j\right)^2}, \quad j = 0, \ldots, n-1 \tag{6.25}$$

In simple linear regression the estimated variance of the slope is [22]:

$$(\delta b)^2 = \frac{n\sum (\tau_j + \varepsilon_j)^2}{(n-2)\left[n\sum j^2 - \left(\sum j\right)^2\right]} \tag{6.26}$$

Since τ_j and ε_j are independent the error δb consists of two components:

$$(\delta b)^2 = (\delta b)_\tau^2 + (\delta b)_\varepsilon^2 \tag{6.27}$$

6.2 Measurement of Velocity and Attenuation of Leaky Waves

where:

$$(\delta b)_\tau = \sqrt{n} \sigma_\tau \Psi_n(0), \quad (\delta b)_\varepsilon = \frac{\sigma_0}{|u_1|\omega_0} \Psi_n(a) \tag{6.28}$$

and:

$$\Psi_n^2(a) = \frac{n \sum \exp\{2ja(\omega_0)\}}{(n-2)\left[n \sum j^2 - \left(\sum j\right)^2\right]} \tag{6.29}$$

The fractional error v_b in the velocity C_R also consists of two terms, $v_b = v_{b\tau} + v_{b\varepsilon}$. Taking a partial derivative these errors can be found according to Equation (6.21):

$$v_{b\tau} = \frac{C_R \cdot \cos\theta_R}{\cos(\theta_R - \theta_0)p} \sigma_\tau \sqrt{n} \Psi_n(0) \tag{6.30}$$

$$v_{b\varepsilon} = \frac{C_R \cos\theta_R}{\cos(\theta_R - \theta_0)p} \cdot \frac{\sigma_0 \Psi_n(a)}{|u_0| H[k_0 \sin(\theta_R - \theta_0)]} \tag{6.31}$$

To estimate the error of the parameter a, the regression Equation (6.19) should be rewritten taking into account the additive noise σ_0:

$$\mu_j = \log(|u_1|\exp\{-aj\} + \sigma_0) = \mu_0 - aj + \log\left(1 + \frac{\sigma_0}{|u_1|\exp\{-aj\}}\right) \tag{6.32}$$

Assuming that the noise is small $\sigma_0 \ll |s_j|$, approximation of the logarithm gives the regression equation in the form:

$$\mu_j = \mu_0 - aj + \frac{\sigma_0}{|u_1|\exp\{-aj\}} \tag{6.33}$$

and the error δa can be derived using the function $\Psi_n(a)$, Equation (6.28):

$$\delta a = \frac{\sigma_0}{|u_1|} \Psi_n(a) \tag{6.34}$$

Thus, this error depends on the attenuation coefficient itself and the signal-to-noise ratio. The corresponding fractional error η_a in the attenuation coefficient can be calculated using Equations (6.11), (6.15), and (6.22):

$$\eta_a = \frac{\delta a}{a} = \frac{\sigma_0 \Psi_n(a)}{a|u_0| H[k_0 \sin(\theta_R - \theta_0)]} \tag{6.35}$$

In the course of analyzing the measurement accuracy, the fractional errors $v_{b\varepsilon}$ and η_a have been calculated as a function of the aperture size $L = pn$ for the various number of elements n. The errors $v_{b\varepsilon}$ calculated for water–steel and water–fused quartz combinations are shown in Figure 6.2a and b, respectively. The typical experimental values of frequency $\omega_0/2\pi = 5\,\text{MHz}$, signal-to-noise ratio $|u_1|/\sigma_0 = 10$, tilt angle of the array $\theta_0 = 30°$ have been assumed in the calculations.

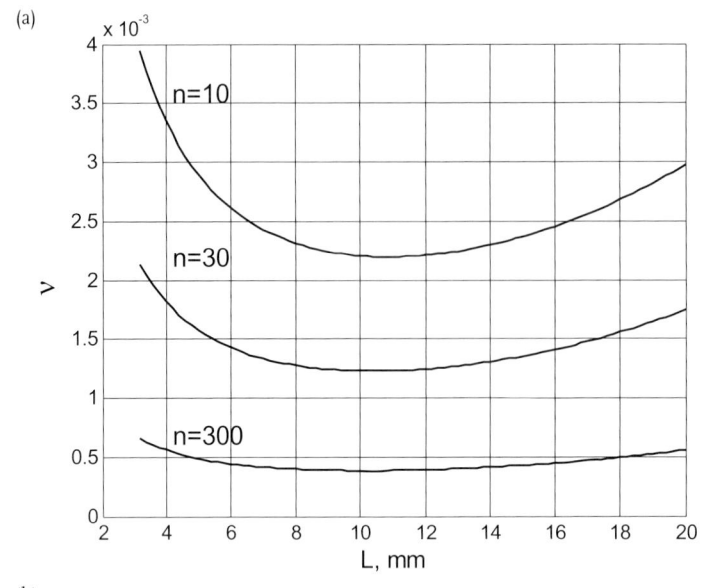

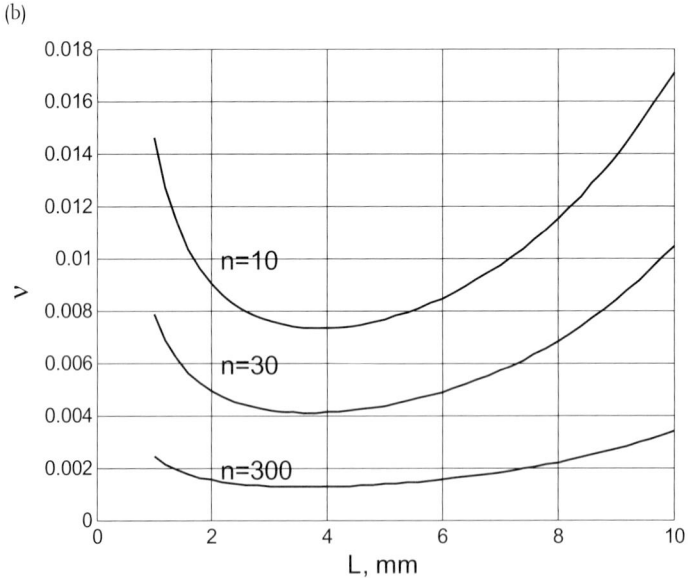

Figure 6.2 Fractional errors $v_{b\varepsilon}$ calculated for water–steel (a) and water–fused quartz (b).

As it should be expected, the error $v_{b\varepsilon}$ monotonically decreases with increasing number of elements n. However, if n is fixed there is an optimal aperture size L where the error reaches the minimal value. The growth of the error can be explained by the fact that due to the attenuation of the leaky wave the signal-to-noise ratio decreases at the far end of the aperture. Thus, the optimal size of the

Table 6.1 Minimal values of errors $v_{b\varepsilon}$ and η_a.

Material	$v_{b\varepsilon}$		η_a		$\eta_a/v_{b\varepsilon}$
	$n = 30$	$n = 300$	$n = 30$	$n = 300$	
Steel ($L \approx 10.5$ mm)	1.25×10^{-3}	0.4×10^{-3}	0.11	0.035	86.8
Fused quartz ($L \approx 4$ mm)	4×10^{-3}	1.5×10^{-3}	0.1	0.04	25.9

aperture and corresponding minimal achievable errors $v_{b\varepsilon}$ depend on the attenuation coefficient of LSAW (Table 6.1). Large attenuation causes fast decay of the signals and gives rise to the elevated error.

In addition, it should be mentioned that the errors $v_{b\varepsilon}$ and η_a depend on the directivity function of the receiving elements H. If the elements are wider than the wavelength of the wave in the liquid, then the directivity function is narrow. In this case, to avoid degradation of the signal-to-noise ratio the array should be tilted at the expected critical angle of LSAW: $\theta_0 \approx \theta_R$.

As follows from Equations (6.30) and (6.35), the errors $v_{b\varepsilon}$ and η_a are proportional because their behavior is determined by the function $\Psi_n(a)$:

$$\eta_a = v_{b\varepsilon} \frac{\omega_0}{\alpha_x C_R} \tag{6.36}$$

The calculated ratios $\eta_a/v_{b\varepsilon}$ are presented in Table 6.1 for selected materials. They are significantly greater than unity, and, generally, the accuracy of the attenuation coefficient measurement is much lower than the accuracy of the velocity measurement.

The fractional error $v_{b\tau}$ calculated for fused quartz is shown in Figure 6.3 as a function of number of elements n. The pitch of the array $p = 0.2$ mm, the tilt angle $\theta_0 = 30°$, and the temporal jitter $\sigma_\tau = 5$ ns were used in the calculation. This error monotonically decreases with increasing n.

The fractional velocity errors v_θ, v_p, and v_C produced by deviations of the tilt angle of the specimen $\delta\theta_0$, the pitch of the array δp, and the velocity of sound in liquid δC, respectively, have been derived according to (6.21) and are presented in Table 6.2. The deviation $\delta\theta_0$ is caused by improper positioning of the specimen or the array, and δp is the pitch bias due to the array technology restrictions. The origins of δC are variations in temperature, pressure, chemical composition of the immersion liquid, and the effect of slowing-down of inhomogeneous plane waves. It follows from Equation (6.2) that the phase velocity of the inhomogeneous plane wave C_E is smaller than the velocity of the homogeneous wave C [23]:

$$C_E = C \left(\sqrt{1 + \frac{\alpha_x^2 + \alpha_z^2}{k_0^2}} \right)^{-1} \tag{6.37}$$

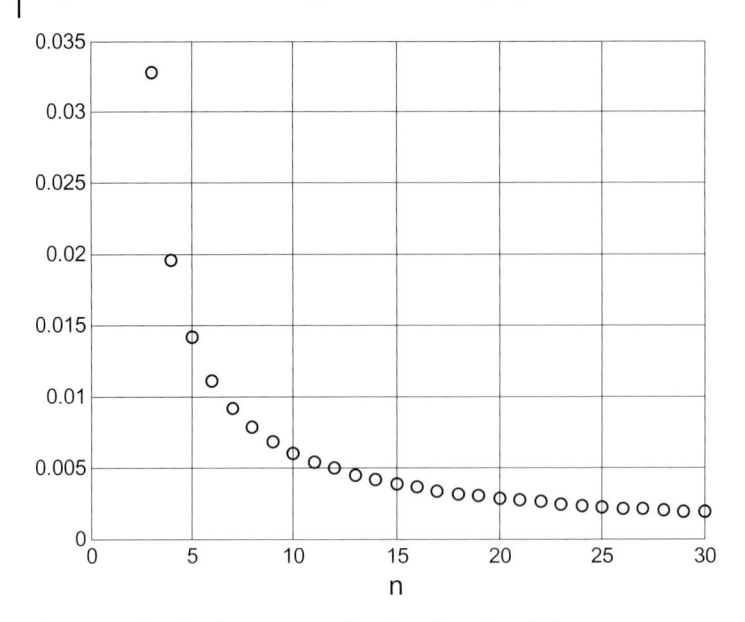

Figure 6.3 Fractional error v_{br} as a function of number of elements n.

Table 6.2 Fractional errors v_k and η_k.

	Velocity errors v_k	Attenuation errors η_k
$\delta\theta_0$	$v_\theta = \dfrac{1}{tg\theta_R} \cdot \delta\theta_0$	$\eta_\theta = \|tg(\theta_R - \theta_0)\|\delta\theta_0$
δp	$v_p = \dfrac{tg(\theta_R - \theta_0)}{tg\theta_R} \cdot \dfrac{\delta p}{p}$	$\eta_p = \dfrac{\delta p}{p}$
δC	$v_C = \dfrac{\sin\theta_0}{\sin\theta_R \cdot \cos(\theta_R - \theta_0)} \cdot \dfrac{\delta C}{C}$	–
$v = \delta(C_R)C_R^{-1}$	–	$\eta_{CR} = \dfrac{\sin\theta_0 \cdot \sin\theta_R}{\cos(\theta_R - \theta_0)\cos^2\theta_R} \cdot v$

Since $\alpha \ll k_0$, the velocity decrease can be estimated:

$$\frac{\delta C}{C} = \frac{C - C_E}{C} = \frac{1}{2}\left(\frac{\alpha_x}{k_0 \cos\theta_R}\right)^2 \tag{6.38}$$

This effect is relatively small. For instance, for the water and fused quartz combination the normalized attenuation factor is $\alpha_x/k_0 = 0.017$ [24]; therefore, $\delta C/C = 1.8 \times 10^{-4}$. Moreover the error v_C vanishes when the array is parallel to the surface of the sample $\theta_0 = 0$.

The fractional attenuation errors η_θ, η_p, and η_{CR} calculated using Equation (6.22) are also shown in the Table 6.2. There is no error in the attenuation measurement

Table 6.3 Numerical estimation of errors.

	Typical values	Angles θ_0, θ_R	v	η
$\Delta\theta$	$\delta\theta \approx 10^{-2}$–$10^{-3}$ rad	$\theta_0 = 0$; $\theta_R = \pi/6$	$v_\theta \approx 2 \cdot \delta\theta$	$\eta_\theta \approx 0.5\delta\theta$
		$\theta_0 = \theta_R = \pi/6$	$v_\theta \approx 2 \cdot \delta\theta$	$\eta_\theta = 0$
δp	$\delta p/p \approx 10^{-3}$	$\theta_0 = 0$; $\theta_R = \pi/6$	$v_p \approx \delta p/p$	$\eta_p \approx \delta p/p$
		$\theta_0 = \theta_R = \pi/6$	$v_p = 0$	$\eta_p \approx \delta p/p$
δC	$\delta C/C \approx 2 \times 10^{-3}$ ($\delta T = 1\,°K$)	$\theta_0 = \theta_R$	$v_C \approx \delta C/C$	–
		$\theta_0 = 0$	$v_C = 0$	–
v	$v \approx 10^{-2}$–10^{-3}	$\theta_0 = \theta_R = \pi/6$	–	$\eta_{CR} \approx 0.3v$
		$\theta_0 = 0$	–	$\eta_{CR} = 0$

caused by the variations in sound velocity δC, since α_x does not explicitly depend on C. On the other hand, the attenuation coefficient depends on the critical angle θ_R; therefore, the velocity error v generates the additional error in the attenuation coefficient η_{CR}.

Table 6.3 presents the fractional errors $v_{\theta,p,C}$ and $\eta_{\theta,p,CR}$ numerically estimated for various combinations of the angles θ_0 and θ_R. The velocity measurement is sensitive to the tilt angle of the receiving aperture, the pitch of the array, and the parameters of the immersion liquid, mostly temperature T. For various experimental arrangements, the errors $v_{\theta,p,C}$ can be larger or smaller than the errors $v_{b\varepsilon}$ and $v_{bт}$. Some errors can be reduced by proper choice of the tilt angle. For instance, the error v_p disappears when the aperture is parallel to the wave front of LSAW $\theta_0 = \theta_R$, and the measurement result is insensitive to the parameters of the liquid when the aperture is parallel to the surface of the specimen $\theta_0 = 0$. The fractional errors of the attenuation measurement $\eta_{\theta,p,CR}$ are all negligibly small in comparison with the error η_a caused by the additive noise.

In the experiment, a lens-less line-focused transducer was used as a transmitter for excitation of the leaky waves (Figure 6.1). In this transducer, the poly(vinylidene fluoride) film (PVDF, Measurement Specialties, Inc.) was in direct contact with the immersion liquid (water) and was capable of radiating a cylindrical converging wave. The film was supported by the cylindrically concave buffer whose acoustical impedance was close to the impedance of the film to achieve a broad frequency band. The central frequency of the transducer was approximately 15 MHz and the relative bandwidth extended up to 100%. The focal distance and the half-aperture angle of the transducer were 20 mm and $\gamma_m = 30°$, respectively, and the line focus was about 10 mm long. The axis of the transducer was tilted at an angle of $\gamma_0 = 45°$; thus the transducer was capable of generating a leaky wave with a critical angle from 15° to 75°.

The receiving linear array has been manufactured using piezocomposite technology (Imasonic Inc., France). The central frequency and relative bandwidth of the array were about 17 MHz and 70%, respectively. The array consisted of 32 rectangular elements having a width of 0.2 mm and a length of 8 mm, and the pitch of the array was $p = 0.25$ mm. The width of the main lobe of the directivity function H between points 3 dB below the peak is approximately $\pm 20°$ for a frequency of 10 MHz.

By using mechanical manual stages, the system was adjusted so that the line focus of the transmitting transducer was positioned on the water–specimen interface, and the long sides of the elements of the array were parallel to the line focus of the transmitter and the surface of the specimen.

The transmitting transducer was excited by the square pulses, which had amplitudes of 200 V and a duration of 50 ns. The pulses were generated by an Ultrasonic Pulser Receiver system (UT 340; Utex Scientific Instruments Inc.). The low and high pass filters of the receiver were adjusted to produce a bandwidth of 1–30 MHz. Sequential connection of the elements of the array to the input of the receiver was performed by an analog multiplexer. The amplified and filtered waveforms $s_j(t)$, where $j = 1, \ldots 32$ is a channel number, were digitized by an oscilloscope (TDS520C; Tektronix) and transmitted via a GPIB interface to a computer. The acquisition time of the full $s_j(t)$ data set was less than 200 ms.

Several materials with known properties were tested using this experimental setup. The $s_j(t)$ data recorded for fused quartz at various tilt angle of the array θ_0 are shown in Figure 6.4 as grayscale images. In the graphs, the vertical axes represent the relative time delay t, and the horizontal axes correspond to the channel number j. The number of elements of the array is relatively small ($n = 32$), therefore the discrete structure of the images over variable j is quite evident.

The incident probing wave generated by the transmitting transducer produces the directly reflected wave D and the leaky wave R. Wave D is a cylindrical wave whose line source is located at the focal point F on the specimen–water interface (Figure 6.1); it demonstrates nonlinear behavior versus j. For all orientations of the array the leaky wave R comes to the receiving elements earlier than wave D. The slopes of the responses R and D and their separation in the time domain depend on the array orientation. The reference dashed lines shown in the images indicate the boundary of the regions where the critical incident LSAW angles θ_R are between $10°$ and $50°$. The mode separation is best at the array tilt angle $\theta_0 = 22.9°$ (Figure 6.4b), when the array is approximately oriented in the direction of the leaky wave propagation. This observation is in agreement with the results of the error analysis since the fractional errors v_{br} (6.30), $v_{b\varepsilon}$ (6.31) are inversely proportional to the factor $\cos(\theta_R - \theta_0)$. In addition, the sensitivity of the receiving elements to the leaky wave is smaller at the tilt angles $\theta_0 = 0°$ and $90°$ since the directivity function H decays when the incident angle of the wave on the aperture plane of the array $|\theta_R - \theta_0|$ increases.

The $s_j(t)$ data presented in Figure 6.4 have been processed to obtain the velocity of the leaky wave C_R. Initially, the responses R have been discriminated and the time positions t_j of the negative peaks were determined as functions of j. Then, by solving the linear regression Equation (6.18), the parameters b have been obtained and the velocities C_R calculated using Equation (6.17). The measured

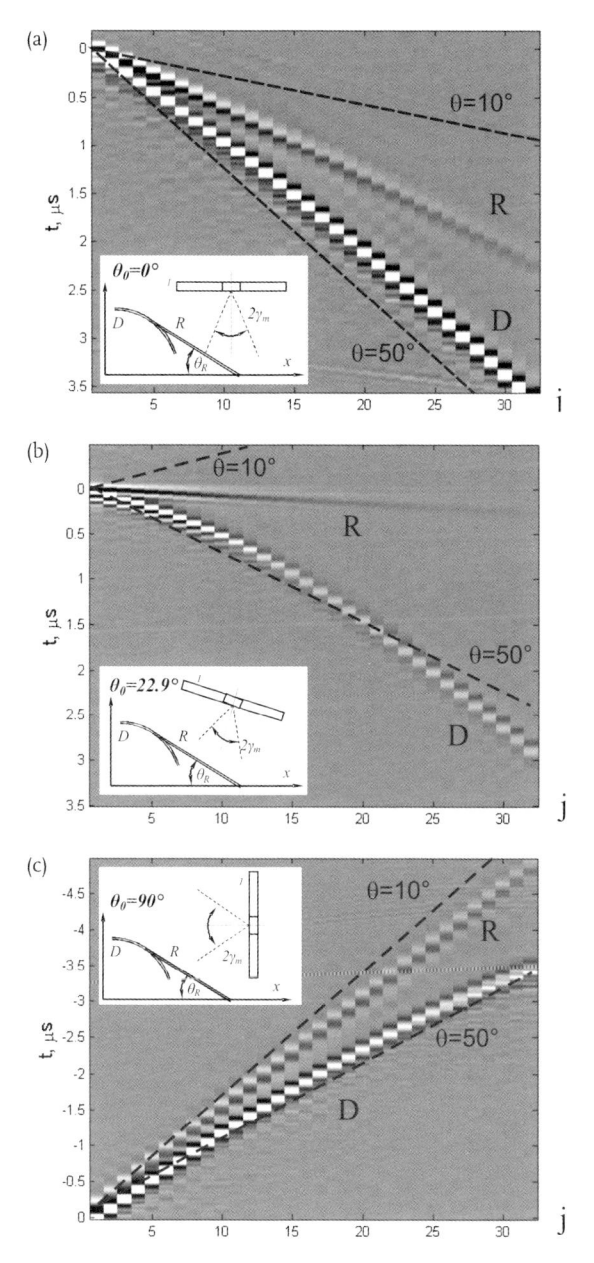

Figure 6.4 $s_j(t)$ data recorded for fused quartz specimen at the array tilt angles $\theta_0 = 0°$, 22.9°, and 90°, respectively.

6 Quantitative Acoustic Microscopy Based on the Array Approach

Table 6.4 Measured and reference LSAW parameters.

Material	Velocity		Normalized attenuation factor	
	Measured	Reference	Measured	Reference [24]
	C_R, C_L (m s^{-1})	C_R^*, C_L^* (m s^{-1})	γ	γ^*
Fused quartz	$C_R = 3440 \pm 26$	$C_R^* = 3426$ [24]	$16 \pm 1.3 \times 10^{-3}$ ($f = 3$–13 MHz)	17×10^{-3}
Steel	$C_R = 3006 \pm 15$	$C_R^* = 2998$ [24]	$5.6 \pm 0.3 \times 10^{-3}$ ($f = 3$–20 MHz)	5.4×10^{-3}
Polystyrene	$C_L = 2355 \pm 20$	2340 [21]	–	–

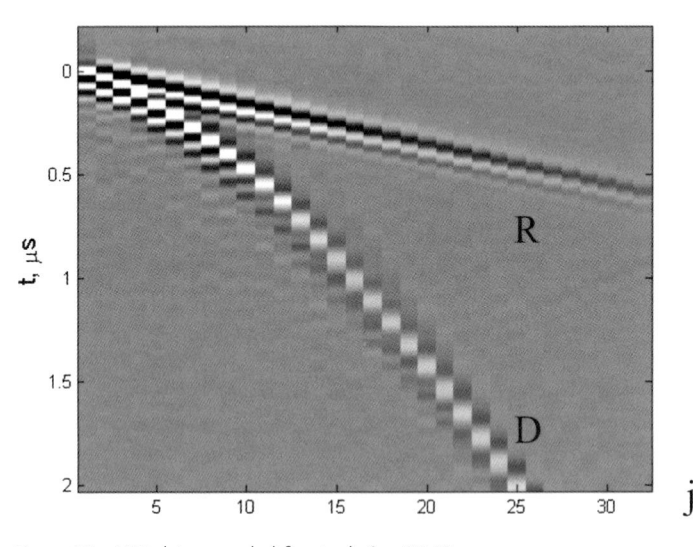

Figure 6.5 $V_j(t)$ data recorded for steel, $\theta_0 = 22.9°$.

values are $C_R = 3461$, 3445, and 3485 m s^{-1} for orientation angles of the array $\theta_0 = 0°$, $22.9°$, and $90°$, respectively. Thus, the tilt angles $0°$ and $90°$ give a large deviation from the measured velocities in comparison with the known reference value $C_R^* = 3426$ m s^{-1} (Table 6.4).

In further measurements, the experimental setup based on the tilt angle $\theta_0 = 22.9°$ was used to test materials with known acoustical parameters. Figure 6.5 shows the data $s_j(t)$ measured for the steel specimen. In comparison with the data recorded for the fused quartz specimen (Figure 6.4b), response R has a different slope and slow amplitude decay. This behavior can be explained by the fact that

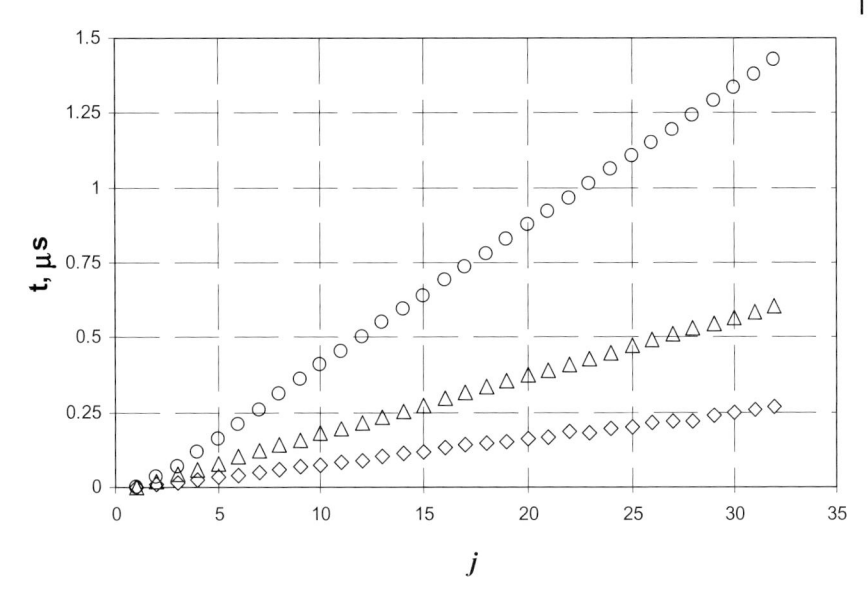

Figure 6.6 Time delay t of the R, L responses versus the channel number j ($\diamond$–fused quartz, $\triangle$–steel, $\bigcirc$–polystyrene).

steel has a lower LSAW velocity and attenuation than fused quartz. Figure 6.6 shows an example of the measured time delays t_j of the responses that were used to obtain the LSAW velocity. The measurements of C_R have been repeated several times to estimate the experimental accuracy (Table 6.4). The confidence intervals of C_R measured for the quartz and steel specimens include the reference values C_R^* obtained for the particular materials used in the separate experiments [24].

To estimate the attenuation factor of the leaky wave, the spectra of the leaky wave responses $s_j(f)$ were calculated. For the steel sample, the amplitude spectra $|s_j(f)|$ are presented in Figure 6.7 for several elements of the receiving array. The graphs show that the frequency of the maximal amplitude decreases with increasing j; thus, attenuation of the spectral components of the wide-band leaky wave is stronger at high frequencies.

Based on the calculated spectra, the attenuation factor of the leaky wave $\alpha_x(f)$ was found using Equation (6.15), where the parameter $a(f)$ was estimated by a linear fitting of the data $\mu_j = \log|s_j(f)|$ (6.19) over the variable j. Figure 6.8 shows the measured attenuation factor $\alpha_x(f)$ for the test materials. For the fused quartz specimens, the signal-to-noise ratio in the spectral domain is high enough to measure $\alpha_x(f)$ within approximately the 3–15 MHz frequency range. The lower boundary of this range is mostly restricted by a cutoff frequency of the system itself, whereas the upper frequency limit is determined by lowering of the signal-to-noise ratio at high frequencies due to attenuation of the leaky Rayleigh wave. For the steel specimen, the attenuation of the wave is significantly lower, and measurement of $\alpha_x(f)$ is possible up to 20 MHz. Assuming linearity of the

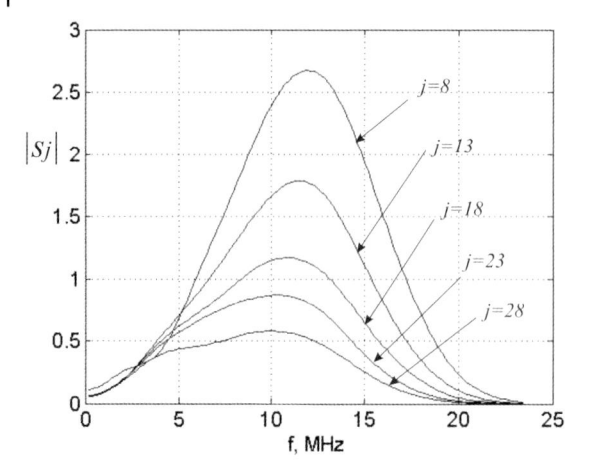

Figure 6.7 Amplitude spectra $|s_j(f)|$ of the leaky wave obtained for a steel specimen.

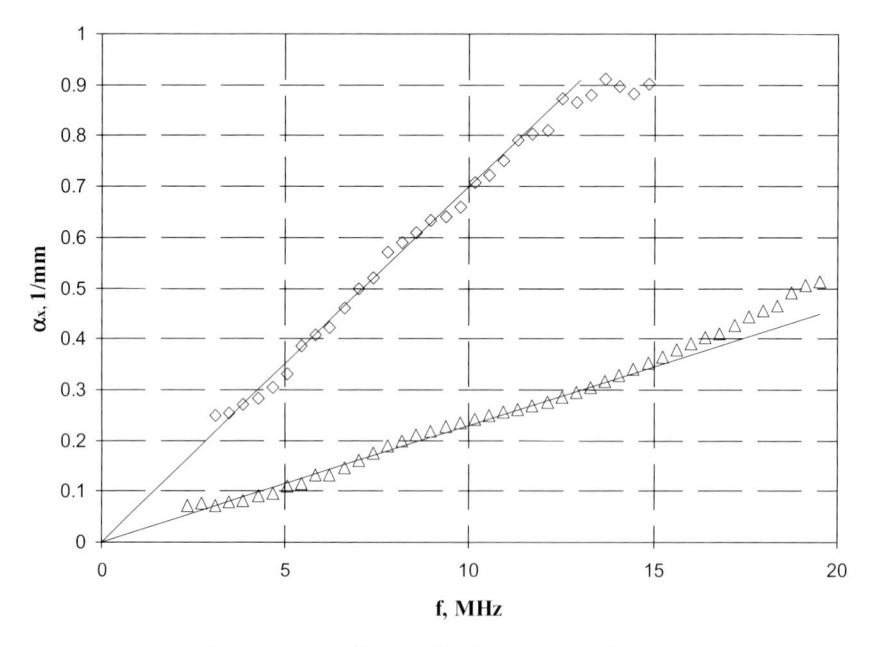

Figure 6.8 Measured attenuation coefficient of the leaky wave $\alpha_x(f)$ ($\Diamond$–fused quartz, $\triangle$–steel).

function $\alpha_x(f)$, the normalized attenuation coefficient $\gamma = \alpha_x C(2\pi f)^{-1}$ was found for each material by fitting of the experimental data and is presented in Table 6.4. The reference attenuation factors γ^* obtained in a separate study and published elsewhere [24] are in reasonable agreement with the measured values.

In addition to leaky Rayleigh waves, head waves have also been observed in the experiments. Figure 6.9 shows the $s_j(t)$ data recorded for the polystyrene specimen.

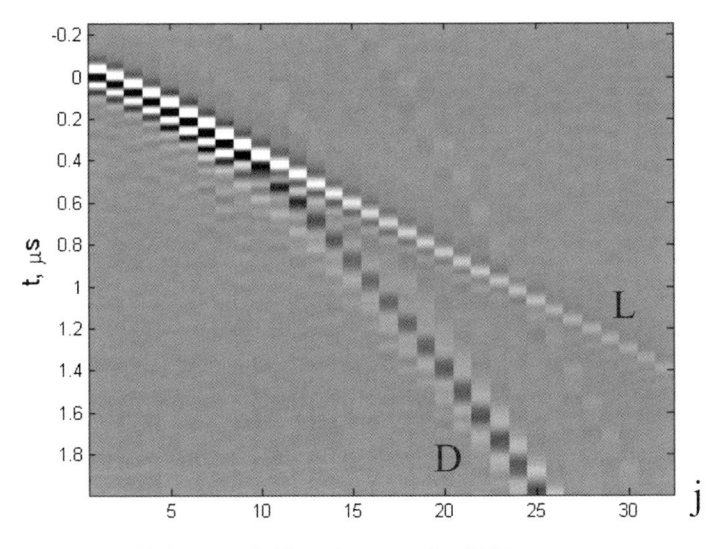

Figure 6.9 $s_j(t)$ data recorded for polystyrene, $\theta_0 = 22.9°$.

The bulk wave velocities are low for polystyrene [21] and the Rayleigh wave cannot be generated at the water–specimen interface. The critical angle of the longitudinal wave $\theta_L \approx 40°$ is within the active angular aperture of the experimental setup, therefore response L is produced by the head wave. The polarity of L is not inverted whereas the leaky Rayleigh wave response R has $-\pi$ phase shift [20]. The velocity measurement method demonstrated for the Rayleigh waves can be applied to the head waves as well. The time delays t_j of response L are plotted in Figure 6.9 as an example. The result of statistical processing of the C_L measurement is presented in Table 6.4, which is in agreement with the published value.

6.3
Measurement of Bulk Wave Velocities and Thickness of Specimen

In the LSAW measurement the tilted transducer is usually used for effective generation of the leaky waves since their critical angles are relatively large. To transmit waves into a solid plate and receive reflected echoes the incidence angles of the probing wave should be smaller corresponding critical angles and can be close to the interface normal. In this case, it is reasonable to position the array parallel to the interface of the specimen and the elements of the array can be used as transmitters and receivers.

Figure 6.10 shows the experimental setup of the measurement system and the main rays in the immersion liquid and solid specimen. There ray D is radiated by the transmitting element Tr, reflected at the liquid–specimen interface, and received by the array element Rs. There are also rays L and T associated with

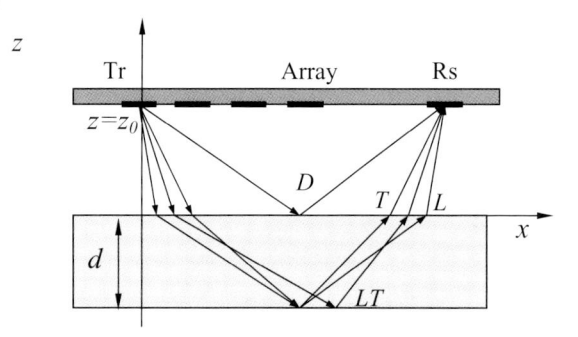

6 Quantitative Acoustic Microscopy Based on the Array Approach

Figure 6.10 Experimental arrangement for bulk wave velocities and thickness of specimen measurement.

propagation of the longitudinal and shear waves in the solid, respectively, and the mixed mode ray LT. The velocities of ultrasound in the solid layer and its thickness can be found using the time delays of the responses produced by these rays. Based on a simple geometrical consideration, the propagation time of the directly reflected wave D can be found as follows:

$$t_D(x) = \frac{\sqrt{x^2 + (2z_0)^2}}{C_0} \tag{6.39}$$

where x is the distance between the transmitting and receiving transducers. Similarly, the time delay of the response L is:

$$t_L(x) = \frac{\sqrt{(x-\xi)^2 + (2z_0)^2}}{C_0} + \frac{\sqrt{\xi^2 + (2d)^2}}{C_L} \tag{6.40}$$

where ξ is the propagation distance of the wave in the solid layer along the axis x and C_L is the velocity of the longitudinal wave in the specimen. Generally, it is possible to estimate d and C_L by fitting of the model Equations (6.39) and (6.40) to the measured $t_D(x)$ and $t_L(x)$ data [25, 26]. However, the distance $\xi = \xi(x)$ in Equation (6.40) is unknown; therefore, the additional relationship based on Snell's law should be included in the model. A more explicit and straightforward technique has been developed in seismology [27, 28] and it can be used for processing the ultrasonic array data. In this approach, the components of the slowness vector $\mathbf{s} = \omega^{-1}\mathbf{k}$ are used to express the time delay as a sum of two parts associated with the wave travel along and across the layer:

$$t_D(x) = s_x x + 2 s_{z0} z_0 = s_x x + \tau_0 \tag{6.41}$$

where $s_x = \omega^{-1} k_x$, $s_{z0} = \omega^{-1} k_z = \sqrt{s_0^2 - s_x^2}$, $s_0 = C_0^{-1}$, and $\tau_0 = 2 s_{z0} z_0$ is the transverse propagation time in the liquid.

The advantage of this representation is that the slowness s_x remains constant when the wave passes the interfaces between the layers. Therefore, the propagation

6.3 Measurement of Bulk Wave Velocities and Thickness of Specimen

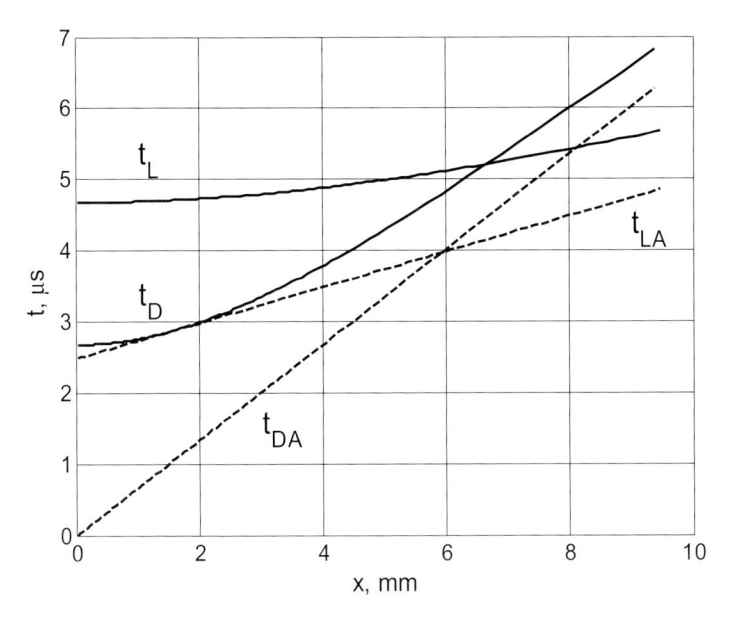

Figure 6.11 Time delays $t_D(x)$ and $t_L(x)$ and their asymptotic lines t_{DA} and t_{LA}, respectively.

time along the layers depends on the overall distance x, and the time delay of the response L is equal to:

$$t_L(x) = s_x x + 2s_{z0} z_0 + 2s_{zL} d = s_x x + \tau_0 + \tau_L \qquad (6.42)$$

where $s_{zL} = \sqrt{s_L^2 - s_x^2}$, $s_L = C_L^{-1}$, and $\tau_L = 2s_{zL} d$ is the transverse propagation time of the longitudinal wave across the solid layer. Similar equations can be written for the shear wave response T and the mixed mode response LT.

Figure 6.11 shows the time delays $t_D(x)$ and $t_L(x)$ calculated for the following set of the parameters: $z_0 = 2\,\text{mm}$, $d = 4\,\text{mm}$, $C_0 = 1485\,\text{m s}^{-1}$, and $C_L = 4000\,\text{m s}^{-1}$. At the normal incidence $t_L(0)$ is larger than $t_D(0)$ on the time of the wave propagation in the solid layer d/C_L. However, starting from some x, $t_D(x)$ becomes larger, $t_L(x)$, because at large x the wave L mostly propagates in the solid layer with the velocity $C_L > C_D$. When x tends to infinity the delay $t_D(x)$ of response D approaches the asymptotic line $t_{DA} = s_0 x$ while the slowness s_x tends to the maximal value s_0. For wave L the slowness s_x cannot exceed s_L, otherwise wave L in the solid layer becomes inhomogeneous. Therefore, as follows from Equation (6.42), the asymptotic line is determined by the equation:

$$t_{LA}(x) = s_L x + 2z_0 \sqrt{s_0^2 - s_L^2} \qquad (6.43)$$

The transverse time delays in the immersion liquid τ_0 and in the solid layer τ_L calculated as functions of normalized slowness s_x/s_0 are shown in Figure 6.12. The delay τ_0 exists for all $s_x < s_0$ whereas the delay τ_L is defined in the interval $(0, s_L/s_0)$.

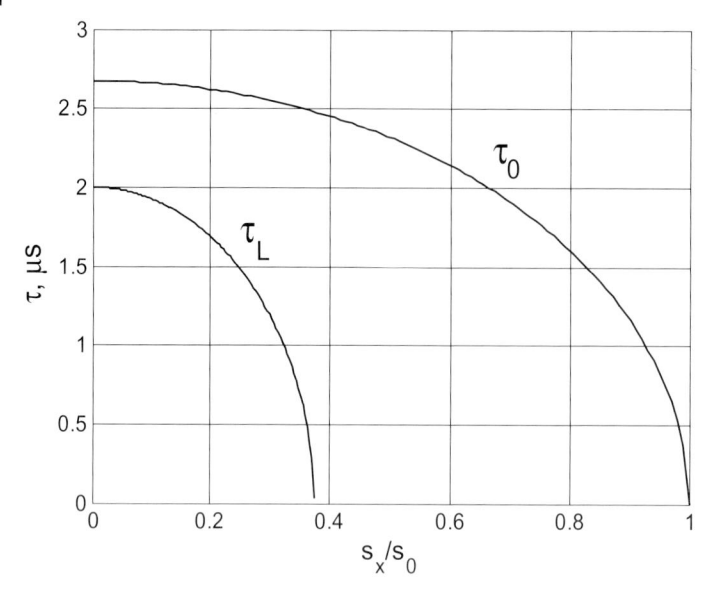

6 Quantitative Acoustic Microscopy Based on the Array Approach

Figure 6.12 Delays τ_0 and τ_L.

The derivatives of the delays τ_0 and $(\tau_0 + \tau_L)$ over s_x are determined by the propagation distances along the array aperture:

$$\frac{d\tau_0}{ds_x} = -2z_0 \frac{s_x}{s_{z0}} = -x, \quad \frac{d(\tau_0 + \tau_L)}{ds_x} = -x \quad (6.44)$$

Based on these important relationships, the derivatives of the times $t_D(x)$ and $t_L(x)$ over x can be found:

$$\frac{dt_D}{dx} = s_x + \frac{ds_x}{dx}\left(x + \frac{d\tau_0}{ds_x}\right) = s_x, \quad \frac{dt_L}{dx} = s_x \quad (6.45)$$

Thus, differentiating the experimental $t_D(x)$, $t_L(x)$ data the parameter s_x can be found as a function of the propagation distance x. Figure 6.13 shows the inverse functions $x_D(s_x/s_0)$ and $x_L(s_x/s_0)$ calculated in the course of the current numerical simulation. The corresponding ray representation of the D and L waves that travel with the same slowness parameter s_x is given in Figure 6.14.

Using established relationships between x_D, x_L, and s_x, the measured $t_D(x)$, $t_L(x)$ data can be expressed as functions of the slowness parameter s_x:

$$t_D(x_D) = s_x x_D + \tau_0 \quad (6.46)$$

$$t_L(x_L) = s_x x_L + \tau_0 + \tau_L \quad (6.47)$$

The time $t_L[x(s_x)]$ is larger than $t_D[x(s_x)]$ due to propagation of the L wave inside the solid layer along the ray ABC (Figure 6.14). This delay consists of the travel time

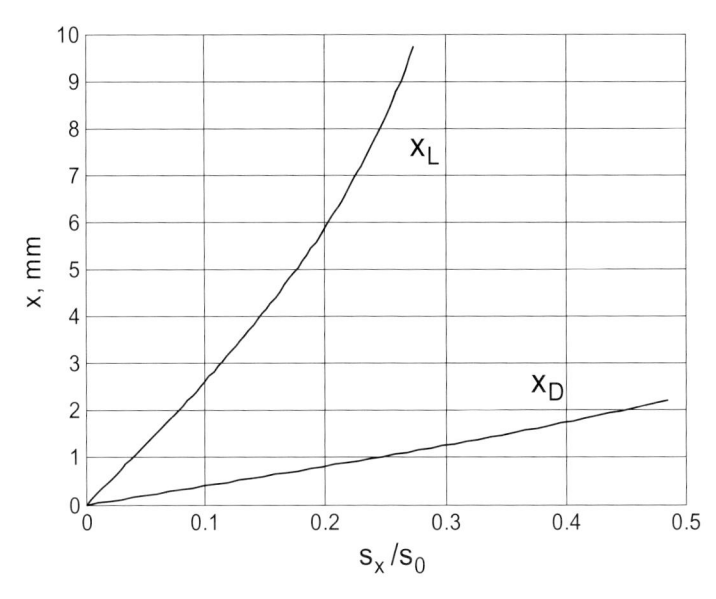

Figure 6.13 Propagation distances x_D, x_L versus s_x/s_0.

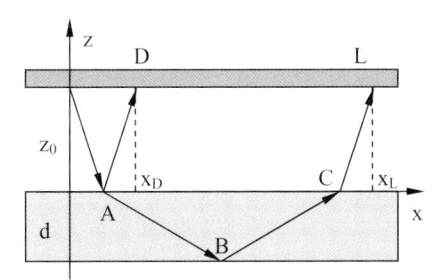

Figure 6.14 D and L rays.

of the L wave along $AC = x_L - x_D$ with the slowness parameter s_x and the transverse propagation time τ_L. Thus the dependence $\tau_L(s_x)$ can be found by subtracting Equation (6.46) from Equation (6.47):

$$\tau_L = t_L(x_L) - t_D(x_D) - s_x(x_L - x_D) \qquad (6.48)$$

Knowing $\tau_L(s_x)$ it is possible to estimate the velocity of the longitudinal wave C_L and the thickness of the layer d. The squared τ_L is a linear function of s_x^2:

$$\tau_L^2 = -(2d)^2 s_x^2 + (2d)^2 s_L^2 \qquad (6.49)$$

This linear regression equation can be solved using the least-squares method and the desired d and C_L can be easily deduced from the coefficients of the estimated

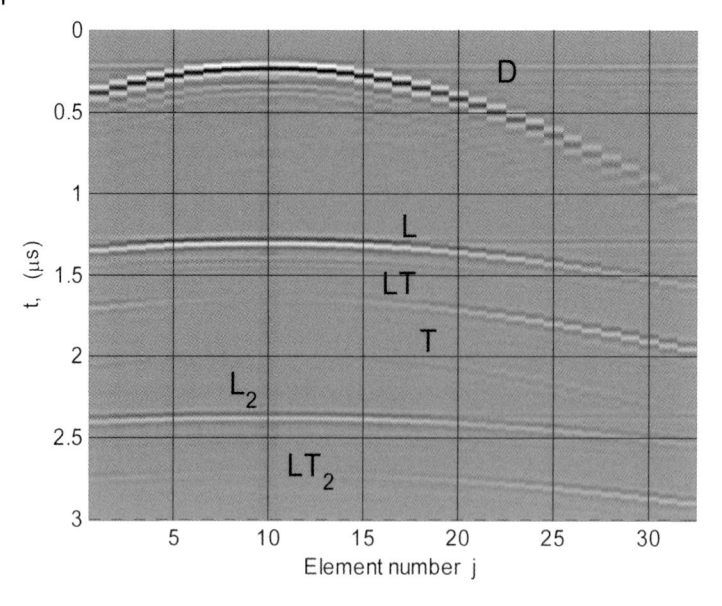

Figure 6.15 $s_{10j}(t)$ data ($k = 10$).

linear function. Notably, similar relationships can be obtained for the shear wave propagation time $\tau_T(s_x)$. For the mixed mode the transverse propagation time of the response LT is $\tau_{LT} = (\tau_L + \tau_T)/2$, and to determine d and $C_L(C_T)$ the velocity $C_T(C_L)$ should be known.

In the experimental testing of this method, the hardware described earlier in this chapter was used for data acquisition. The linear array was adjusted to be parallel to the immersed specimen, and all elements of the array were used for transmitting and receiving of ultrasound waves to record a full data set $s_{kj}(t)$, where k and j are the transmitting and receiving element numbers, respectively. Figure 6.15 shows the signals reflected for a fused quartz plate with a thickness of $d = 3.2 \pm 0.01$ mm. The data set $s_{10j}(t)$, presented in the picture, was acquired by all ($1 \leq j \leq 32$) elements of the array when the element $k = 10$ was used as a transmitter.

There are several responses produced by the directly reflected wave (D): longitudinal (L) and shear (T) waves reflected from the back side of the specimen and the longitudinal–shear mixed mode response LT. In addition, the responses L_2 and LT_2 caused by double reflection in the solid layer are presented in the image. To find the thickness and velocity in the layer according to the proposed technique it is sufficient to use the partial data recorded with a single transmitting element since the signals depend only on the distances between the transmitter and receiver, $x = p(j - k)$. However, full data processing provides a better signal-to-noise ratio, compensation of array–specimen non-parallelism, and variations of element parameters.

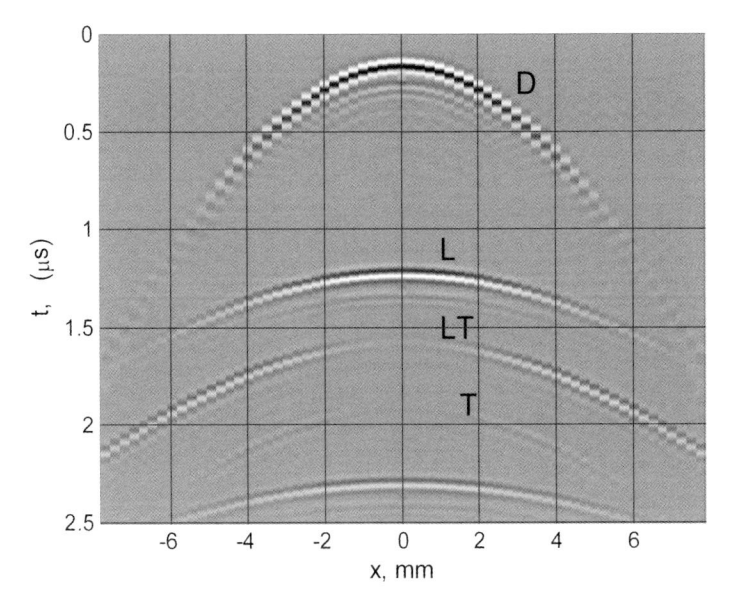

Figure 6.16 $r_m(t)$ data.

For these purposes, 32 signals $s_{ij}(t)$ are selected from the full data set to determine the amplitudes a_j and delays Δt_j of the pulses reflected in the pulse–echo mode from the front interface of the specimen. Then the output data $r_m(t)$ ($-N \leq m \leq N$) are generated by summing all time corrected and amplitude normalized waveforms for which the transmitter–receiver distance is constant:

$$r_m(t) = \frac{1}{N-|m|} \sum_{j,k=1}^{N} s_{jk}\left(t - \frac{\Delta t_j + \Delta t_k}{2}\right)(a_j \cdot a_k)^{-\frac{1}{2}} \cdot \delta_{(k-j)m} \qquad (6.50)$$

where $\delta_{km} = 1$ if $k = m$ but otherwise is 0. The number of terms summed to form r_m depends on m; therefore, the coefficient $(N - |m|)^{-1}$ in Equation (6.50) is used to compensate this effect. The result of the described processing is shown in Figure 6.16 as a function of $x = pm$. Obviously, the data are even: $r_{-m}(t) = r_m(t)$; this is a consequence of the acoustic reciprocity principle.

The calculated $r_m(t)$ data were used to determine the delays $t(x)$ for the responses D, L, LT, and T (Figure 6.17). The polarity of the L, LT ultrasonic pulses is inversed with respect to the polarity of D and T; therefore, the minimal values of the D, T and maximal values of the L, LT were detected to calculate the delays. Then the MATLAB function *polyfit* was applied to the experimental curves for polynomial fitting of the 4th degree. Having the data in the polynomial form it is possible to employ symbolic differencing and avoid noisy direct finite difference calculations. Figure 6.18 shows the result of the evaluation with Equation (6.45) and inversion of the obtained $s_x(x)$ functions.

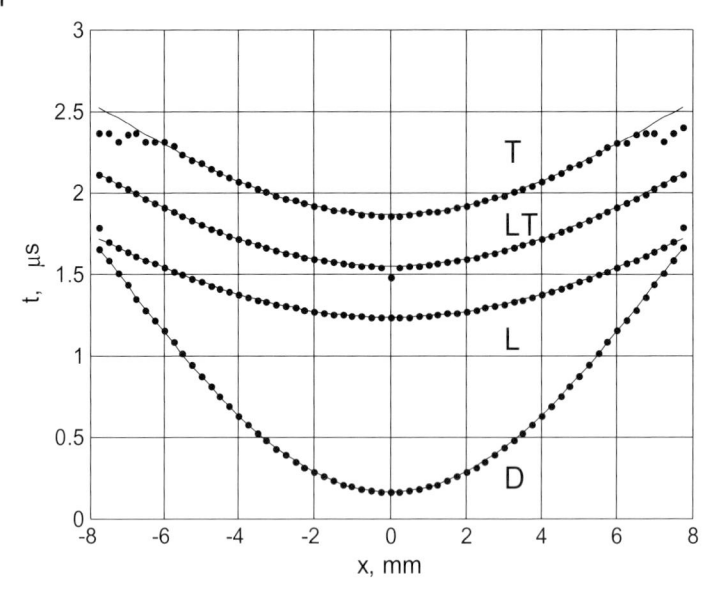

148 | *6 Quantitative Acoustic Microscopy Based on the Array Approach*

Figure 6.17 Measured (•) and fitted (solid lines) delays $t(x)$.

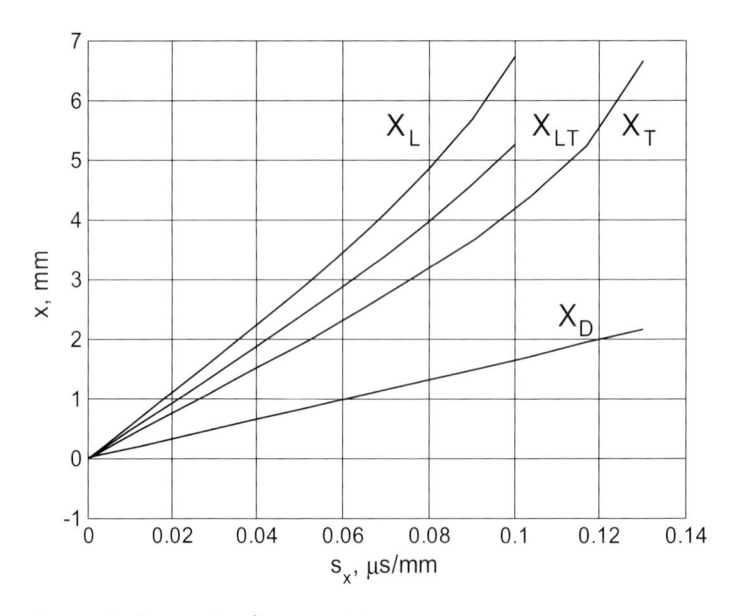

Figure 6.18 Propagation distances $x(s_x)$.

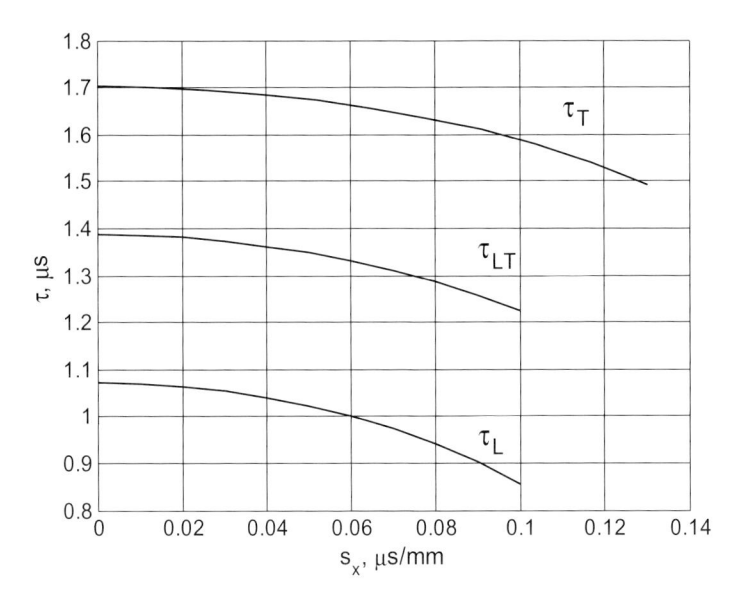

Figure 6.19 Delays $\tau(s_x)$.

Table 6.5 Measured and reference thicknesses and bulk velocities.

	Thickness d (mm)	Velocity C_L (m s^{-1})	Velocity C_T (m s^{-1})
Reference values	3.20 ± 0.01	5970 [21]	3765 [21]
Response L	3.215 ± 0.015	5999 ± 26	
Response T	3.17 ± 0.035		3741 ± 37
Response LT (known C_L)	3.06 ± 0.03		3757 ± 34
Response LT (known d, C_L)			3744 ± 18

Using the calculated $x_D(s_x)$, $x_L(s_x)$, $x_{LT}(s_x)$, and $x_T(s_x)$ functions the corresponding delays $\tau(s_x)$ were determined in accordance with Equation (6.48) (Figure 6.19). Finally, the thickness of the plate d and the bulk wave velocities C_L, C_T were found using Equation (6.49). The results of statistical analysis of the measured values in the form of 95% confidence intervals are presented in Table 6.5. The longitudinal wave response L has a good signal-to-noise ratio (Figure 6.16) and it spreads over a wide range of x; therefore, the accuracy of the thickness and velocity determination using this response is the best. The amplitude of the shear wave response T is smaller; it occupies a narrower range of x. This can be considered as a reason for higher errors in d and C_T determined by processing the response T. The mixed mode response LT was also used to evaluate d and C_T. In these calculations, the velocity C_L determined earlier by processing of the response L from the same measured data set $s_{kj}(t)$ was taken as the known value. If the thickness d was assumed to be known as well, the error in C_T decreases. The results presented in Table 6.5 show that the measured velocities are in agreement with the values published for fused quartz, and that ultrasonically determined d matches the directly measured plate thickness. The relative errors achieved with the described technique and experimental setup can be roughly estimated to be 0.5–1% in thickness and 0.5% and 1% in longitudinal and shear velocities, respectively.

6.4
Conclusions

This chapter has considered the measurement method of acoustic parameters of a flat specimen immersed in a liquid. The main idea of the method is to use an ultrasonic array to record the wave reflected by the specimen and process the acquired spatiotemporal data to determine the desired parameters of the specimen. It was shown that the pitch–catch arrangement of the single-element transmitting transducer and receiving array can be used to measure the velocity and attenuation of leaky surface waves. This arrangement allows for the effective generation and detection of leaky waves. Moreover, as follows from the measurement error analysis, to optimize the signal-to-noise ratio the array should be tilted at the expected critical angle of LSAW. For measurement of the bulk wave velocities and the thickness of the plate specimen the incident angles of the probing and reflected waves are relatively small. Thus, the aperture of the array should preferably be aligned along the specimen and the elements of the array can serve as transmitters and receivers. In comparison with the existing quantitative acoustic microscopy method, mechanical scanning of the transducers is no longer used in the considered technique, and the measurement time is only limited by the time of wave propagation and the speed of the electronic data acquisition system.

References

1 Briggs, A. (1992) *Acoustic Microscopy*, Clarendon Press, Oxford, pp. 105–152.
2 Weglein, R.D. and Wilson, R.G. (1978) Characteristic material signature by acoustic microscopy. *Electron. Lett.*, **14**, 352–354.
3 Liang, K.K., Kino, G.S., and Khuri-Yakub, B.T. (1985) Material characterization by the inversion of V(z). *IEEE Trans. Sonics Ultrason.*, **32**, 213–224.
4 Kushibiki, J. and Chubachi, N. (1985) Material characterization by line-focus-beam acoustic microscope. *IEEE Trans. Sonics Ultrason.*, **32**, 189–212.
5 Atalar, A., Koymen, H., Bozkurt, A., and Yaralioglu, G. (1995) Lens geometries for quantitative acoustic microscopy, in *Advances in Acoustic Microscopy*, vol. 1 (ed. A. Briggs), Plenum Press, New York, pp. 117–151.
6 Lee, Y. and Cheng, S. (2001) Measuring lamb wave dispersion curves of a bi-layered plate and its application on

material characterization of coating. *IEEE Trans. Ultrason. Ferroelectr. Freq. Control.*, **48**, 830–837.

7 Nadal, M.-H., Lebrun, P., and Gondard, C. (1998) Prediction of the impulse response of materials using a SAM technique in the MHz frequency range with a lensless cylindrical-focused transducer. *Ultrasonics*, **36**, 505–512.

8 Maev, R.G. and Levin, V.M. (1997) Principles of local sound velocity and attenuation measurement using transmission acoustic microscope. *IEEE Trans. Ultrason. Ferroelectr. Freq. Control.*, **44** (6), 1224–1231.

9 Hanel, V. and Kleffner, B. (2000) Double focus technique for simultaneous measurement of sound velocity and thickness of thin samples using time-resolved acoustic microscopy, in *Proceedings of the 24th International Symposium on Acoustical Imaging* (eds P. Tortoli and L. Masotti), Kluwer Academic/Plenum, New York, pp. 187–192.

10 Nakaso, N., Ohira, K., Yanaka, M., and Tsukahara, Y. (1994) Measurement of acoustic reflection coefficients by an ultrasonic microspectrometer. *IEEE Trans. Ultrason. Ferroelectr. Freq. Control.*, **41**, 494–502.

11 Vines, R.E., Tamura, S., and Wolfe, J.P. (1995) Surface acoustic wave focusing and induced Rayleigh waves. *Phys. Rev. Lett.*, **74**, 2729–2732.

12 Every, A.G., Maznev, A.A., and Briggs, G.A.D. (1997) Surface response of a fluid loaded anisotropic solid to an impulsive point source: application to scanning acoustic microscopy. *Phys. Rev. Lett.*, **79** (13), 2478–2481.

13 Hauser, M.R., Weaver, R.L., and Wolfe, J.P. (1992) Internal diffraction of ultrasound in crystals: phonon focusing at long wavelengths. *Phys. Rev. Lett.*, **68** (17), 2604–2607.

14 Pluta, M., Schubert, M., Jahny, J., and Grill, W. (2000) Angular spectrum approach for the computation of group and phase velocity surface of acoustic waves in anisotropic materials. *Ultrasonics*, **38**, 232–236.

15 Titov, S., Maev, R., and Bogatchenkov, A. (2003) Wide-aperture, line-focused ultrasonic material characterization system based on lateral scanning. *IEEE Trans. Ultrason. Ferroelectr. Freq. Control.*, **50**, 1046–1056.

16 Lobkis, O.I. and Chimenti, D.E. (1999) Three-dimensional transducer voltage in anisotropic materials characterization. *J. Acoust. Soc. Am.*, **106**, 36–45.

17 Alleyne, D. and Cawley, P. (1991) A two-dimensional Fourier transform method for the measurement of propagating multimode signals. *J. Acoust. Soc. Am.*, **89**, 1159–1168.

18 Mah, M. and Schmitt, D.R. (2001) Near point-source longitudinal and transverse mode ultrasonic arrays for material characterization. *IEEE Trans. Ultrason. Ferroelectr. Freq. Control.*, **48** (3), 691–698.

19 Bossy, E., Talmant, M., and Laugier, P. (2002) Effect of bone cortical thickness on velocity measurements using ultrasonic axial transmission: a 2D simulation study. *J. Acoust. Soc. Am.*, **112**, 297–307.

20 Brekhovskikh, L.M. and Godin, O.A. (1990) *Acoustics of Layered Media I*, Springer, Berlin, pp. 98–112.

21 Birks, A.S., Green, R.E., Jr., and McIntire, P. (eds) (1991) *Ultrasonic Testing*, Nondestructive Testing Handbook, 2nd edn, vol. 7. American Society for Nondestructive Testing, Columbus, OH, p. 325, pp. 836–841.

22 Montgomery, D.C. and Runger, G.C. (2003) *Applied Statistics and Probability for Engineers*. 3rd edn, John Wiley & Sons, Inc., Hoboken.

23 Victorov, I.A. (1967) *Rayleigh and Lamb Waves: Physical Theory and Applications*, Plenum Press, New York.

24 Titov, S.A., Maev, R.G., and Bogatchenkov, A.N. (2006) Measurement of velocity and attenuation of leaky waves using an ultrasonic array. *Ultrasonics*, **44** (2), 182–187.

25 Sousa, A.V.G., Pereira, W.C.A., and Machado, J.C. (2007) An ultrasonic theoretical and experimental approach to determine thickness and wave speed

in layered media. *IEEE Trans. UFFC*, **54** (2), 386–393.

26 Titov, S.A., Maev, R.G., and Bogatchenkov, A.N. (2009) Measuring the acoustic wave velocity and sample thickness using an ultrasonic transducer array. *Tech. Phys. Lett.*, **35**, 1029–1031.

27 Bessonova, E.N., Fishman, V.M., Ryaboi, V.A., and Sitnikova, G.A. (1974) The tau method for inversion of travel times – I. Deep seismic sounding data. *Geophys. J. Roy. Astr. Soc.*, **36**, 377–398.

28 Stoffa, P.L., Buhl, P., Diebold, J.B., and Wenzel, F. (1981) Direct mapping of seismic data to the domain of intercept time and ray parameter – a plane-wave decomposition. *Geophysics*, **46** (3), 255–267.

Part Three
Advanced Biomedical Applications

7
Study of the Contrast Mechanism in an Acoustic Image for Thickly Sectioned Melanoma Skin Tissues with Acoustic Microscopy

Bernhard R. Tittmann, Chiaki Miyasaka, Elena Maeva, and David Shum

7.1
Introduction

Since contrast in an acoustic image (i.e., C-scan image) for skin tissue including a cancerous portion is formed by attenuation differences within the tissue, to predict the amplitude of a transducer output may be important for the field of medical ultrasound. We developed a mathematical model for this prediction employing frequencies from 25 to 250 MHz in pulse wave mode. In addition, we developed an algorithm that is a component of the model to determine Young's modulus at frequencies from 200 to 600 MHz in tone burst wave mode. In the present study, an abnormal skin tissue (i.e., a tissue including a melanoma portion) was selected as a specimen, and compared to normal skin tissue. The measurements gave approximate values of 2.73 MPa and 0.22 dB mm^1 for normal tissue and 2.80 MPa and 6.67 dB mm^1 for abnormal tissue for the modulus and attenuation, respectively. The technique suggests the use of this approach as a diagnostic tool and opens the door for more sophisticated analysis.

7.1.1
What Is Melanoma?

Melanoma is a form of cancer that begins in melanocytes (i.e., cells that make the pigment melanin). It may start predominantly in a mole (i.e., skin melanoma), but also may present itself in other pigmented tissues, such as in the eye or in the intestines. It is one of the less common types of skin cancer but causes the majority (75%) of skin cancer related deaths.

The skin is the body's largest organ. It protects the organs inside your body from injury, infection, heat, and ultraviolet light from the sun. The skin has two main layers. The layer at the surface is called the *epidermis*. Below the epidermis is the inner layer, the *dermis*. Deep in the epidermis are cells called melanocytes. Melanocytes make melanin, which gives color to your skin. When skin is exposed to the sun, the melanocytes make more melanin and cause the skin to tan or darken. Sometimes melanocytes cluster together and form moles (i.e., *nevi*).

There are three types of skin cancer:

- squamous cell cancer (SCC), which starts in the squamous cells (i.e., thin flat cells found on the surface of the skin);
- basal cell cancer (BCC) starts in the basal cells (i.e., round cells that lie under the squamous cells);
- melanoma starts in the melanocytes.

Melanoma is less common than squamous cell and basal cell skin cancers (sometimes called non-melanoma skin cancers). Melanoma can start in other places in the body where melanocytes are found, such as the eyes, mouth, or under the fingernails.

Early signs of melanoma are changes in the shape or color of existing moles. A mole, whose medical name is a melanocytic nevus, is a common benign growth of the color cells of the skin called melanocytes. Moles that are irregular in color or shape are suspicious of malignant or premalignant melanoma. Melanoma needs to be detected as early as possible for patients to have the best possible prognosis due to its aggressive tendency to invade other tissues of the body. Treatment is much less successful if the cancer is well advanced or has already metastasized to other areas.

7.1.2
How Is Melanoma Diagnosed?

To detect melanomas it is recommended to be aware of moles and check for changes (e.g., shape, size, color, itching, bleeding, and the like). Furthermore, a popular method for remembering the signs and symptoms of melanoma is the so-called mnemonic "ABCDE:"

- **A**symmetrical skin lesion; if the lesion was folded in half, the two sides would not match.
- **B**order of the lesion is irregular;
- **C**olors: melanomas usually have multiple colors;
- **D**iameter: moles greater than 6 mm are more likely to be melanomas than smaller moles;
- **E**nlarging: enlarging or evolving; a mole or skin lesion that looks different from the rest or is changing in size, shape, or color.

Therefore, any moles found that are notably altered, or asymmetric, more than one-quarter inch in diameter, have irregular borders or coloration, should be of special concern and be brought to the attention of a medical professional.

A biopsy is necessary to make a definite diagnosis of melanoma. The doctor will try to remove all of the unusual-looking growth or mole. The entire tumor along with a margin of tissue that is not a visible part of the tumor is removed. This type of biopsy is called an *excisional biopsy*. If the doctor cannot remove all of the growth, then a sample of the tissue will be removed. This is called an *incisional biopsy*. The removed tissue sample is examined under a microscope to determine if cancer cells are present and, if so, which kind. If an excisional biopsy is performed, the physician examining the sample should be able to determine how deeply the cancer has penetrated the skin. If the sample reveals melanoma cells then further testing is usually carried out to determine the extent of the cancer growth before a specific treatment plan is developed.

7.1.3
Present Problems for Biopsy

The excisional biopsy is the preferred method for removing lesions suspected to be melanoma. The main limitation of excisional biopsy is the problem of sampling error. The assumption that pathological examination of a few samples of tissue is representative of the pathological process involving the entire organ is not always correct. This is particularly true when there is a localized pathological process. Furthermore, when performing incisional biopsy, large volumes of the tissues are required to reduce the sampling error. It is unrealistic to take such amounts from patients suffering from diseases. A technique that enables visualization of *in situ* cellular detail in multiple areas would therefore provide more reliable information than that obtained by biopsy.

Conventional optical microscopes are often used for observing such data from tissues, including cells taken from patients. However, the cells within the tissues need to be chemically stained and fixed for optical microscopic observation. The staining and/or fixation usually affects the cells. Therefore, when those techniques are applied, it is difficult to understand the real effects of the treatment or the medicine, as the transformation in the tissue may only be visible in living cells. Further, it takes at least one week to prepare the specimen for pathological diagnosis.

The contrast mechanism of a mechanical scanning acoustic reflection microscope (SAM) is different from that of a conventional optical microscope [1]. Since staining and/or fixation to the specimen are not required for the SAM, the living cells activities may be observed directly [2] and the sample preparation process may be effectively reduced. Furthermore, a SAM can nondestructively observe not only the surface but also the internal structure of the specimen with sub-micron resolution when it is operated at a frequency near 1.0 GHz [3]. The SAM can also measure the mechanical properties (e.g., loss factor and modulus) of tissues [4]. Therefore, a high frequency ultrasonic diagnostic apparatus could provide pathological information about multiple areas of skin (e.g., epidermis and dermis) and enable a more accurate assessment of pathology. If sampling error is effectively reduced, the transformation of living cells and the real effects of treatment within the tissue can be observed more readily.

7.1.4
Objective of Present Study

Our present study aims is to investigate the feasibility of applying *in vivo* acoustic microscopy to the analysis of cancerous tissue. However, observation of an internal structure of a thickly sectioned biological tissue with the ultrahigh-frequency SAM is relatively difficult for the following reasons [4–26]. First, an ultrasonic wave is attenuated in proportion to the square of its frequency. Therefore, when the high frequency ultrasonic wave is emitted from the acoustic lens to the tissue through the coupling medium (e.g., water), the wave may not penetrate into the appropriate internal focal plane of the tissue. Even if the wave can penetrate into the internal focal plane, the wave may not reflect from the tissue to the acoustic lens with enough amplitude to form the acoustic image. Second, when a low frequency ultrasonic wave is used, the image of the appropriate internal focal plane of the specimen may be formed, but the resolution of the image may not be very useful. Most importantly, if the surface has irregularities or is rough, subsurface imaging is sub-optimal. Third, a biological tissue is acoustically transparent so that we cannot apply a conventional "time-of-flight" method.

To overcome the limitations, in the present study, first, imaging of thickly sectioned cancer tissues with acoustic lenses operating at frequencies in the range 25–50 MHz was found to provide useful information without some of the disadvantages of the ultrahigh-frequency systems. Lower frequency acoustic lenses provide better tissue penetration and stronger backscattered signals. However, to determine the ideal trade-off between resolution and penetration, it is important to establish the lowest frequency at which cellular imaging for identification of tissue pathology is feasible. The resultant images were then discussed to find appropriate techniques to improve scanning acoustic microscopy for melanoma cancer tissues with high frequency microscopy. Second, a thin cover plate was used to remove the effect of surface roughness of the specimen. Third, a sapphire (c-cut) was used as a substrate to optimize acoustical refection from the specimen.

Here, we present an approach for systematically predicting contrast in an acoustic image using thin and thick specimens sectioned from the same melanoma skin tissue with the SAM. We visualized the specimen with the shear wave polarized lens [7]. As a result, there is no change in the contrast due to different angles. Since the specimen is isotropic we can use an angular spectrum approach [5]; we developed a mathematical model to analyze amplitude of waveform of the transducer output for a five layer system. In addition, part of approach [i.e., $V(z)$ simulation algorithm] was found useful for quantitative estimates of the longitudinal velocity of the tissue, which provide the key quantity for estimating the modulus.

7.2
Physical and Mathematical Modeling for Five Layer Wave Propagation in an Acoustic Microscope

Figure 7.1 shows a schematic diagram of the geometry and coordinate system used in the mathematical modeling. The planes labeled 1 and 2 represent the back and

7.2 Physical and Mathematical Modeling for Five Layer Wave Propagation

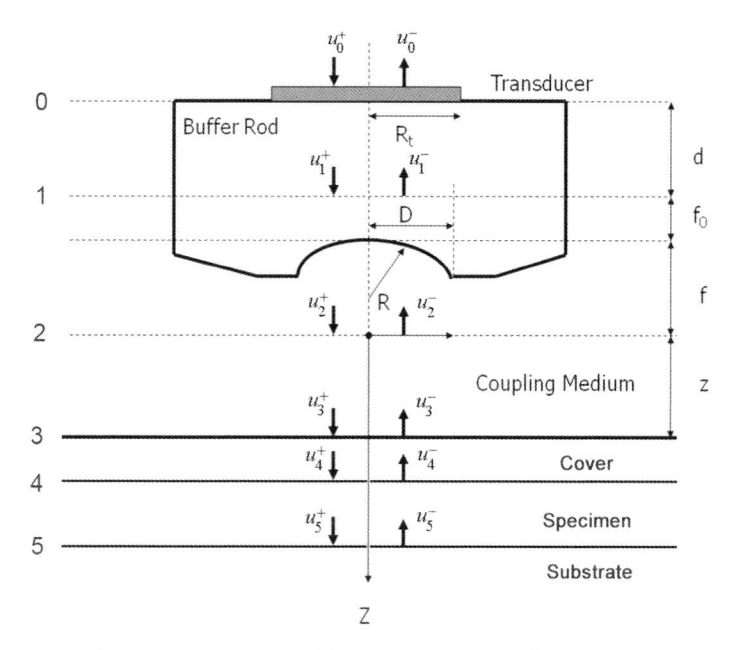

Figure 7.1 Schematic diagram of the geometry and coordinate system.

front *focal planes of the acoustic lens.* They are *not* at the same distance from the lens because different media are involved. Plane 3 is the plane of the surface of the cover. Plane 4 is the 1st interface between the cover and the front surface of the specimen. Plane 5 is the 2nd interface between the back surface of the specimen and the front surface of the substrate. Superscripts + or − indicate the direction of field travel, that is, in the +z or −z direction, respectively.

Since we are using a "time-of-flight" method, we focus a spherical acoustic wave onto an interface between a specimen and a substrate, and gate an electronic signal converted from the acoustic wave. Therefore, we must obtain the transducer output. The output is expressed as the following equation:

$$V(z) = \iint_{-\infty}^{\infty} u_0^+(x, y) u_0^-(x, y) dx dy \tag{7.1}$$

The transducer excites a uniform field [i.e., $u_0^+(x, y)$] when a unit voltage is applied at its terminals. Therefore, the transducer input may be determined when we know the properties of the transducer. Further, it is experimentally approximated to the Gaussian distribution; $u_1^+(x, y)$ is an acoustic field incident at the back focal plane (i.e., plane 1). The transducer plane (i.e., plane 0) is a distance "d" from plane 1. If we know d, the field $u_1^+(x, y)$ can be mathematically determined through implementing Fast Fourier transformation (FFT) of $u_0^+(x, y)$; $u_1^-(x, y)$ is an acoustical field reflected from the back focal plane. The transducer plane (i.e., plane 0) is a distance "d" away from plane 1.

Then, we need to calculate an acoustical field on each critical plane:

$$u_2^+(x, y) = \frac{e^{ik_0 f(1+\bar{c}^2)}}{i\lambda_0 f} \int\int_{-\infty}^{\infty} u_1^+(x, y) P_1(x, y) e^{-i\frac{2\pi}{\lambda_0 f}(x, y)} dx dy$$

$$= \frac{e^{ik_0 f(1+\bar{c}^2)}}{i\lambda_0 f} \mathcal{F}\{u_1^+(x, y) P_1(x, y)\}\bigg|_{\substack{k_x = \frac{k_0 x}{f} \\ k_y = \frac{k_0 y}{f}}}$$

(7.2)

where $k_0 = 2\pi/\lambda_0$ is the wave number in the coupling medium (i.e., deionized water), λ_0 is the wavelength of the coupling medium, f is the focal length of the lens, $\bar{c}$ is the ratio of the velocity of sound in the coupling medium to that in the solid (i.e., material of the buffer rod such as fused quartz), and $P_1(x, y)$ is the pupil function expressed as the following equation:

$$P_1(x, y) = \text{circ}\left(\frac{\sqrt{x^2 + y^2}}{R}\right) \quad (7.3)$$

where R is the radius of the pupil function of the lens.

Propagation of the wave beyond the focal plane is easily calculated if the angular spectrum representation is utilized:

$$U_2^+(k_x, k_y) = \mathcal{F}\{u_2^+(x, y)\}$$

$$= \mathcal{F}\left[\frac{e^{ik_0 f(1+\bar{c}^2)}}{i\lambda_0 f} \mathcal{F}\{u_1^+(x, y) P_1(x, y)\}\bigg|_{\substack{k_x = \frac{k_0 x}{f} \\ k_y = \frac{k_0 y}{f}}}\right] \quad (7.4)$$

$$= -i\lambda_0 f e^{ik_0 f(1+\bar{c}^2)} u_1^+\left(-\frac{f}{k_0}k_x, -\frac{f}{k_0}k_y\right) P_1\left(-\frac{f}{k_0}k_x, -\frac{f}{k_0}k_y\right)$$

$$U_3^+(x, y) = U_2^+(x, y) e^{ik_z Z} \quad (7.5)$$

$$k_z = \sqrt{k_0^2 - k_x^2 - k_y^2} \quad (7.6)$$

The lens is focused at the front surface of the substrate via the cover and the specimen. This means that the amplitude of the reflected signal from the 2nd interface is maximized, and gated for acoustical visualization:

$$U_4^+(x, y) = U_3^+(x, y) T_1^+\left(\frac{k_x}{k_1}, \frac{k_y}{k_1}\right) e^{ik_{z1} L_1} \quad (7.7)$$

$$k_{z1} = \sqrt{k_1^2 - k_x^2 - k_y^2} \quad (7.8)$$

7.2 Physical and Mathematical Modeling for Five Layer Wave Propagation

where T_1^+ is the transmittance function within the cover including attenuation factor experimentally obtained, $k_1 = 2\pi/\lambda_1$ is the wave number in the cover, and λ_1 is the wavelength of the cover:

$$U_5^+(x,y) = U_4^+(x,y)T_2^+\left(\frac{k_x}{k_2},\frac{k_y}{k_2}\right)e^{ik_{z2}L_2} \tag{7.9}$$

$$k_{z2} = \sqrt{k_2^2 - k_x^2 - k_y^2} \tag{7.10}$$

where T_2^+ is the transmittance function within the cover including the estimated attenuation factor, $k_2 = 2\pi/\lambda_2$ is the wave number in the specimen, and λ_2 is the wavelength of the tissue:

$$U_5^-(x,y) = U_5^+(x,y)\mathcal{R}\left(\frac{k_x}{k_2},\frac{k_y}{k_2}\right) \tag{7.11}$$

$$U_4^-(x,y) = U_5^-(x,y)T_2^-\left(\frac{k_x}{k_2},\frac{k_y}{k_2}\right)e^{ik_{z2}L_2} \tag{7.12}$$

$$U_3^-(x,y) = U_4^-(x,y)T_1^-\left(\frac{k_x}{k_1},\frac{k_y}{k_1}\right)e^{ik_{z1}L_1} \tag{7.13}$$

$$U_2^-(x,y) = U_3^-(x,y)e^{ik_z Z} \tag{7.14}$$

$$u_2^-(x,y) = \mathcal{F}^{-1}\{U_2^-(k_x,k_y)\} \tag{7.15}$$

$$u_1^-(x,y) = \frac{e^{ik_0 f(1+\bar{c}^2)}}{i\lambda_0 f}\iint_{-\infty}^{\infty} u_2^-(x,y)P_2(x,y)e^{-i\frac{2\pi}{\lambda_0 f}(x,y)}dxdy$$

$$= \frac{e^{ik_0 f(1+\bar{c}^2)}}{i\lambda_0 f}P_2(x,y)\mathcal{F}\{u_1^+(x,y)\}\bigg|_{\substack{k_x=\frac{k_0 x}{f}\\k_y=\frac{k_0 y}{f}}} \tag{7.16}$$

$$= \frac{e^{ik_0 f(1+\bar{c}^2)}}{i\lambda_0 f}P_2(x,y)U_2^-\left(\frac{k_0}{f}x,\frac{k_0}{f}y\right)$$

By using Equation (7.1), one can carry out simulation calculations to predict the nature of the received signal. However, we need to input the longitudinal wave of the specimen. Since the specimen is compressed by the cover and the substrate it is difficult to measure the wave velocity. We will describe how to obtain the wave velocity in a later section.

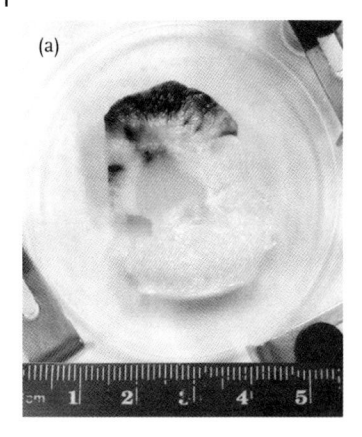

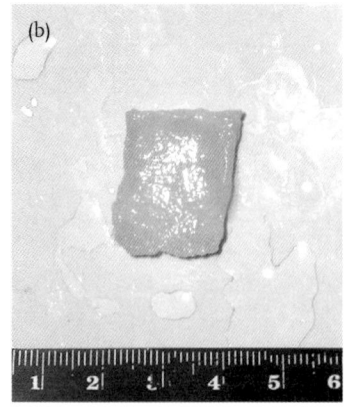

Figure 7.2 Optical images of thickly sectioned abnormal and normal skin tissue. The images were formed by a digital camera (Olympus): (a) thickly sectioned melanoma tissue including groups of cancer cells; (b) thickly sectioned normal skin tissue.

Table 7.1 Comparison of acoustic impedance, Z.

Material	Z (kg m^{-2})
Polystyrene	2.52×10^6
Silica glass	13.0×10^6
Sapphire	44.3×10^6

7.3
Sample Preparation

Figure 7.2 shows a thickly sectioned skin tissue specimen located on the surface of a substrate made of c-cut sapphire with a thin transparent polysilicone (thickness: 500 μm) cover pressed onto the tissue by a holder made of aluminum. The cover makes the surface of the specimen flat, thereby reducing effects on the acoustic image caused by surface roughness. The cover also prevents the specimen from floating into the water. The chamber is filled with a coupling medium (i.e., deionized water). To obtain better acoustic reflection, the specimen is located on a substrate made of a highly reflective material such as a sapphire (Table 7.1).

Figure 7.2b shows a thickly sectioned normal skin tissue. The tissue is consistent in color and softness. The tissue is 3.0 mm thick. Figure 7.2a shows a thickly sectioned abnormal skin tissue. The abnormal tissue has a firmer consistency, is substantially flatter, and has less surface roughness. When compared to normal tissue, it is also 3.0 mm thick. The tissue includes, in fact, black, white, and yellow

Figure 7.3 Setup of the specimen in the chamber.

colored portions. The black colored portions have been diagnosed as melanoma carcinoma. In addition, the carcinoma area has less surface roughness than the normal sections.

7.4
Digital Imaging – Optical and Ultrasonic

The thickly sectioned melanoma tissue is placed on the sapphire substrate; it is then covered by a transparent cover made of polysilicone, and is pressed by the holder toward the substrate to avoid artifacts in the acoustic image caused by surface roughness (Figure 7.3).

7.4.1
Optical Image

Figure 7.4a and b are comparisons of the optical images of thickly sectioned melanoma and normal skin tissues. These images are formed with a stereoscopic microscope (Olympus, Model: SZX16). The optical image shows only the surface information of the thick tissues. The surface of melanoma tissue varies in contrast, which indicates the specimen has elastic discontinuities. However, the normal skin tissue is consistent in contrast, meaning that the internal structure of the specimen is continuous.

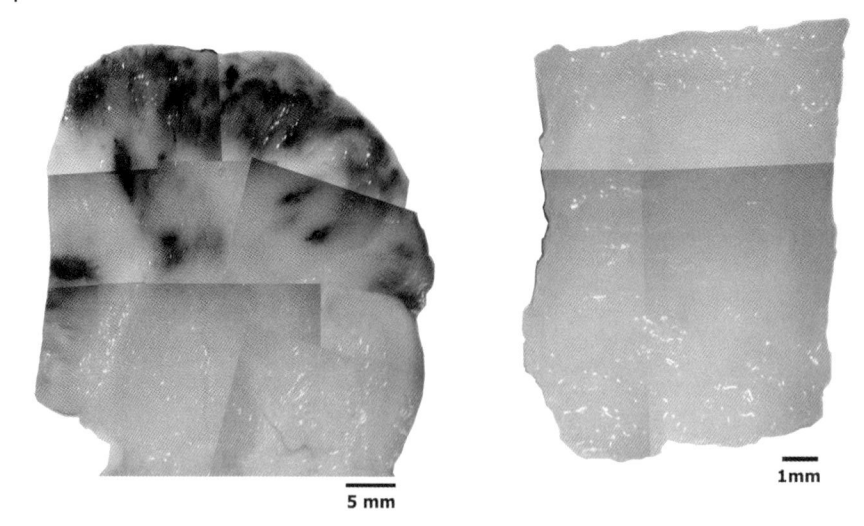

Figure 7.4 Optical mosaic images of both thickly sectioned melanoma and normal skin tissues (thickness: 3 mm): (a) optical image of the melanoma skin tissue; (b) optical image of the normal skin tissue.

By comparison with the normal tissues, the appearance of the black coloration in the abnormal tissue is evidently recognized as melanoma. The optical image of the abnormal tissue shows that the full extent of the carcinoma tumors spread mostly in the epidermis and the dermis, as well as some that have scattered into the hypodermis and appeared around the cartilage, which is present in the yellow zone (bottom left of the figure) that is surrounded by mostly fat. Within this zone, cartilage is merged with adjacent supporting tissue containing adipocytes, capillaries, and small nerves.

7.4.2
Acoustic Imaging Principle (Pulse-Wave Mode)

Figure 7.5 shows the schematic diagram of the SAM (pulse-wave mode); the imaging mechanism of the pulse-wave mode is described below.

An electrical signal (i.e., pulse wave) generated by a RF source is transmitted via a circulator to a piezoelectric transducer (i.e., $LiNbO_3$) located on the top of a long buffer rod designed for the time-of-flight method. The input voltage from the source to the transducer is approximately 100 V. The electrical signal is converted into an acoustic signal (i.e., ultrasonic plane wave) by the transducer. The ultrasonic plane wave travels through the buffer rod made of fused quartz to a spherical recess (hereinafter called simply the "lens") located at the bottom of the buffer rod. The lens converts the ultrasonic plane wave into an ultrasonic spherical wave (i.e., ultrasonic beam).

7.4 Digital Imaging – Optical and Ultrasonic

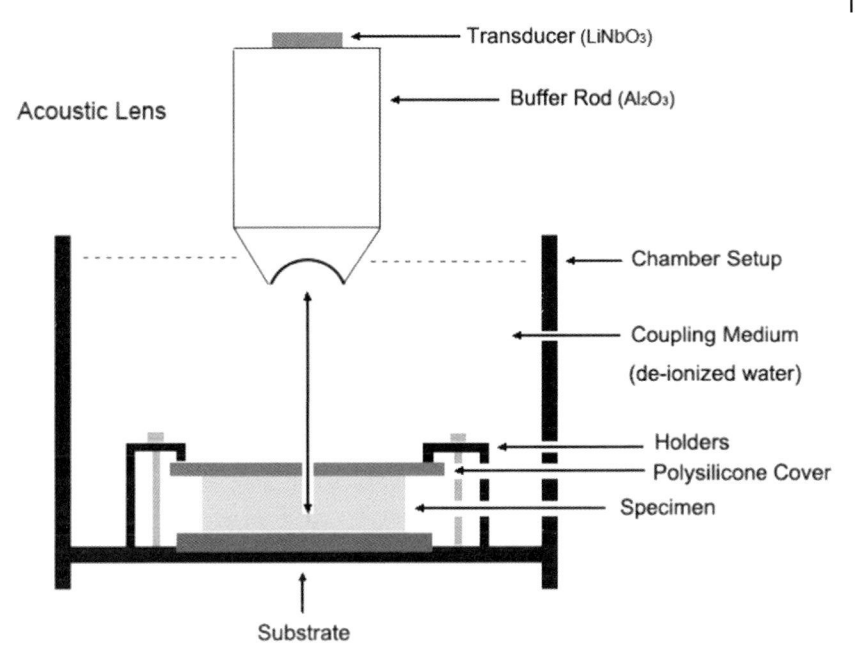

Figure 7.5 Schematic diagram of the experimental setup for the SAM (pulse-wave mode).

The specimen is located on a c-cut sapphire substrate (thickness: 1.78 mm), and is submerged in deionized water. A biological tissue with acoustic impedances close to that of water has virtually no contrast because differences in reflection coefficients are displayed. In other words, the contrast in the acoustic image of the tissue is generated from the variation in attenuation from tissue structure to tissue structure as the acoustic waves propagate within the tissues. Therefore, it was necessary for the specimens to be mounted onto substrates composed of highly reflective materials (e.g., sapphire, fused quartz, or the like) as a background so as to maximize the ultrasound signal reflection.

The ultrasonic beam is focused on the back surface of the specimen via a thin transparent poly-silicon cover, and reflected from the back surface. It has various properties such as good electrical insulation, thermal stability, low toxicity, low chemical reactivity, etc., which are suitable for medical preparation and applications. In addition, it is stable for months and even years and, unlike other materials, insensitive to rough handling. The cover is substantially 0.5 mm thick. The cover (i.e., Cover Well™ imaging chamber gasket with adhesive) is manufactured by Molecular Probes. The cover was fixed by an aluminum holder. When the specimen does not have remarkable surface roughness it is not necessary to place the cover for visualization. The reflected ultrasonic beam, which carries acoustic information of the specimen, is again converted into an ultrasonic plane wave by the lens. The ultrasonic plane wave returns to the transducer through the buffer rod. The ultrasonic plane wave is again converted into an electric signal by the transducer.

Suppose that a pulse wave emitted from an acoustic lens via a coupling medium (i.e., deionized water) is focused onto the back surface of the specimen. In addition, suppose that the pulse wave is strong enough to travel through the specimen and reflect back to the acoustic lens. The timing of each reflection is different because the distance traveled is different. The waveform includes information such as amplitude, phase, or delay, and can be monitored by an oscilloscope (Figure 7.6). The reflected pulse wave is electronically gated out to visualize the portion of a horizontal cross-sectional image by horizontally scanning the acoustic lens.

The acoustic lens is able to translate axially along the z-direction, varying the distance between the specimen and the lens for clearly focused subsurface focusing so as to obtain clear visualization – that is, when the acoustic lens is focused on the surface of the specimen, the amplitude of the reflected signal from the surface is optimized (we denote $z = 0\,\mu m$). Then, we look for a reflected signal from the specimen. When focusing the acoustic signal reflected from the subsurface plane of the specimen, the acoustic lens is mechanically defocused toward the specimen (we denote $z = -x\,\mu m$, where x is the defocused distance) till the amplitude of the reflected signal is optimized.

To form a two-dimensional acoustic image, an acoustic lens is mechanically scanned across a certain area of the specimen.

An appropriate acoustic lens must be selected by considering its penetration depth in a specimen and the resolution in an image obtained from the interior plane of the specimen.

In this setup, a thickly sectioned biological specimen is sandwiched between a thin transparent polysilicone cover and a sapphire substrate. Four reflections of the acoustic waves are shown in Figure 7.6. The reflections are from the surface of the cover, interface between the cover and the front surface of the specimen, the interface between the back surface of the specimen, and the surface of the substrate, respectively. The thickness of specimen is measured as substantially 2.1 mm.

The 1st signal is the reflected wave from the surface of the cover. The 2nd signal is the reflected wave from the 1st interface between the back surface of the cover and the surface of the specimen. The 3rd signal is the reflected wave from the 2nd interface between the back surface of the specimen and the front surface of the substrate. The 4th signal is the reflected wave from the back surface of the substrate.

To form an interior acoustic image, a gate is set on an appropriate reflected signal. Selecting the appropriate reflected signal from the specimen is an important step in the experiment. This step can be implemented by observing a change of signal amplitude when moving the acoustic lens along its z-axis. When the acoustic lens is focused on the interior plane of the specimen, the amplitude of the wave from the plane is maximized.

In this study, the 3rd reflected wave is gated to visualize the specimen interior. Since the specimen is acoustically transparent, no image can be formed by gating the reflected signal from the 1st interface.

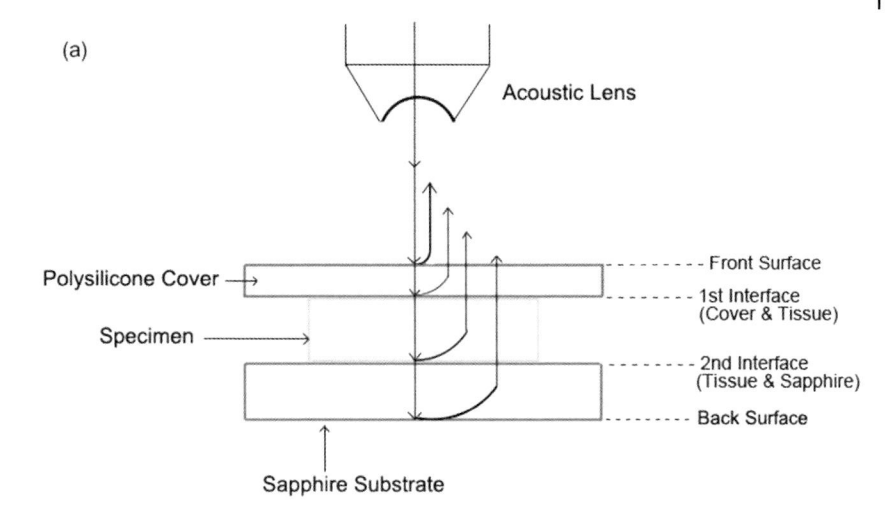

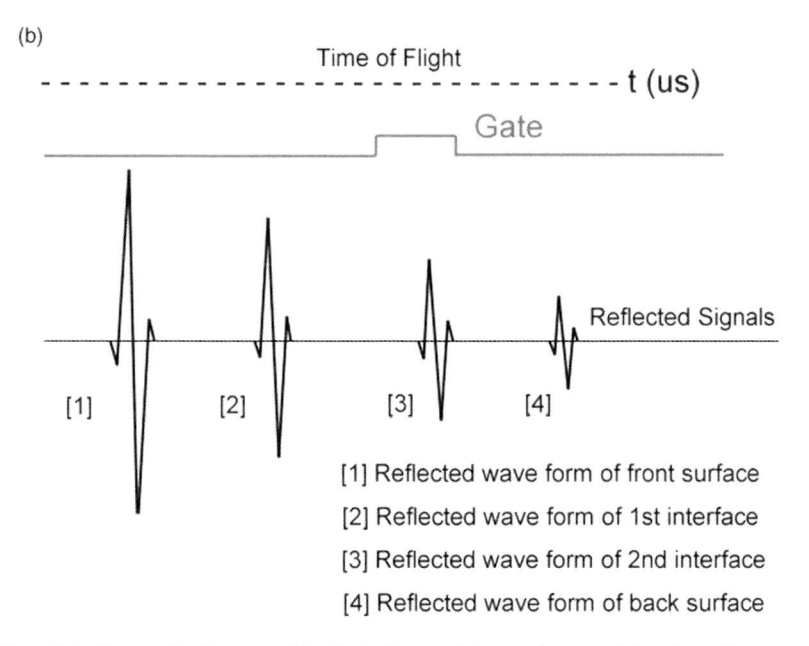

Figure 7.6 Schematic diagrams: (a) reflected acoustic beams from each interface of a specimen system; (b) acoustic waves from a specimen located on the substrate via a cover – the diagram shows the timings of reflected pulse waves reflected from a specimen.

7.4.3
Resolution

Two types of resolution must be considered for the SAM. One is a lateral resolution (i.e., Δr), and the other is a vertical resolution (i.e., $\Delta \rho$). They are expressed as follows:

$$\Delta r = F\lambda = F\left(\frac{v_w}{f}\right) \tag{7.17}$$

$$\begin{aligned}\Delta \rho &= 2F^2\lambda = (2F^2)\left(\frac{v_w}{f}\right) \\ &= 2\left(\frac{f_0}{D}\right)^2\left(\frac{v_w}{f}\right) \\ &= 2\left(\frac{1}{2\tan\theta}\right)^2\left(\frac{v_w}{f}\right) \\ &= \frac{1}{2\tan^2\theta}\left(\frac{v_w}{f}\right)\end{aligned} \tag{7.18}$$

where F is a constant related to lens geometry (i.e., F-number), λ is the wavelength in the coupling medium (i.e., deionized water), f is the frequency of the wave generated by the transducer, v_w is the longitudinal wave velocity in the coupling medium, f_0 is the focal distance of the lens, D is the diameter of the lens aperture, and θ is one-half of the aperture angle of the lens.

To observe cellular details within the tissue, the maximum lateral resolution, Δr, has to be at least the same size as a cell, which is about $10\,\mu m$. For example, suppose the acoustic wave velocity in water is $1500\,m\,s^{-1}$, and the F-number is 0.7. Then the lowest frequency for pulse-wave mode to visualize the cells is approximately 100 MHz. Therefore, when the frequency of the transducer output signal is less than 100 MHz, the pulse-wave mode apparatus with the conventional C-scan technique is not capable of visualizing the cells.

Referring to Equation (7.17), we know that the resolution of the image formed by the SAM is determined by the frequency of the ultrasonic wave, the velocity of the coupling medium, and the lens geometry. Therefore, one approach to increase resolution is to increase the acoustic frequency. However, as can be appreciated, the ultrasonic wave is attenuated in proportion to the square of its frequency; therefore, when a high-frequency ultrasonic wave is applied as a probe, the wave will not penetrate with a significant depth inside the specimen, and the internal information may not be obtained. Consequently, we need to know the upper limitation of the frequency that can be applied to the biomedical imaging of thickly sectioned tissues.

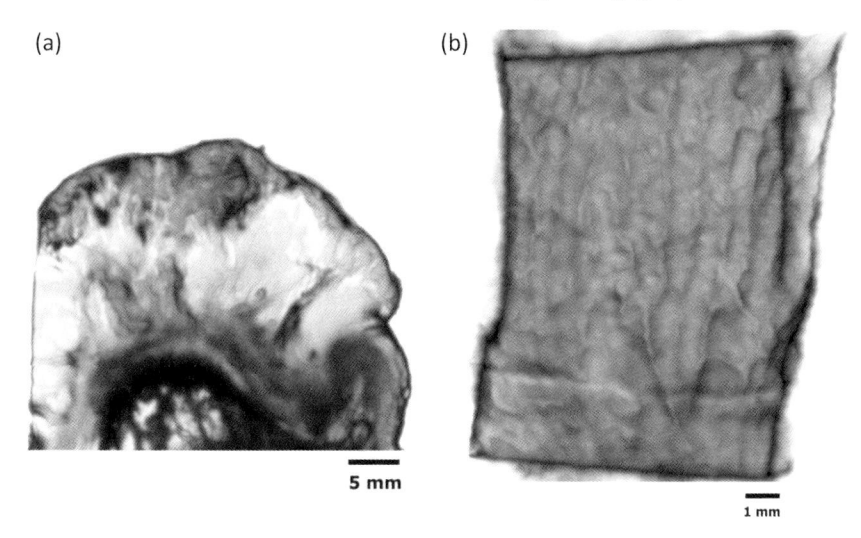

Figure 7.7 (a) Acoustic image (C-scan) of a thickly sectioned melanoma skin tissue; (b) normal skin tissue. The tissues are 3 mm thick. Operating frequency for pulse-wave mode is 50 MHz.

7.4.4
Acoustic Images

Figure 7.7a and b show acoustic C-scan images of thickly sectioned abnormal melanoma and normal skin tissues (horizontal cross-sections; thickness: 3 mm), respectively, taken with an acoustic lens (Olympus NDT) with an operating frequency at 50 MHz. The input frequency was 50 MHz, which was the maximum frequency emitted from the acoustical lens to the specimen. The acoustic beam was focused at the interface between the specimen and the substrate [i.e., $Z = 0\mu m$ or $V(0)$]. The resolution of the image is limited due to the frequency of the acoustical lens. At this point we are unable to diagnose the melanoma cancer cells from the acoustic images of the thick tissues. However, they show both the surface and the internal structures of the thickly sectioned tissue. Since the images cannot review cellular details of the thickly sectioned tissue, we need to select an acoustic lens capable of operating at a frequency at 50 MHz or more.

Furthermore, since acoustic properties (i.e., reflection coefficient, attenuation, and velocity of acoustic wave) and the surface condition (i.e., surface roughness and discontinuities) of the specimen are factors in forming acoustic images, it is also significantly important to study the acoustic images to obtain the acoustic properties and surface condition. Therefore, waveforms, FFT analysis, and B-scan images need to be performed and reviewed attentively.

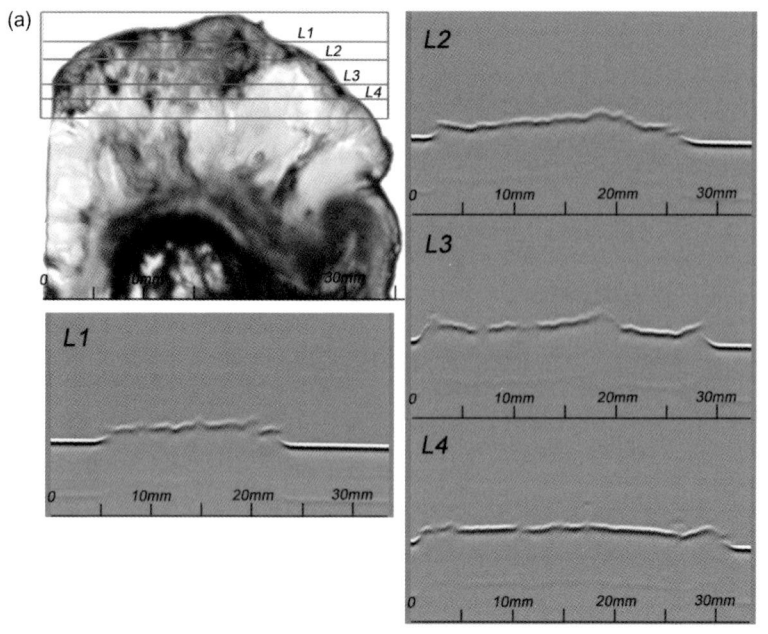

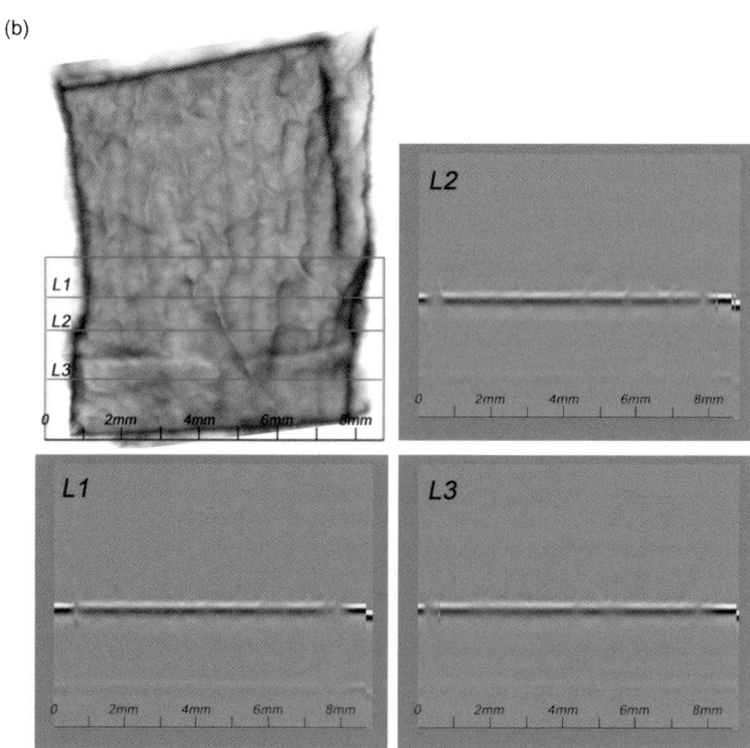

Figure 7.8 B-scan images of thickly sectioned abnormal and normal skin tissues, wherein scanning positions are selected from the C-scan image of the tissues. (a) Melanoma tissue (frequency: 50 MHz; mode: pulse-wave mode); (b) normal skin tissue (frequency: 50 MHz; mode: pulse-wave mode).

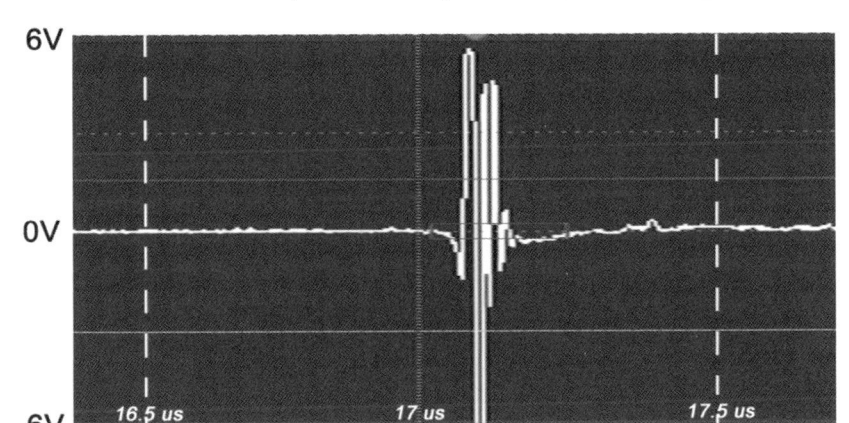

Pulse Width: 17.098us, 12.678 mm Depth: 0.005us, 0.021mm Amp: 100% FSH

Figure 7.9 Reference for waveform analysis. Waveform of the reflected wave from the surface of the sapphire substrate via water shows that its amplitude is 100% FSH (full screen height; 12 V_{pp}).

A SAM B-scan image can form a vertical cross-sectional 2D image of a specimen. Figure 7.8a and b are B-scan images of melanoma and normal skin tissues, respectively, and show the surface condition of the normal and abnormal tissues. Figure 7.8a clearly illustrates the surface is much rougher. The surface structure of the normal tissue is more uniform and consistent

7.4.5
Waveform Analysis

The insertion technique [21] is used to measure the acoustic properties of different materials. This technique is a relative measurement method that employs water as the reference to study the transmission of longitudinal ultrasonic waves through solid media embedded in an aqueous environment (i.e., water). Therefore, the waveform of the wave reflected from the substrate via water was taken as a reference waveform (Figure 7.9). In this case, the acoustic beam is focused on the sapphire substrate, and because of high reflectivity the reflected signal carries relatively high energy and its waveform has an amplitude of 100% full screen height (FSH). The echo amplitudes represent the product of the local scattering strength and the attenuation loss factor, which takes into account scattering in all directions and absorption. Based on the difference in contrast, three points were chosen (i.e., P1, P2, and P3 in Figure 7.10). The waveforms corresponding to these points were acquired (Figure 7.10) and the Fast Fourier transforms obtained (Figure 7.11) [8, 9]. The amplitudes of the waveforms are given in Table 7.2.

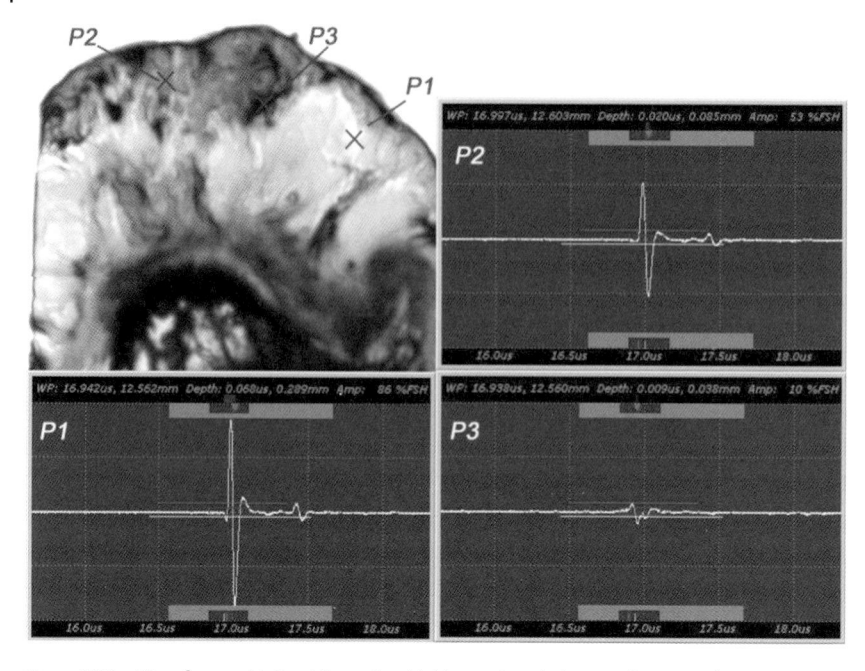

7 Study of the Contrast Mechanism in an Acoustic Image

Figure 7.10 Waveforms obtained from the thickly sectioned abnormal tissue. The tissue includes melanoma cancer cells that were sandwiched between the polysilicone cover and the sapphire substrate.

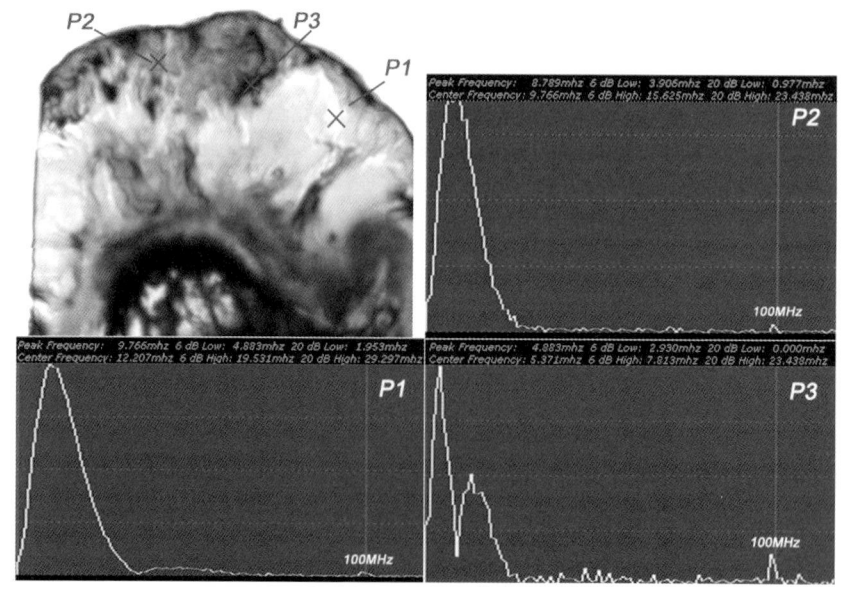

Figure 7.11 FFT of thickly sectioned specimen including melanoma cancer cells located between the polysilicone cover and the sapphire substrate.

Table 7.2 Amplitude of the waveform (A) obtained at the points shown in Figure 7.10.

	Amplitude A
P1	86% FSH
P2	53% FSH
P3	10% FSH

Table 7.3 Location, signal amplitude, relative attenuation, and center frequency from fast Fourier transform of waveform (A) and frequency shift due to attenuation.

Location	Signal amplitude A	Relative attenuation (dB)	Center frequency (MHz) (shift, MHz)
P1 normal	86% FSH	−1.32	12.2 ($\Delta = 14.1$)
P2 transitional	53% FSH	5.52	9.77 ($\Delta = 16.5$)
P3 abnormal	10% FSH	−40.0	5.37 ($\Delta = 20.9$)
Reference	100% FSH	0	26.3 ($\Delta = 0$)

Before imaging and measurements were performed the specimens were checked for elastic anisotropy. A method, previously described [7], using shear wave transducers was employed to observe any angular dependence of the images. No anisotropy was found and the measurements could proceed on the basis that the tissue was essentially isotropic.

It can be seen (Figure 7.11) that the frequency of the acoustic beam returned from the thickly sectioned abnormal tissue is less than 15 MHz because of the high attenuation of the tissue including cancer cells. Table 7.3 shows the location, signal amplitude, relative attenuation, center frequency from the fast Fourier transform of waveform (A), and the frequency shift due to attenuation. It can be concluded that emitting a 50 MHz frequency acoustic beam is not an appropriate way to form highly resolved acoustic image in the "time-of-flight" method. Since the skin tissues are acoustically transparent, there is no ultrasonic reflection from the surfaces of the tissues. The contrast of the tissue in an acoustic image is primarily generated from differences in attenuation. High attenuation causes significant signal loss; therefore, by comparing the signal amplitudes of the waveforms, the highest attenuation occurs at P1 and the lowest at P3 (Table 7.3).

7.5
High Frequency Acoustic Microscopy

7.5.1
Normal Control Skin Tissue

The acoustic image was formed by the acoustic lens (Olympus, Model: AL4M631, frequency: 400 MHz, transducer: ZnO, buffer rod: sapphire, aperture angle: 120°, working distance 310 μm). The acoustic lens was focused onto the surface of a soda-lime glass substrate. Therefore, the acoustic image includes information stemming from the region ranging from the front surface to the back surface.

As noted above, two of the main layers of the human skin are the epidermis (i.e., an outer keratinizing stratified squamous epithelium) and the dermis (i.e., an underlying touch supporting and nourishing layer of fibroelastic tissue). As can be seen in Figure 7.12, both layers are clearly identifiable. The top layer, which is presented in light gray, is the thin epidermis with an overlying layer of loose keratin. The external surface of the epidermis is fairly smooth and flat, but the junction between epidermis and thick dermis is marked by downward folds of epidermis. The thick dermis, which just beneath the epidermis, contains large compact fine collagen and elastic fibers and small blood vessels. Melanocytes, as seen in the optical micrograph, are responsible for the synthesis and release of

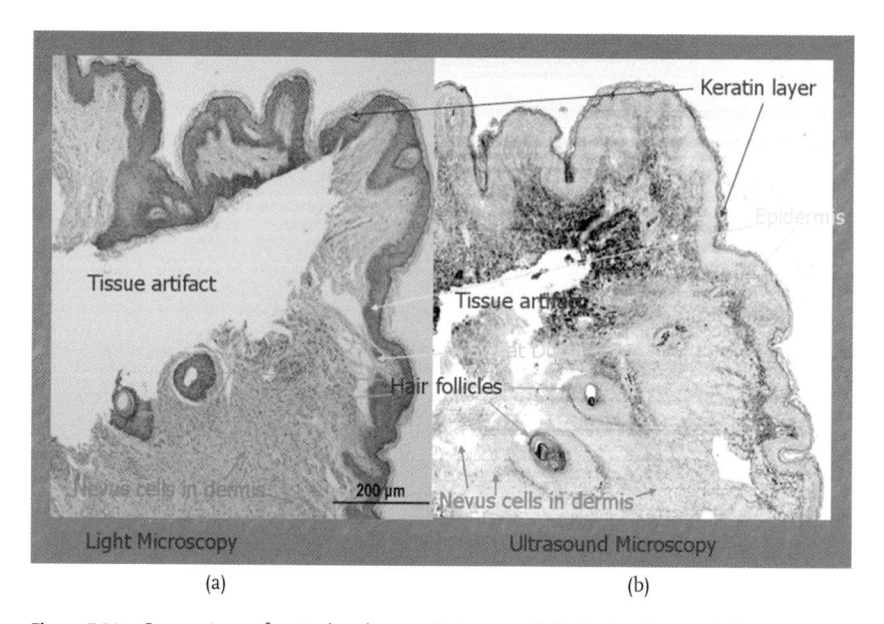

Figure 7.12 Comparison of optical and acoustic images of thinly sliced normal skin specimen: (a) optical image formed at 4× objective lens; (b) acoustic image scanned at 400 MHz.

the brown pigment melanin, which is largely responsible for skin coloration. They are located in the basal layer of the epidermis and scattered infrequently inside of the dermis that appear as dark blue round cells with pale-staining cytoplasm. The skin appendages such as sweat glands and hair follicles reside within the dermis. The hair follicles are essentially cylindrical downgrowths of the surface epithelium ensheathed by collagenous tissue; and sweat glands are simple, coiled tubular glands that secrete a watery fluid onto the skin surface by the process of merocrine secretion. Both of them are easily discernible from the acoustic images by comparison with the optical image as a reference. However, the melanocytes (i.e., nevus cells) and other cellular details in all skin layers of the acoustic image are poorly reviewed.

7.5.2
Abnormal Skin Tissue

For abnormal skin tissue, the series of micrographs in Figure 7.13 compare optical and acoustical images made at 40×, 100×, and 400× original magnifications. Because the melanocytes and other cellular details were not fully observed in acoustic image of normal skin tissue above (Figure 7.12), the acoustic images of abnormal skin tissues were formed with a higher input frequency of 600 MHz.

Figure 7.13a reviews the general structure of the thin skin tissue; all skin layers can be easily identified due to the contrast differences for both optical and acoustic images. The cornified layer (C) that appears in the darkest outer layer in the acoustic image–also called the keratin layer–is mainly composed of fibrous protein and loose keratin. The external surface of the underlying layer, which is the epidermis (E), is seen to be fairly smooth and flat. The dermis (D) consists of two layers: the papillary dermis (PD) just beneath the epidermis, which contains fine collagen and elastic fibers and small blood vessels, and the thicker reticular dermis (RD), which contains large compact collagen fibers and thick elastin fibers. The carcinoma tumors are readily distinguishable at a higher magnification in Figure 7.13b. The stratum basale (or the basal layer) (B) in the epidermis can also be discerned in the acoustic image at this level, where the cells are small uniform and darkly stained in the basal epidermis shown in the optical image. Figure 7.13c shows the carcinoma tumor residing underneath the epidermis, which is surrounded by loose connective tissues. The melanocytes, which are located in the basal layer and appear as round cells with cytoplasm scattered infrequently between the low columnar basal cells, are seen in both images.

7.5.3
Acoustic Velocity

Figure 7.11 shows the reflected waves from the three layered structure (i.e., the polysilicone cover, the tissue, and the sapphire substrate) of the specimen system

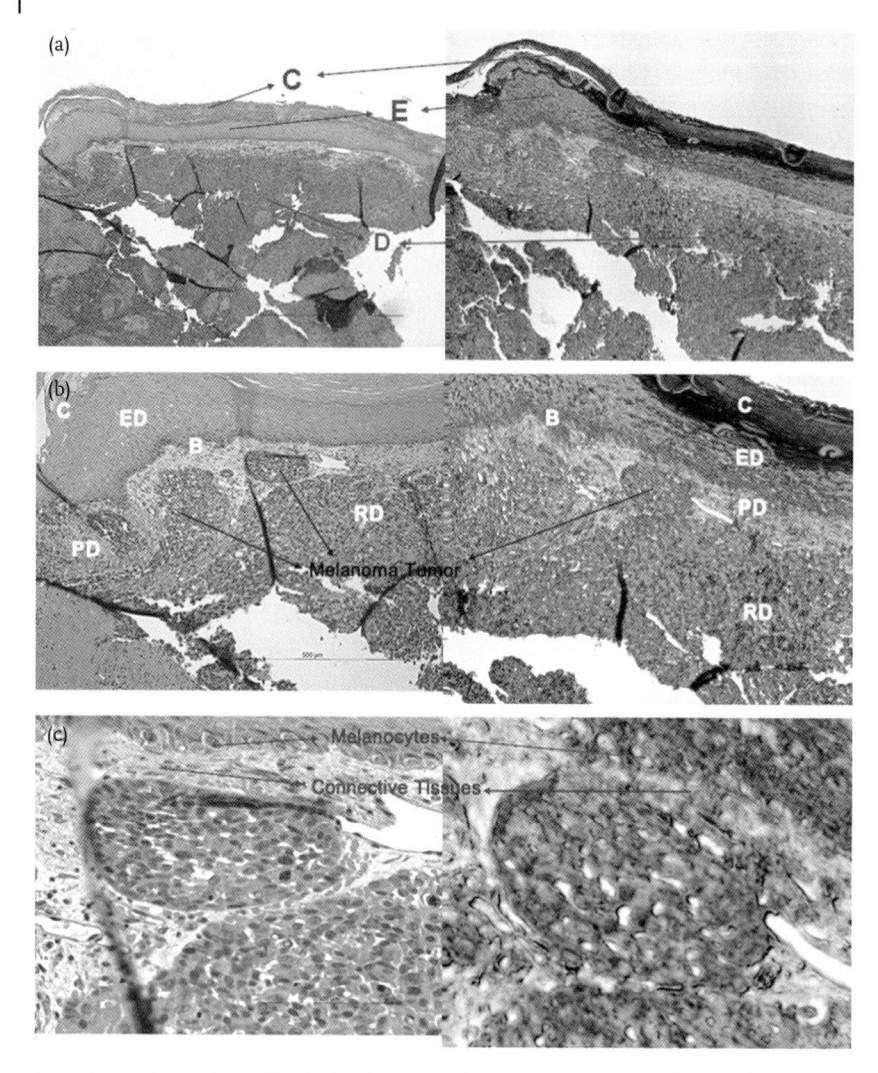

Figure 7.13 Comparison of optical and acoustic images of thinly sliced *abnormal* specimen. Left images are optical; right hand images are acoustic images. Operating frequency: 600 MHz; defocusing distance: $z = -3.5$ um; temperature: 21.5 °C. Optical: (a) 400×, (b) 100×, and (c) 40× objective lens; acoustic: scanning area $x = 0.25$ mm.

(i.e., A-scan). The longitudinal wave velocity of the tissue (denoted simply as "V_{tissue}") can be calculated by the following equation:

$$V_{tissue} = \left(\frac{1}{V_{water}} - \frac{\Delta t}{\Delta x} \right)^{-1} \tag{7.19}$$

where V_{water} is the longitudinal wave velocity in water, Δt is the time of flight from 1st interface to 2nd interface, and Δx is the thickness of the tissue.

Note that the tissue is compressed during the experiment. Therefore, the real thickness of the tissue is relatively difficult to measure precisely. Furthermore, without the cover, no signal from the surface of the tissue can be seen. The longitudinal wave velocity of the tissue can be more accurately obtained with $V(z)$ analysis and is described below.

Ever since the advent of mechanical scanning acoustic reflection microscope (SAM) [12], a key objective has been quantitative data acquisition besides the enhancement of resolution in the acoustic images. For quantitative data acquisition, the $V(z)$ curve techniques have been developed and applied to various materials with fruitful results [13–19]. However, it is not easy to characterize soft materials by the $V(z)$ curve technique. First, the critical angle of the Rayleigh wave generation of the soft materials is generally high. Therefore, the Rayleigh wave is often not generated within a specimen, even though an acoustic lens having a high numerical aperture (e.g., 120°) is used. This means that no $V(z)$ curve is formed. Second, attenuations of the soft materials are high. Therefore, the $V(z)$ curve may not have enough oscillations for the FFT analysis to measure the Rayleigh wave velocities accurately, even though the Rayleigh wave is generated. When an acoustic lens, having a high numerical aperture and a long working distance, with a low frequency (e.g., 10 MHz or less) is used for soft materials, the $V(z)$ curve might be formed. However, in this case, an advantage of the $V(z)$ curve technique for characterizing a small area of a material is lost.

There may be a simple solution for applying the $V(z)$ curve techniques to the soft materials when the soft material is treated as a thin film coated on an isotropic substrate (e.g., sapphire, fused quartz, silica glass, or the like). Then, the reflectance function can be obtained by using the theory of ultrasonic propagation in layered media [20]. Using the reflectance function, the $V(z)$ curve for the thin soft material mounted onto the substrate can be simulated. In the simulation, only the velocity of the longitudinal wave of the soft material is set by estimation. The actual velocity of the longitudinal wave of the soft material can be obtained by matching the $V(z)$ curve obtained from the experiment in an iterative procedure. This chapter presents analyses and laboratory results for the above case.

7.5.4
Computer Simulation

7.5.4.1 Experimental $V(z)$ Curve

The main procedures for characterizing the surface acoustic wave (SAW) velocity of the specimen by the $V(z)$ curve method is illustrated in Figure 7.14. Figure 7.14a and b show the major components contributing to the $V(z)$ curve in Figure 7.14c conceptually. We now assume that the entire transducer output $V(z)$ can be approximately represented by combing ray theory for the leaky SAW component and field theory for the directly reflected component in the negative z region. Once the $V(z)$ curve has been smoothed, we window the smoothed curve by selecting

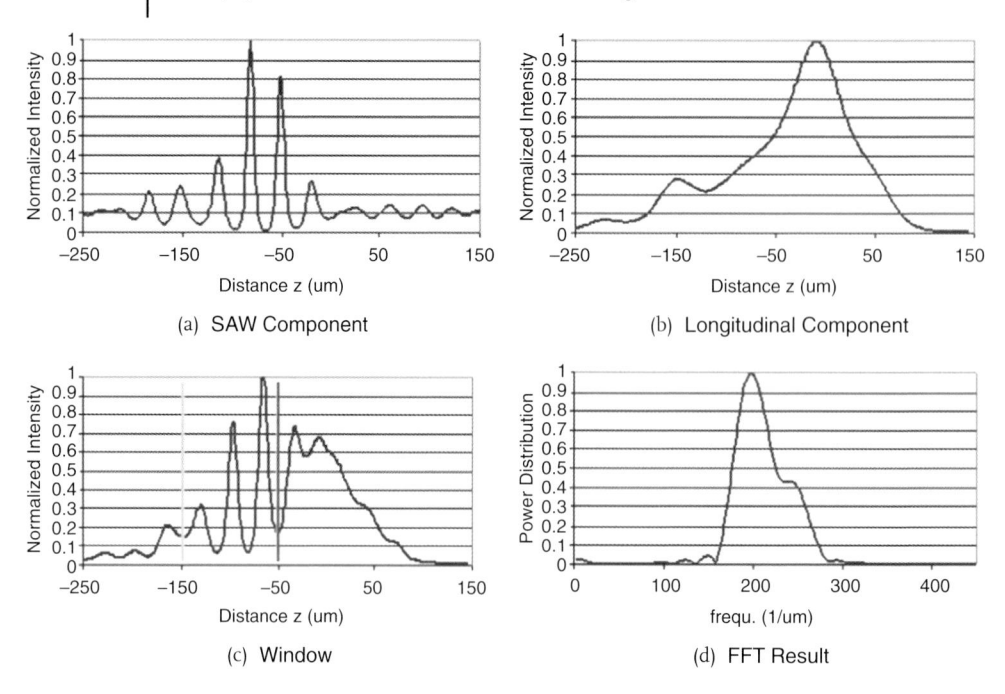

Figure 7.14 SAW velocity acquisition with FFT.

(a) SAW Component
(b) Longitudinal Component
(c) Window
(d) FFT Result

its first 3–5 oscillations. The SAW velocity is obtained from the power FFT result by transforming the V(z) curve into a frequency domain.

Figure 7.15 shows the original V(z) curves obtained at both normal and abnormal points of the thin melanoma tissue mounted on the soda-lime glass substrate, while the corresponding FFT analysis are shown in Figure 7.16.

7.5.4.2 Theoretical V(z) Curve (Simulation of V(z) Curve)

Following well-known procedures to obtain, measure, and interpret V(z) curves [27–30] the longitudinal wave velocities were obtained by implementing a computer parameter-fitting technique. The algorithm of the V(z) curve simulation is described in the following steps. First, initialize parameters of the acoustic lens, specimen (thinly sectioned biological tissue), and soda-lime substrate. Second, calculate parameters of the acoustic field at the back focal plane, pupil function of the lens, and reflectance function. Third, calculate and draw the V(z) curve. For the simulation, the parameters of the acoustic lens are shown in Table 7.4.

The velocity of the longitudinal wave of water was set as $1487\,\mathrm{m\,s^{-1}}$. Based on previous research [21], the speed of sound in most normal tissues of the human body is quite constant and close to that in water, and the ultrasound velocities in

(a)

Abnormal Point V(z) Curve

(b)

Normal Point V(z) Curve

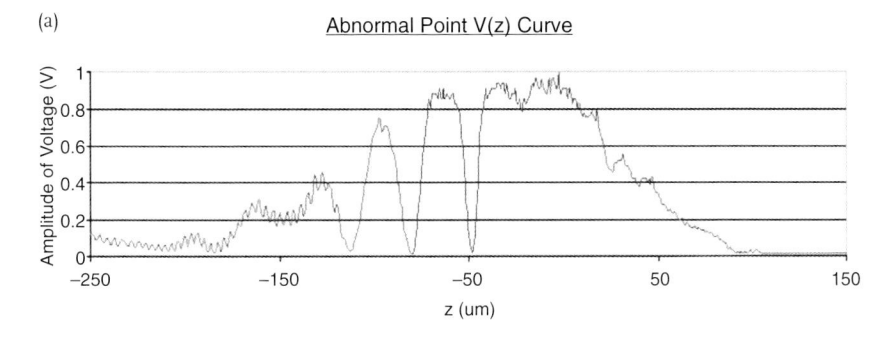

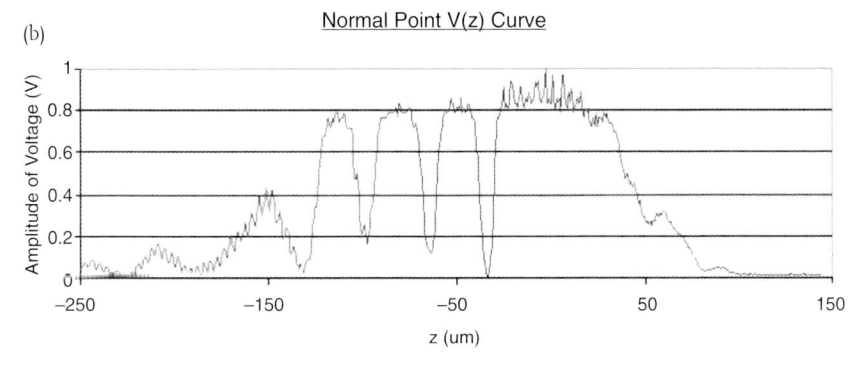

Figure 7.15 V(z) curves obtained at (a) abnormal and (b) normal points of the melanoma thin tissue on soda-lime glass substrate. (Input frequency: 200 MHz; thickness of the specimen: 5 μm; temperature: 21.5 °C.)

malignant melanoma tissues varied between 1553 and 1588 m s^{-1} with a mean of 1564 m s^{-1}.

Therefore, the longitudinal velocity of the normal/abnormal skin tissue (thickness: 5 μm) was set in the range 1540–1590 m s^{-1}. Moreover, the human skin density is set within the range of $1.11–1.19 \times 10^3$ kg m^{-3} [22]. Soda-lime glass was chosen as the substrate; the velocities of its longitudinal and shear waves are 6000 and 3200 m s^{-1}, respectively. Table 7.5 summarizes the physical parameters of the biological tissue used for the computer simulation.

As can be seen in Figure 7.17, from the FFT analysis, the SAW velocities of the specimen system for abnormal and normal skin tissues/soda-lime glass were obtained as 3248.02 and 3162.90 m s^{-1}, respectively. Changing the values of the longitudinal wave velocity and the density, the simulations were continued till the surface acoustic velocities substantially coincided with the experimental values. The longitudinal wave velocities were obtained as 1560 and 1540 m s^{-1} at a density 1.15×10^3 kg m^{-3} for abnormal and normal specimens, respectively. These values are in approximate agreement with the longitudinal wave velocities and densities referred to in the literature.

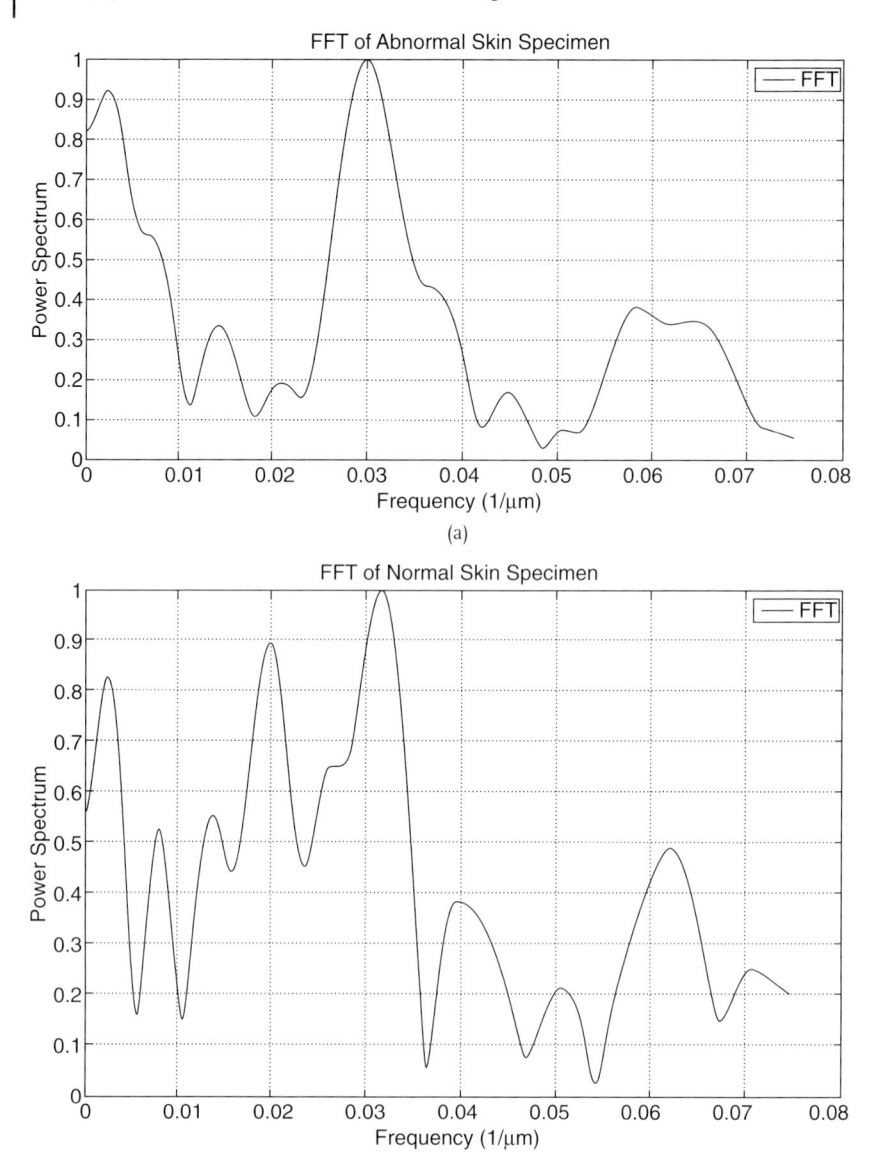

Figure 7.16 FFT analysis obtained at (a) abnormal and (b) normal points of the melanoma thin tissue on soda-lime glass substrate. (Input frequency: 200 MHz; thickness of the specimen: 5 μm; temperature: 21.5 °C.)

Table 7.4 Parameters of the acoustic lens.

Parameter	Value
Name	Acoustic lens Olympus
Model	AL2M631
Transducer	ZnO
Buffer rod	Al_2O_3
Radius of transducer	383.000 μm
Longitudinal wave velocity in buffer rod	11 175.00 m s^{-1}
Shear wave velocity in buffer rod	6950 m s^{-1}
Density in buffer rod	3.980 g cm^{-3}
Frequency	200.00 MHz
Focal length	577.52 μm
Half aperture angle	60.00°
Acoustic ZL	6122.00 μm

Table 7.5 Physical parameters of biological tissue.

Parameter	Value
Lens	AL2M631
Liquid	Deionized water
Temperature of water	22 °C
Substrate	Soda-lime glass
Radius of lens	500.6207 μm
Wavelength in water	7.4400 μm
Longitudinal wave velocity of tissue	1540.00–1590.00 m s^{-1}
Density of tissue	1.11–1.19 g cm^{-3}
Thickness of tissue	5.0000 μm

7.6 Conclusions

In this chapter we have presented the results of a study to use broadband frequency acoustic microscopy to carry out digital imaging and computer simulation to characterize tissue. Since contrast in an acoustic image (i.e., C-scan image) for skin tissue including a cancerous portion is formed by attenuation differences within the tissue, prediction of the amplitude of a transducer output is important for the field of medical ultrasound. We have developed a mathematical model for this prediction, employing frequencies from 25 to 600 MHz. In addition, we have developed an algorithm that is a component of the model to determine Young's modulus. In the present study, an abnormal skin tissue (i.e., a tissue including a melanoma portion) was selected as a specimen, and compared to a normal skin

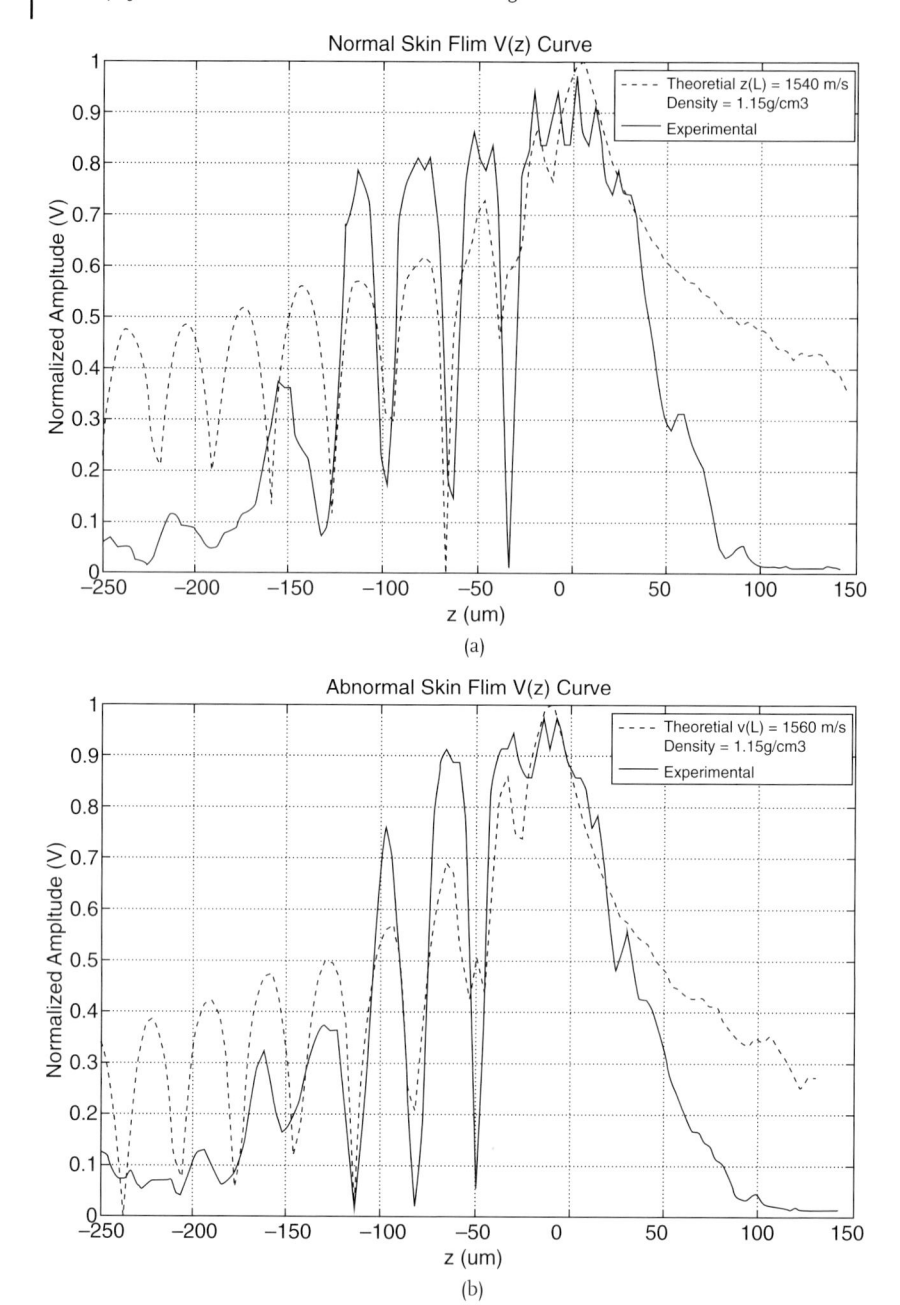

7 Study of the Contrast Mechanism in an Acoustic Image

Figure 7.17 Matched up result of the V(z) curve simulation with the experimental data for (a) normal and (b) abnormal skin tissue.

Table 7.6 Approximate values for the modulus and attenuation in normal and abnormal melanoma tissue.

	Normal	Abnormal
Modulus E (MPa)	2.73	2.80
Attenuation (dB mm^{-1})	0.22	6.67

tissue. The measurements gave approximate values of 2.73 MPa and 0.22 dB mm^1 for normal tissue but 2.80 MPa and 6.67 dB mm^1 for abnormal tissue for the modulus and attenuation, respectively. The technique suggests the use of this approach as a diagnostic tool and opens the door for more sophisticated analysis. In particular, we have developed a mathematical model for a five-layer acoustic wave propagation system. We then used as an example melanoma skin tissue to obtain digital images at low frequencies, that is, 10–50 MHz, which we processed by FFT waveform analysis to obtain estimates of ultrasonic attenuation in normal and abnormal tissue. We then went to high frequencies, that is, 200–600 MHz, to measure the V(z) signatures of normal and abnormal tissue. We used the model to carry out simulations to match the V(z) curves, which allowed us to estimated ultrasonic velocities for both. Thus, by a combination of low and high frequency acoustic microscopy we were able to measure two important acoustic parameters, giving quantitative estimates of the acoustic properties of tissues. The results are summarized in Table 7.6.

The results indicate both qualitative and quantitative differences between normal and abnormal melanoma tissue, thus paving the way for a potentially useful diagnostic medical tool without the need for staining tissue as is most often required in optical imaging.

Acknowledgment

The authors thank Yihan Tian for her assistance in the experiments.

References

1 Atalar, A., Quate, C.F., and Wickramasinge, H.K. (1977) Phase imaging in reflection with acoustic microscope. *Appl. Phys. Lett.*, **31**, 791.

2 Johnston, R.N., Atalar, A., Heiserman, J., Jipson, V., and Quate, C.F. (1979) Acoustic microscopy: resolution of subcellular detail. *Proc. Natl. Acad. Sci. USA*, **76** (7), 3325–3329.

3 Parmon, W. and Bertoni, H.L. (1979) Ray interpretation of the material signature in the acoustic microscope. *Electron. Lett.*, **15**, 684–686.

4 Lamarque, J.L., Djoukhadar, A., Rodiere, M.J., Attal, J., and Boubals, E. (1981) Acoustic microscopy in the study of breast tissue, in *1981 Ultrasonic Symposium Proceedings* (ed. B.R. McAvoy), IEEE, pp. 565–567.

5 Bennett, S.D. and Ash, E.A. (1981) Differential imaging with the acoustic microscope. *IEEE Trans. Sonic Ultrason.*, **SU-28** (2), 59–64.

6 Kolosov, O.V., Levin, V.M., Maev, R.G., and Senjushkina, T.A. (1987) The use of acoustic microscopy for biological tissue characterization. *Ultrasound Med. Biol.*, **13** (8), 477–483.

7 Tittmann, B.R. and Miyasaka, C. (2011) Visualization of anisotropy of biomedical tissue with an acoustic microscopy lens with a shear wave transducer. Imaging Conference in Malibu, CA; also Tittmann, B.R., Miyasaka, C., Maeva, E., and Maev, R. (2009). Acoustic imaging of isotropic and anisotropic thick tissue. Presented at the 30th International Acoustical Imaging Symposium, March 1–4, 2009, Monterey, CA.

8 Hildebrand, J.A., Rugar, D., Johnston, R.N., and Quate, C.F. (1981) Acoustic microscopy of living cells. *Proc. Natl. Acad. Sci. USA*, **78** (3), 1656–1660.

9 Chubachi, N., Kushibiki, J., Sannomiya, T., Akashi, N., Tanaka, M., Okawaki, H., and Dunn, F. (1987) Scanning acoustic microscope for quantitative characterization of biological tissues. *Acoust. Imag.*, **16**, 1–9.

10 Maev, R.G. (1988) Scanning acoustic microscopy of polymeric and biological substances. *Tutorial Arch. Acoust.*, **13** (1–2), 13–43.

11 Chandraratna, P.A.N., Awaad, M.I., Chandrasoma, P., and Khan, M. (1995) High frequency ultrasound: determination of the lowest frequency required for cellular imaging and detection of myocardial pathology. *Am. Heart J.*, **129**, 15–19.

12 Daft, C.M.W. and Briggs, G.A.D. (1989) The elastic microstructure of various tissues. *J. Acoust. Soc. Am.*, **85**, 416.

13 Itoh, K., Gosung, G., Jeno, E., Kasahara, K., and Zhao, L. (1983) Studies of the relationship between acoustic patterns produced by liver carcinoma in ultrasonography and in scanning acoustic microscopy. *Asian Med. J.*, **26** (9), 585–597.

14 Marmor, M., Wickramasinghe, H., and Lemons, R. (1977) Acoustic microscopy of the human retina and pigment epithelium. *Invest. Ophthalmol. Vis. Sci.*, **16** (7), 660–666.

15 D'Astous, F. and Foster, F. (1986) Frequency dependence of ultrasound attenuation and backscatter in breast tissue. *Ultrasound Med. Biol.*, **12** (10), 795–808.

16 Jones, J.P. (1997) Applications of acoustical microscopy in diagnostic medicine. *Int. J. Imag. Syst. Technol.*, **8**, 61–68.

17 Tittmann, B.R., Miyasaka, C., Maestro, A.M., and Mercer, R.R. (2007) Study of cellular adhesion with scanning acoustic microscopy. Special issue on high resolution ultrasonic imaging in industrial, material and biomaterial applications. *IEEE Trans. Ultrason. Ferroelectr. Freq. Control*, **54** (8), 1502–1513.

18 Tittmann, B.R. and Miyasaka, C. (2003) Imaging and quantitative data acquisition of biological cells and soft tissues with scanning acoustic microscopy, in *Science, Technology and Education of Microscopy: An Overview* (ed. A. Mendez-Vilas), Formatex, Badajoz, Spain, pp. 325–344.

19 Kumon, R.E., Bruno, I., Heartwell, B., and Maeva, E. (2004) Breast tissue characterization with high-frequency scanning acoustic microscopy. *J. Acoust. Soc. Am.*, **115**, 2376(A).

20 Maeva, E., Severin, F., Miyasaka, C., Tittmann, B.R., and Maev, R.G. (2009) Acoustic imaging of thick biological tissue. *IEEE Trans. Ferroelectr. Ultrason. Freq. Control*, **56** (7), 1352–1358.

21 Bamber, J.C. (1997) *Acoustical Characterization of Biological Media Encyclopedia of Acoustics*, 4th edn, John Wiley & Sons, Inc., New York.

22 Goodman, J.W. (1968) *Introduction to Fourier Optics*, McGraw Hill, New York, p. 48.

23 Champeney, D.C. (1973) *Fourier Transformations and their Physical Applications*, Academic Press, London, p. 142.

24 Atalar, A. (1978) An angular-spectrum approach to contrast in reflection acoustic microscopy. *J. Appl. Phys.*, **49** (10), 5130–5139.

25 Weglein, R.D. (1979) A model for predicting acoustic materials signatures. *Appl. Phys. Lett.*, **34**, 179–181.

26 Atalar, A. (1979) A physical model for acoustic signature. *J. Appl. Phys.*, **50** (12), 8237.

27 Kushibiki, J., Horii, K., and Chubachi, N. (1983) Velocity measurement of multiple leaky waves on germanium by line-focus-beam acoustic microscope using FFT. *Electron. Lett.*, **19**, 404–405.

28 Liang, K., Kino, G.S., and Khuri-Yakub, B.T. (1985) Material characterization by the inversion of V(z). *IEEE Trans.*, **32**, 213–224.

29 Endo, T., Sasaki, Y., Yamagishi, T., and Sakai, M. (1992) Determination of sound velocities by high frequency complex V(z) measurement in acoustic microscopy. *Jpn. Appl. Phys.*, **31**, 160–162.

30 Kulik, A., Gremaud, G., and Sathish, S. (1989) Continuous wave reflection scanning acoustic microscope, in *Acoustic Imaging*, vol. 17 (eds H. Shimizu, N. Chubachi, and J. Kushibiki), Plenum Press, New York, pp. 71–78.

8
New Concept of Pathology – Mechanical Properties Provided by Acoustic Microscopy

Yoshifumi Saijo

8.1
Introduction

The biomedical application of scanning acoustic microscopy (SAM) began early on in the history of acoustic microscopy. SAM introduced a new form of contrast that is based on the mechanical properties of what is imaged. There are three unique features of SAM compared with other microscopies such as optical, electron, and atomic force. First, SAM can be applied for easy and simple histopathological examinations because it does not require special staining techniques because the contrast observed in SAM images depends on the acoustic properties (i.e., density, stiffness, and attenuation) and on the topographic contour of the tissue.

Second, microscopic acoustic properties obtained with high frequency ultrasound can be used for assessing echo intensity and texture in clinical echography with lower frequency ultrasound. The density ρ and sound speed c determine the characteristic acoustic impedance Z of the material as:

$$Z = \rho c$$

On the assumption that the interface between two fluid-like media is infinite and plane, the relative reflected sound power, in dB, can be determined by the specific acoustic impedance of each medium if the material is assumed to be isotropic:

$$dB = 10\log_{10}\frac{P_r}{P_i} = 10\log_{10}\frac{(Z_a - Z_b)^2}{(Z_a + Z_b)^2}$$

where P_r is the sound power reflected at the interface, P_i is incident sound power, Z_a is the acoustic impedance of medium a, and Z_b is the acoustic impedance of medium b.

Third, SAM data can be used as the basic data for assessing the biomechanics of tissues and cells. Especially, it is useful for microscopic targets where direct mechanical measurements cannot be applied. In its simplest form, the

---

Advances in Acoustic Microscopy and High Resolution Imaging: From Principles to Applictaions, First Edition.
Edited by Roman Gr. Maev.
© 2013 Wiley-VCH Verlag GmbH & Co. KGaA. Published 2013 by Wiley-VCH Verlag GmbH & Co. KGaA.

relation between the sound speed and the elastic bulk modulus of liquid-like media is:

$$c = \sqrt{\frac{K}{\rho}}$$

where c is the sound speed, K the elastic bulk modulus, and ρ the density.

As a biological soft tissue may be considered as a liquid-like material, this equation may be applied to assess its elastic properties. Recent biomechanical studies have suggested that the mechanical properties of tissues may not be sufficiently similar to liquids and should be treated as soft solid materials. However, the acoustical relations of solid materials can be described by the following equation:

$$c = \sqrt{\frac{E(1-\sigma)}{\rho(1+\sigma)(1-2\sigma)}}$$

where c is the sound speed, E the Young's modulus, σ the Poisson's ratio, and ρ the density.

This relation shows that the Young's modulus of tissue and the sound speed are closely related.

Soft materials are sometimes considered as viscoelastic material. The viscosity can also be derived from acoustic properties, although it is complicated [1]:

$$\alpha = \frac{2f^2\pi^2}{3\rho c^3}\left(\eta_v + \frac{4}{3}\eta_s\right)$$

where α is the absorption, f the frequency, η_v the volumetric viscosity, η_s the shear viscosity, ρ_0 the density, and c the speed of sound.

8.2
Principle of Acoustic Microscopy

Figure 8.1 shows a schematic illustration of reflections from the tissue surface and from the interface between tissue and substrate in acoustic microscopy. The soft biological material is attached on a substrate. Normal slide glass or high-molecular polymer materials used in dishes for cell culture can be used as the substrates. The biological material is sectioned as an appropriate thickness to separate the reflections from the tissue surface and those from the interface between tissue and substrate. Single-layered cultured cells are also appropriate objects for SAM.

The ultrasound is transmitted through a coupling medium and focused on the surface of the substrate. Transmitted ultrasound is reflected at both the surface of the biological material (S_s) and the interface between the biological material and the substrate (S_d). The transducer receives the sum of these two reflections. The

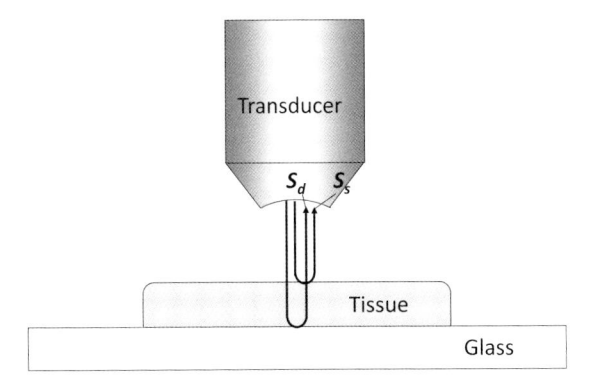

Figure 8.1 Schematic illustration of reflections from the tissue surface and from the interface between tissue and substrate in acoustic microscopy.

interference of these two reflections is determined by acoustic properties of the biological material. The determinants of the interference in the frequency are thickness and sound speed of the sample. The determinant of the interference of the intensity is the amplitude of the surface reflection and attenuation of ultrasound propagating through the tissue. The concept of quantitative measurement of the speed of sound is based on the analysis of the interference of frequency-dependent characteristics. In our previous SAM system, the frequency-dependent characteristics were obtained by serial measurements with varying frequencies from 100 to 200 MHz with 10 MHz steps. The newly proposed ultrasound sound microscope obtains the frequency-dependent characteristics by fast Fourier transform of a single broad-band pulse.

8.3
Application to Cellular Imaging

The application of SAM for cellular imaging began very early on in its development [2]. Johnston et al. used SAM for the analysis of subcellular components [3]. They could detect such features as nuclei and nucleoli, mitochondria, and actin cables of fixed cells. Hildebrand et al. applied SAM for the observation of living cells [4]. Their analysis of acoustic images of actively motile cells indicated that leading lamella were less dense or stiff than the quiescent trailing processes of the cells.

Following the Stanford group, Bereiter-Hahn at Frankfurt a performed series of important work on the biomechanics of living cells by observations with SAM. He proposed the hypothesis that the shape and locomotion of tissue cells depended on the interaction of elements of the cytoskeleton, adhesion to the substrate, and an intracellular hydrostatic pressure [5]. His group also found that higher values of impedance and attenuation coefficients were found in the cell periphery than in the central part of the cell. The phenomenon was suggested to be due to the different organization of cytoskeletal elements [6]. Veselý et al. developed

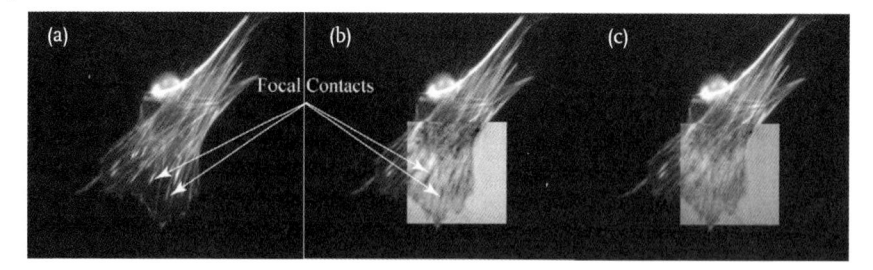

Figure 8.2 (a) Optical and (b), (c) acoustical images of embryonic chicken heart muscle cells taken at room temperature.

subtraction of the SAM images (SubSAMs) of live cells as a method for investigating minimal changes in cellular topography and elasticity. SubSAM opened up an approach to the characterization of cell motility *in vitro* and to an understanding of early cellular reactions to various stimuli [7]. They assumed that migration was due to an extension of the cell into the direction of minimum stiffness, and they were consistent with the hypothesis that local release of hydrostatic pressure provided the driving force for the flux of cytoplasm [8].

Briggs at Oxford University is also one of the pioneers of acoustic microscopy. His group measured the waveform of very short pulses, and derived the thickness of the cell. They thus calculated the acoustic velocity, impedance, and attenuation by analyzing two separate signals reflected from the top and the bottom of the cell [9].

Recently, time-resolved acoustic microscopy with GHz frequency ultrasound has been developed for cellular imaging. Weiss *et al.* compared the acoustical images of chicken heart muscle cells, and fluorescence optical images of the same cells after staining showed that the actin fibers ended inside the dark streaks in the acoustical images and thus represented the focal contacts (FCs) [10]. Figure 8.2 shows optical and acoustical images of embryonic chicken heart muscle cells taken at room temperature. The acoustical images (65 µm × 65 µm) were taken with a center frequency of 860 MHz in focus ($z = 0$). They also measured quantitative acoustical properties of a single HeLa cell *in vivo* and derived elastic parameters of subcellular structures. The value of the sound velocity inside the cell ($1534.5 \pm 33.6 \, \text{m s}^{-1}$) was only slightly higher than that of the cell medium ($1501 \, \text{m s}^{-1}$) [11].

A group based in Toronto has investigated ultrasound backscatter from leukemia cells for monitoring treatment. After treatment, backscatter increased by 400% compared with estimates obtained from control samples. Changes in spectral parameters were hypothesized to be linked to structural cell changes during apoptosis [12]. The group clarified the mechanism of backscatter change by comparing high frequency ultrasound spectroscopy (10–60 MHz) and SAM (0.9 GHz) on HeLa cells that were exposed to the chemotherapeutic agent cisplatin [13].

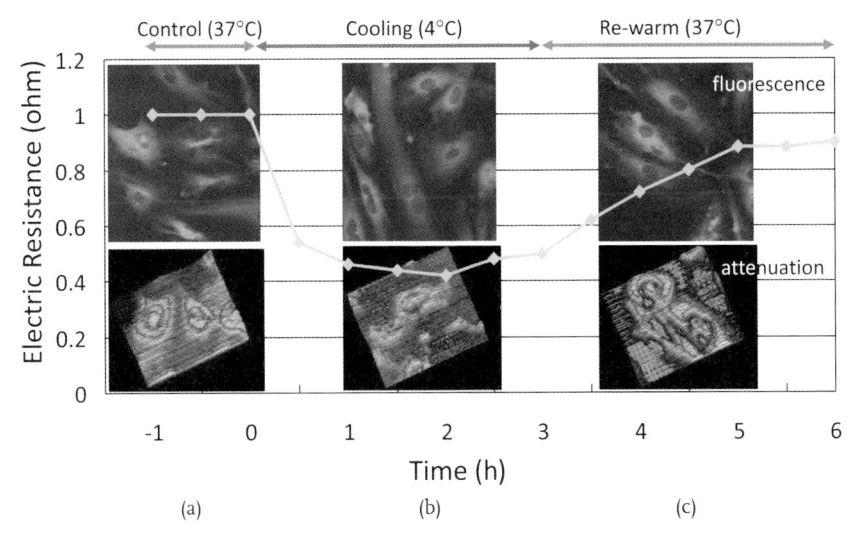

Figure 8.3 (a) Ultrasonic attenuation, (b) endothelial electric resistance, and (c) microtubule formation during and after rapid cooling of pulmonary endothelial cell.

We have also applied SAM for cellular imaging. Cold preservation is the most practical method to maintain the viability of isolated lungs in clinical lung transplantation. However, rapid cooling may affect pulmonary endothelial function. Human pulmonary arterial endothelial cells were incubated at 4 °C for 2 h. Microtubules were visualized using immunocytochemical techniques. Ultrasonic attenuation was measured with scanning acoustic microscopy. The endothelial barrier integrity was measured as transendothelial electric resistance. The low temperature caused a reversible microtubule disassembly [14]. Figure 8.3 shows the ultrasonic attenuation, endothelial electric resistance, and microtubule formation during and after rapid cooling of pulmonary endothelial cell.

A two-dimensional distribution of the ultrasonic intensity, which is closely related to the mechanical properties, was visualized to analyze cell organs, such as the nucleus at the central part and the cytoskeleton at the peripheral zone. After stimulation with TGF-β1, the ultrasonic intensity at the actin zone was significantly increased compared with the control [15].

Figure 8.4 shows a 1.2-GHz SAM image of a cultured renal vascular smooth muscle cell. The fringe shift indicates the difference of the cellular thickness.

8.4
Application to Hard Tissues

Special treatment is needed for the application of SAM to hard material. For soft tissues including cells, ultrasound penetrates through the specimen and is reflected at the interface between the tissue and the substrate such as slide glass. For hard

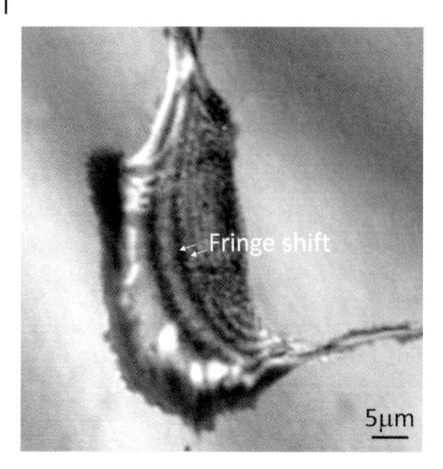

Figure 8.4 1.2-GHz SAM image of a cultured renal vascular smooth muscle cell.

materials, ultrasound is assumed to be reflected at the surface of the tissue. Thus, the surface of a hard material is polished to ensure complete reflection. Peck and Briggs applied SAM for imaging dental caries in their early phase in a biomedical application of SAM [16, 17]. As well as visualization, they developed quantitative data analysis methods in SAM measurement. One method is the $V(z)$ curve analysis that has a gradually decayed background signal due to the transducer's defocusing. They applied $V(z)$ curve analysis to the enamel of a tooth and found that the variation of velocity indicated a substantial reduction of elastic stiffness in the lesion [18].

Katz and coworkers at Case Western Reserve University applied SAM to hard biological materials such as bone. They developed a simple reflection acoustic microscope with 20 MHz and investigated inhomogeneities in the surface acoustic properties of mineralized tissues and implant materials [19]. They then used 400–600-MHz SAM for quantitative measurement on tissue elasticity. Elastic moduli for both trabecular and cortical bone were obtained by means of a series of calibration curves correlating SAM gray levels of known materials with their elastic moduli; specimens included polypropylene, PMMA [poly(methyl methacrylate)], Teflon, aluminum, Pyrex glass, titanium, and stainless steel [20, 21].

A group at the Université Paris VI has developed 50-MHz SAM and applied it to the ultrasonic tissue characterization of cartilage tissues. They found that the acoustic parameters and thickness of the articular cartilage excised from Wistar male rats changed significantly as the animal matures [22]. Collaborating with the Paris group, Raum's group at the Universitätsmedizin Berlin has applied SAMs with various frequencies for bone characterization. They investigated the chemical composition and anisotropic elasticity of individual lamellae in secondary osteons by comparing site-matched Raman microspectroscopy, 50-MHz SAM, and a nanoindentation method [23]. They also employed 200-MHz SAM and synchrotron radiation microCT (SR-microCT) to assess microstructural parameters, acoustic impedance Z, and tissue degree of mineralization of bone (DMB) in site-matched regions of interest in mouse femoral bone [24]. Measurements with two microscopes operating in the pulse-echo mode, either with frequencies up to 200 MHz and time-resolved detection or between 100 MHz and 2 GHz and amplitude detection, have been reviewed. The methods were compared and their application potentials and limitations were discussed with respect to the hierarchical structure of cortical bone [25].

For the application of SAM to tendon or ligament insertions it was necessary to determine the role of decalcification in SAM measurements since mineralized tissues including bone or mineralized fibrocartilage were present at the insertion site. To assess whether decalcification alters the tissue sound speed, supraspinatus tendon insertion of six Japanese white rabbits were measured with SAM operating in the frequency range 50–150 MHz. The sound speed in non-mineralized fibrocartilage was 1544 m s^{-1} in the undecalcified specimens, while a value of 1541 m s^{-1} was determined in the decalcified ones. On the other hand, it decreased 2–3% after decalcification in the mineralized tissue including mineralized fibrocartilage and bone (mineralized fibrocartilage: undecalcified = 1648 m s^{-1}, decalcified = 1604 m s^{-1}; bone: undecalcified = 1716 m s^{-1}, decalcified = 1677 m s^{-1}). However, no significant differences were found between the undecalcified and the decalcified specimens (non-mineralized fibrocartilage: $p = 0.84$, mineralized fibrocartilage: $p = 0.35$, bone: $p = 0.28$) [26].

8.5
Application to Soft Tissues

As described before, soft tissues should be sliced thinly so that ultrasound can penetrate through the tissue and be reflected at the interface between tissue and substrate. In our experience, a specimen thickness of approximately 10 μm was appropriate for 100-MHz SAM in terms of signal amplitude and image quality. If the speed of sound is 1500 m s^{-1} in the specimen, the wavelength of the 100-MHz ultrasound is approximately 15 μm. Thus, a time-of-flight method cannot be applied for the thickness measurement of such thin materials. Our group has proposed a unique method for calculating the thickness and sound speed of thinly sliced tissues using the interference between surface and bottom reflections [27, 28].

8.5.1
Gastric Cancer [27]

The gastric cancer tissues were classified into five groups according to pathological findings: papillary adenocarcinoma, well-differentiated tubular adenocarcinoma, moderately differentiated tubular adenocarcinoma, poorly differentiated adenocarcinoma, and signet-ring cell carcinoma. Figure 8.5 shows an example of optical and acoustical images of papillary adenocarcinoma. Figure 8.6 presents a

(a) (b) (c)

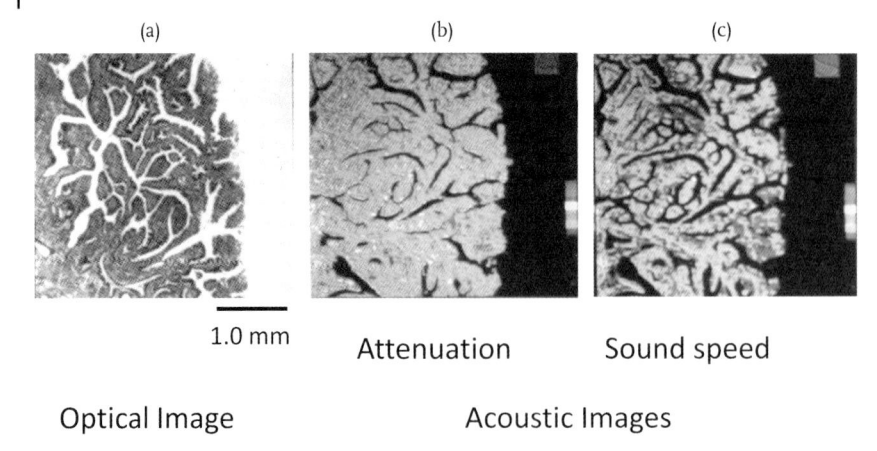

1.0 mm   Attenuation   Sound speed

Optical Image   Acoustic Images

Figure 8.5 Optical (a), attenuation (b), and sound speed (c) images of papillary adenocarcinoma.

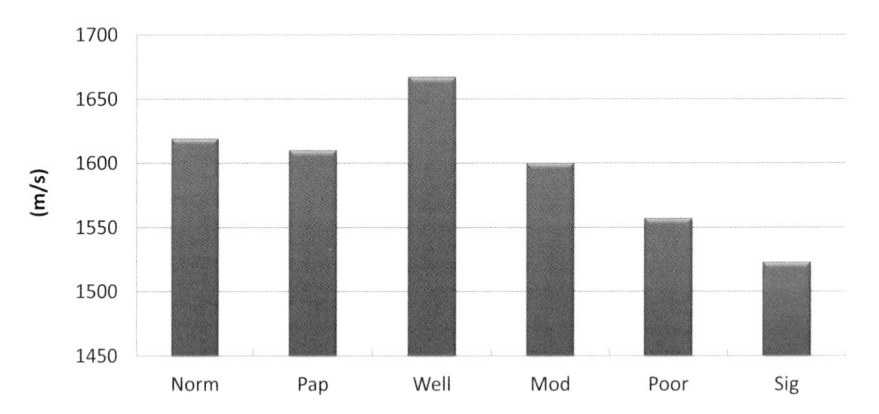

Figure 8.6 Bar graph showing sound speed in normal mucosa and five kinds of cancer tissues. Norm: normal mucosa, Pap: papillary adenocarcinoma, Well: well-differentiated tubular adenocarcinoma, Mod: moderately-differentiated tubular adenocarcinoma, Poor: poorly-differentiated adenocarcinoma, Sig: signet-ring cell carcinoma.

bar graph showing the sound speed in normal mucosa and five kinds of cancer tissues.

The values of the sound speed increased as the cellular differentiation proceeded through the three kinds of tubular adenocarcinoma. As the density of the biological soft tissues can be assumed to be nearly constant, increased sound speed can thus be interpreted to mean that tubular adenocarcinoma tissues become acoustically stiffer as the differentiation of the tissue proceeds. Electron microscopy has shown that the number of desmosomes, which are considered to attach cell-to-cell, is significantly decreased in poorly differentiated adenocarcinoma. Well-differentiated tubular adenocarcinoma specimens exhibit nearly the same number of desmosomes as in normal mucosal tissue. This increasing trend was thus regarded as the result of tightening of the intercellular attachment. Both the attenuation constant and the sound speed were significantly lower in the signet-ring cell carcinoma than in the adenocarcinoma. The intracellular component of the signet-ring cell carcinoma is the periodic acid, Schiff stain (PAS) positive substrate. The lower values of the attenuation and sound speed may be accounted for by the intracellular chemical components of the tumor tissues. These data on gastric cancer tumors clearly show that the SAM system can be used to classify the types of cancer tissues, as revealed by measurement of the acoustic parameters of the pathologies.

8.5.2
Myocardial Infarction

Since the cardiac B-scan was developed at Tohoku University in the early 1960s [29], the origin of the strong echo in myocardium has been investigated by various methods. The "sensitivity varying method" in which the relative echo intensity of the tissue is compared with those of left ventricular cavity (defined as zero) and pericardium (defined as the strongest) has been used for semi-quantitative analysis of the echo intensity of myocardium. The echo intensity of myocardium and histopathology were compared in hypertrophic cardiomyopathy. The study showed a relationship between strong echo portion and collagen fiber distribution [30]. However, there have been several reports that collagen content and myocardial echo amplitude showed only weak correlation [31].

The discrepancy may be caused by differences between optical histology and acoustic properties; thus, the acoustic microscopy was equipped for quantitative measurement of acoustic properties of tissue components at the microscopic level. The acoustic properties of the tissue elements in myocardial infarction were measured and the elastic bulk modulus of the normal and pathological myocardium was assessed from the acoustic parameters. Four kinds of tissue elements – normal myocardium, degenerated myocardium, granulation, and fibrosis – were observed in the specimens. Figure 8.7 is an acoustic image of the myocardial infarction tissue. The fibrotic lesion represents high attenuation in this image. The sound speeds were $1620\,\mathrm{m\,s^{-1}}$ in the normal myocardium, $1572\,\mathrm{m\,s^{-1}}$ in the degenerated myocardium, and $1690\,\mathrm{m\,s^{-1}}$ in the fibrosis (Figure 8.8).

The density of each tissue element was measured by the graded $CuSO_4$ solution method, and the specific acoustic impedance was calculated by the sound speed and the density of each tissue element. The values were $1.75 \times 10^6\,\mathrm{N\,s\,m^{-3}}$ in the normal myocardium, $1.69 \times 10^6\,\mathrm{N\,s\,m^{-3}}$ in the degenerated myocardium, and $1.85 \times 10^6\,\mathrm{N\,s\,m^{-3}}$ in the fibrosis. The dB level of the relative reflected sound power was calculated on the assumption that the interface between two kinds of tissue elements was infinite and plane. The level at the interface between degenerated myocardium and fibrosis was calculated as 15.4 dB. The clinical echocardiography literature showed that the strong echo of 15 dB was observed at the area of the scar

(a)	(b)	(c)
1.0 mm	Attenuation	Sound speed
Optical Image	Acoustic Images	

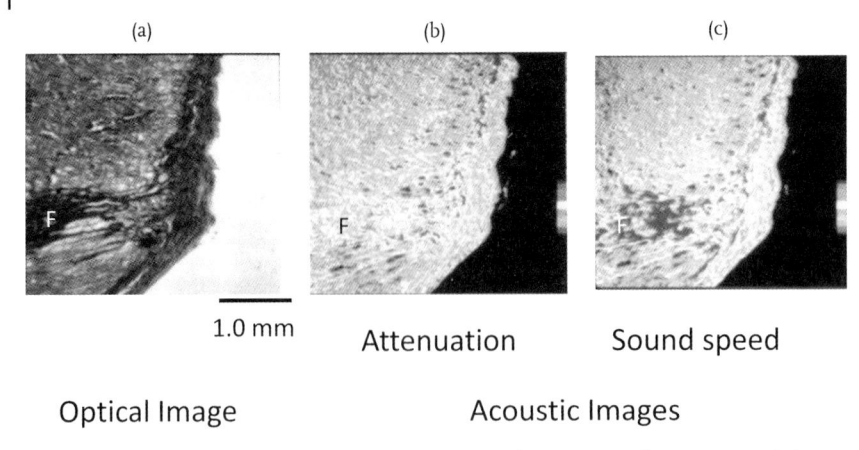

Figure 8.7 Optical (a), attenuation (b), and sound speed (c) images of acute myocardial infarction.

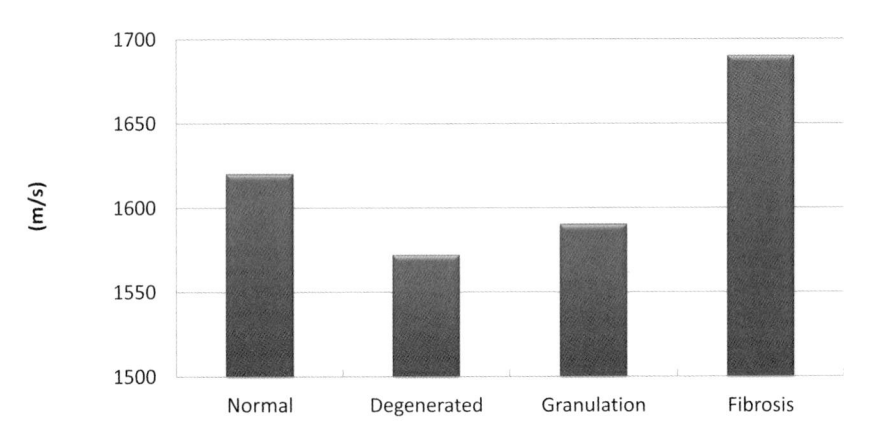

Figure 8.8 Bar graph showing sound speed in four kinds of tissue elements observed in acute myocardial infarction.

in myocardial infarction. The origin of the strong echo was clarified by the acoustic microscopy measurement [28].

Chandraratna et al. also assessed the echo-bright area in myocardium by using 600-MHz SAM [32]. They also showed that the echo intensity was affected by collagen fiber morphology [33].

As the biological tissues are modeled as fluid, the bulk modulus can also be calculated from the values of sound speed and density. The value of bulk modulus was $2.84 \times 10^9 \, \text{N m}^{-2}$ in the normal myocardium, $2.65 \times 10^9 \, \text{N m}^{-2}$ in the degenerated myocardium, and $3.12 \times 10^9 \, \text{N m}^{-2}$ in the fibrosis. One of the roles of collagen fibers in acute myocardial infarction is to prevent expansion of the infarction, and the frequency of left ventricular rupture has been reported as higher in a group that exhibits no remarkable increment of the scar in myocardium [34]. From the measurements, the bulk modulus of fibrosis was highest in the tissue components in myocardial infarction. This suggested that the fibrosis formation soon after myocardial infarction may prevent the infarct expansion and the cardiac rupture.

8.5.3
Kidney

SAM investigation of kidney was started by Dunn and Kessler at the dawn of acoustic microscopy [35]. However, quantitative data on kidney tissues were presented 20 years later when our group re-started investigations on kidney. Regarding renal cell carcinoma, the values of attenuation constant and sound speed were lower in both kinds of cancer cells than those in normal kidney, although a significant difference was not found between the clear cell and granular cell. In addition, both acoustic parameters of cancer cells were significantly lower than those in hemorrhage and fibrosis. These data suggest that the elasticity of renal cell carcinoma tissue may be lower than that of normal kidney. Moreover, the high intensity echo in clinical echography may be related to the heterogeneity of the micro-acoustic field in the carcinoma tissue [36]. We also applied SAM to differentiate renal angiomyolipoma from renal cell carcinoma [37].

For the acoustic properties of dialyzed kidney, the attenuation constant for inflammatory granulation tissue was significantly higher than that for hyaline degeneration tissue ($P < 0.001$). The sound speed was high for granulation tissue, but tended to diminish gradually for hyaline degeneration. The sound speed increased again with progression to cystic degeneration ($P < 0.001$), but the attenuation constant remained low. When a cystic kidney contained a malignant lesion, the previously low attenuation constant rose at that site ($P < 0.001$), and the previously high sound speed was diminished ($P < 0.001$). Our data suggest that the physical properties of dialyzed kidneys at different stages of pathology can be classified by their acoustic properties [38].

8.5.4
Atherosclerosis

The normal coronary artery consists of three parts and the structure is represented in optical microscopic images. The intima consists of the endothelium, the inner elastic membrane, and thin collagen fibers. The media consists of the elastic fibers and smooth muscle. The adventitia consists of the collagen fibers. The atherosclerotic intima consisted of thick collagen fibers, calcification, and lipid pool. In one study, the acoustic properties of five kinds of tissue elements, viz., the intimal collagen fiber, calcification, normal media, adventitia, and fatty plaque were measured. The cross-sectional images of coronary arteries were observed by both optical and acoustic microscopy [39].

Based on the sound speed distribution, regions with different elastic parameters were divided. Geometric and compositional information for each acoustic image was digitized using NIH Image 1.60 software (free software from NIH) on a PC (Power Macintosh 9600/233). Finite element meshes were generated using ANSYS 5.5 (SAS IP, Inc., PA) software on a workstation (Ultra10, Sun Microsystems, CA). The finite element models were solved for an intraluminal pressure load of 110 mmHg–14.6 kPa. In the calculation process, the region in the images was considered as part of a circumferential vessel wall and the central angle was 30°. The sound speed distribution was inhomogeneous and the discontinuity of the elastic property in the fibrous cap was observed in the atherosclerotic coronary artery. Then, the fibrotic region was divided into four regions according to the elasticity distribution. The first principle stress distribution of the normal coronary artery showed that the stress was dominant in intima but the stress distribution in the intima was uniform. The stress distribution in the atherosclerotic plaque showed that the mean value of the stress was smaller than that found in the intima of the normal coronary artery, but the peak stress was very high and concentrated into the crack-like structure of the fibrous cap (Figure 8.9a–d). The results indicated that the pathophysiology of coronary plaque rupture was strongly

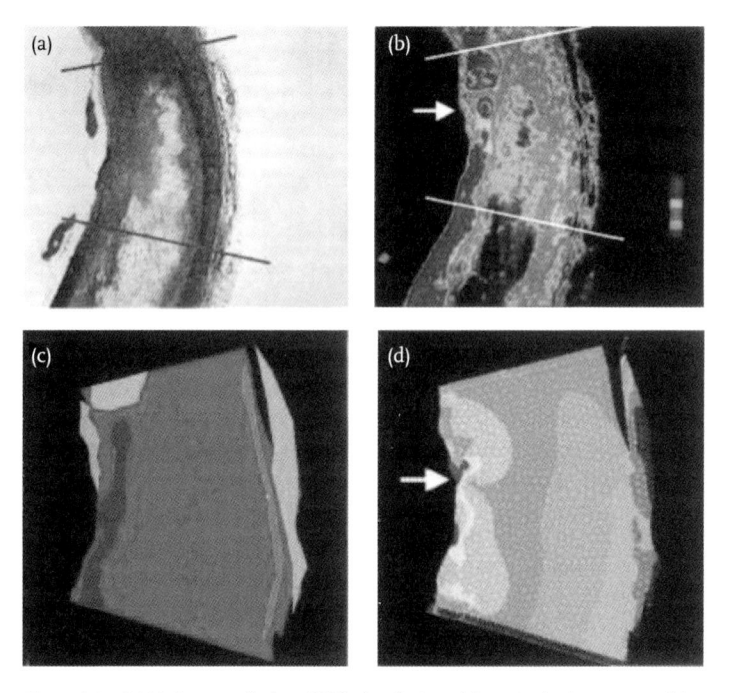

Figure 8.9 SAM data applied to FEM simulation: (a) optical microscope, (b) acoustic microscope (sound speed), (c) FEM model, and (d) stress distribution of atherosclerotic coronary artery.

| (a) | (b) | (c) |

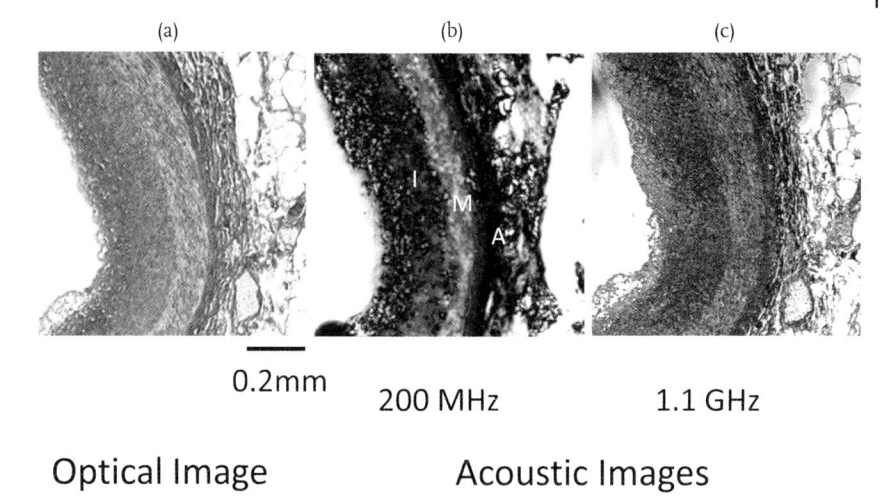

0.2mm 200 MHz 1.1 GHz

Optical Image   Acoustic Images

Figure 8.10 Comparison of resolution according to the ultrasonic frequency: (a) optical, (b) 200-MHz SAM, (c) 1.1-GHz SAM images of normal human coronary artery.

correlated with biomechanical properties of the tissue components in the coronary artery [40].

Figure 8.10 shows the optical (a), 200 MHz (b), and 1.1 GHz (c) acoustic microscopy images of normal human coronary artery. The image quality is better in 1.1 GHz image. For example, each elastic fiber in the media can be observed in the 1.1 GHz image while the media is almost homogeneous in the 200 MHz image. However, the three-layered appearance of coronary artery is more obvious in the 200 MHz image. The intima and adventitia show higher ultrasonic attenuation than that of media.

For the assessment of plaque rupture in a vulnerable plaque, SAM investigations on atherosclerosis-prone mice were performed. The acoustic properties of the normal vessel wall and plaques, particularly fibrous caps of lipid-rich plaques, were evaluated in the aortic roots of six normal C57BL mice and 12 atherosclerosis-prone apoE-deficient [apoE(-/-)] mice by SAM. After processing, the attenuation of high-frequency (1.1 GHz) focused ultrasound was measured in unstained tissue sections by SAM followed by quantification of the amount and type of collagen in picrosirius red stained sections by means of polarized light microscopy (PLM). Collagen appeared green in thin fibrous caps and bright orange in thick caps by PLM. The attenuation of ultrasound was significantly higher in the collagen fibers with orange color compared to those with green color (17.2 versus 6.6×10^3 dB mm^{-1}) [41].

Human carotid atherosclerotic lesions were observed by GHz-range scanning acoustic microscopy (SAM). The atherosclerotic lesions were characterized by either thickened fibrosis with dense collagen fibers or lipid accumulation with sparse collagen network by optical microscopy. SAM revealed that the fibrosis was classified into type I and III collagen by attenuation of ultrasound (US) and that the sound field of lipid accumulation lesions became inhomogeneous. The results would provide the scientific basis for the images of vulnerable plaques being produced in diagnostic ultrasound [42].

8.6
Ultrasound Speed Microscopy (USM) [39]

A single ultrasound pulse with a pulse width of 2 ns was emitted and received by the same transducer above the specimen. Saline was used as the coupling medium between the transducer and the specimen. The reflections from the tissue surface and those from the interface between the tissue and glass were received by the transducer and were introduced into a Windows-based PC with a fast digitizer card (Acqiris DP210, 2 GSa s^{-1}, 8-bit, onboard memory 16 MB, Geneva, Switzerland). The frequency range was 500 MHz, and the sampling rate was 2 GS s^{-1}. Eight consecutive values of the signal taken for a pulse response were averaged to reduce random noise.

The transducer was mounted on an X-Y stage with a microcomputer board that was driven by the PC through RS232C. Both the X-scan and Y-scan were driven by linear servo motors and the position was detected by an encoder. The scan was controlled to reduce the effects of acceleration at the start and deceleration at the end of the X-scan. Finally, two-dimensional distributions of ultrasonic intensity, sound speed, attenuation coefficient, and thickness of a specimen measuring 2.4 × 2.4 mm were visualized using 300 × 300 pixels. The total scanning time was 63 s.

Denoting the standardized phase of the reflection wave at the tissue surface as φ_{front} and the standardized phase at the interference between the tissue and the substrate as φ_{rear}:

$$2\pi f \times \frac{2d}{c_o} = \varphi_{front}$$

$$2\pi f \times 2d \left(\frac{1}{c_o} - \frac{1}{c} \right) = \varphi_{rear}$$

where d is the tissue thickness, c_o is the sound speed in the coupling medium, and c is the sound speed in the tissue.

The thickness is obtained as:

$$d = \frac{c_o}{4\pi f} \varphi_{front}$$

Finally, the sound speed is calculated as:

$$c = \left(\frac{1}{c_o} - \frac{\varphi_{rear}}{4\pi f d} \right)^{-1}$$

(a) (b)

0.5mm

Attenuation　　　　Sound speed

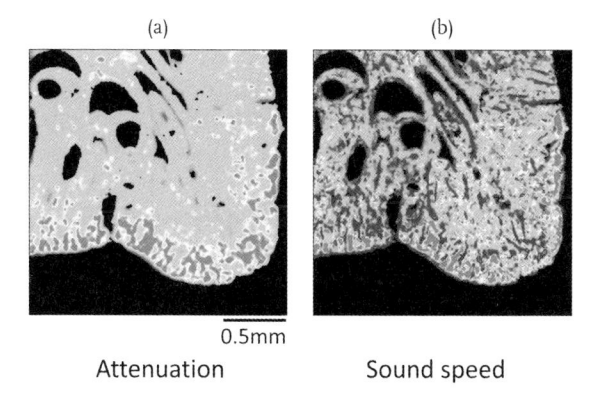

Figure 8.11 USM images – (a) attenuation and (b) sound speed – of well-differentiated tubular adenocarcinoma of stomach.

(a) (b)

0.5mm

Attenuation　　　　Sound speed

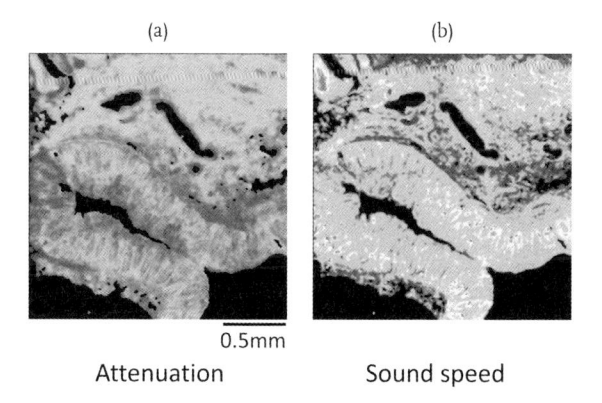

Figure 8.12 USM images – (a) attenuation and (b) sound speed – of moderately differentiated adenocarcinoma of colon.

After determination of the thickness, attenuation of ultrasound was then calculated by dividing the amplitude by the thickness and frequency.

Figure 8.11 shows a well-differentiated tubular adenocarcinoma of the stomach. The adenocarcinoma shows a clear tubular structure. The tubular structure lesion shows higher attenuation and sound speed than the surrounding zone. The sound speed values were similar to our previous measurement [27].

Figure 8.12 shows a moderately differentiated adenocarcinoma of the colon. An adenocarcinoma with desmoplasia (fibrosis) is typical. This case is classified as an early stage because the tumor is localized in the submucosal layer. Part of the cancer tissue contains rich fibrosis and the lesion shows high attenuation and sound speed.

8.7
Articular Tissues

As described before, SAM data can be used as the basic data for assessing the biomechanics of tissues and cells. Research on biomechanics is widely performed in the field of orthopedic surgery. The acoustic properties of rabbit supraspinatus tendon insertions were measured by SAM. In the tendon proper and the non-mineralized fibrocartilage, the sound speed and attenuation constant gradually decreased as the predominant collagen type changed from I to II. In the mineralized fibrocartilage, they increased markedly with mineralization of the fibrocartilaginous tissue. These results indicate that the non-mineralized fibrocartilage shows the lowest elastic modulus among four zones at the insertion site, which could be interpreted as an adaptation to various types of biomechanical stress [43].

Our group also attempted to determine the changes of articular cartilage of the knee joint during immobilization in a rat model. The knee joints of adult male rats were immobilized at 150° of flexion using an internal fixator for 3 days, and for 1, 2, 4, 8, and 16 weeks. The articular cartilage from the medial midcondylar region of the knee was obtained and divided into three areas (non-contact area, transitional area, contact area); in each area, a degree of degeneration was evaluated by gross observation, histomorphometric grading, and measurements of thickness and number of chondrocytes. Figure 8.13 shows the results.

Degeneration of the articular cartilage was mainly observed in the contact and transitional areas. Matrix staining intensity by safranin-O and the number of chondrocytes were decreased in these two areas. The thickness of the articular cartilage in the non-contact and contact areas was unchanged, but it was increased in the transitional area. A decrease in sound speed was observed in the transitional area of both the femoral and tibial cartilage, indicating the softening of the articular cartilage. The changes in articular cartilage became obvious as early as 1 week after immobilization. These changes may be due to a lack of mechanical stress or a lack of joint fluid circulation during immobilization [44, 45].

8.8
Summary

The principle and brief history of scanning acoustic microscopy (SAM) for medicine and biology have been described here. SAM was able to visualize high quality microscopy images of tissues and cells suitable for histopathological examinations.

For bone, cartilage, tendon, and cardiovascular tissues, micro-acoustic properties provided important information on biomechanical properties. The biomechanics of these tissues are especially important for assessing the pathophysiology.

Cells are considered to consist of viscoelastic materials and SAM has revealed information on viscosity by attenuation and information on elasticity by sound

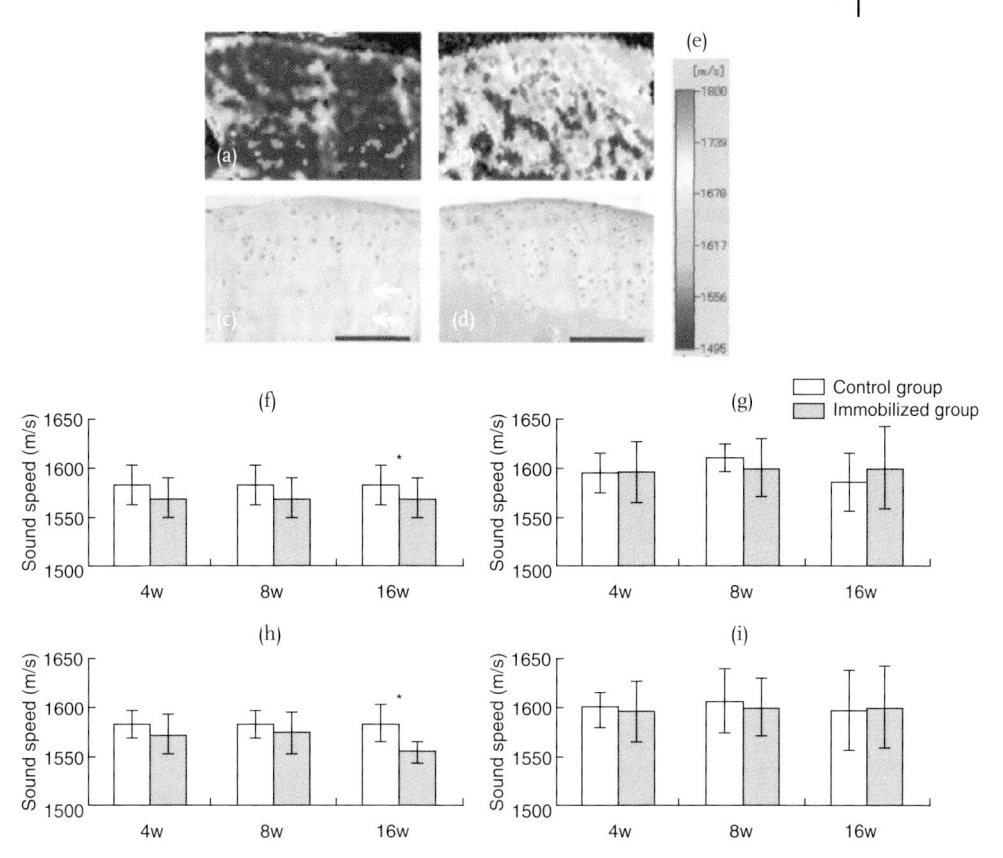

Figure 8.13 Gradation images and sound speed changes of articular cartilage assessed by scanning acoustic microscopy (SAM). Upper row: (a) and (b) show the gradation images of the tibial articular cartilage at 16 weeks; the second row, (c) and (d), show each corresponding area to each upper row with hematoxylin and eosin staining, respectively. (a) and (c) The immobilized group, (b) and (d) the control group, and (e) gradation scale bar. The low sound speed area gradually expanded from the surface of the articular cartilage in the immobilized group. The third row, (f) and (g), shows the femoral cartilage; the lower row, (h) and (i), shows the tibial cartilage of the sound speed. (f) and (h) Are the transitional area, (g) and (i) the noncontact area.

speed. Instead of stretching cells or using atomic force microscopy to measure biomechanical properties, SAM can be used to measure a precise mechanical property distribution without contact to the cells.

Thus, SAM has introduced a new concept of pathology that is based on the mechanical properties of what is imaged. Recent developments such as ultrasound speed microscopy, 3D ultrasound microscopy, and high frequency array transducer may realize a clinically applicable SAM in the near future.

References

1 Mikhailov, I.G., Soloviev, V.A., and Syrnikov, Y.P. (1964) *Basics of Molecular Acoustics*, Khimia Publ, Moscow.
2 Lemons, R.A. and Quate, C.F. (1975) Acoustic microscopy: biomedical applications. *Science*, **188** (4191), 905–911.
3 Johnston, R.N., Atalar, A., Heiserman, J., Jipson, V., and Quate, C.F. (1979) Acoustic microscopy: resolution of subcellular detail. *Proc. Natl Acad. Sci. USA*, **76** (7), 3325–3329.
4 Hildebrand, J.A., Rugar, D., Johnston, R.N., and Quate, C.F. (1981) Acoustic microscopy of living cells. *Proc. Natl Acad. Sci. USA*, **78** (3), 1656–1660.
5 Bereiter-Hahn, J. (1985) Architecture of tissue cells. The structural basis which determines shape and locomotion of cells. *Acta Biotheor.*, **34** (2–4), 139–148.
6 Litniewski, J. and Bereiter-Hahn, J. (1990) Measurements of cells in culture by scanning acoustic microscopy. *J. Microsc.*, **158** (Pt 1), 95–107.
7 Veselý, P., Lücers, H., Riehle, M., and Bereiter-Hahn, J. (1994) Subtraction scanning acoustic microscopy reveals motility domains in cells *in vitro*. *Cell Motil. Cytoskeleton*, **29** (3), 231–240.
8 Bereiter-Hahn, J. and Lüers, H. (1998) Subcellular tension fields and mechanical resistance of the lamella front related to the direction of locomotion. *Cell Biochem. Biophys.*, **29** (3), 243–262.
9 Briggs, G.A., Wang, J., and Gundle, R. (1993) Quantitative acoustic microscopy of individual living human cells. *J. Microsc.*, **172** (Pt 1), 3–12.
10 Weiss, E.C., Lemor, R.M., Pilarczyk, G., Anastasiadis, P., and Zinin, P.V. (2007) Imaging of focal contacts of chicken heart muscle cells by high-frequency acoustic microscopy. *Ultrasound Med. Biol.*, **33** (8), 1320–1326.
11 Weiss, E.C., Anastasiadis, P., Pilarczyk, G., Lemor, R.M., and Zinin, P.V. (2007) Mechanical properties of single cells by high-frequency time-resolved acoustic microscopy. *IEEE Trans. Ultrason. Ferroelectr. Freq. Control*, **54** (11), 2257–2271.
12 Taggart, L.R., Baddour, R.E., Giles, A., Czarnota, G.J., and Kolios, M.C. (2007) Ultrasonic characterization of whole cells and isolated nuclei. *Ultrasound Med. Biol.*, **33** (3), 389–401.
13 Brand, S., Weiss, E.C., Lemor, R.M., and Kolios, M.C. (2008) High frequency ultrasound tissue characterization and acoustic microscopy of intracellular changes. *Ultrasound Med. Biol.*, **34** (9), 1396–1407. Epub 2008 April 24.
14 Suzuki, S., Bing, H., Sugawara, T., Matsuda, Y., Tabata, T., Hoshikawa, Y., Saijo, Y., and Kondo, T. (2004) Paclitaxel prevents loss of pulmonary endothelial barrier integrity during cold preservation. *Transplantation*, **78** (4), 524–529.
15 Hagiwara, Y., Saijo, Y., Ando, A., Chimoto, E., Suda, H., Onoda, Y., and Itoi, E. (2009) Ultrasonic intensity microscopy for imaging of living cells. *Ultrasonics*, **49** (3), 386–388. Epub 2008 Oct 31.
16 Peck, S.D. and Briggs, G.A. (1986) A scanning acoustic microscope study of the small caries lesion in human enamel. *Caries Res.*, **20** (4), 356–360.
17 Peck, S.D. and Briggs, G.A. (1987) The caries lesion under the scanning acoustic microscope. *Adv. Dent. Res.*, **1** (1), 50–63.
18 Peck, S.D., Rowe, J.M., and Briggs, G.A. (1989) Studies on sound and carious enamel with the quantitative acoustic microscope. *J. Dent. Res.*, **68** (2), 107–112.
19 Meunier, A., Katz, J.L., Christel, P., and Sedel, L. (1988) A reflection scanning acoustic microscope for bone and bone-biomaterials interface studies. *J. Orthop. Res.*, **6** (5), 770–775.
20 Katz, J.L. and Meunier, A. (1993) Scanning acoustic microscope studies of the elastic properties of osteons and osteon lamellae. *J. Biomech. Eng.*, **115** (4B), 543–548.
21 Bumrerraj, S. and Katz, J.L. (2001) Scanning acoustic microscopy study of human cortical and trabecular bone. *Ann. Biomed. Eng.*, **29** (12), 1034–1042.
22 Chérin, E., Saïed, A., Pellaumail, B., Loeuille, D., Laugier, P., Gillet, P., Netter, P., and Berger, G. (2001) Assessment of

rat articular cartilage maturation using 50-MHz quantitative ultrasonography. *Osteoarthritis Cartilage*, **9** (2), 178–186.

23 Hofmann, T., Heyroth, F., Meinhard, H., Fränzel, W., and Raum, K. (2006) Assessment of composition and anisotropic elastic properties of secondary osteon lamellae. *J. Biomech.*, **39** (12), 2282–2294. Epub 2005 September 6.

24 Raum, K., Hofmann, T., Leguerney, I., Saïed, A., Peyrin, F., Vico, L., and Laugier, P. (2007) Variations of microstructure, mineral density and tissue elasticity in B6/C3H mice. *Bone*, **41** (6), 1017–1024. Epub 2007 September 7.

25 Raum, K. (2008) Microelastic imaging of bone. *IEEE Trans. Ultrason. Ferroelectr. Freq. Control*, **55** (7), 1417–1431.

26 Sano, H., Hattori, K., Saijo, Y., and Kokubun, S. (2006) Does decalcification alter the tissue sound speed of rabbit supraspinatus tendon insertion? *In vitro* measurement using scanning acoustic microscopy. *Ultrasonics*, **44** (3), 297–301. Epub 2006 April 18.

27 Saijo, Y., Tanaka, M., Okawai, H., and Dunn, F. (1991) The ultrasonic properties of gastric cancer tissues obtained with a scanning acoustic microscope system. *Ultrasound Med. Biol.*, **17** (7), 709–714.

28 Saijo, Y., Tanaka, M., Okawai, H., Sasaki, H., Nitta, S.I., and Dunn, F. (1997) Ultrasonic tissue characterization of infarcted myocardium by scanning acoustic microscopy. *Ultrasound Med. Biol.*, **23** (1), 77–85.

29 Tanaka, M., Neyazaki, T., Kosaka, S., Sugi, H., Oka, S., Ebina, T., Terasawa, Y., Unno, K., and Nitta, K. (1971) Ultrasonic evaluation of anatomical abnormalities of heart in congenital and acquired heart diseases. *Br. Heart J.*, **33** (5), 686–698.

30 Tanaka, M., Nitta, S., Nitta, K., Sogo, Y., Yamamoto, A., Katahira, Y., Sato, N., Ohkawai, H., and Tezuka, F. (1985) Non-invasive estimation by cross sectional echocardiography of myocardial damage in cardiomyopathy. *Br. Heart J.*, **53** (2), 137–152.

31 Lythall, D.A., Bishop, J., Greenbaum, R.A., Ilsley, C.J., Mitchell, A.G., Gibson, D.G., and Yacoub, M.H. (1993) Relationship between myocardial collagen and echo amplitude in non-fibrotic hearts. *Eur. Heart J.*, **14** (3), 344–350.

32 Chandraratna, P.A., Whittaker, P., Chandraratna, P.M., Gallet, J., Kloner, R.A., and Hla, A. (1997) Characterization of collagen by high-frequency ultrasound: evidence for different acoustic properties based on collagen fiber morphologic characteristics. *Am. Heart J.*, **133** (3), 364–368.

33 Tabel, G.M., Whittaker, P., Vlachonassios, K., Sonawala, M., and Chandraratna, P.A. (2006) Collagen fiber morphology determines echogenicity of myocardial scar: implications for image interpretation. *Echocardiography*, **23** (2), 103–107.

34 Uusimaa, P., Risteli, J., Niemelä, M., Lumme, J., Ikäheimo, M., Jounela, A., and Peuhkurinen, K. (1997) Collagen scar formation after acute myocardial infarction: relationships to infarct size, left ventricular function, and coronary artery patency. *Circulation*, **96** (8), 2565–2572.

35 Kessler, L.W., Fields, S.I., and Dunn, F. (1974) Acoustic microscopy of mammalian kidney. *J. Clin. Ultrasound*, **2** (4), 317–320.

36 Sasaki, H., Tanaka, M., Saijo, Y., Okawai, H., Terasawa, Y., Nitta, S., and Suzuki, K. (1996) Ultrasonic tissue characterization of renal cell carcinoma tissue. *Nephron*, **74** (1), 125–130.

37 Sasaki, H., Saijo, Y., Tanaka, M., Nitta, S., Yambe, T., and Terasawa, Y. (1997) Characterization of renal angiomyolipoma by scanning acoustic microscopy. *J. Pathol.*, **181** (4), 455–461.

38 Sasaki, H., Saijo, Y., Tanaka, M., Nitta, S., Terasawa, Y., Yambe, T., and Taguma, Y. (1997) Acoustic properties of dialysed kidney by scanning acoustic microscopy. *Nephrol. Dial. Transplant.*, **12** (10), 2151–2154.

39 Saijo, Y., Santos Filho, E., Sasaki, H., Yambe, T., Tanaka, M., Hozumi, N., Kobayashi, K., and Okada, N. (2007) Ultrasonic tissue characterization of atherosclerosis by a speed-of-sound microscanning system. *IEEE Trans. Ultrason. Ferroelectr. Freq. Control*, **54** (8), 1571–1577.

40 Saijo, Y., Ohashi, T., Sasaki, H., Sato, M., Jorgensen, C.S., and Nitta, S. (2001) Application of scanning acoustic microscopy for assessing stress distribution in atherosclerotic plaque. *Ann. Biomed. Eng.*, **29** (12), 1048–1053.

41 Saijo, Y., Jørgensen, C.S., and Falk, E. (2001) Ultrasonic tissue characterization of collagen in lipid-rich plaques in apoE-deficient mice. *Atherosclerosis*, **158** (2), 289–295.

42 Saijo, Y., Jørgensen, C.S., Mondek, P., Sefránek, V., and Paaske, W. (2002) Acoustic inhomogeneity of carotid arterial plaques determined by GHz frequency range acoustic microscopy. *Ultrasound Med. Biol.*, **28** (7), 933–937.

43 Sano, H., Saijo, Y., and Kokubun, S. (2006) Non-mineralized fibrocartilage shows the lowest elastic modulus in the rabbit supraspinatus tendon insertion: measurement with scanning acoustic microscopy. *J. Shoulder Elbow Surg.*, **15** (6), 743–749.

44 Hagiwara, Y., Ando, A., Chimoto, E., Saijo, Y., Ohmori-Matsuda, K., and Itoi, E. (2009) Changes of articular cartilage after immobilization in a rat knee contracture model. *J. Orthop. Res.*, **27** (2), 236–242.

45 Ando, A., Suda, H., Hagiwara, Y., Onoda, Y., Chimoto, E., Saijo, Y., and Itoi, E. (2011) Reversibility of immobilization-induced articular cartilage degeneration after remobilization in rat knee joints. *Tohoku J. Exp. Med.*, **224** (2), 77–85.

9
Quantitative Scanning Acoustic Microscopy of Bone

Pascal Laugier, Amena Saïed, Mathilde Granke, and Kay Raum

9.1
Introduction

9.1.1
Hierarchical Structure of Bone and Properties

Musculoskeletal mineralized tissues (MMTs), for example, bone, mineralized tendons, and teeth, are natural examples of tissues that achieve unique combinations and also great ranges of stiffness and strength. These tissues have a common basic building block – a collagen type I fibril that is reinforced by small hydroxyapatite mineral crystals. One of the striking features of these tissues is their ability to adapt to variable loading conditions by multiple structural arrangements of this building block at several levels of hierarchical organization.

In contrast to other MMTs bone tissues have high adaptation and regeneration potentials. This is realized by tissue remodeling, that is, two synchronized cellular processes of tissue resorption and synthesis, which leads to an organ that is permanently adapted to its mechanical environment and maintained throughout its lifetime by an incremental tissue repair.

At the nano-scale, collagen Type I molecules with a specific tertiary structure, having a 67-nm periodicity and 40-nm gaps or holes between the ends of the molecules, assemble into collagen fibrils. The diameter of an individual fibril is of the order of 100–200 nm. After synthesis, plate-like crystals of hydroxyapatite are deposited either within the discrete spaces within the fibrils (Figure 9.1) or between adjacent fibril walls. This intra- and interfibrillar deposition of mineral platelets and crystal growth leads to an anisotropic and time-dependent stiffening of the mineralized fibrils. The fibrils aggregate to form structural units, for example, woven type, plexiform, or lamellar bone. Up to seven hierarchical levels of organization have been proposed for MMTs [2]. With respect to the levels of acoustic assessment described in the next sections, four levels of hierarchy, from the nano-scale to the macro-scale, are used (Table 9.1).

Advances in Acoustic Microscopy and High Resolution Imaging: From Principles to Applictaions, First Edition.
Edited by Roman Gr. Maev.
© 2013 Wiley-VCH Verlag GmbH & Co. KGaA. Published 2013 by Wiley-VCH Verlag GmbH & Co. KGaA.

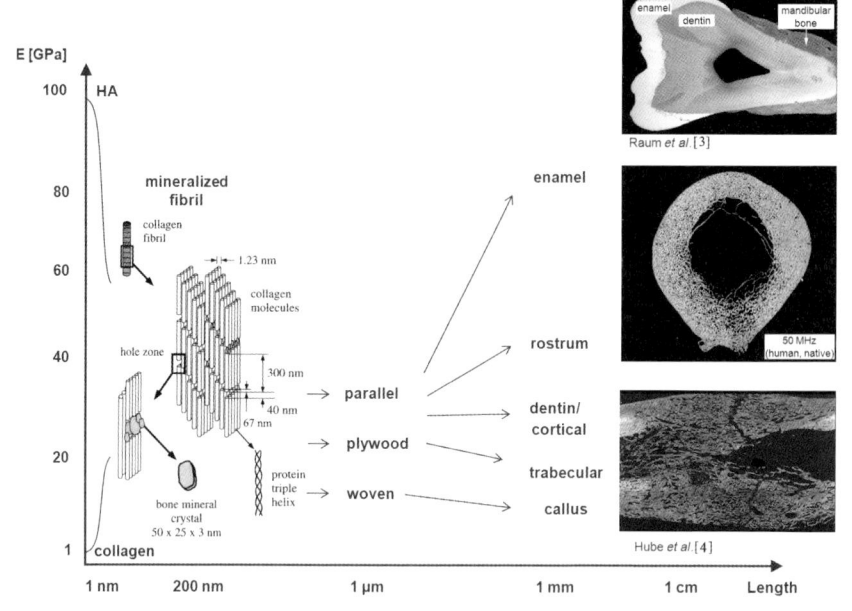

Figure 9.1 Different material properties are realized in various musculoskeletal mineralized tissues with a single building block – the mineralized collagen fibril, that is, a compound of collagen type I molecules, hydroxyapatite (HA) mineral crystals, and water. Acoustic images of enamel, dentin, mandibular bone, cortical and trabecular tissues of a human femoral cross section and a healing sheep callus are shown on the right. From Reference [1, 3, 4].

Table 9.1 Definition of characteristic length scales.

Length scale	Composition	Compound	Nomenclature
1–200 nm	Hydroxyapatite, collagen, water, other	Mineralized collagen fibril	Nano-scale
200 nm–~10 μm	Mineralized collagenFibrils, pores (osteocyte lacunae, canaliculi)	Mineralized tissue matrix (variable structures)	Micro-scale
10 μm–~1 mm	Mineralized tissue matrix, pores (Haversian canals, resorption lacunae)	MMTs	Meso-scale
>1 mm	MMTs, macrostructure	Organs	Macro-scale

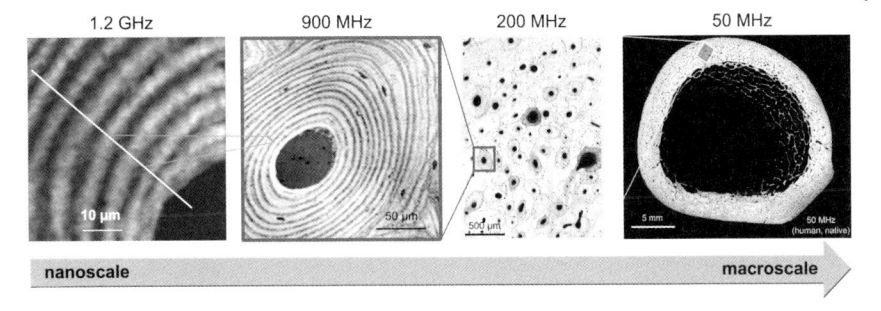

Figure 9.2 Acoustic impedance images of a human cortical bone cross-section from the femoral mid-diaphysis. Measurements with different frequencies resemble the hierarchical structure of the tissue.

The peripheral skeleton of mature mammals and humans consists of trabecular and cortical bone. The latter is composed of a highly ordered system of Haversian and Volkmann canals with typical diameters in the range 10–200 μm, canaliculi (diameter: 0.2–0.3 μm), and other pores, for example, osteocyte and resorption lacunae with diameters of the order of 2–8 μm and up to about 200 μm, respectively (Figure 9.2).

9.1.2
Relevance of Multiscale Elastic Properties

The elastic behavior of lamellar bone cannot be simply described by a set of unique material constants. The complex hierarchical structure results in spatial, temporal, and directional variations of elastic properties leading to an almost perfect adaptation to their functional demands. Macroscopic mechanical properties, for example, stiffness or resistance to failure, are determined by composition, structure, and elastic properties at all lower hierarchical levels. Many attempts have been made to downscale well established macroscopic mechanical testing methods to the smaller hierarchy levels. For example, nanoindentation has emerged from macroscopic indentation measurements and has become the most accepted micromechanical tool for the direct assessment of elastic properties of mineralized tissues at the tissue level (micro-scale). The spot-limited and destructive measurement principle, however, prohibits comprehensive evaluations of the heterogeneous microstructure, which can only partially be overcome by scanning approaches [5].

Multiscale assessment of bone elastic properties is possible using ultrasonic methods. Ultrasound waves are mechanical waves that are inherently affected by the material and the structural characteristics of the propagation medium. The scalability of the acoustic wavelength (from 6 mm at 500 kHz down to 0.5 μm at 2 GHz) makes ultrasound a suitable modality to investigate bone properties at different scales: low frequency methods (500 kHz to 10 MHz) are widely used *in vitro* and in clinical devices to assess bone quality and elastic properties of bone specimens at the macro-scale [6–10]. Scanning acoustic microscopy (SAM) with frequencies between 50 MHz and 2 GHz is adapted to the investigation of structure and elastic properties of the bone tissue matrix, that is, the linear elastic interaction of acoustic waves with the material under investigation can be used to visualize the microarchitecture, and also to measure sound velocities and acoustic impedances at various length scales. When combined with local density estimates (e.g., derived from quantitative X-ray micro-computed tomography data), acoustic impedance estimates can be used to derive tissue stiffness [11–13]. By combining *multiscale* experimental data with numerical and continuum mechanical homogenization approaches, structural properties can be decoupled from material properties and their respective impacts on the elastic functional behavior of the compound can be studied [14, 15]. The underlying principles and application will be described in the following sections.

9.1.3
History of Measurement Principles

Various high frequency ultrasound techniques have been developed during the last two decades with the intention of assessing the elastic properties of bone at the tissue level. The basic measurement principles can be divided into (i) measurement of the compressional wave velocity in thin tissue sections, (ii) measurement of surface acoustic wave velocities in thick sections, and (iii) derivation of the acoustic impedance from the confocal reflection amplitude in thick sections.

Early quantitative studies [16–18] have measured the compressional wave velocity v_p in 500-µm thick sections with a 50-MHz pulse–echo microscope. In these studies, the sound velocity was estimated from the time-of-flight difference (ΔTOF) of front and back side reflections and the a priori determined sample thickness d (Figure 9.3):

$$v_p = \frac{2d}{\Delta \text{TOF}} \quad (9.1)$$

The elastic coefficient C was obtained using the relation:

$$C = \rho v_p^2 \quad (9.2)$$

where ρ is the mass density.

However, several limitations hindered a broad application of this method. First, the sample has to be thick enough so that front and back side echoes can be separated (i.e., approximately twice the wavelength). If the sample becomes thicker than the depth of focus of the lens, the spatial resolution degrades. Second, two parallel and smooth surfaces have to be prepared and the thickness has to be determined with another method. Therefore, any thickness variations within the sample directly affect the estimated parameters.

Surface acoustic waves (SAWs) can be generated when the sound field is focused inside a stiff and thick material (Figure 9.4). These SAWs leak waves back into the coupling fluid and they can eventually be detected by the same transducer. The

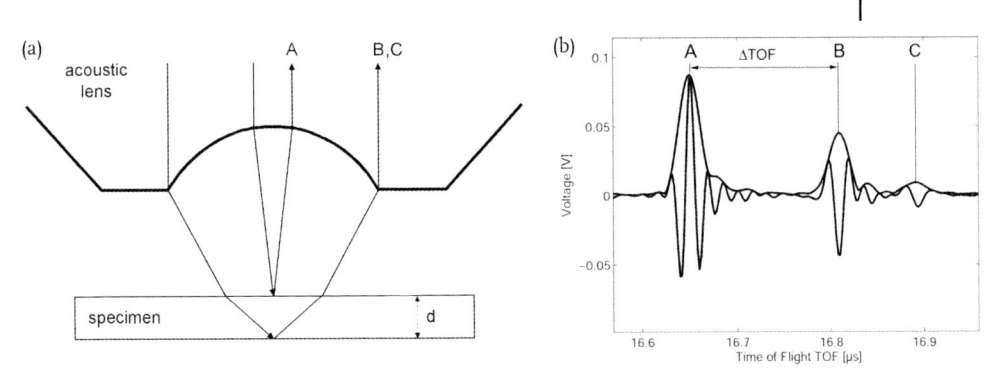

Figure 9.3 Principle of the measurement of compressional and shear wave velocities in thin tissue sections (a). The time delay between waves reflected at the front (A) and back side (B and C) of a thin specimen can be used to calculate the wave velocities (b). Mode conversions can result in the generation of compressional (B) and shear waves (C). Because of the lower sound velocities, shear waves have larger time delays than compressional waves. From Reference [19].

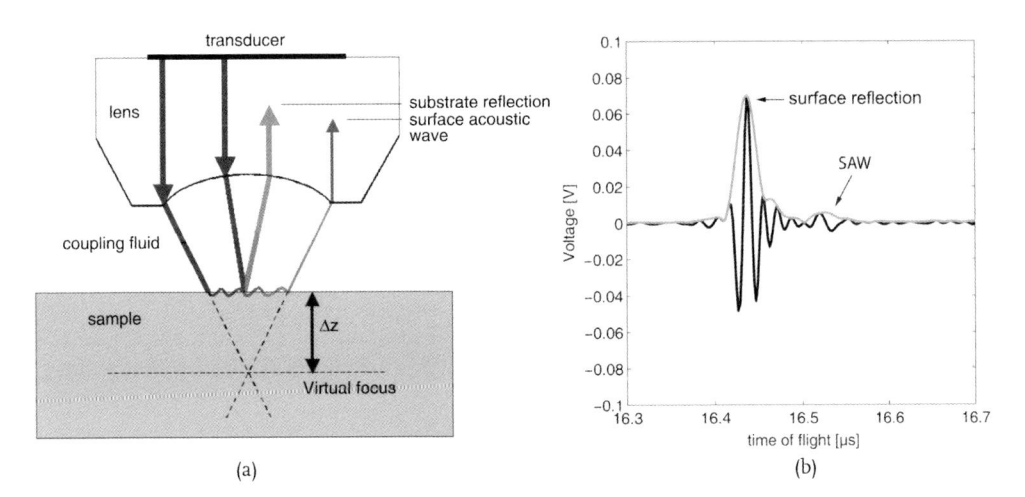

Figure 9.4 (a) Principle of the SAW measurement; (b) a typical 50-MHz pulse–echo measured at 800 μm defocus distance in human cortical bone.

different propagation velocities in coupling fluid and sample and the different propagation paths of the directly reflected waves and the SAW waves lead to a time delay Δt between reflected and SAW waves at the transducer that depends on the SAW velocity v_{SAW} and the defocus distance Δz:

$$v_{SAW} = \left[\frac{\Delta t}{v_0 \Delta z} - \frac{1}{4} \left(\frac{\Delta t}{\Delta z} \right)^2 \right]^{-\frac{1}{2}} \tag{9.3}$$

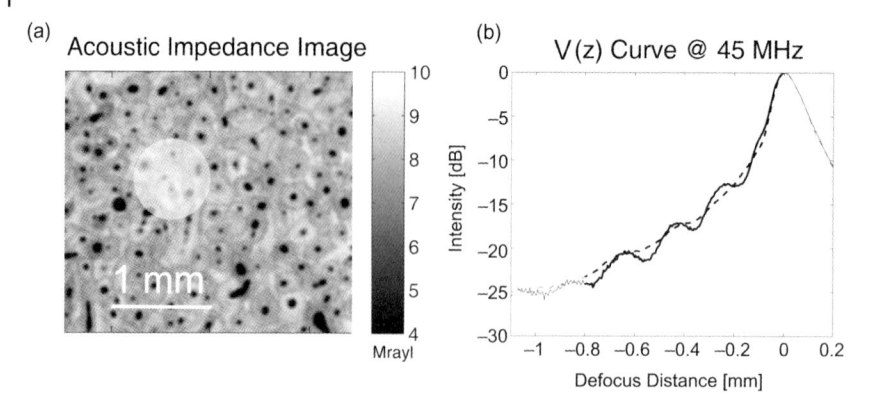

9 Quantitative Scanning Acoustic Microscopy of Bone

Figure 9.5 50-MHz acoustic impedance image (a) with a circular bright spot indicating the region for a V(z,t) measurement (b). The acoustic impedance was determined from the confocal reflection amplitude (z = 0 mm) measured in x and y directions. The V(z)-curve in (b) shows oscillation at negative defocus. The dashed line is from the reference signal measured in Teflon. The bold section was used for the estimation of the surface skimming compressional wave. From Raum [21], © 2008 IEEE.

In a quasi-monochromatic system the interference between SAW and reflected waves results in phase differences $\Delta\phi$ between the surface reflection and leaking surface wave, which successively increases with increasing Δz:

$$\frac{\Delta\phi}{\Delta z} = 2k(1-\cos\theta_{SAW}) \tag{9.4}$$

where $k = 2\pi/\lambda$ is the wave number, λ is the acoustic wavelength, and θ_{SAW} is the critical angle for the generation of a surface wave [20]. The interference of these waves lead to characteristic oscillations of the detected reflection amplitude in a so-called V(z) measurement (Figure 9.5). The spatial oscillation frequency due to this interference is:

$$\frac{1}{\Delta z} = \frac{2f}{v_0}(1-\cos\theta_{SAW}) \tag{9.5}$$

where f and v_0 denote the ultrasound frequency and the compressional wave velocity in the coupling fluid, respectively. The SAW velocity v_{SAW} is obtained by Snell's law:

$$v_{SAW} = \frac{v_0}{\sin\theta_{SAW}} \tag{9.6}$$

Rayleigh wave and surface skimming compressional wave (SSCW) velocities can easily be measured in homogeneous stiff biological materials, for example, dentin and tooth enamel, either with quasi-monochromatic burst excitation using the well-established V(z) technique [22–24] or with broad-band excitation, time-resolved V(z,t) acquisition and spectral analysis [3].

However, the heterogeneity and anisotropy of bone tissue render measurement and interpretation of SAW velocities in bone more difficult. The necessary defocusing increases the interrogated surface area and the spatial resolution is dramatically reduced compared to a confocal reflection measurement (Figure 9.5a). The radius of the illuminated surface area r_{max} is:

$$r_{max} = -z_{max} \tan \theta_0 \tag{9.7}$$

where θ_0 is the semi-aperture angle of the transducer and z_{max} is the maximum defocus position used for the analysis. The surface diameter contributing to the SAW measurement in Figure 9.5b was approximately 670 μm. It can be seen that the tissue contains several pores within this area.

Even with a 2 GHz lens (and a typical lens aperture of 100°), for which the maximum defocus is approximately 20 μm, the corresponding sampled surface diameter would be about 40 μm. Therefore, most of the recent SAM studies on bone tissue have measured the confocal reflection amplitude with frequencies between 50 MHz and 1.2 GHz.

9.2
Quantitative SAM-Based Impedance of Bone

9.2.1
Theory

The reflected amplitude of a plane wave incident at a boundary between a fluid and a homogeneous isotropic elastic material is proportional to the angular dependent reflectance function $R(\theta)$:

$$R(\theta) = \frac{Z_P \cos^2 2\theta_S + Z_S \sin^2 2\theta_S - Z_1}{Z_P \cos^2 2\theta_S + Z_S \sin^2 2\theta_S + Z_1} \tag{9.8}$$

$$Z_1 = \frac{\rho_1 v_1}{\cos \theta}, \quad Z_P = \frac{\rho_2 v_P}{\cos \theta_P}, \quad Z_S = \frac{\rho_2 v_S}{\cos \theta_S} \tag{9.9}$$

where θ is the angle of the incident wave with the normal vector of the surface, Z_1 is the acoustic impedance value of the coupling fluid. Z_P and Z_S in the solid material are related to the product of mass density ρ and compressional (v_P) and shear (v_S) velocities [25]. The characteristic acoustic impedance Z of a material is defined as the ratio of tensile stress σ_T to particle displacement velocity $\partial \mathbf{u}/\partial t$:

$$Z = -\frac{\sigma_T}{\partial \mathbf{u}/\partial t} \tag{9.10}$$

and is usually expressed in Mrayl (1 rayl = $1 \text{ kg m}^{-2} \text{s}^{-1}$). The transmission and reflection of plane waves at plane boundaries of anisotropic materials is described by the acoustic impedance $Z^{\hat{n}}$, which relates traction force T_{in} to particle velocity v_j [21, 26, 27]:

$$-T_{in} = \left(Z^{\hat{n}}\right)_{ij} v_j \qquad (9.11)$$
$$i, j = x, y, z,$$

where $\hat{n}$ is the direction in which the impedance is measured. Equation (9.11) can be written in matrix notation:

$$-T_{in} = \frac{n_{iK} C_{KL} k_{Lj}}{\omega} v_j \qquad (9.12)$$

where:

$$n_{iK} = \begin{bmatrix} n_x & 0 & 0 & 0 & n_z & n_y \\ 0 & n_y & 0 & n_z & 0 & n_x \\ 0 & 0 & n_z & n_y & n_x & 0 \end{bmatrix} \qquad (9.13)$$

and:

$$k_{Lj} = \begin{bmatrix} k_x & 0 & 0 \\ 0 & k_y & 0 \\ 0 & 0 & k_z \\ 0 & k_z & k_y \\ k_z & 0 & k_x \\ k_y & k_x & 0 \end{bmatrix} \qquad (9.14)$$

The acoustic impedance matrix elements for the direction $\hat{n}$ are:

$$\left(Z^{\hat{n}}\right)_{ij} = \frac{n_{iK} C_{KL} k_{Lj}}{\omega} \qquad (9.15)$$

where c_{KL} are the components of the stiffness tensor C. For a compressional wave propagation in the x-direction with $k_x = (\omega/v_{Px})$, $k_y = k_z = 0$, $n_x = 1$, $n_y = n_z = 0$, v_{Px} is the phase velocity of the longitudinal wave and Equation (9.15) becomes:

$$\omega\left(Z^{n_x}\right)_{xx} = c_{11} k_x \qquad (9.16)$$

which can be written in the form:

$$\left(Z^{n_x}\right)_{xx} = \sqrt{c_{11} \rho} \qquad (9.17)$$

Similarly, the impedance for a compressional wave propagating in the z-direction is:

$$\left(Z^{n_z}\right)_{zz} = \sqrt{c_{33} \rho} \qquad (9.18)$$

Equations (9.17) and (9.18) show that if the wave propagation direction and particle displacement are normal to the interface and the propagation direction is parallel to the direction i, the acoustic impedance normal to the surface $\left(Z^{n_i}\right)_{ii}$ is directly proportional to the square root of the elastic coefficient c_{ii} and the mass density ρ. The impedance for the propagation not parallel to the elastic symmetry axes can easily be obtained by rotation of the elastic stiffness tensor [26]. For the transverse isotropic case rotation in the xz plane yields [28]:

$$c(\theta) = c_{33}\cos^4\phi + 2(c_{13} + 2c_{44})\sin^2\phi\cos^2\phi + c_{11}\sin^4\phi \qquad (9.19)$$

where ϕ is the rotation angle; $c(\phi)$ is the elastic coefficient c_{33} of the rotated tensor. It can be seen that $c(0°) = c_{33}$ and $c(90°) = c_{11}$. Combining Equation (9.18) with (9.19) gives:

$$(Z^{n\theta})_\theta = \sqrt{c(\phi)\rho} \qquad (9.20)$$

hereinafter simply referred to as $Z(\phi)$.

With a focusing system, the received signal is composed of multiple waves incident at the boundary with angles between $0°$ and θ_{max}, that is, the semi-aperture angle of the lens. The theoretical description of the V(z) response of a spherically focused acoustic lens has been described in detail in Reference [25]. For isotropic materials the V(z) response is determined by a pupil function $P(\theta)$, the complex reflectance function $R(\theta)$, and an exponential function expressing the phase delay with respect to the defocus distance z:

$$V(z) = \int_0^{\pi/2} P(\theta)R(\theta)e^{-i2zk\cos\theta}\sin\theta\cos\theta d\theta \qquad (9.21)$$

However, for spherically focused sound fields the condition of plane wave propagation can be approximated in the focal point. Under the condition of normal incidence, that is, the surface of the sample is perpendicular to the sound beam axis, the generation of shear waves is not possible and the reflectance function can be replaced by the reflection coefficient R:

$$R = \frac{Z_2 - Z_1}{Z_2 + Z_1} \qquad (9.22)$$

where Z_1 and Z_2 are the acoustic impedance values of the coupling fluid and the material under investigation, respectively. Under this condition, the measured voltage is proportional to the reflection coefficient:

$$V(z_0) \sim R \qquad (9.23)$$

Hirsekorn et al. [29, 30] have shown that Equations (9.22) and (9.23) can be applied for highly focused beams, if the interface is positioned perpendicular to the beam axis in the focal plane of the acoustic beam. The major limitation of Equation (9.23) is the assumption that the solid material is homogeneous and flat within the area interrogated by the sound beam. Special care should therefore be taken to ensure that the dimensions of the structures to be measured are either much larger or much smaller than the wavelength and that the surface roughness is much smaller than the wavelength.

The spatial mapping of the confocal reflection amplitude has become the most straightforward way of high frequency assessment of bone, as it combines numerous advantages, for example:

- only one smooth sample surface of a thick or thin section has to be prepared;
- for a given transducer the highest spatial resolution is obtained in the focal plane of the sound field;
- a large lens aperture can be used that provides a spatial resolution of the order of the acoustic wavelength;
- the confocal reflection amplitude can be assessed with all common excitation and detection modes;
- rapid scans in the plane parallel to the sample surface enable a fast two-dimensional data acquisition;
- the normal stress components of the incoming wave front induce a uniaxial strain at the interface; in anisotropic and heterogeneous materials this uniaxial response is very sensitive to the directional compressional response of the material;
- easy adjustment of wavelength and interaction volume to the hierarchical structural dimension; for example, by variation of the ultrasound frequency from 1 MHz to 1 GHz the acoustic wavelength passes several structural levels of organization (Figure 9.2).

9.2.2
Time-Resolved Measurements

The analysis of time-resolved pulse–echo data has several advantages compared to measurements with amplitude-detected signals. The major benefit is that all the information is kept in the signal and can be used for analysis. The amplitude of the reflected signal can be determined in the time-domain or in the frequency-domain. However, some processing steps are necessary for a reliable estimation of the confocal reflection amplitude, that is:

- band-pass filtering,
- amplitude detection,
- time-of-flight based defocus correction.

Band-pass filtering is necessary to remove DC and other signal components outside of the signal band-width. It is important to preserve both amplitude and phase information for reliable time-of-flight estimates. For that reason, phase preserving forward and backward filtering approaches, for example, using a type II Chebyshev filter, have most often been applied. The amplitude can then be detected from the Hilbert-transformed envelope signal (Figure 9.6a). However, if the sampling frequency of the digitized signal is close to the Nyquist limit, this

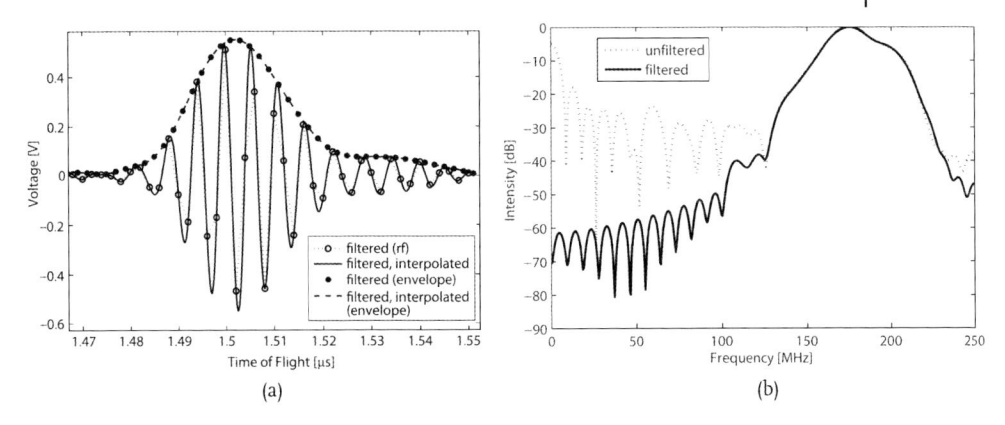

Figure 9.6 Time-resolved signal processing. Confocal pulse echo (a) and power spectrum (b).

approach can be quite inaccurate (Figure 9.6a). In this case, the signal should be up-sampled using an FFT-based interpolation prior to the Hilbert transformation. In the frequency domain, the square root of the integrated spectral intensity (ISI) from the power spectrum $S(f)$ (Figure 9.6b) can be computed:

$$\text{ISI} = \int_{f_1}^{f_2} S(f) df \tag{9.24}$$

where f_1 and f_2 are the $-6\,\text{dB}$ bandwidth limits. The time-of-flight (TOF) can be determined from either the position of the maximum of the Hilbert transformed envelope signal (Figure 9.6a) or the slope of the unwrapped phase spectrum within the bandwidth of the transducer:

$$\text{TOF} = t_0 + t_{\text{ph}} = t_0 + \frac{\partial \phi}{\partial \omega} = t_0 + \frac{\partial N}{\partial f} \tag{9.25}$$

where ϕ is the phase, $\omega = 2\pi f$ is the angular frequency, t_{ph} is the time relative to the start time t_0 of the digitized sequence, and N is the number of phase rotations [12].

TOF is a measure of the two-way pulse travel time from the transducer towards the reflecting surface. If the surface is located in the focal plane, the maximum amplitude is obtained. For a given combination of transducer, coupling fluid, and temperature the confocal time-of flight ($\text{TOF}_{\text{focus}}$) is invariant. Therefore, the TOF of a measured pulse echo can be used to estimate the distance of the surface from the focal plane (defocus) and to estimate the relative decrease of the reflection amplitude relative to the confocal reflection amplitude (Figure 9.7).

9.2.3
Measurements with Time-Gated Amplitude Detection

Acoustic impedance mapping above 200 MHz has often been performed with systems that use burst excitation and time-gated amplitude detection. With these

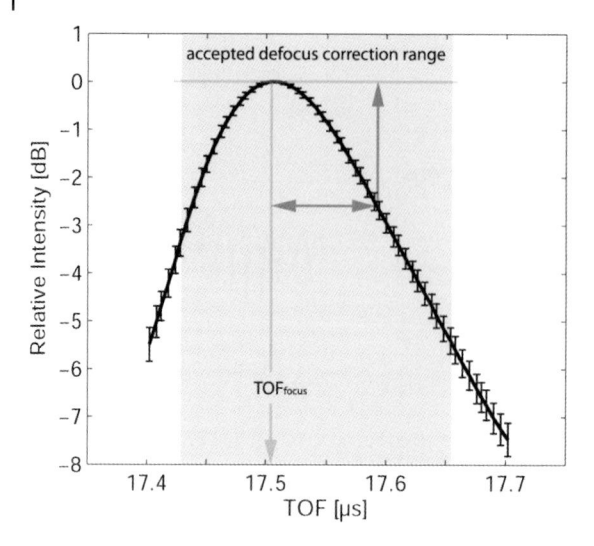

Figure 9.7 Defocus correction function for a 200-MHz transducer. Mean and standard error of the normalized intensity as a function of TOF. The accepted TOF range corresponds to a defocus range from 27 µm to +15 µm.

systems, the radio-frequency data are usually not accessible. Therefore, careful adjustment of all hardware settings is mandatory for a reliable assessment of the reflection coefficient. Proper selection of excitation frequency and time-gate position are crucial to obtain a good signal-to-noise ratio without saturation or truncation of the measured signal.

The confocal reflection amplitude is determined by a 2D analysis of the $V(z)$ signatures. This requires the acquisition of a set of digital xy-scans at successively decreasing transducer–sample distances [31]. Starting from a z-position, where the focus of the lens is well above the sample surface (positive defocus), images are captured with a successively decreasing lens–surface distance. The image acquisition is stopped when the focus is well below the surface everywhere in the scanned image (negative defocus). The z-increment between two adjacent x,y-scans should be small enough to fulfill the Nyquist limit, that is, the sampling frequency in the z direction should be at least two-times the highest expected oscillation frequency in the $V(z)$ curve.

For each x,y-coordinate, the confocal position z_0 and value of the maximum of $V_{xy}(z_0)$ can be determined (Figure 9.8). These values are used to compute the two-dimensional surface topography $z_0(x,y)$ and a topographically corrected confocal amplitude map $V_0(x,y)$. Moreover, for each x,y position the surface normal vector can be computed, which allows the estimation and correction for the local inclinations [21].

9.2.3.1 Calibration
The confocal reflection amplitudes $V_0(x,y)$ can be converted into values of the reflection coefficient by calibration with homogeneous isotropic and non-dispersive

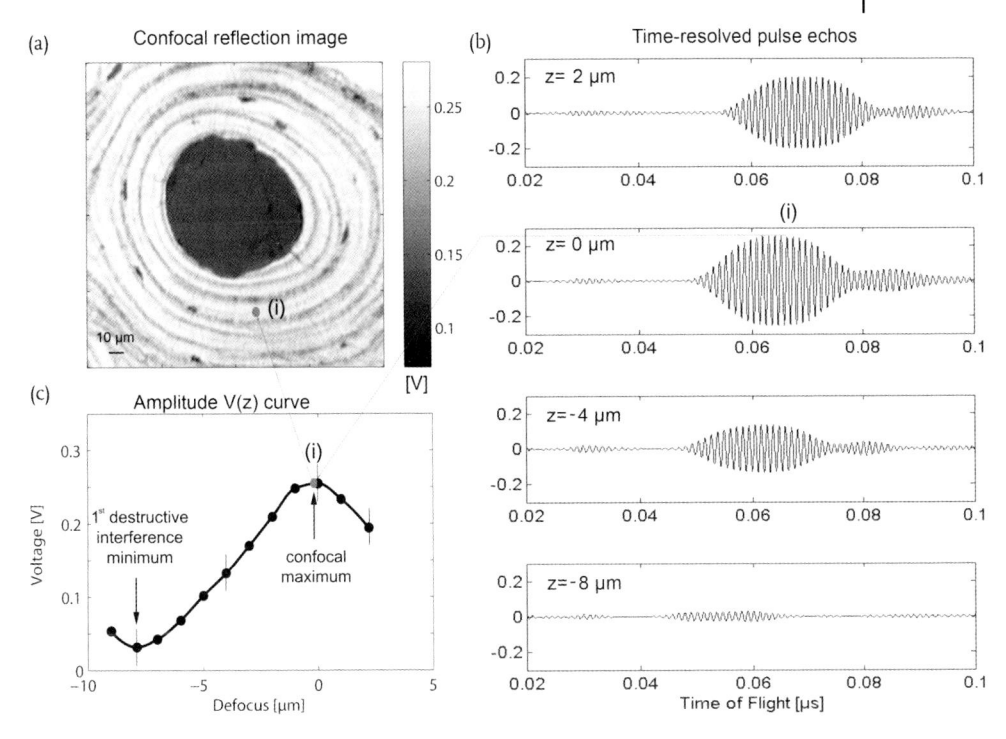

Figure 9.8 Confocal reflection amplitude image of an osteon with a central Haversian canal (a). The image was scanned with a time-resolved system and an acoustic center frequency of 905 MHz. The gray levels correspond to the maximum pulse echo amplitude at each x,y scan position. The corresponding radio frequency echo signals at a single x,y position indicated by the dot labeled (i) are shown in (b). Systems with amplitude detection only provide the variation of the signal amplitude with respect to the defocus position, which is determined by the characteristic V(z) curve (c).

materials. The speed of sound and mass density of these materials and the coupling fluid of these materials can be determined by a low frequency substitution method and by Archimedes' principle, respectively. From Equation (9.22) the corresponding reflection coefficients can be calculated and the relation between the calculated reflection coefficients R and the measured voltages is obtained by linear regression (Figure 9.9).

9.3
Tissue Mineralization, Acoustic Impedance, and Stiffness

Extensive studies have been conducted to clarify the relations between tissue mineralization and resulting elastic and acoustic properties. Equation (9.20) suggests that the acoustic impedance is determined by both mass density ρ and the apparent stiffness $c(\phi)$. It can also be seen from Equation (9.22) that the relation

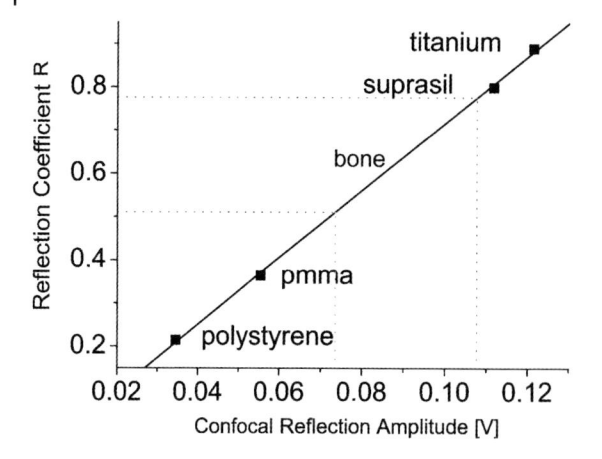

Figure 9.9 Impedance calibration. For homogeneous reference materials confocal reflection amplitude V_0 is proportional to the reflection coefficient R. The range of values measured in the bone tissue is marked with dotted lines.

between the amplitude of the reflected wave and the acoustic impedance is not linear. A good sensitivity to varying acoustic properties is only obtained for materials with intermediate or low acoustic impedance values. As the impedance increases, the reflection coefficient converges towards one and the confocal image contrast decreases. Within the typical impedance range of bone tissue (5–12 Mrayl) the average variation of the reflection coefficient R is approximately 3.3% Mrayl^{-1}.

Bone consists of three basic components: hydroxyapatite, collagen, and water. The total mass density of the tissue is therefore given by:

$$\rho_{tissue} = vf_{HA}\rho_{HA} + vf_{col}\rho_{col} + vf_{H_2O}\rho_{H_2O} \tag{9.26}$$

where vf_j is the volume fraction of the component j and ρ_j is the density. The subscript is HA for mineral, col for collagen, and H$_2$O for water. The mass densities for the three components are $\rho_{HA} = 3.0\,g\,cm^{-3}$, $\rho_{col} = 1.41\,g\,cm^{-3}$, and $\rho_{H_2O} = 1.0\,g\,cm^{-3}$. The volume fraction of mineral can be assessed in 3D by synchrotron radiation micro-computed tomography (SRµCT) [32] and is usually expressed as tissue degree of mineralization of bone (DMB):

$$DMB = vf_{HA}\rho_{HA} \tag{9.27}$$

Raum et al. [11] have shown that the relation of DMB with mass density can be approximated by a second-order polynomial:

$$\rho_{tissue} = 1.12\,g\,cm^{-3} + 0.73\,DMB - 0.033\,cm^3\,g^{-1} \cdot DMB^2 \tag{9.28}$$

Thus, by site-matched measurements of DMB and Z_i by SRµCT and SAM, respectively (Figure 9.10), the apparent elastic coefficient $c(\phi)$ can be determined using Equation (9.20).

9.3 Tissue Mineralization, Acoustic Impedance, and Stiffness

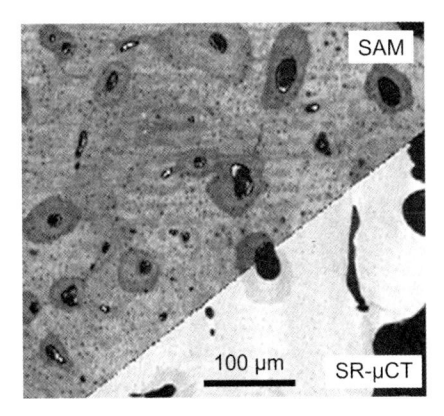

Figure 9.10 Detail of a fused acoustic impedance Z (SAM) and DMB (SR-µCT) image of a native cortical bone sample from a human forearm. For a better illustration the diagonal line separates regions for which either the SAM or the SR-µCT image is in the foreground. Osteons appear darker with both imaging modalities, indicating that this tissue is softer and less mineralized than the surrounding interstitial tissue. From Reference [12].

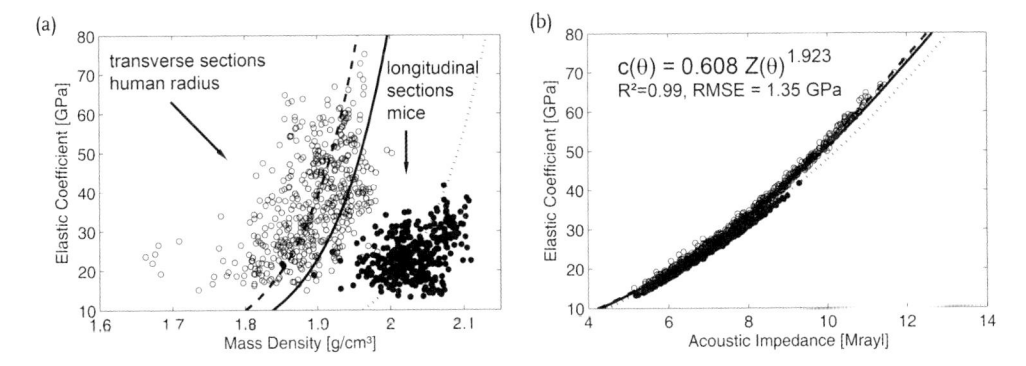

Figure 9.11 Relationships between acoustic impedance, mass density, and elastic stiffness for cortical bone from different species, anatomical sites, measurement directions, and sample preparation. The lines indicate exponential functions of the form $c = f(\rho) = \lambda \rho^\alpha$, and $c = f^*(Z) = \mu \rho^\beta$, where $\alpha = \beta(2 - \beta)^{-1}$ and $\lambda = \mu^{\alpha+1}$ that were fitted to the data (dashed: native transverse sections of human radius, dotted: embedded longitudinal sections of mice femora). From Reference [33].

Figure 9.11a shows the relation between mass density ρ and the apparent elastic coefficient $c(\phi)$ for mature human radius and mice femur samples. There is clearly an increase of the elastic coefficient with increasing mass density, but the relations differ between the tissues. This can be due to different fibril orientations, embedding, or other differences in the collagenous organic matrix [12]. Remarkably, this difference is much less apparent when $c(\phi)$ is plotted as a function of $Z(\phi)$ (Figure

9.11b). Therefore, the apparent elastic coefficient $c(\phi)$ can be approximated from $Z(\phi)$ by a single regression function:

$$c(\phi) = 0.608 Z(\phi)^{1.923} \tag{9.29}$$

Hereinafter, we call $Z(\phi)$ and $c(\phi)$ apparent impedance and apparent stiffness, respectively.

Notably, to obtain the Young's modulus it is necessary to know additional parameters, for example, the Poisson ratios.

9.4
Elastic Anisotropy at the Nanoscale (Lamellar) Level

With frequencies in the GHz range a characteristic lamellar pattern with alternating impedance values between adjacent lamellae is usually observed (Figures 9.2, 9.8, and 9.12). However, since the diameter of a single mineralized fibril (~0.1 µm) is still approximately one order of magnitude smaller than the wavelength, the individual fibrils are not resolved. Therefore, interpretation of the data obtained in the GHz range requires further ultrastructural model assumptions. Hofmann et al. [34] have evaluated osteonal tissue by site-matched SAM at 911 MHz, nanoindentation, and 2D Raman spectroscopy. They found that the relative mineral concentration within individual osteons is relatively homogeneous and concluded that the alternating impedance pattern observed with GHz ultrasound arises from mineralized collagen fibrils with relatively equal transverse isotropic elastic properties (with c_{33} parallel and c_{11} perpendicular to the fibril long axis; $c_{33} > c_{11}$) that are arranged in an asymmetric twisted plywood structure [35, 36]. According to this model fibril bundles are tilted progressively layer by layer with an angle of rotation between adjacent layers of around 30° (Figure 9.12d). One layer consists of a variable number of parallel fibrils and a lamellar unit is composed of six layers with fibril orientations from 0 to 180°. For example, the thickness of a lamellar unit in Figure 9.12 can be considered as the space between two adjacent low-impedance regions. The average lamellar unit thickness can be estimated from the oscillation period along the line drawn in Figure 9.12 to be 6.9 ± 0.1 µm. It is reasonable to argue that the fibrils oriented perpendicular to the osteon long axis are located in the regions with the local minima (corresponding to c_{11}), while the fibrils oriented parallel to the osteon long axis are at the regions of local maxima (corresponding to c_{33}). The layers with orientations of 30° and 60° should exhibit impedance values corresponding to the rotated elastic coefficients $c(30°)$ and $c(60°)$ of the fibrils. Using a transverse isotropic model, the remaining nanoscale coefficients can be approximated (Table 9.2). Apparently, due to the spatial resolution limit of approximately 1 µm at 1.2 GHz the six individual sublayers of the lamellar unit cannot be distinguished. However, the asymmetric shape of the oscillation pattern in Figure 9.12d supports an asymmetric arrangement of individual layer thicknesses within a lamellar unit, which explains the tissue anisotropy at the next (micro-scale) level of organization.

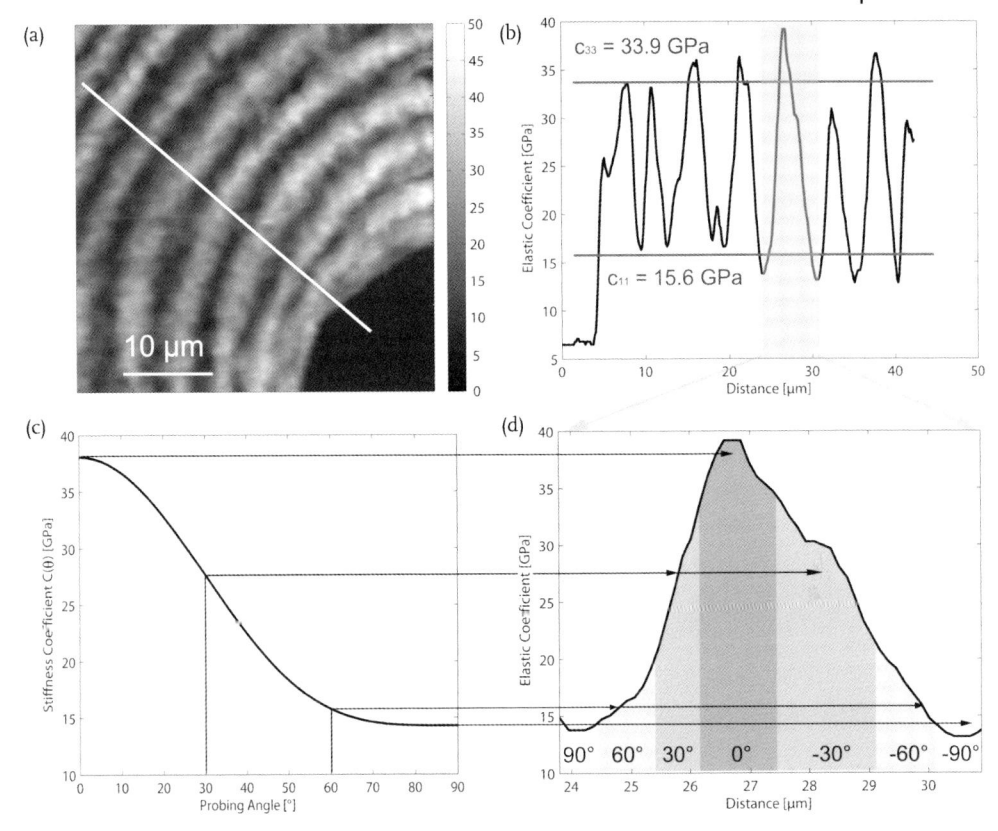

Figure 9.12 (a) Typical alterations of the elastic coefficient in osteonal lamellae. 1.2 GHz image, image plane perpendicular to the osteonal long axis, one quarter of an osteon. The Haversian canal is in the lower right part of the image. (b) Elastic coefficient measured along the line in (a). The mean maxima and minima correspond to c_{33} and c_{11} of the individual fibrils, respectively. A representative lamellar unit is highlighted in the gray box. (c) Theoretical $c(\phi)$ with estimated coefficients from Table 9.2. (d) Schematic illustration of a six-layer lamellar unit. The different gray scales indicate fibril layers with parallel alignment and variable thickness, but distinct orientations (0°: parallel to the osteon long axis).

9.5
Elastic Anisotropy at the Microscale (Tissue) Level

The elastic anisotropy of the tissue matrix has been measured by Lakshmanan et al. [27] in small cylindrically shaped cortical bone sections of a human femoral shaft (Figure 9.13). This method allowed direct assessment of c_{33}, c_{11}, and $c^* = 2(c_{13} + 2c_{44})$ from $c(\phi)$. The remaining elastic coefficients of a transverse isotropic stiffness tensor were derived using continuum micro-mechanical model constraints. The means and standard deviations of the derived elastic coefficients

Table 9.2 Stiffness coefficients (GPa) of human cortical tissue at several length scales.

Stiffness tensor	$[C]_{fibril}$	$[C]_{osteon}$	$[C]_{meso}$
Source (Reference)	[1]	[1]	[37]
Origin	Human femur, proximal shaft, 1 osteon, 1 male, age: 63 years	Human femoral shaft, 56 locations, osteonal & interstitial tissue, 1 female, age: 72 years	Human femur, mid-diaphysis, osteonal & interstitial tissue, ten female, age: 66–98 years
Frequency	1.2 GHz	50 MHz	1–2.25 MHz
c_{11}	15.6	23.7 ± 1.9	19.3 ± 2.2
c_{22}	$=c_{11}$	$=c_{11}$	19.8 ± 2.2
c_{33}	33.9	33.0 ± 3.0	29.2 ± 3.2
c_{44}	3.4	6.6 ± 3.0	5.8 ± 0.8
c_{55}	$=c_{44}$	$=c_{44}$	5.6 ± 0.8
c_{66}	$=(c_{11}-c_{12})/2$	$=(c_{11}-c_{12})/2$	4.2 ± 0.6
c_{12}	5.2	9.5 ± 1.2	$=c_{11}-2c_{66}$
c_{13}	8.8	10.0 ± 1.3	15.3
c_{23}	$=c_{13}$	$=c_{13}$	$=c_{13}$

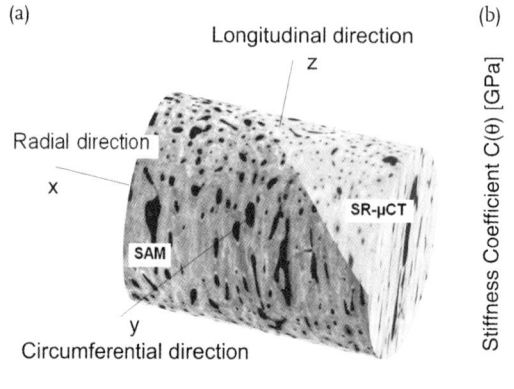

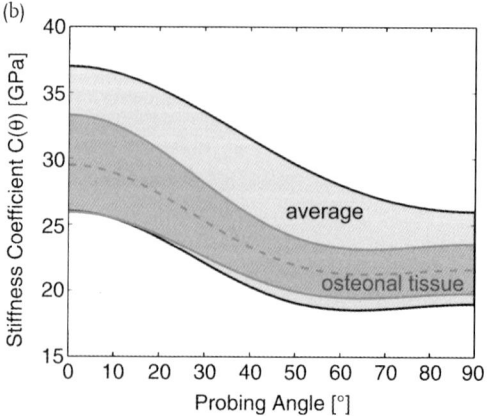

Figure 9.13 (a) Site-matched evaluation of DMB and $Z(\phi)$ in small cylindrically shaped cortical bone tissue sections (human femur: 4.4 mm in diameter); (b) typical course of $c(\phi)$ averaged over the entire mineralized matrix (osteonal and interstitial tissue) and separately for osteonal tissue.

for osteonal tissue are summarized in Table 9.2 and the range and course of the apparent stiffness in osteonal and interstitial tissue is shown in Figure 9.13b.

The conversion between elastic coefficients and engineering constants is straightforward, if all independent coefficients are known. For example, the conversion between c_{33} and E_3 is:

$$E_3 = \frac{(1+v_{12})(1-v_{12}-2v_{13}v_{31})}{1-v_{12}^2} \cdot c_{33} \tag{9.30}$$

where v_{12} is the Poisson's ratio in the cross-sectional plane ($x_1 x_2$-plane) and $v_{13} = v_{23}$ and $v_{31} = v_{32}$ are the Poisson ratios in the longitudinal section (i.e., $x_1 x_3$- and $x_1 x_2$-) planes.

An experimental approach is to assess the Poisson ratio at this length scale is a site-matched analysis of the acoustic impedance and the indentation modulus E_{IT} by SAM and nanoindentation, respectively [34, 38, 39]. Interestingly, in all studies consistent, rather moderate correlations between Z and the indentation modulus E_{IT} ($0.61 \leq R^2 \leq 0.67$) have been observed. Although at least some of the unexplained variances have to be attributed to experimental artifacts, for example, caused by surface roughness, viscous, and contact effects, matched interaction volumes, a considerable amount has been suggested to be caused by variations and directional dependence of the Poisson ratios [39].

9.6
Applications in Musculoskeletal Research

In vitro scanning acoustic microscopy (SAM) is currently the only nondestructive micro-elastic imaging technique that provides large-scale (cm range) structural and elastic properties at the tissue level with a spatial resolution down to the μm range (Figure 9.14). Two-dimensional impedance maps are particularly suitable for investigations of tissue healing in animal models. For example, Hube *et al.* [4] have shown that the combined assessment of structural and anisotropic elastic tissue properties in a sheep callus distraction model by 50-MHz SAM allowed the prediction of the fracture force, determined by four-point bending, with a very high accuracy ($R^2 = 0.86$, $p < 0.0005$). Multivariate regression analysis also revealed the relative contributions of structural (37.5%) and elastic matrix properties (48.5%) to the predicted variations of the fracture force. A very important finding of that study was that not only structural and elastic properties of the newly formed callus tissue but also those of the adjacent old cortical tissue contributed significantly to the macroscopic resistance to fracture.

In a similar animal model, Preininger *et al.* [33] have investigated the spatial and temporal elastic variations in cortical and callus tissues in the course of healing. Most importantly, it has been demonstrated that the gradual softening of the periosteal cortical tissue at the later consolidation time-points results in an elastic "handshake" with the mineralized callus, which is essential for the minimization of stress concentrations at the cortex–callus interface (Figure 9.14). The

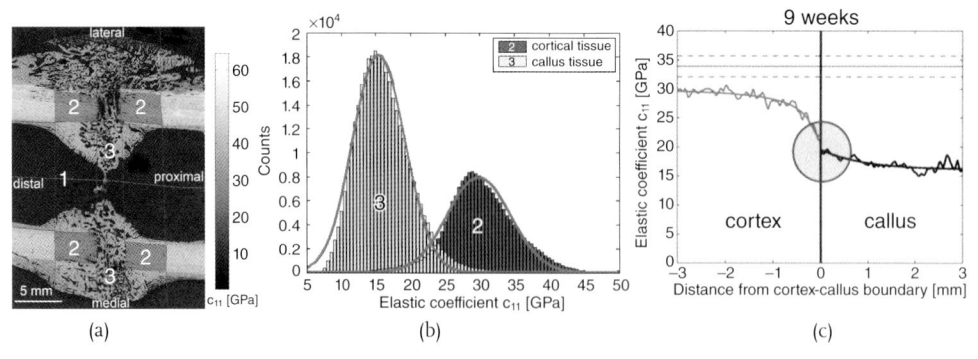

Figure 9.14 Stiffness map of a transverse section of a sheep osteotomy after 9 weeks of consolidation, measured by 50-MHz SAM (a). The overlaid colors indicate segmented evaluation regions (1: excluded non-mineralized tissue and embedding material; 2: cortical tissue; 3: callus tissue). Histograms of the elastic coefficient c_{11} obtained within the ROIs indicated in (a) reveal remarkable differences between the two tissues (b). Gradients of elastic coefficients relative to the periosteal cortex–callus boundary exhibit distinct steps at the interface at the early consolidation time points (not shown), while after 9 weeks a smooth elastic transition has been developed (c). The solid and dashed lines indicate mean ± standard deviation of c_{11} in intact cortical tissue. From Reference [33].

degree of this transformation from a step-like towards a smooth interface may be a future indicator for the biomechanical competence of the healing bone, since many fractures occur at the interface between callus and cortical tissues.

In recent studies, Rupin et al. [40] have demonstrated that quantitative impedance mapping at 200 MHz reveals intrinsic microelastic properties of bone induced by adaptive remodeling process in response to mechanical loading. On the other hand, genetic influences on the bone phenotype between C57BL/6J@Ico and C3H/HeJ@Ico mice have been proposed to be more pronounced in acoustic impedance variations (up to 13.2%) than in variations of DMB (up to 3.8%) [13].

9.7
Conclusions

Ultrasound offers various possibilities for the evaluation of bone. The acoustic wavelength can be varied over more than four orders of magnitude. Acoustic parameters, for example, acoustic impedance and sound velocities are directly linked with elastic parameters of the material interrogated by the acoustic wave. Owing to the hierarchical organization of bone the elastic properties at each level are determined by the compound properties of the preceding level. The mechanical function and resistance to fracture of cortical bone are predominantly determined by the intrinsic elastic properties of the mineralized collagen matrix and by the porous microstructure. While the porous microstructure can be assessed with high accuracy in three dimensions with other imaging modalities, for example, µCT, the target of ultrasound with frequencies between 50 MHz and 2 GHz is to assess the heterogeneous anisotropic elastic properties of the mineralized collagen matrix. Because of the small spatial dimensions of the characteristic structural units, for example, osteonal and interstitial tissue or lamellar units with rapid alterations of fibril composition and orientation, the requirements on the spatial resolution are demanding. The measurement of sound velocity in thin samples or the surface acoustic waves in thick samples requires defocusing of the sound field and, consequently, an increase in the interrogated sample volume. Therefore, the applicability of these methods is limited to relatively homogeneous tissue regions that are sufficiently far from structural boundaries.

Two-dimensional mapping of the confocal reflection amplitude has emerged from being a semi-quantitative method to the modality of choice for ultrasonic investigations of bone at the tissue level. A reliable estimation of the confocal reflection amplitude with defocus correction and surface tilt control is possible with time-resolved or amplitude detection microscopes. By adjusting the ultrasound frequency rapid scans can be performed either with large scan fields, to map the effective elastic coefficient of the tissue matrix, or with small scan fields and frequencies in the GHz range to map the anisotropic tissue properties at the lamellar level. In addition, the high-resolution imaging capability allows a precise estimation of microstructural properties.

The potential of a combined assessment of structural and tissue elastic properties in musculoskeletal research has already been demonstrated in several studies. Raum et al. [41] have used 50-MHz impedance maps in conjunction with synchrotron radiation µCT data to predict the velocity of the first arriving signal measured with diagnostic ultrasound (bi-axial transmission) in human radius sections. In the low megahertz range ultrasonic propagation in cortical bone depends on anisotropic elastic tissue properties, porosity, and on the cortical geometry, for example, thickness. Based on the SAM data a new model was derived that accounts for the nonlinear dispersion relation with the cortical thickness and predicts the velocity of the first arriving signal by a nonlinear combination of fracture-determining parameters, that is, porosity, cortical thickness, and tissue impedance ($R^2 = 0.69$, $p < 10^{-4}$, RMSE = 52 m s^{-1}). High-resolution acoustic impedance maps in combination with the locally derived average elastic stiffness tensor are perfectly suited for numerical deformation or sound propagation analyses on "real life" models (Figure 9.15). Such models are crucial for the development and validation of new non-invasive diagnostic tools dedicated to the prediction of an individual fracture risk. Moreover, assessment of changes of local tissue anisotropy at the lamellar level with ultrasound in the GHz range may provide new insight in studies of bone remodeling, for example, in the course of fracture healing, bone pathologies, aging, or adaptation to modified loading conditions at the bone–implant interface after endoprosthetic surgeries.

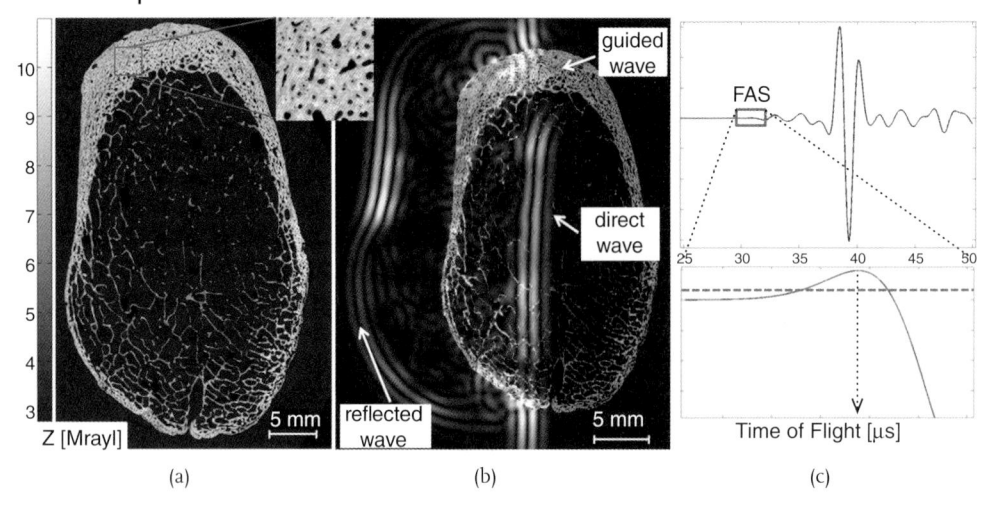

Figure 9.15 A 50-MHz impedance map of a human femoral neck (a) is directly meshed into a finite difference time domain (FDTD) simulation model (b). The snapshot shows a 0.5 MHZ wave traveling from the anterior (left) to the posterior (right) side. A part of the wave has been reflected at the cortical shell and scattered in the trabecular structure. Two transmitted waves can be seen. The direct wave propagates directly through the medullar canal, while the guided wave is directed through the cortical shell. A receiver on the right-hand side detects the transmitted waves (c). In most *in vivo* systems the time of flight of the first arriving signal is analyzed.

References

1 Raum, K., Grimal, Q., Laugier, P., and Gerisch, A. (2011) Multiscale structure-functional modeling of lamellar bone. *Proc. Meet. Acoust.*, **9**, 1–15.
2 Weiner, S. and Wagner, H.D. (1998) The material bone: structure mechanical function relations. *Annu. Rev. Mater. Sci.*, **28**, 271–298.
3 Raum, K., Kempf, K., Hein, H.J., Schubert, J., and Maurer, P. (2007) Preservation of microelastic properties of dentin and tooth enamel *in vitro*–a scanning acoustic microscopy study. *Dent. Mater.*, **23**, 1221–1228.
4 Hube, R., Mayr, H., Hein, W., and Raum, K. (2006) Prediction of biomechanical stability after callus distraction by high resolution scanning acoustic microscopy. *Ultrasound Med. Biol.*, **32**, 1913–1921.
5 Gupta, H.S., Stachewicz, U., Wagermaier, W., Roschger, P., Wagner, H.D., and Fratzl, P. (2006) Mechanical modulation at the lamellar level in osteonal bone. *J. Mater. Res.*, **21**, 1913–1921.
6 Ashman, R.B., Cowin, S.C., Rho, J.Y., Van Buskirk, W.C., and Rice, J.C. (1984) A continuous wave technique for the measurement of the elastic properties of cortical bone. *J. Biomech.*, **17**, 349–361.
7 Bensamoun, S., Ho Ba Tho, M.C., Luu, S., Gherbezza, J.M., and de Belleval, J.F. (2004) Spatial distribution of acoustic and elastic properties of human femoral cortical bone. *J. Biomech.*, **37**, 503–510.
8 Pithioux, M., Lasaygues, P., and Chabrand, P. (2002) An alternative ultrasonic method for measuring the elastic properties of cortical bone. *J. Biomech.*, **35**, 961–968.

9 Rho, J.Y. (1996) An ultrasonic method for measuring the elastic properties of human tibial cortical and cancellous bone. *Ultrasonics*, **34**, 777–783.
10 Xu, J., Rho, J.Y., Mishra, S.R., and Fan, Z. (2003) Atomic force microscopy and nanoindentation characterization of human lamellar bone prepared by microtome sectioning and mechanical polishing technique. *J. Biomed. Mater. Res.*, **67A**, 719–726.
11 Raum, K., Cleveland, R.O., Peyrin, F., and Laugier, P. (2006) Derivation of elastic stiffness from site-matched mineral density and acoustic impedance maps. *Phys. Med. Biol.*, **51**, 747–758.
12 Raum, K., Leguerney, I., Chandelier, F., Talmant, M., Saied, A., Peyrin, F., and Laugier, P. (2006) Site-matched assessment of structural and tissue properties of cortical bone using scanning acoustic microscopy and synchrotron radiation μCT. *Phys. Med. Biol.*, **51**, 733–746.
13 Raum, K., Hofmann, T., Leguerney, I., Saied, A., Peyrin, F., Vico, L., and Laugier, P. (2007) Variations of microstructure, mineral density and tissue elasticity in B6/C3H mice. *Bone*, **41**, 1017–1024.
14 Baron, C., Talmant, M., and Laugier, P. (2007) Effect of porosity on effective diagonal stiffness coefficients (cii) and elastic anisotropy of cortical bone at 1 MHz: a finite-difference time domain study. *J. Acoust. Soc. Am.*, **122**, 1810.
15 Grimal, Q., Raum, K., Gerisch, A., and Laugier, P. (2008) Derivation of the mesoscopic elasticity tensor of cortical bone from quantitative impedance images at the micron scale. *Comput. Methods Biomech. Biomed. Eng.*, **11**, 147–157.
16 Hasegawa, K., Turner, C.H., Recker, R.R., Wu, E., and Burr, D.B. (1995) Elastic properties of osteoporotic bone measured by scanning acoustic microscopy. *Bone*, **16**, 85–90.
17 Turner, C.H., Chandran, A., and Pidaparti, R.M. (1995) The anisotropy of osteonal bone and its ultrastructural implications. *Bone*, **17**, 85–89.
18 Turner, C.H., Rho, J.Y., Takano, Y., Tsui, T.Y., and Pharr, G.M. (1999) The elastic properties of trabecular and cortical bone tissues are similar: results from two microscopic measurement techniques. *J. Biomech.*, **32**, 437–441.
19 Raum, K. (2011) Microscopic elastic properties, in *Bone Quantitative Ultrasound* (eds P. Laugier and G. Haiat), Springer Science + Business Media B.V., Dordrecht, Heidelberg, London, New York, pp. 409–440.
20 Raum, K. (2003) Ultrasonic characterization of hard tissues, in *Ultrasonic Nondestructive Evaluation: Engineering and Biological Material Characterization* (ed. T. Kundu), CRC Press, Boca Raton, FL, pp. 761–781.
21 Raum, K. (2008) Microelastic imaging of bone. *IEEE Trans. Ultrason. Ferroelectr. Freq. Control*, **55**, 1417–1431.
22 Peck, S.D., Rowe, J.M., and Briggs, G.A. (1989) Studies on sound and carious enamel with the quantitative acoustic microscope. *J. Dent. Res.*, **68**, 107–112.
23 Peck, S.D. and Briggs, G.A. (1986) A scanning acoustic microscope study of the small caries lesion in human enamel. *Caries Res.*, **20**, 356–360.
24 Peck, S.D. and Briggs, G.A. (1987) The caries lesion under the scanning acoustic microscope. *Adv. Dent. Res.*, **1**, 50–63.
25 Briggs, G.A. (1992) *Acoustic Microscopy*, Clarendon Press, Oxford.
26 Auld, B.A. (1990) *Acoustic Fields and Waves in Solids*, Krieger Publishing Company, Malabar, Florida.
27 Lakshmanan, S., Bodi, A., and Raum, K. (2007) Assessment of anisotropic tissue elasticity of cortical bone from high-resolution, angular acoustic measurements. *IEEE Trans. Ultrason. Ferroelectr. Freq. Control*, **54**, 1560–1570.
28 Jones, R.M. (1984) *Mechanics of Composite Materials*, McGraw-Hill, New York.
29 Hirsekorn, S., Pangraz, S., Weides, G., and Arnold, W. (1995) Measurement of elastic impedance with high spatial resolution using acoustic microscopy. *Appl. Phys. Lett.*, **67**, 745–747.
30 Hirsekorn, S., Pangraz, S., Weides, G., and Arnold, W. (1996) Erratum: Measurement of elastic impedance with high spatial resolution using acoustic microscopy. *Appl. Phys. Lett.*, **69**, 2138.

31 Raum, K., Jenderka, K.V., Klemenz, A., and Brandt, J. (2003) Multilayer analysis: Quantitative scanning acoustic microscopy for tissue characterization at a microscopic scale. *IEEE Trans. Ultrason. Ferroelectr. Freq. Control*, **50**, 507–516.

32 Nuzzo, S., Peyrin, F., Cloetens, P., Baruchel, J., and Boivin, G. (2002) Quantification of the degree of mineralization of bone in three dimensions using synchrotron radiation microtomography. *Med. Phys.*, **29**, 2672–2681.

33 Preininger, B., Checa, S., Molnar, F.L., Fratzl, P., Duda, G.N., and Raum, K. (2011) Spatial-temporal mapping of bone structural and elastic properties in a sheep model following osteotomy. *Ultrasound Med. Biol.*, **37**, 474–483.

34 Hofmann, T., Heyroth, F., Meinhard, H., Franzel, W., and Raum, K. (2006) Assessment of composition and anisotropic elastic properties of secondary osteon lamellae. *J. Biomech.*, **39**, 2284–2294.

35 Giraud-Guille, M.M. (1988) Twisted plywood architecture of collagen fibrils in human compact bone osteons. *Calcif. Tissue Int.*, **42**, 167–180.

36 Giraud-Guille, M.M., Besseau, L., and Martin, R. (2003) Liquid crystalline assemblies of collagen in bone and *in vitro* systems. *J. Biomech.*, **36**, 1571–1579.

37 Granke, M., Grimal, Q., Saied, A., Nauleau, P., Peyrin, F., and Laugier, P. (2011) Change in porosity is the major determinant of the variation of cortical bone elasticity at the millimeter scale in aged women. *Bone*, **49**, 1020–1026.

38 Rupin, F., Saied, A., Dalmas, D., Peyrin, F., Haupert, S., Barthel, E., Boivin, G., and Laugier, P. (2008) Experimental determination of Young modulus and Poisson ratio in cortical bone tissue using high resolution scanning acoustic microscopy and nanoindentation. *J. Acoust. Soc. Am.*, **123**, 3785.

39 Rupin, F., Saied, A., Dalmas, D., Peyrin, F., Haupert, S., Raum, K., Barthel, E., Boivin, G., and Laugier, P. (2009) Assessment of microelastic properties of bone using scanning acoustic microscopy: a face-to-face comparison with nanoindentation. *Jpn J. Appl. Phys.*, **48**, 07GK01.

40 Rupin, F., Bossis, D., Vico, L., Peyrin, F., Raum, K., Laugier, P., and Saied, A. (2010) Adaptive remodeling of trabecular bone core cultured in 3-D bioreactor providing cyclic loading: an acoustic microscopy study. *Ultrasound Med. Biol.*, **36**, 999–1007.

41 Raum, K., Leguerney, I., Chandelier, F., Bossy, E., Talmant, M., Saied, A., Peyrin, F., and Laugier, P. (2005) Bone microstructure and elastic tissue properties are reflected in QUS axial transmission measurements. *Ultrasound Med. Biol.*, **31**, 1225–1235.

Part Four
Advanced Materials Applications

10
Array Imaging and Defect Characterization Using Post-processing Approaches
Alexander Velichko, Paul D. Wilcox, and Bruce W. Drinkwater

10.1
Introduction

A typical ultrasonic array consists of several individually addressable piezoelectric transducer elements in a single package. The most common type of array is a linear, one-dimensional (1D) array with a row of rectangular elements in a straight line as shown in Figure 10.1a. This type of array enables control of the transmitted and received sound field in the two-dimensional (2D) imaging-plane indicated in the figure. In the simplest type of 2D array the array elements are distributed in a grid pattern (Figure 10.1b) and control of the sound field is possible in a three-dimensional (3D) imaging-volume.

There has been rapid growth in the use of ultrasonic array technology in non-destructive evaluation applications over the last decade. This growth has been driven by two key benefits of arrays over traditional single element transducers. Firstly, an array is able to perform a range of different inspections from a given location and so is significantly more flexible than a single element transducer. For example, the same array can produce plane, focused, and steered beams as well as more complex inspections such as those involving multiple reflections and/or mode conversions. In this respect, an array can be deployed in the same way as a single-element transducer in standard inspection configurations, such as the direct contact and wedge-coupled modes shown in Figure 10.1c and d, respectively. Secondly, arrays can be used to produce detailed images at each test location. This allows rapid visualization of the internal structure of a component and the mapping of results onto engineering drawings in an intuitive way. These advantages have led to a significant take-up of ultrasonic array technology by industry and the sector is still expanding. At the time of writing there are already thousands of array systems in use in industry and it now seems possible that array inspections will completely replace the present industry standard single element measurements over the course of the next decade.

There are now a large number of successful demonstrations of the use of commercially available array systems for specific industrial application. For example,

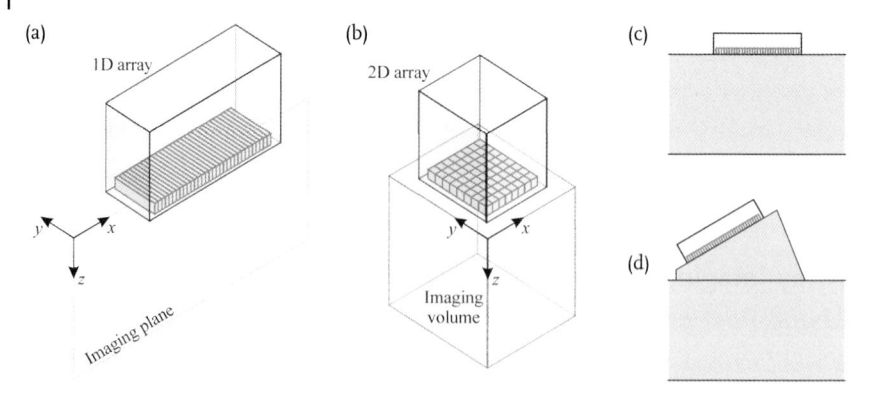

Figure 10.1 Schematic diagrams of array geometries and common configurations: (a) 1D array for 2D imaging; (b) 2D array for 3D imaging; (c) contact configuration; (d) wedge or immersion coupled array.

Long et al. [1] and Brekow et al. [2] both describe the use of ultrasonic arrays for the inspection of thick section pressure vessels commonly found in the power generation industry. Long et al. showed that array images can be corrected to account for known anisotropy, such as found in austenitic steel welds. Both publications demonstrate the use of an array to steer the beam through a range of angles and so ensure good sensitivity to a range of crack orientations. Lane et al. [3] adaptively compensated for the unknown anisotropy found in an aircraft turbine blade manufactured from a single crystal nickel alloy. This compensation then allowed highly focused beams to be synthetically produced and various crack-like defects detected reliably. This paper touches on the emerging topic of adaptive imaging in which initially unknown array imaging parameters are extracted as part of the imaging process. Recent publications in the array field describe increasingly challenging inspections, such as small defects, complex geometries, and accurate sizing/characterizing. For example, Johnston [4] has explored the use of arrays to detect micron thickness oxide inclusions in friction stir welds and Satyarnarayan et al. [5] have demonstrated how surface breaking cracks due to fatigue can be accurately sized in two dimensions.

Array imaging and defect characterization generally follow one of two distinct philosophies: beamforming or post-processing. In the beamforming philosophy multiple elements are fired almost simultaneously according to a predefined set of delays (often termed a delay law) such that a physical ultrasonic beam of the desired characteristics (e.g., focal depth and steering direction) is formed within the test structure. On reception, the signals from multiple elements are delayed and summed in a similar fashion, leading to a single time-domain signal (called an A-scan) that typically contributes one line in the final image. The process is then repeated with a different set of delays, for example to rotate or translate the beam and generate another line in the image. This process continues until a complete image (a B-scan) is formed. In the beamforming approach, the array directly emulates the operation of a monolithic single element device. In this approach, the delay laws must be predefined and the result is a single image. If a different imaging method is required then the whole process must be repeated. Conversely, in the most general version of the post-processing philosophy the array elements are fired individually and the data received from each element is captured and stored. In this way the full matrix of all possible transmit–receive combinations can be obtained and this is referred to as full matrix capture (FMC) [6]. To form an image this data matrix is post-processed and it is only in this post processing step that the imaging method (or algorithm) must be defined. In this sense the post-processing approach splits the data capture and imaging into two independent actions. With the assumption that all the ultrasonic phenomena in the sample are governed by a linear differential equation (which is a very accurate assumption at the power levels used in non-destructive evaluation, NDE), the FMC data matrix contains all possible data from the array. Consequently, any conceivable imaging algorithm can be realized by processing the contents of the FMC data matrix. For example, if an image based on a conventional beamforming algorithm is generated from FMC data it is impossible to discern if it has been formed through physical beamforming or post-processing [7].

The post-processing philosophy has some significant advantages over the beamforming philosophy. Arguably the most important advantage comes from the separation of the data capture from the imaging, which means that a wide range of imaging algorithms can be applied to the same data, either immediately after capture or at some later date. To some extent this future proofs the measurement and means that as new and improved imaging methods emerge they can be deployed on historic data. Indeed the idea of long-term storage of the raw data matrix data in this way is itself appealing in safety critical applications where auditability is increasingly important. The second major advantage is that various algorithms that would be prohibitively slow as beamforming operations take little or no extra time as post-processing operations. The most obvious example is the total focusing method, in which the beam is focused, in turn, on every pixel in the image to achieve very high-resolution imaging [6]. As a beamforming operation this would require as many firings as there are pixels in the image. Conversely, as a post-processing algorithm it takes a fraction of a second on a standard desktop computer. Given the full matrix of data it is also possible interrogate the data and examine, for example, the angular reflectivity of a scatterer to aid characterization [8–10]. The characterization potential of array post-processing methods is one of the central themes of this chapter. Post-processing also opens the door to a range of adaptive imaging techniques or those based on inverse modeling. Such algorithms often involve computation times an order of magnitude greater than the data capture time [11, 12] and so are better performed off-line.

Although arrays are more commonly viewed as a means of obtaining high-resolution images, often at the expense of scanning speed, several approaches have been developed that allow for rapid inspection. For example, Freemantle *et al.* [13] and Smith *et al.* [14] have described array scanning systems with application to large aerospace components. The device in Freemantle *et al.* uses an array mounted in a wheel probe that can be rapidly scanned across the component whereas the device in Smith *et al.* uses a confined water column and is mounted in a more traditional scanning system. In both cases the array is used to produce a predetermined beam and this is swept across the array perpendicular to the scan direction to form a c-scan type image. Ultimately, the speed of any array inspection is limited by the number of transmit cycles required at a given location. Often the electronics is capable of extremely fast repetitions but the maximum usable scanning speed is governed by the time taken for the ultrasonic waves to attenuate within the sample as this determines the minimum time between the transmit cycles. For this reason various authors have considered approaches in which images can be formed with the minimum number of transmit cycles. For example, Verkooijen and Boulavinov [15] use a single simultaneous firing on transmission, receive on all elements (in parallel), and then post-process the data using the synthetic aperture focusing technique (SAFT) to reconstruct an image. Moreau *et al.* [16] used two or four transmissions and attempted to compensate for the missing data using the effective aperture post-processing technique. Inevitably both approaches involve a trade-off between speed and resolution and, therefore, do not suit applications that demand the highest possible resolution.

A large proportion of the current array applications and research involve 1D arrays that generate 2D images. The reasons for this are historic and based on the high cost per channel of arrays and array controllers. However, the use of 2D arrays to perform 3D imaging is probably the most rapidly expanding aspect of array development at the present time. The benefits, in terms of visualization and characterization [17, 18] are obvious so it seems reasonable to assume this trend will continue. For this reason we look to the future and present several applications of 2D array technology in this chapter.

Because of the wide range of possible array geometries and inspection modalities, modeling plays a crucial part in the use of arrays for NDE. Typically the challenge is to select an array to achieve a required level of performance. This performance requirement could include a specification of minimum detectable defect size, defect sizing accuracy, maximum steering angle, depth of penetration, range of inspection angles, volume inspected from each position, or speed of inspection. The modeling of ultrasonic array data is therefore described in detail in Section 10.2. In Section 10.3, the key imaging algorithms are described for 1D arrays. This section includes a comparison of the algorithms in terms of both their performance and computational burden. The section concludes with some model and experimental examples. Section 10.4 investigates the special considerations associated with 2D array imaging, one of the most important of which is the layout of elements within a 2D aperture. Linear imaging algorithms of the type discussed in Sections 10.3 and 10.4 can be used to detect but not characterize sub-wavelength defects because of the diffraction limit. To characterize and size defects below the diffraction limit requires more advanced techniques. One such method is based on extraction of the so-called scattering matrix from the ultrasonic array data. The scattering matrix encodes all the available ultrasonic information about a defect and techniques for performing the extraction are the subject of Section 10.5. Once extracted, the scattering matrix must be probed to deduce the salient information about a defect, such as its shape, orientation, and size. This is covered in Section 10.6. Finally, Section 10.7 presents some concluding remarks concerning the future use of arrays in NDE.

10.2 Modeling Array Data

10.2.1 Introduction

An ultrasonic array coupled to a component under inspection either directly or indirectly via an intermediate coupling medium produces a collection of time-domain signals or time-traces. The analytical modeling or numerical simulation of these time-traces for a particular component containing known defects defines the forward problem; the recovery of information about the interior of the component from the time-traces is the inverse problem. This section considers two methods of describing the forward modeling problem. The first is based on an intuitive ray-tracing approach and the second on a more formal mathematical description of the elastodynamic scattering problem.

The formulation of the forward model itself informs the development of solutions to the inverse problem and enhances understanding of the physical phenomena associated with arrays. For these reasons, the two methods of modeling array data described in this section should not be regarded simply as alternative ways of arriving at the same result in terms of numerically simulated data; they should also be considered as complementary tools for gaining insight into the overall process. Throughout, this section, the simple contact inspection geometry shown in Figure 10.2 is used as an example. Here a 5 MHz center frequency, 32-element

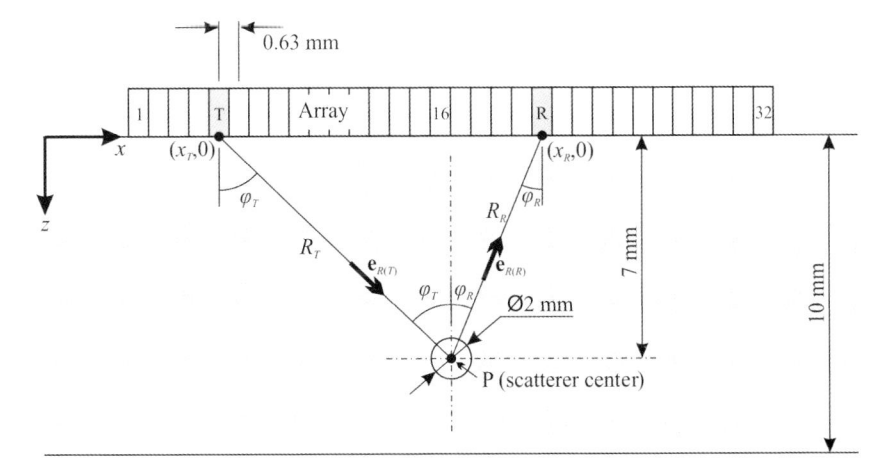

Figure 10.2 Geometry of contact configuration used for simulation examples. The material is assumed to be aluminum (Young's modulus = 70 GPa, Poisson's ratio = 1:3, density = 2700 kg m^{-3}).

array with 0.63 mm element pitch in the top surface of a 10 mm thick aluminum plate is considered. Directly below the center of the plate is a reference reflector in the form of a circular cylindrical cavity (i.e., a side-drilled hole) 2 mm in diameter. The geometry is 2D and a 1D linear array is assumed. This means that scatterers and array elements are assumed to be infinite in the third dimension. However, all results can be easily extended for a 3D system.

10.2.2
Ray-Based Description of Ultrasonic Array Data

In a ray-based description, the ultrasonic propagation along particular ray-paths between a transmitting element and a receiving element is considered, rather than the complete elastodynamic wave-field. A ray-based description of array data has several benefits, not least in its intuitiveness. It leads directly to equivalent ray-based imaging algorithms such as the total focusing method (TFM) described in Section 10.3.2. It is also the basis of most efficient numerical simulations of ultrasonic array data.

After an element in an array has emitted a pulse, the overall signal recorded at a given receiver element is the superposition of signals along all possible ray-paths connecting the transmitter to the receiver. For practical purposes the infinite number of possible ray-paths considered is limited to the finite number that result in arrivals within a certain time-frame of interest. In numerical simulations the number of rays considered may be further limited, both for the purposes of reducing computational load and also to aid understanding of how different signal paths contribute to different features in the final ultrasonic image obtained from an array.

10.2.2.1 Determining the Ray-Paths

The first part of modeling ultrasonic array data using rays is to identify the possible ray-paths between a particular transmitter–receiver pair of array elements. An array element is considered to be a point source or receiver of rays from any direction. In a homogeneous material rays propagate as straight lines; at discontinuities some form of scattering occurs, possibly resulting in one ray splitting into multiple rays. Each ray-path is therefore a continuous line from transmitter to receiver, composed of one or more straight segments with changes of direction occurring whenever the ray interacts with a scatterer. As an illustrative example, the first four longitudinal (L) ray-paths associated with the simple contact configuration of Figure 10.2 are shown in Figure 10.3a. In Figure 10.3a, T denotes a transmitting element, R a receiving element, P is the location of a small scatterer such as a defect, and W is the back wall of the component. Ray-paths 1 and 2 are the direct reflections from P and W, respectively. Ray-paths 3 and 4 are secondary scattering paths and it is destructive interference from contributions along these paths that gives rise to "shadows" of P on W in subsequent image formation. Note that these four paths do not cover all the signals that will be physically observed over the time window of interest as shear, S, wave paths have not been considered.

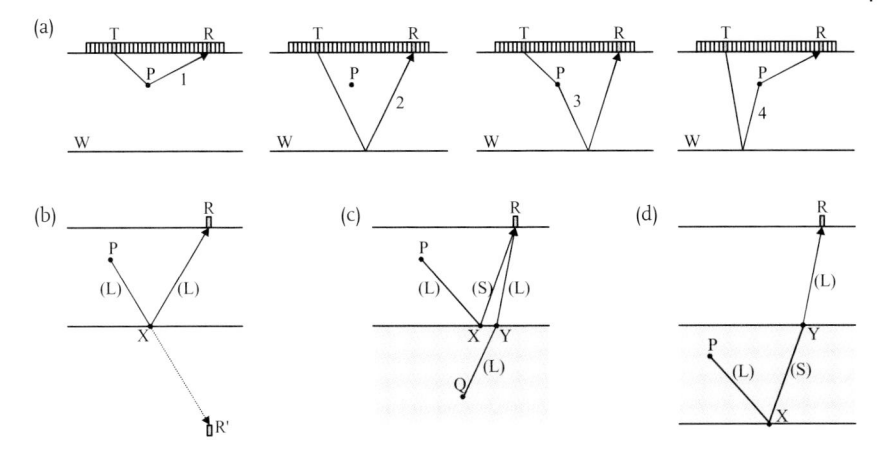

Figure 10.3 (a) Longitudinal mode ray-paths for a simple contact array configuration shown in Figure 10.2; (b) ray-path for a reflection without mode conversion between two known points; (c) ray-paths for reflection with mode conversion and transmission at interface between media; (d) ray-path involving multiple interfaces.

When computing the physical ray-path, it is necessary to distinguish between geometrically large and small scatterers. A large scatterer is one such as the back wall of a specimen that is many times larger than the spatial size of an ultrasonic wavepacket. Consequently, as far as an incoming ray is concerned, the discontinuity is infinite in spatial extent and the scattered ray or rays continue in straight lines at certain directions according to Snell's law. Conversely, for a scatterer that is comparable in size to the incident wave-packet, diffraction effects become significant and, from the point of view of ray-tracing, it should be assumed that rays can potentially leave the scatterer in any direction. This means that ray tracing when only small scatterers are considered is very straightforward as it only involves line segments that have start or end points at predefined locations (i.e., the transmitter and receiver element positions and the scatterer positions). The two segments of the first ray-path shown in Figure 10.3a are of this type. On the other hand, large scatterers present more difficulty as the point at which a ray is scattered is not predefined and must be calculated. Consider reflection and transmission at a plane surface on the portion of a ray between two defined points. If the ray of interest undergoes reflection without mode conversion then the calculation of the reflection point, X, is trivial trigonometry. In the example shown in Figure 10.3b one end point (in this case the receiver R) is mirrored in the reflector to form a virtual receiver at R' and a straight line drawn between P and R' defines the ray-path. However, if the ray of interest is subject to mode conversion in reflection or if refraction occurs then the calculation of the interaction point involves the solution of a quartic equation (arising out of the application of Snell's law). These cases are illustrated in Figure 10.3c. If there is interaction with more than one large scatterer between defined start and end points the calculation of the interaction points and hence the ray-path generally requires iterative techniques. A practical situation where this arises is in certain types of immersion testing, as illustrated in Figure 10.3d. Here scattered longitudinal waves from a defect at P are mode-converted by reflection at point X on the back wall of the component to shear waves. The shear waves are themselves mode-converted again to longitudinal waves by refraction into the coupling fluid at Y before detection at the receiver R. In this case, the locations of X and Y must be determined using an iterative method in order for the complete ray-path to be determined.

10.2.2.2 Predicting the Signal Associated with a Ray-Path

Once a complete path from a transmitter element to a receiver element associated with a particular ray is known, it is necessary to predict the corresponding received signal. It is assumed that the signal injected into the sample from a transmitter element is known. The effects that need to be taken into account when predicting the received signal are:

- the directivity of the transmitting and receiving elements;
- a delay due to the transit time along the path;
- a reduction in amplitude due to geometric attenuation (beam spreading) effects;
- a possible exponential reduction in amplitude with propagation distance from attenuation due to absorption and scattering from material microstructure;
- modifications to amplitude and phase due to interaction of the ray with discrete features, such as reflections from boundaries and scattering from defects.

All these effects can be exactly modeled in the frequency-domain and Fourier transformed to provide a time-domain signal. Modeling entirely in the time-domain is also possible, but with the limitation that frequency-dependent phenomena (e.g., element directivity, dispersion, attenuation, some scattering effects) have to be approximated as being frequency-independent. These approximations can have some significant effects, as illustrated by the following example.

10.2.2.3 Simple Example

The basic features of a ray-based simulation will now be demonstrated by considering the simulation of signals along the first of the ray-paths shown in Figure 10.3a using both time and frequency-domain expressions. The starting point is the time-domain signal injected into the sample, $u_0(t)$, and its frequency-domain equivalent $U_0(\omega)$. Let R_T be the length of the ray-path segment between T and P and let R_R be the length of that between P and R. In addition, let $\mathbf{e}_{R(T)}$ and $\mathbf{e}_{R(R)}$ be unit vectors directed along these segments of the ray-path, as shown in Figure 10.2. The total propagation distance is $R_T + R_R$ and in the frequency domain this is manifested as a phase delay equal to $\omega\tau$, where $\tau = (R_T + R_R)/v$ and v is the wave speed. Along the first segment, the beam is spreading two dimensionally and it can be shown that this results in a loss in amplitude proportional to the square root of R_T. As the scatterer at P is small, there is a secondary decay of the scattered signal proportional to the square root of R_R along the second segment of the ray-path. The directivity of the transmit and receive elements can be expressed as functions, f_T, of the direction of the segment of the ray-path at the element and frequency. These modify the amplitude of the received signal. There is another modification to the amplitude of a signal propagating along this ray-path due to the scattering from the defect itself. This can be described by the so-called scattering matrix, S, which is a function of the incident and scattered directions at the defect and frequency. Conceptually, the scattering matrix provides the amplitude and phase of scattered waves in the far-field of the scatterer as a function of scattering direction, $e_{R(R)}$, for unit amplitude plane wave incident on the scatterer from direction, $e_{R(T)}$. Therefore, in the frequency domain, the exact expression for the received signal that has propagated along this ray-path is:

$$G(\omega) = AU_0(\omega) f_T(e_{R(T)}, \omega) S(e_{R(T)}, e_{R(R)}, \omega) f_T(e_{R(R)}, \omega) \frac{\exp(-i\omega\tau)}{\sqrt{R_T R_R}} \quad (10.1)$$

where A is a constant.

The equivalent approximate time-domain expression is:

$$g(t) = Au_0(t-\tau) \frac{f_T(e_{R(T)}, \omega_c) S(e_{R(T)}, e_{R(R)}, \omega_c) f_T(e_{R(R)}, \omega_c)}{\sqrt{R_T R_R}} \quad (10.2)$$

where ω_c is the nominal center frequency of the signal. Note that a crucial piece of information lost in the time-domain model is the changing phase of S with frequency. For scatterers larger than a wavelength, it is the phase-gradient of S with frequency that encodes the fact that signals reflected from the near-side of the scatterer arrive sooner than if they are assumed to be reflected from the center of the scatterer. Hence although the amplitude of the signals from a large scatterer may be predicted with reasonable accuracy using Equation (10.2), their arrival time is not. This can be seen in the example time-traces shown in Figure 10.4a and b, where the reflected signal from the scatterer predicted in the time-domain appears later in time than it should do.

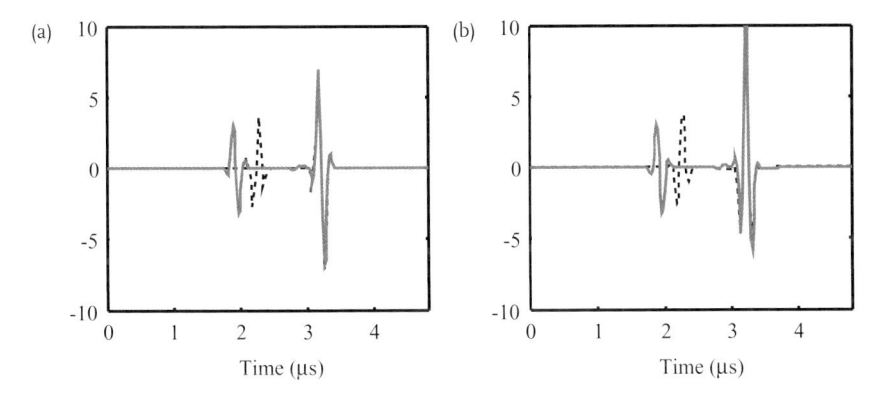

Figure 10.4 Example simulated signals for transmission and reception on element 16 for the example shown in Figure 10.2 using both time-domain (dashed line) and frequency-domain (solid line) simulations: (a) ignoring secondary scattering; (b) including secondary scattering.

The contributions along other ray-paths can be developed in a similar manner. An important point to note is that where there is a reflection from W there is no secondary beam-spreading. Hence, the beam-spreading term associated with the two legs of a ray-path of length r_1 and r_2 that have a reflection from W in between is $\sqrt{r_1 + r_2}$ not $\sqrt{r_1 r_2}$ as in the case of the scattering from the defect on ray-path 1.

The effect of inclusion of contributions along ray-paths 3 and 4 is to significantly alter the back-wall signal, as can be seen by comparison of Figure 10.4a and b.

10.2.3
Mathematical Model of Ultrasonic Array Data

In this section a formal elastodynamic model of the ultrasonic transmit–receive array data is developed for the same example. Such a model starts from a model of an individual element. There are various techniques for modeling array element output including finite element analysis (see, for example, Reference [19]) and Huygens' principle (see, for example, Reference [20]). Finite element analysis has been used to model both the electromechanical effects within the element and the propagation of ultrasound into the surrounding media. For example, Robertson et al. [21] used finite elements to relate the voltage applied across a piezocomposite array element to the displacement field in the surrounding media. This approach is time consuming to set-up but lends itself to modeling more complex geometries [22], near field effects, and details such as inter-element cross-talk [23].

In the Huygens' principle approach the output of a transducer is described in terms of the summation of point (for a 3D model) or line (for a 2D model) sources representing the surface of the transducer. Huygens' principle only models the mechanical aspects of the element (or transducer) and so the distribution of pressure across the element must be know a priori.

Any time-domain signal $u(t)$ can be expressed as a linear superposition of its spectral components $u(\omega)$:

$$u(t) = \frac{1}{2\pi} \int u(\omega) e^{i\omega t} d\omega \qquad (10.3)$$

In this section the model is developed in the frequency domain and hence only time-harmonic wave fields are considered. As is common practice, the factor $\exp(i\omega t)$ is omitted for reasons of brevity.

For simplicity, the 2D example shown in Figure 10.2 is again considered. The Cartesian coordinate axes (x, z) are defined with the z axis normal to the surface of the half-space. A transmitter element is modeled by a time-harmonic load q applied to the free surface $z = 0$ in a contact area with its center at $x = x_T$.

For comparison with the ray-based approach, only longitudinal waves are considered although this method too can be extended to include S-waves. In the case of L waves, the scalar potential ϕ can be used instead of the displacement vector $\mathbf{u}$, so $\mathbf{u} = \nabla \phi$. Using the integral transform method [24] and taking the part of the solution associated with the longitudinal wave, the resulting time-harmonic potential field φ_T can be written in the form:

$$\varphi_T(\mathbf{r}) = \frac{1}{2\pi}\int \Phi_T(k_{x(T)}) e^{-i\mathbf{k}_{(T)} \cdot (\mathbf{r}-\mathbf{r}_{(T)})} dk_{x(T)} \qquad (10.4)$$

where $\mathbf{k}_{(T)} = \{k_{x(T)}, k_{z(T)}\}$ is the wave vector, k is the wavenumber, and $k_{z(T)} = \sqrt{(k^2 - k_{x(T)}^2)}$. The branch of the function $k_{z(T)}$ in the complex $k_{x(T)}$ domain is chosen by the condition [24] $\operatorname{Re} k_{z(T)} \geq 0$, $\operatorname{Im} k_{z(T)} \leq 0$, $\omega > 0$. The function $\Phi_T(k_{x(T)})$ is the angular spectrum of the excited field and $\mathbf{r} = \{x,z\}^T$, $\mathbf{r}_{(T)} = \{x_T, 0\}$.

In the far-field of the transmitter element $k|\mathbf{r} - \mathbf{r}_{(T)}| \gg 1$ the integral (10.4) can be calculated using the stationary phase method. The local polar coordinate system R_T, φ_T is defined with its origin at the center of the transmitter element as illustrated in Figure 10.2. The resulting far-field expression for the displacements $\mathbf{u}_T$ is given by:

$$\mathbf{u}_T = \mathbf{e}_{R(T)} f_T(\varphi_T, \omega) \sqrt{\frac{\lambda}{R_T}} e^{-ikR_T}, \quad \mathbf{e}_{R(T)} = \{\sin\varphi_T, \cos\varphi_T\}^T \qquad (10.5)$$

where $\lambda = 2\pi/k$ is the wavelength and the directivity function of the transmitter element, f_T, has the form:

$$f_T(\varphi_T, \omega) = \frac{1}{2\pi} e^{-i\pi/4} k^2 \cos(\varphi_T) \Phi_T(k \sin\varphi_T) \qquad (10.6)$$

In the particular case of a unit line load normal to the surface, $\mathbf{q} = \{0,1\}\delta(\mathbf{r} - \mathbf{r}_T)$, the function Φ_T is given by (Glushkov et al. [24]):

$$\Phi_T(k_x) = -\frac{k_S^2 - 2k_x^2}{\mu\left[(k_S^2 - 2k_x^2)^2 + 4k_x^2 k_z k_{S,z}\right]} \qquad (10.7)$$

Here μ is the shear Lame elastic constant, k_S is the wavenumber of the shear wave, and $k_{S,z} = \sqrt{k_S^2 - k_x^2}$. In this case Equation (10.6) for the directivity function agrees with the result obtained by Miller and Pursey [25]. In the case of a finite width element the wave-field (10.5) for a point source has to be integrated along the element width. As a result the far field directivity function of a finite width source can be obtained by multiplying the point source directivity of Equations (10.6) and (10.7) by the function $D(\varphi_T,\omega) = \operatorname{sinc}[\tfrac{1}{2}ka \sin(\varphi_T)]$, where a is the element width [7].

The wave-field from any transmitter element propagates into the material and interacts with the scatterers. Each scatterer acts as a secondary source and generates a scattered wave field. This secondary source can be modeled by a distribution of forces that are applied in the area occupied by the scatterer.

It is assumed that the scatterer is located in the half-space $z \geq z_{\min} > 0$ and occupies the vicinity of the point $\mathbf{r}_0 = \{x_0, z_0\}$. Let the wave incident on a scatterer be a plane wave $\exp[-i\mathbf{k}_{(T)}(\mathbf{r} - \mathbf{r}_0)]$. The response of the scatterer is described by the force distribution $\mathbf{f}(\mathbf{k}_{(T)}, \mathbf{r} - \mathbf{r}_0)$. Then the scattered wave field is given by the convolution of the Green's function with the function f. Using the integral representation of the Green's function [26] and taking the part of the solution associated with the longitudinal wave, the scattered wave field in the region $z \leq z_{\min}$ can be written as:

$$\varphi_{sc}(k_{x(T)}, r) = \frac{1}{2\pi} \int k_{z(R)}^{-1} \Phi_{sc}(k_{x(T)}, k_{x(R)}) e^{-ik_{(R)}(r-r_0)} dk_{x(R)} \tag{10.8}$$

Here $k_{(R)} = \{k_R, -k_{z(R)}\}^T$ is the wave vector, $k_{z(R)} = \sqrt{k^2 - k_{x(R)}^2}$, and the function $\Phi_{sc}(k_{x(T)}, k_{x(R)})$ is given by:

$$\Phi_{sc}(k_{x(T)}, k_{x(R)}) = \frac{1}{2\rho\omega^2} \int k_{(R)} \cdot f(k_{(T)}, r') e^{ik_{(R)} r'} dr' \tag{10.9}$$

where ρ is the material density.

In the far-field from the reflector, $k|r - r_0| \gg 1$, and the integral (10.9) can be evaluated by the stationary phase method. Let the local polar coordinate system R_R, φ_R be defined with the origin at the center of the receiver element, as shown in Figure 10.2. Then the far-field expression for the scattered field is given by:

$$u_{sc} = e_{R(R)} S(e_{k(T)}, e_{R(R)}, \omega) \sqrt{\frac{\lambda}{R_R}} e^{-ikR_R}, \quad e_{k(T)} = \frac{k_{(T)}}{k}, \quad e_{R(R)} = -\{\sin\phi_R, \cos\phi_R\}^T \tag{10.10}$$

The function S is the scattering matrix of the scatterer and is related to the function Φ_{sc} by:

$$S(e_{k(T)}, e_{R(R)}, \omega) = \frac{1}{2\pi} e^{-i\pi/4} k \Phi_{sc}(k_{x(T)}, -k\sin\phi_R) \tag{10.11}$$

The scattered wave propagates back and is detected by the transducer element located at the point $r_{(R)} = \{x_R, 0\}$. The action of the transducer as a receiver can be defined using the electromechanical reciprocity argument of Auld [27, 28]. Therefore, the signal $g(\omega, x_T, x_R)$ measured by the receiver element at $r_{(R)}$ when the element at $r_{(T)}$ is the transmitter can be written in the form:

$$g(\omega, x_T, x_R) = \frac{i\omega}{4P} \int_C (t_{1n} \cdot u_2 - t_{2n} \cdot u_1) dS \tag{10.12}$$

where C is an arbitrary contour enclosing the scatterer, u is the displacement vector, t_n is the normal stress on the contour C, and P is some quantity that is proportional to the squared amplitude of the electrical signal in the transmitter [27]. The index 1 refers to the wave field generated by the transmitter element when the scatterer is present and the index 2 refers to the wave-field generated by the receiver element when the scatterer is absent. Finally, the measured signal can be written as [29]:

$$g(\omega, x_T, x_R) = \frac{1}{4\pi^2} \iint G(\omega, k_{x(T)}, k_{x(R)}) e^{-i(k_{x(T)} x_T + k_{x(R)} x_R)} dk_{x(T)} dk_{x(R)} \tag{10.13}$$

or alternatively in the $(k_{x(T)}, k_{x(R)})$ domain as:

$$G(\omega, k_{x(T)}, k_{x(R)})$$
$$= ik^3 \Phi_0(\omega) \Phi_T(-k_{x(T)}) \Phi_{sc}(-k_{x(T)}, k_{x(R)}) \Phi_T(-k_{x(R)}) e^{i(k_{x(T)}+k_{x(R)})x_0} e^{-i(k_{z(T)}+k_{z(R)})z_0}$$
(10.14)

where $\Phi_0 = \rho v^3 k U_0(\omega)/(4P)$.

If the transmitter and receiver are in the far-field of the scatterer, $kR_T \gg 1$, $kR_R \gg 1$, then integral (10.13) can be evaluated by the stationary phase method and the transmit–receive signal G can be written as:

$$G(\omega) = \Phi_0(\omega) f_T(e_{R(T)}, \omega) S(e_{R(T)}, e_{R(R)}, \omega) f_T(e_{R(R)}, \omega) e^{-ik(R_T+R_R)} \left(\frac{4\pi^2 \lambda^2}{R_T R_R} \right)^n c$$
(10.15)

Here the unit vectors $e_{R(T)} = (r_0 - r_{(T)})/R_T$ and $e_{R(R)} = (r_{(R)} - r_0)/R_R$ define incident and scattered directions, respectively, and coefficients $c = \exp(-i\pi/4)$, $n = \frac{1}{2}$ in the 2D case, and $c = i/k$, $n = 1$ in the 3D case. It can be seen that this is exactly the same as Equation (10.1); the intuitive ray-based result has been deduced from a rigorous elastodynamic analysis. However, it is now clear exactly which assumptions are necessary for this to be valid.

10.3 Imaging with 1D Arrays

It is assumed throughout this section that all possible data from an array is available and that the array is spatially fully sampled. It is more elegant to work with continuous functions than data that is discrete in both time and space. Therefore, here the time-domain data from the array is described by the function $g(t, x_T, x_R)$. The objective of an imaging process is to map this function to a 2D image that represents, in some sense, the ultrasonic reflectivity as a function of position in the sample. In the following subsections, various schemes for performing this mapping will be discussed, starting with the classical beamforming operations.

10.3.1 Classical Beam-Forming Imaging Methods in Post-processing

In what is arguably the simplest imaging algorithm that can be implemented in beamforming, an active aperture consisting of a subset of the elements in the complete array is swept across the full array aperture to produce a B-scan. At each position the elements in the active aperture are fired simultaneously and the complete process therefore emulates an unfocused transducer being scanned over

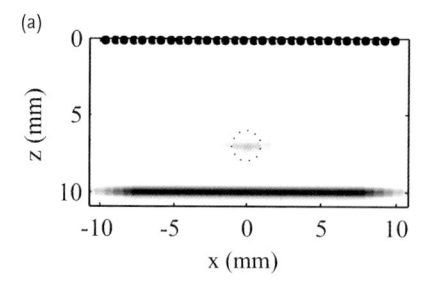

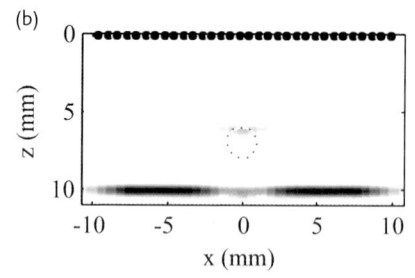

246 | 10 Array Imaging and Defect Characterization Using Post-processing Approaches

Figure 10.5 B-scan images generated from simulated data for configuration shown in Figure 10.2: (a) ignoring secondary scattering using time-domain simulated data; (b) including secondary scattering using frequency-domain simulated data. The scale is linear and normalized to the peak amplitude in each image.

the surface of a sample. As a post-processing operation applied to the full matrix of array data, the result can be expressed as:

$$I(x, z) = \iint B(x_T - x) B(x_R - x) g\left(\frac{2z}{v}, x_T, x_R\right) dx_T dx_R \qquad (10.16)$$

where the aperture function B is defined as:

$$B(\chi) = \begin{cases} 1, & |\chi| \leq L/2 \\ 0, & |\chi| > L/2 \end{cases} \qquad (10.17)$$

where L is the aperture size. Similar expressions for other classical beamforming algorithms can be derived fairly easily and are detailed elsewhere [7].

Examples of the application of a B-scan processing to the simulated data for the configuration shown in Figure 10.2 are shown in Figure 10.5, both with and without the inclusion of secondary scattered signals in the simulated data. The effect of inclusion of the secondary scattered signals is clear; without them the shadowing artifact on the back-wall signal is not observed. Again, the shortcoming of time-domain simulation can be seen by the incorrect spatial position of the near-side reflection from the side-drilled hole in Figure 10.5a.

10.3.2
Total Focusing Method

The concept of the total focusing method is simple and intuitive: the entire array is focused at every image point on transmission and reception. This can be very efficiently written as:

$$I(x, z) = \iint g\left(\frac{R_T + R_R}{v}, x_T, x_R\right) dx_T dx_R \qquad (10.18)$$

In the TFM the signal $g(t)$ is mapped to a function of propagation distance, r, by the relationship $r = vt$. This is a simple back-propagation operation. For the

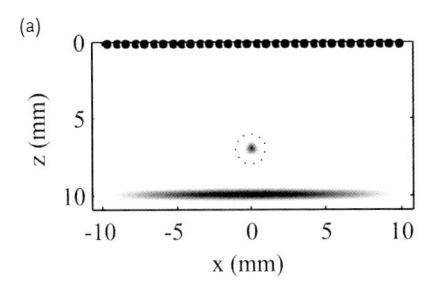

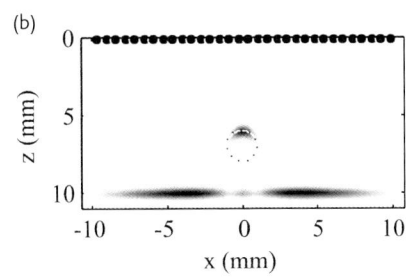

Figure 10.6 TFM images generated from simulated data for configuration shown in Figure 10.2 with $a = 2$ mm: (a) ignoring secondary scattering using time-domain simulated data; (b) including secondary scattering using frequency-domain simulated data. The scale is linear and normalized to the peak amplitude in each image.

transmit–receive signal $g(t,x_T,x_R)$ the propagation distance is equal to the sum of the distances between the transmitter and receiver positions and the imaging point, $R_T + R_R$. By comparison with the time-domain expression of array data simulation (10.2) it is clear why this is effective: the TFM inverts the time-delays associated with a scatterer at each pixel position.

Figure 10.6 shows examples of the application of TFM processing to the simulated data for the configuration shown in Figure 10.2. The effect of inclusion of omitting the scattered signals and shortcomings of time-domain simulation are again clear. Comparison of the B-scan and TFM results from the correctly simulated data shown in Figures 10.5b and 10.6b shows that the amplitude from the side-drilled hole is significantly higher in the latter. In addition, the shape of the signal from the hole in the TFM image can be seen to be more closely matched to the profile of the near side of the hole than that of the equivalent B-scan.

10.3.3
Wavenumber Method

The formulation of the wavenumber method for imaging is rather more rigorous than the TFM and is based on taking a model of the expected scattering process and then inverting it to obtain the scatterer distribution from the measured data.

Equation (10.9) for the reflected wave field represents the superposition of responses from point sources distributed within the scatterer with amplitudes $f(k_{(T)},r)$. If the vector field f is differentiable then according to Helmholtz's theorem it can be decomposed into the sum of irrotational (curl-free) and solenoidal (divergence-free) vector fields. Using Gauss's theorem it can be shown that the integral (10.9) for the solenoidal component of f is equal to zero. Therefore, the vector field f can be considered as irrotational and can be written as a gradient of a scalar potential $f_1(k_{(T)},r)$. To obtain the function f_1, the problem of interaction of a plane elastic wave with the scatterer must be solved.

If the effect of the scatterers on the incident wave field is small then the Born approximation can be applied. This means that the amplitude of each point source is proportional to the amplitude of the incident wave at the location of that point. If in addition each point scatterer is omnidirectional with respect to the incident wave, then the function f can be written as:

$$f(k_{(T)}, r) = c_1(\omega)\nabla\left[f_1(r+r_0)e^{-ik_{(T)}r}\right] \tag{10.19}$$

where c_1 is the constant of proportionality, the point r_0 is the nominal center of the scatterer, and the scalar function $f_1(r)$ is the object function of a scatterer that describes the scatterer distribution. Consequently, in this case the imaging can be considered as an inverse problem of reconstructing the object function $f_1(r)$ from the array data.

Substituting Equation (10.19) into Equation (10.9) and evaluating the spatial integral, yields the following expression:

$$\Phi_{sc}(k_{x(T)}, k_{x(R)}) = c_2 F_1(k_{(R)} - k_{(RT)})e^{i(k_{(T)} - k_{(R)})r_0} \tag{10.20}$$

where $c_2 = -ic_1(\omega)/(2\rho v^2)$, and F_1 is the 2D spatial Fourier transform of the function f_1.

For further calculations it is assumed that the constant c_2 is equal to unity. Equation (10.4) for the measured signal also contains the characteristics of the transmitted signal, Φ_0, and the beam-pattern of the transducer element, Φ_T. However, these effects can be compensated for in a preprocessing step as shown in Reference [30]. Therefore, below it is assumed that the function $\Phi_0 = 1$ and the array elements are omnidirectional. If the directivity function of transducer is equal to unity, then from Equation (10.4) it follows that the transducer angular spectrum $\Phi_T(k_x) = 2\pi \exp(i\pi/4)/(kk_z)$. Using Equations (10.2) and (10.20) the Fourier domain array data $G(\omega, k_{x(T)}, k_{x(R)})$ can now be written as:

$$G(\omega, k_{x(T)}, k_{x(R)}) = 4\pi^2 ik^2 F_1(k_{x(T)} + k_{x(R)}, -k_{z(T)} - k_{z(R)})(k_{z(T)}k_{z(R)})^{-1} \tag{10.21}$$

The relationship (10.21) between the array data and the object function agrees with the results obtained earlier for a 2D medium [30] and a 3D medium [31].

From Equation (10.21) it can be seen that there is significant redundancy in the array data. Indeed, the object function can be fully reconstructed by using only diagonal data, $k_{x(T)} = k_{x(R)} = k_x/2$, in the Fourier $(\omega, k_{x(T)}, k_{x(R)})$ domain and applying the 2D inverse Fourier transform with respect to the variables k_x, k_z to the function $-ik_z^2 G(w, k_x/2, k_x/2)/(16\pi^2 k^2)$:

$$f_1(x, z) = \frac{-i}{(8\pi^2)^2} \iint G\left(\omega, \frac{k_x}{2}, \frac{k_x}{2}\right) \frac{k_z^2}{k^2} e^{-i(k_x x - k_z z)} dk_x dk_z \tag{10.22}$$

However, it should be stressed that this data redundancy is caused by the approximate model of the scatterer (10.19) only (see also Reference [32]). In this case each point (k_x, k_z) corresponds to some curve $l(k_x, k_z)$ in the $(\omega, k_{x(T)}, k_{x(R)})$ domain. In practice the effect of noise and side lobes can be reduced by additional averaging of the array data over the curve l. If the variable $k_{x(T)}$ is chosen as a parameter of the curve l then the resulting image can be written in the form [30]:

$$f_1(x,z) = \frac{-i}{(4\pi^2)^2} \iint G(\omega, k_{x(T)}, k_{x(R)}) \frac{k_{z(T)} k_{z(R)}}{k^2} N(k_{x(T)}, k_x, k_z) e^{-i(k_x x - k_z z)} dk_{x(T)} dk_x dk_z \quad (10.23)$$

where $N = N_0(k_x, k_z) dl/dk_{x(T)}$. The coefficient N_0 is the normalization factor and can be taken in the form:

$$N_0 = \int_{-L}^{L} dl_0 \bigg/ \int_{-L}^{L} dl, \ L \to \infty$$

where l_0 is the reference curve corresponding to some point (k_{x0}, k_{z0}). Note that Equation (10.22) can be obtained from Equation (10.23) by taking $N = \delta(k_{x(T)} - k_{x(R)})$.

10.3.4
Back-Propagation Method

The wavenumber imaging method is based on the approximate model (10.19) of the scatterer. This model can be justified if scatterers are considered as a collection of frequency-independent point scatterers [30] or for the scattering from an elastic inclusion with properties that only slightly differ from the host material [31, 33]. However, the approximation (10.19) is quite restrictive and is not valid, for example, in the case of a crack-like defect.

Therefore, if the distribution of the scatterers is described by a general function $f(k_{(T)}, r)$ then a different imaging procedure is needed. One of the possible ways to construct an image is by back-propagation of the detected wave field. In fact, it could be argued that this is the implicit basis of the total focusing method (TFM) although it is not usually described in that way. An alternative to the TFM is to back-propagate the angular spectrum of the transmitted and received wave fields rather than the individual signals. From the expressions, Equations (10.13) and (10.14), for the array data it is seen that the angular spectrum $E(\omega, k_{x(T)}, k_{x(R)}, z)$ of the transmit–receive array data at the depth z is given by:

$$E(\omega, k_{x(T)}, k_{x(R)}, z) = G(\omega, k_{x(T)}, k_{x(R)}) e^{ik_z z} \quad (10.24)$$

where $G(\omega, k_{x(T)}, k_{x(R)})$ is the angular spectrum at $z = 0$. Note: because $k_z = k_{z(T)} + k_{z(R)}$ this equation can be considered as two consecutive back-propagations in transmission and reception. Then the time domain signal, $e(t, x_T, x_R, z)$, measured by the receiver element located at $\{x_R, z\}$ when the transmitter element located at $\{x_T, z\}$ is activated can be written as:

$$e(t, x_T, x_R, z) = \frac{1}{8\pi^3} \iiint E(\omega, k_{x(T)}, k_{x(R)}, z) e^{i(\omega t - k_{x(T)} x_T - k_{x(R)} x_R)} dk_{x(T)} dk_{x(R)} d\omega$$

$$(10.25)$$

To estimate the location of the scatterers the back-propagated array data corresponding to $x_T = x_R$ are considered. At the moment of time t the signal $e(t,x,x,z)$ is localized in the vicinity of the points (x,z) such that the time required for the wave propagation from the point (x,z) to the scatterer location and back is equal to t. Therefore, at the moment of time $t = 0$ the data $e(t,x,x,z)$ are nonzero only in the vicinity of the areas occupied by the scatterers and to form an image $I(x,z)$ the function $e(0,x,x,z)$ can be taken. From Equation (10.25) it follows that:

$$I(x,z) = \frac{1}{8\pi^3} \iiint G(\omega, k_{x(T)}, k_{x(R)}) e^{-i(k_x x - k_z z)} dk_{x(T)} dk_{x(R)} d\omega \qquad (10.26)$$

10.3.5
Theoretical Comparison of Imaging Methods

An imaging method can be considered as an operator, L, that transforms the transmit–receive array data set $g(t, \mathbf{r}_{(T)}, \mathbf{r}_{(R)})$ into a new data set $I(\mathbf{r})$ usually called an image:

$$I(\mathbf{r}) = L[g(t, \mathbf{r}_{(T)}, \mathbf{r}_{(R)})] \qquad (10.27)$$

where $\mathbf{r}_{(T)}$ and $\mathbf{r}_{(R)}$ are the position vectors of the transmitter element and receiver element, respectively, and $\mathbf{r}$ is the imaging point.

In the previous subsections, the TFM, wavenumber, and back-propagation imaging methods for processing of the full matrix of transmit–receive array data were described. These methods are conceptually different, use different approximations, and have different expressions for the image construction. However, they are all linear and can all be expressed as an operator of the following form:

$$L[g(t, \mathbf{r}_{(T)}, \mathbf{r}_{(R)})] = \frac{1}{2\pi} \iiint s(\omega, x_T, x_R, x, z) g(\omega, x_T, x_R) d\omega dx_T dx_R \qquad (10.28)$$

where the function s represents focusing coefficients. Note that there also exist nonlinear imaging methods with regard to the data, such as decomposition of the time reversal operator (DORT) [34] and MUSIC [35, 36], and a linear sampling methods [37] has been developed. DORT and MUSIC are eigenvalue approaches and the linear sampling method provides a solution to the inverse problem of reconstructing the shape of a scatterer from knowledge of the far-field reflection pattern. These are not considered here.

Detailed derivations to cast the TFM, wavenumber, and back-propagation algorithms into this form may be found in Reference [29]; here only the final results for the corresponding focusing coefficients, s_{tfm}, s_{wn}, and s_{bp}, are presented:

$$s_{tfm} = e^{ik(R_T + R_R)} \qquad (10.29)$$

$$s_{wn} = -c \frac{k}{2\pi} \frac{\cos\phi_T \cos\phi_R (\cos\phi_T + \cos\phi_R)}{2\pi v (R_T R_R)^n} e^{ik(R_T + R_R)} \qquad (10.30)$$

$$s_{bp} = -c \frac{k}{2\pi} i \frac{\cos\phi_T \cos\phi_R}{(R_T R_R)^n} e^{ik(R_T + R_R)} \qquad (10.31)$$

Note that back-propagation based on Equation (10.28) has also been termed the inverse wave field extrapolation (IWEX) method [38].

The above expressions show that the three methods all share the same phase factor but have different amplitude factors. In terms of conventional beamforming this means the algorithms have the same focal law delay profile for a given imaging point but different apodizations. The amplitude factors depend on transmitter and receiver positions and the imaging point and can be considered as frequency-independent spatial filters. Because of these filters the wavenumber and the back-propagation methods generally provide lower side lobes in the image than the TFM. However, as will be demonstrated in Section 10.3.8, the actual improvement in the signal to coherent noise ratio given by the wavenumber and back-propagation methods relative to the TFM depends on the position of the scatterer relative to the array and its scattering behavior.

10.3.6
Computational Burden

In practice an important property of an imaging algorithm is its computational efficiency. In this section the numerical implementation of the different imaging techniques and their computational performance are discussed.

As has been shown in the previous section, different imaging algorithms can be expressed in the form of a linear superposition of transmit–receive signals in the frequency domain with some focusing coefficients. Equation (10.28) leads to the TFM-type implementation of the imaging methods and is ideally suited for parallel processing. Additionally, the TFM-type processing (10.28) can be performed while data acquisition is still in progress.

On the other hand, the wavenumber method operates in the Fourier domain, which naturally leads to its implementation using the fast Fourier transform algorithm [30]. This makes the wavenumber method in general computationally more efficient in terms of the number of floating point operations required than the TFM-type approach. As shown in the literature [30], for large data sets the computational cost of the wavenumber method is N times less than the computational cost of the TFM approach, where N is the number of array elements. It can be shown [10, 29] that the back-propagation method can also be implemented in the Fourier domain as a series of fast Fourier transforms and has the same advantages in terms of computational performance as the wavenumber algorithm. However, the Fourier domain implementations require simultaneous processing of the full matrix of array data, which results in a heavy memory load compared with the TFM-type approach. Moreover, if a scatterer is located outside the array aperture then additional coherent noise appears in the image because of aliasing. To avoid this, the original array aperture data must be padded with zeros to perform its Fourier transform with more x_T, x_R-positions. Consequently, the number of sampling points in the spatial Fourier domain is generally greater than the number of array elements.

Consider firstly 2D imaging using 1D array. In practical testing the total number of sampling points in the x-domain (i.e., the number of array elements and the number of padded zeros) is typically of order of $N \sim 10^2$. Therefore, the number of sampling points in the Fourier ($k_{x(T)}$, $k_{x(R)}$) domain is $N^2 \sim 10^4$. Consider now 3D imaging using 2D array. It is assumed that the array is located in the (x,y)-plane. To cover the imaging area of the same aperture as for the 1D array the number of sampling points in the (x,y)-domain is N^2 and the number of sampling points in the 4D Fourier ($k_{x(T)}$, $k_{y(T)}$, $k_{x(R)}$, $k_{y(R)}$) domain is $N^4 \sim 10^8$. Therefore, for 2D arrays the computer memory requirements make the Fourier domain implementation of the imaging algorithms impractical. Moreover, the Fourier domain implementation using fast Fourier transform algorithm implies that the array element layout represents a regular sampling scheme (Section 10.4.1.2). However, in some cases for a 2D array an irregular element distribution is more beneficial. Consequently, in the case of a 2D array the TFM type implementation based on Equation (10.28) with the equivalent focusing coefficients is preferable.

10.3.7
Focusing Performance

In this section the performance of the wavenumber, TFM, and back-propagation imaging algorithms are compared in terms of their resolution. The resolution can be quantified by the array performance indicator [6] (API), which is defined in 2D as:

$$P = \frac{A}{\lambda^2} \qquad (10.32)$$

where A is the area over which the amplitude of the point-spread function (i.e., the image of an omnidirectional point scatterer) is greater than some threshold below its maximum value and λ is the wavelength at the center frequency. Therefore, the function P depends on the imaging point and the threshold value.

The specifications of a commercial 64-element array (manufactured by Imasonic, Besançon, France) were used for the modeling. The array has an element pitch of $\Delta x = 0.63$ mm and a total length of $L = 40$ mm. The simulated specimen was mild steel with a longitudinal wave velocity of 5900 m s^{-1}. The directivity function of the array elements was modeled using Equations (10.6) and (10.7). The transmitted signal was a Hanning windowed toneburst with a center frequency of 5 MHz.

The API maps corresponding to a −20 dB threshold for the TFM and the wavenumber algorithm were calculated in Reference [30]. It was shown [30] that with this threshold the wavenumber algorithm provides a marginally better resolution than the TFM. Below the API as a function of the threshold value is now investigated.

Figure 10.7 shows the dependence of the API for different imaging algorithms on the threshold value for two different point omnidirectional scatterers located

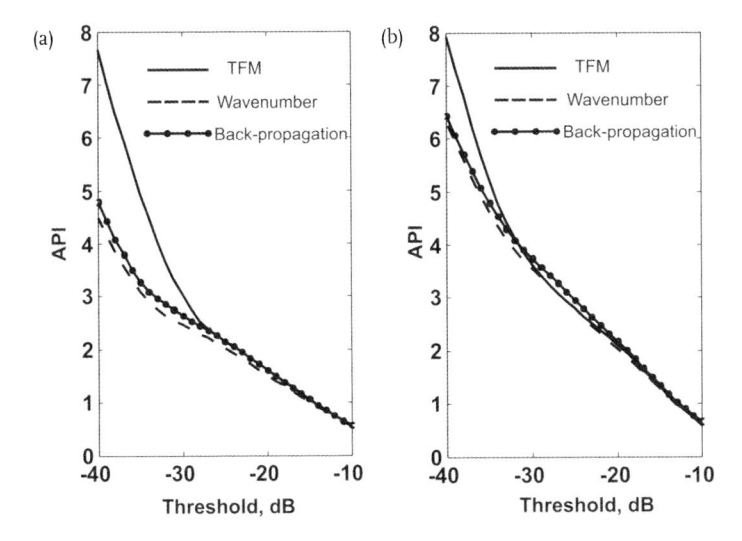

Figure 10.7 Dependence of the array performance indicator on the threshold level for the point scatterers with x-coordinates corresponding to the (a) center of the array, (b) end of the array.

at the same depth $z = 20$ mm from the array. The x-coordinate was taken equal to the center of the array ($x_0 = 0$) for the first scatterer and the end of the array ($x_0 = L/2$) for the second scatterer. It is seen that the wavenumber algorithm and back-propagation method have approximately the same API for all threshold values. Moreover, all three imaging methods provide very similar resolution for threshold values greater than -26 dB for the first scatterer and for threshold values greater than -32 dB for the second scatterer. For lower thresholds the API corresponding to the TFM increases much faster than APIs for the wavenumber and back-propagation methods. For example, at a threshold of -40 dB the resolution of the wavenumber and back-propagation methods is about 1.7 times better than the resolution of TFM for the scatterer in the middle of the array and about 1.2 times better for the scatterer at the end of the array. Note that in both cases the wavenumber algorithm has slightly better resolution than back-propagation. This is expected since the scatterers are omnidirectional and this is consistent with the assumption of the wavenumber algorithm.

10.3.8
Experimental Example

To illustrate the performance of different imaging methods the experiment described below was set up. The 64-element array with 5-MHz center frequency described above was used on a mild steel block with dimensions $160 \times 80 \times 20$ mm. The velocity of the bulk longitudinal wave is approximately $5900 \, \text{m s}^{-1}$, so the element pitch is $0.53\lambda_0$, where $\lambda_0 = 1.18$ mm is the wavelength at the center

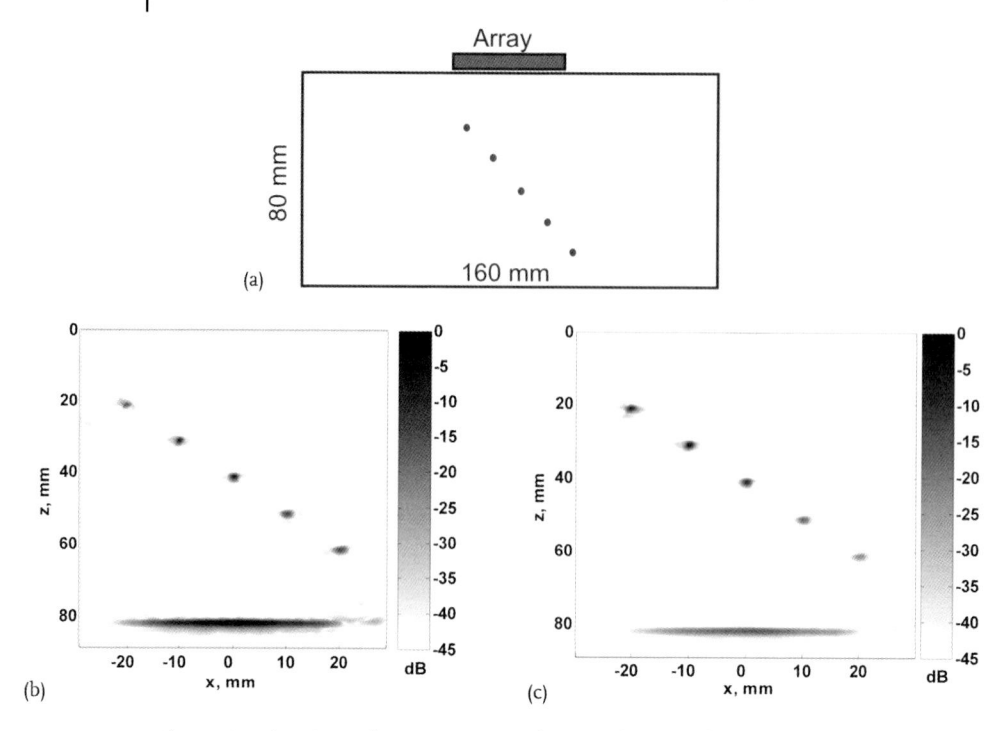

Figure 10.8 Imaging with a 1D array: (a) schematic diagram of experimental sample with five side drilled holes of 1 mm diameter; (b) TFM image; (c) back-propagation image.

frequency. Five circular holes 1 mm in diameter were drilled through the side of the specimen as schematically illustrated in Figure 10.8a. The TFM image is shown in Figure 10.8b and the image obtained by the back-propagation method is shown in Figure 10.8c. The images obtained by the wavenumber algorithm have approximately the same signal-to-noise ratio and resolution as the back-propagation images and are not shown. Note that all processing was performed using positive frequencies only and then the absolute value of the image function was taken. This is equivalent to the extraction of the envelope of the image.

For this example the Nyquist frequency is about 10 MHz and the element pitch is equal to the half wavelength at the center frequency of 5 MHz. Therefore, at frequencies greater than the center frequency the Nyquist criterion is not satisfied and grating lobes should appear in the image. However, it can be seen that both methods (and also the wavenumber method) give a similar signal-to-noise ratio of about 45 dB, which can be explained by the scattering behavior of the scatterers. In this case the scattered energy is contained mainly in the non-wrapped part of the wavenumber spectra and, hence, the amplitude of the grating lobes is below −45 dB for all imaging methods. Therefore, in this situation the wavenumber and back-propagation methods give no improvement over the TFM in terms of the signal-to-noise ratio. However, as has been predicted, there is some improvement in the resolution. For example, the −40 dB API of the scatterer located in the middle of the array is 8.5 for the TFM and 7 and 6.5 for the asymptotic back-propagation and asymptotic wavenumber methods, respectively.

10.4
Imaging with 2D Arrays

10.4.1
Optimization of 2D Array Layout

Two-dimensional arrays offer the potential to image in three dimensions. This has clear benefits as real defects and engineering structures are three-dimensional – for example, defects such as cracking and inclusions in welds are of arbitrary shape and can occur in arbitrary orientations. Two-dimensional arrays can "view" a given defect from a range of angles leading to the possibility of obtaining characterization detail far beyond what is currently achievable. In the case of an array with a finite-sized aperture, each defect can be illuminated only over a finite range of angles that is determined by the size of array aperture. The resolution of a 3D image and accuracy of defect characterization can be improved by increasing the array aperture. However, there is only a limited number of array elements and therefore it is necessary to consider different configurations of a limited number of array elements within an aperture [17].

Consider the frequency spectrum of array data $g(\omega, r_{(T)}, r_{(R)})$. The array data represent samples of the continuous transmitted and received wave-field $g(\omega, r_1, r_2)$ at points $r_1 = r_{(T)}$ and $r_2 = r_{(R)}$. The best performance of the array can be achieved if the function $g(\omega, r_1, r_2)$ can be reconstructed from its sample values $g(\omega, r_{(T)}, r_{(R)})$, otherwise there is loss of information. Generally, the function $g(\omega, r_1, r_2)$ is an arbitrary function but one that is spatially bandlimited (meaning that there cannot be any spatial variations of shorter wavelength than the ultrasonic wavelength at the frequency ω). In this case the problem of optimum element layout can be formulated as follows: what is the most efficient sampling scheme requiring the minimum number sampling points per unit area for reconstruction a continuous bandlimited 2D function?

10.4.1.1 Optimization Criterion
This section describes the general procedure for quantifying the performance of any array layout. Any function $f(r)$ can be represented as an inverse Fourier transform:

$$f(r) = \frac{1}{4\pi^2} \int F(k) e^{-ik \cdot r} dk \qquad (10.33)$$

where $F(\mathbf{k})$ is the Fourier transform of the function $f(\mathbf{r})$. If $f(\mathbf{r})$ is known at the sampling points $\mathbf{r}_n$ only, then instead of $F(\mathbf{k})$ the discrete Fourier transform, $F_d(\mathbf{k})$, can be calculated:

$$F_d(\mathbf{k}) = \frac{1}{N} \sum_{n=1}^{N} f(\mathbf{r}_n) e^{i\mathbf{k}\cdot\mathbf{r}_n} \quad (10.34)$$

The function $F_d(\mathbf{k})$ is related to the spectrum $F(\mathbf{k})$ by:

$$F_d(\mathbf{k}) = \int F(\mathbf{k}') F_0(\mathbf{k}-\mathbf{k}') d\mathbf{k}' \quad (10.35)$$

where $F_0(\mathbf{k})$ is the spatial 2D Fourier transform of the array layout:

$$F_0(\mathbf{k}) = \frac{1}{N} \sum_{n=1}^{N} e^{i\mathbf{k}\cdot\mathbf{r}_n} \quad (10.36)$$

Consider the bandlimited function $f(\mathbf{r})$ with spectrum $F(\mathbf{k})$, which is equal to zero outside the circle $k = k_0 = 2\pi/\lambda$. Equation (10.35) shows that $F_d(\mathbf{k}) \approx F(\mathbf{k})$ if the Fourier transform of the array layout, $F_0(\mathbf{k})$, has a single narrow peak at $\mathbf{k} = (0,0)$ and no other peaks anywhere in the circle $k < 2k_0$. In this case applying the inverse Fourier transform (10.33) to $F_0(\mathbf{k})$ shows that the function $f(\mathbf{r})$ can be fully reconstructed from its sampled values $f(\mathbf{r}_n)$. Otherwise, high-frequency components of the function f appear as low-frequency components of the reconstructed function. This effect is known as aliasing.

10.4.1.2 Regular Sampling

Any regular (i.e., periodic) sampling pattern has a corresponding sampling criterion. If the distance between sampling points is greater than some maximum allowable value then aliasing occurs.

The most straightforward generalization of the 1D regular sampling to the 2D case is rectangular sampling, which is schematically shown in Figure 10.9a for the case of an element pitch $d = 1.1$ mm. In this case the function $F_d(\mathbf{k})$ shown in Figure 10.9c has a periodic pattern with the distance between peaks equal to $2\pi/d = 5.71$ mm^{-1}, where d is the element pitch. Therefore, to avoid aliasing the spacing between sampling points in a rectangular pattern must be less than $\lambda/2$.

However, in contrast to the 1D case there are an infinite number of other possible regular sampling patterns in the 2D case. It can be shown [39] that the optimum regular sampling scheme with the lowest density of sampling points is hexagonal sampling. Here the elements are located on a triangular grid. In this case the element spacing can be taken as equal to $\lambda/\sqrt{3}$ rather than $\lambda/2$ for the rectangular sampling scheme. Overall, hexagonal sampling can be shown to result in a 2D array with an aperture that is approximately 20% larger than that of one with the same number of elements in a rectangular sampling pattern.

10.4.1.3 Non-uniform Sampling

If the sampling criterion for the regular sampling scheme is not satisfied then the reconstructed function contains aliasing peaks. As a result artifacts due to grating lobes appear in the image. Note that these localized artifacts are caused by the periodicity of the array element layout only. Irregular distribution of the array elements suppresses aliasing peaks but at the expense of increased uniform noise throughout image. The problem of finding the optimal non-uniform sampling distribution is well known in computer graphics [40]. One of the best solutions is a Poisson disk distribution – random element locations with constraint on minimum separation distance. Interestingly, the distribution of the photoreceptors in a human eye is a Poisson disk distribution [41].

An example of an array with a Poisson disk element distribution and the corresponding Fourier transform of array layout are shown in Figure 10.9b and its 2D spatial Fourier transform can be seen in Figure 10.9d. The average inter-element spacing, d, in this example is ~2.3 mm. It can be seen that there is only

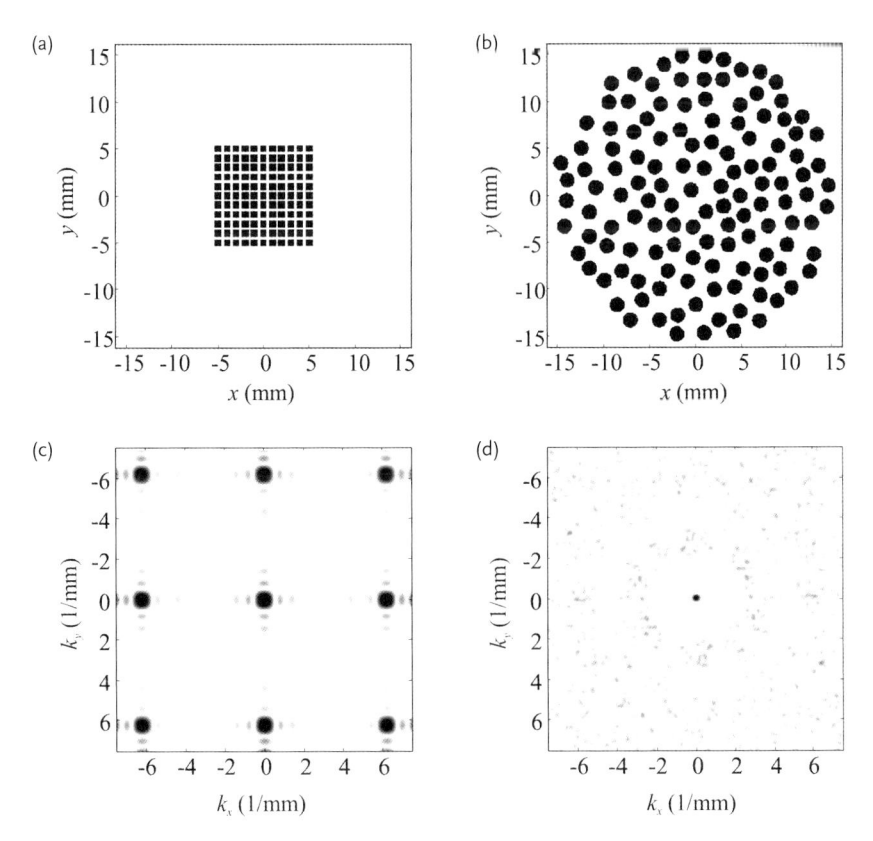

Figure 10.9 2D array layouts: (a) rectangular sampling; (b) Poisson disk sampling. The images in (c) and (d) show the 2D spatial Fourier transforms of (a) and (b) on 20 dB scales.

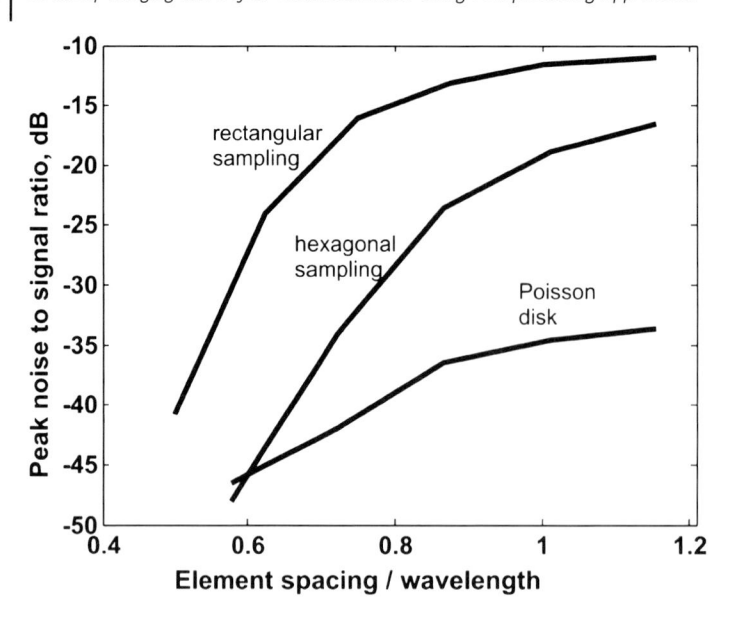

Figure 10.10 Peak noise-to-signal ratio as a function of the element pitch for rectangular, hexagonal, and Poisson disk sampling schemes.

one main peak in the Fourier domain at $k = (0,0)$. All other peaks are effectively suppressed, but at the expense of uniformly distributed noise when $|k|$ is greater than $2\pi/d \sim 2.7\,\text{mm}^{-1}$ in this case.

The limiting factor in using an irregular sampling pattern is how high a uniform noise level can be tolerated in the image. The peak noise-to-signal ratio as a function of the element pitch for rectangular, hexagonal, and Poisson disk sampling schemes with 128 elements is presented in Figure 10.10. In this case one point scatterer was placed at $x = 7.7\,\lambda$, $y = 0$, $z = 11.5\,\lambda$. It can be seen that, to keep the noise level below −35 dB, the Poisson disk configuration enables the mean element pitch to be made approximately twice as big as the best regular configuration (hexagonal).

10.4.2
Experimental Comparison of 2D Array Layouts

Here experimental results obtained using two different 2D arrays are presented [18]. Both arrays are manufactured by Imasonic, France with 3-MHz center frequency. The first array is an 11 × 11 2D matrix array with an element pitch of 1.1 mm, which satisfies the sampling criterion for imaging in steel. The second array is a 128 element sparse circular array with Poisson disk element distribution and a minimum inter-element distance of 1.9 mm, resulting in a mean inter-element spacing of 2.3 mm. These correspond to the element layouts shown in Figure 10.9a and b. The aperture size (length of one side) of the 11 × 11 matrix array is 12 mm and the aperture size (diameter) of the 128 element sparse array is 30 mm.

10.4.2.1 Spherical Inclusion

The first examples are from measurements made on a cylindrical steel block containing a spherical inclusion of diameter 2.1 mm. The images obtained using the TFM are shown in Figure 10.11. Figure 10.11a and b shows isosurfaces at −30 dB (relative to the back wall signal) of 3D images. Figure 10.11c and d shows the 2D

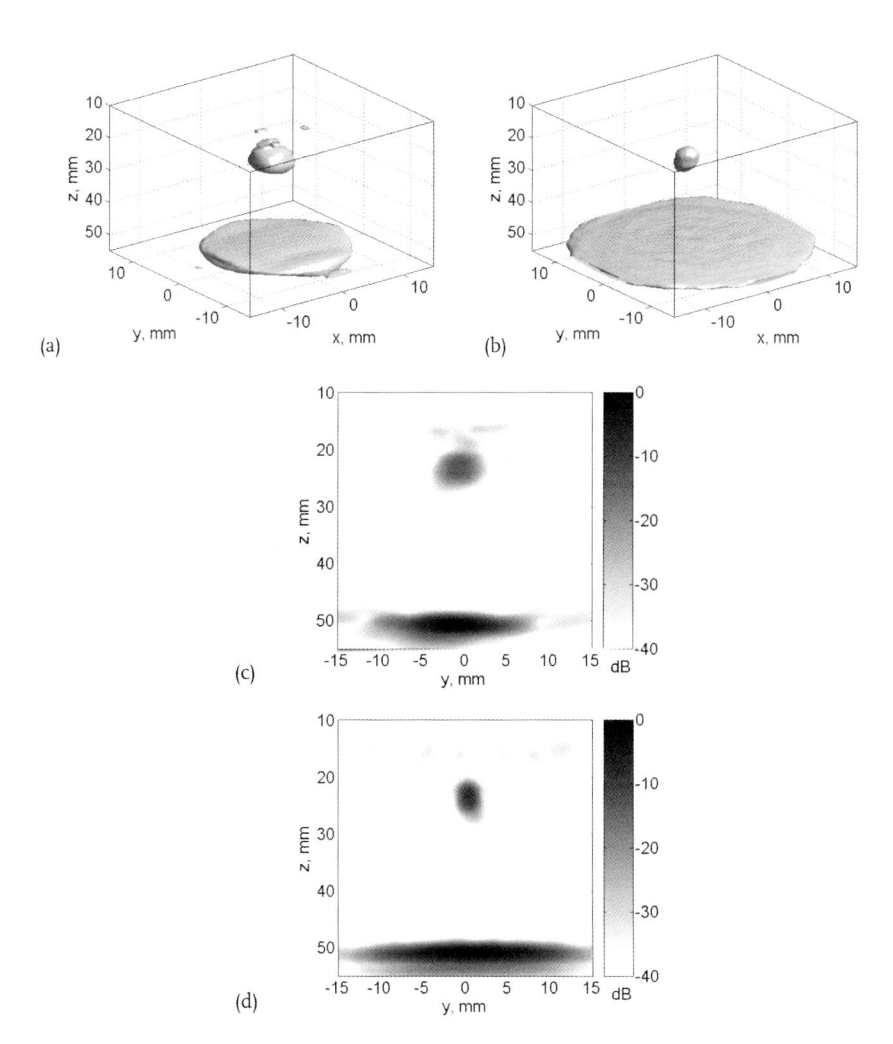

Figure 10.11 Experimental results for a steel sample with 2.1 mm diameter spherical inclusion: (a, b) −30 dB isosurface images; (c, d) 2D cross-sectional images in the y–z plane. Images (a, c) were obtained using 11 × 11 matrix array and (b, d) were obtained using a 128 element sparse array.

images in the y–z plane. It can be seen that the image produced using the sparse array does not contain any localized artifacts. On the other hand, because the aperture of the sparse array is larger than that of the matrix array, the lateral resolution of the defect image obtained by the sparse array is about two times better than the resolution achieved using the matrix probe.

10.4.2.2 Aluminum Block with Flat Bottom Holes

The second example is an aluminum test block with three flat bottom holes of diameter 2 mm at different depths and a 1 mm wide through-thickness slot (Figure 10.12a). The 3D array image obtained from the 128 element sparse array is shown in Figure 10.12b, with a cross-section in the x–z plane in Figure 10.12c. The slot, tips of the holes, and back wall can be readily identified in Figure 10.12b.

10.4.2.3 Surface-Breaking Fatigue Crack

The third sample is a steel block with a surface-breaking fatigue crack with a penetration depth of about 5 mm. Figure 10.13 shows the experimental arrangement and results using the 128 element sparse array. The protrusion of the crack into the volume of the specimen from the back-wall is clearly visible in the 3D image, while the 2D sectional views enable more quantitative estimations of its size to be made.

10.5
Scattering Matrices and Their Experimental Extraction

In many cases defect detection alone is not enough and quantitative information about defect shape, size, and orientation is needed. This information helps to estimate how a defect affects the structural integrity. For defects greater in size than the wavelength this information can be obtained directly from the image. For example, the size of a crack can be estimated using the full width at half maximum (FWHM) parameter [42]. However, sub-wavelength defect characterization is still a major challenge for nondestructive evaluation.

Wilcox *et al.* [8] have developed the technique called the vector total focusing method (VTFM) for characterizing sub-wavelength defects. The array is divided into sub-arrays and transmit–receive data corresponding to each sub-array is processed using the total focusing method (TFM). The amplitude of a scatterer in the image obtained from a sub-array data approximately gives the pulse–echo ultrasonic response of this scatterer in the direction defined by the position of the scatterer relative to the sub-array. It has been shown that the signals from a sub-wavelength slot and hole can be distinguished and that the orientation of a slot also can be determined. However, better characterization accuracy requires a smaller sub-array, but this results in a higher noise level and worse spatial resolution.

10.5 Scattering Matrices and Their Experimental Extraction

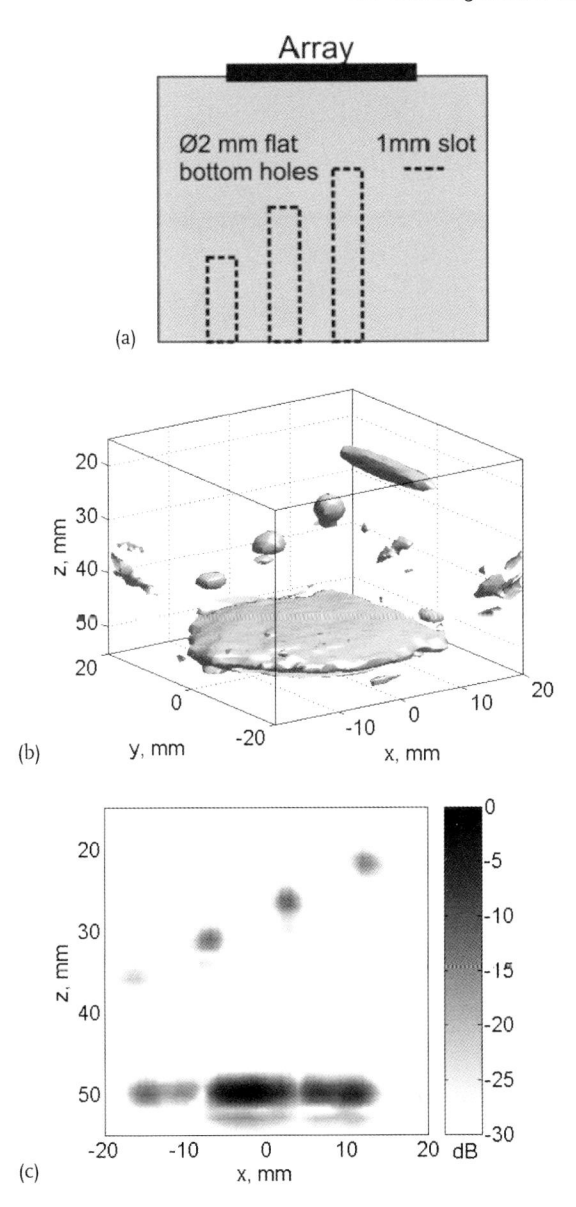

Figure 10.12 Experiment on an aluminum test block using a 128 element sparse array: (a) sample geometry; (b) 3D image; (c) 2D cross-sectional image in the x–z plane.

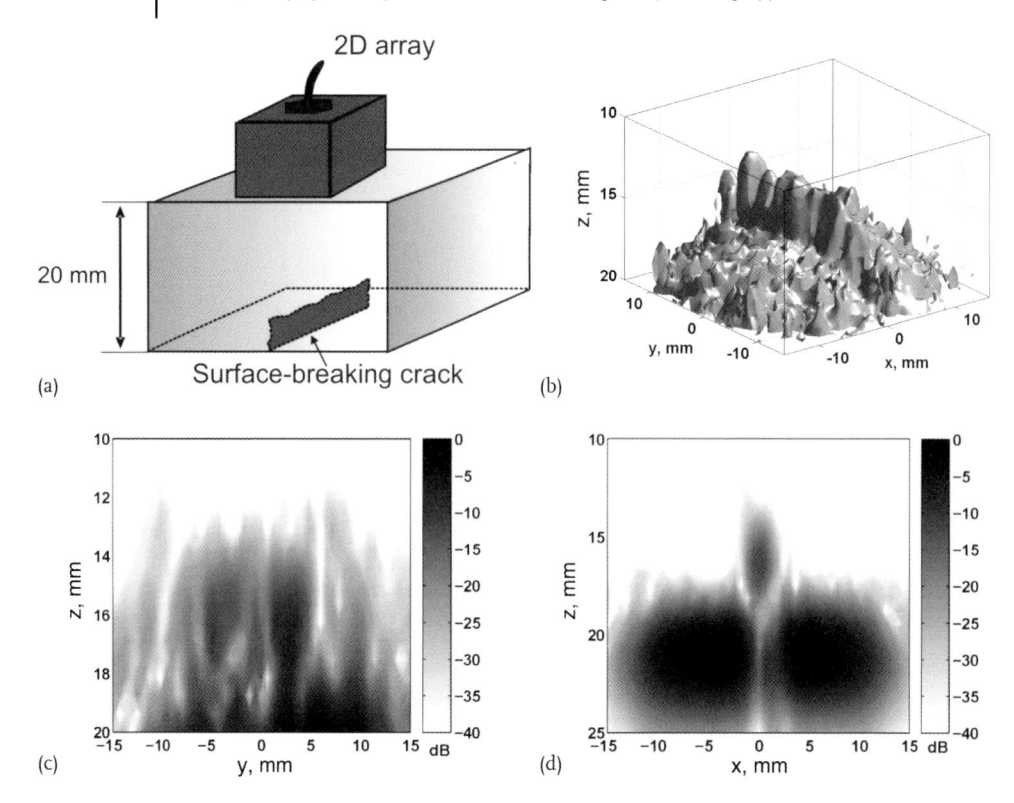

Figure 10.13 Experiment on a steel block with surface-breaking fatigue crack using a 128 element sparse array: (a) sample geometry; (b) 3D image; (c) 2D cross-sectional image in y–z plane; (d) 2D cross-sectional image in x–z plane.

In this section a different approach is described. The problem of characterizing a defect can be divided into two parts. From an array "point of view" all full available information about a scatterer is contained in the ultrasonic response (in the full matrix of transmitter–receiver array data) associated with this scatterer. In fact, this is the maximum possible amount of data corresponding to the scatterer that can be collected by the array in a particular position. Therefore, the first part of characterization is extraction of the raw array data for a particular scatterer from the full transmit–receive data matrix. The second part is processing the array data for a particular scatterer in order to characterize it.

10.5.1
Feature Extraction from Array Data

10.5.1.1 Concept
The real transmit–receive array data contains a superposition of responses from all scatterers including defects and structural features. In the case when the signals

from each scatterer can be identified and separated in time within each transmitter–receiver time-trace, the problem of individual feature extraction is trivial. However, in practice the signals from different scatterers are usually overlapped and a more advanced method is needed.

Any synthetic focusing technique is a procedure that transforms the raw array data set into a new data set usually called an image. The important property of the image is that data associated with each scatterer are approximately localized within some finite area of the image. The degree of this localization depends on the focusing capability of the imaging method. For a finite aperture array the point spread function always has a finite width and non-zero side-lobe level, which means that every image point contains some data corresponding to each scatterer. However, the main information about a scatterer is concentrated in the vicinity of the scatterer's location in the image.

If the distance between scatterers is greater than the resolution limit of the imaging method, most of the data corresponding to the different scatterers can be assumed to be separated in the image. Therefore, the image can be spatially filtered to extract the data associated with a particular scatterer. However, it is very difficult to interpret the data in the "image" representation for characterization of sub-wavelength defects.

On the other hand, the transmitter–receiver array data provides greater flexibility for the characterization of the scatterer than the image. To obtain the array response for each individual scatterer the filtered image data needs to be transformed back into the raw array data. This transformation is only possible if the imaging method used is reversible. Consequently, reversibility of the imaging method is the key to feature extraction from the array data [10].

10.5.1.2 Inverse Imaging

The imaging operator is defined as an operator that transforms the transmit–receive array data set $g(t,r_{(T)},r_{(R)})$ into an image $I(r)$, as described by Equation (10.28). The domain of the imaging operator L is a linear space of array data functions $g(t,r_{(T)},r_{(R)})$. Note that the properties of this space depend on an assumed model of wave propagation and scattering. The reversibility of an imaging method means that the operator L has an inverse operator, L^{-1}, which maps the image $I(r)$ back into the array data $g(t,r_{(T)},r_{(R)})$. The operator L^{-1} exists if and only if there is a one-to-one relation between array datasets, $g(t,r_{(T)},r_{(R)})$, and images, $I(r)$. Note that the operator L^{-1} is defined for the images $I(r)$ that belong to the range of the operator L.

Intuitively it seems a plausible conjecture that, for any imaging operator that produces an image of scatterer locations, two different array datasets lead to two different images. However, the construction of an inverse imaging operator requires the range of the imaging operator to be known, which in turn requires a specific model of the scattering mechanism. For example, if approximation (10.19) for the secondary sources is valid, then the wavenumber imaging method can be used, which reconstructs the scatterer distribution function. In this case the inverse imaging operator is represented by the forward model of wave scattering by a known distribution of point scatterers.

If the imaging operator is linear then for an ideal array with continuous element distribution the operator can be written in the form (10.28). Initially mapping Equation (10.28) has been defined only for the functions g corresponding to the array data. However, if the focusing coefficients s is a bounded function then formally Equation (10.28) defines the linear operator on the set of all absolutely integrable functions g. In this section an imaging method that is reversible in such a global sense is defined [10].

For simplicity, a 2D imaging using a 1D array is considered, but all results are easily extended to 3D imaging with 2D arrays. In Section 10.3.4 it was shown that the back-propagation imaging method can be represented in operator form by Equations (10.25) and (10.26). The function $b(x_T, x_R, z) = e(0, x_T, x_R, z)$ is referred to as a generalized image. The generalized image contains more information than is necessary for localization of the scatterers and the basic 2D image of scatterer position $I(x, z)$ is given by the pulse–echo data $x_T = x_R$. However, the extra information corresponding to the non-diagonal data $x_T \neq x_R$ is crucial for the inverse imaging. Below, the inverse imaging operator will be defined that allows the array data for each individual scatterer to be extracted from the generalized image.

It can be shown that the back-propagation imaging algorithm to create a generalized image can be written as a product of three operators: (i) a Fourier transform, $F_{k_{x(T)} k_{x(R)}}$, to the frequency-wavenumber domain; (ii) back-propagation, H, of the angular spectrum; and (iii) inverse Fourier transform, $F_{x(T)x(R)}$, to the spatial domain. It can be further shown that the operator H can also be expressed as a Fourier transform. Therefore, each operator in the expression for the back-propagation operator B is reversible and the generalized image $b(x_T, x_R, z)$ can be converted back into the array data $g_n(t, x_{(T)}, x_{(R)})$:

$$g(t, x_T, x_R) = B^{-1}[b(x_T, x_R, z)] \tag{10.37}$$

where B^{-1} is the inverse imaging operator:

$$B^{-1} = F^{-1}_{k_{x(T)} k_{x(R)}} H^{-1} F_{x_T x_R} \tag{10.38}$$

From Equation (10.38) it is seen that the operator H^{-1} maps the function $h(z, k_{x(T)}, k_{x(R)})$ into the angular spectrum $G(t, k_{x(T)}, k_{x(R)})$. Using Equation (10.39), H^{-1} can be written as:

$$G(t, k_{x(T)}, k_{x(R)}) \equiv H^{-1}[h(z, k_{x(T)}, k_{x(R)})] = \frac{1}{2\pi} \int v_z^{-1} \left[\int h(z, k_{x(T)}, k_{x(R)}) e^{-ik_z z} dz \right] e^{i\omega t} d\omega \tag{10.39}$$

Note that the group velocity v_z is given by:

$$v_z = \left(\frac{dk_z}{d\omega} \right)^{-1} = \frac{v}{k} \left(\frac{1}{k_{z(T)}} + \frac{1}{k_{z(R)}} \right)^{-1} \tag{10.40}$$

It can be seen that $v_z = 0$ at the frequencies $\omega = vk_{x(T)}, vk_{x(R)}$. Therefore, in practice the expression v_z^{-1} in integral (10.39) should be implemented as a Wiener filter.

10.5 Scattering Matrices and Their Experimental Extraction

If there are several scatterers the time-trace $g(t,x_{(T)},x_{(R)})$ is the superposition of scattered signals:

$$g(t, x_T, x_R) = \sum g_n^0(t, x_T, x_R) \qquad (10.41)$$

where the time-trace $g_n^0(t, x_{(T)}, x_{(R)})$ is the transmit–receive response from n-th scatterer. As the back-propagation operator is a linear operator, then the generalized image $b(x_T, x_R, z)$ is represented by the sum:

$$b(z, x_T, x_R) = \sum b_n^0(z, x_T, x_R) \qquad (10.42)$$

where the function $b_n^0(z, x_{(T)}, x_{(R)})$ is the generalized image of the n-th scatterer. The back-propagation operator focuses the energy from each scatterer into the vicinity of its location. This fact can be expressed in the form:

$$\left| b_n^0(z, x_T, x_R) \right| \leq 10^{-L_{dB}/20} \max \left| b(z, x_T, x_R) \right|, (z, x_T, x_R) \notin D_n \qquad (10.43)$$

here the area D_n defines the location of the n-th scatterer and depends on the chosen dB level L_{dB}. Note that for a point scatterer the spatial size of the area D can be used to quantitatively describe the array focusing performance (Holmes et al. [6]).

It is assumed that for some value of L_{dB} areas D_n for the different scatterers do not intersect. In this case the functions b_n defined as:

$$b_n(z, x_T, x_R) = \begin{cases} b(z, x_T, x_R), & (z, x_T, x_R) \in D_n \\ 0, & (z, x_T, x_R) \notin D_n \end{cases} \qquad (10.44)$$

are satisfied to the following condition:

$$b_n(z, x_T, x_R) = b_n^0(z, x_T, x_R) + \delta_n, \; \delta_n = O(10^{-L_{dB}/20} \max |b(z, x_T, x_R)|) \qquad (10.45)$$

Then using the inverse imaging operator the functions g_n can be calculated as $g_n = B^{-1} b_n$ and it follows:

$$g_n(t, x_T, x_R) = g_n^0(t, x_T, x_R) + O(\delta_n) \qquad (10.46)$$

Therefore, the function $g_n(t,x_{(T)},x_{(R)})$ approximately represents the transmit–receive array data for the n-th scatterer. The accuracy of this approximation is determined by the focusing ability of the array. The better the resolution of different scatterers in the generalized image that can be achieved the more accurate transmit–receive array data for each scatterer can be extracted from the full array data.

The inverse imaging approach described above requires a Fourier domain implementation and as explained in Section 10.3.6 this becomes impractical for 2D arrays because of the computer memory load (Hunter et al. [30]) and the requirement of a regular array element distribution. In this case a different implementation of the inverse imaging is needed that is based on the approximate TFM type formulation of the imaging algorithm. The calculations in this case are similar to that in Section 10.3.2 and only the final expressions are given below.

The generalized image $b(z,x_1,x_2)$ is calculated using Equation (10.28). The far-field back-propagation focusing coefficients s are given by Equation (10.31), where distances R_T and R_R are calculated as:

$$R_T = \sqrt{z^2 + r_1^2} \text{ and } R_R = \sqrt{z^2 + r_2^2} \quad (10.47)$$

where $r_1 = |x_1 - x_T|$ and $r_2 = |x_2 - x_R|$. Note that in contrast to conventional imaging, different focusing points x_1 and x_2 are used for transmitter and receiver array elements. The extracted part of array data, $g_n(t,x_{(T)},x_{(R)})$, is given by:

$$g_n(t, x_T, x_R) = -\left(\frac{\partial}{\partial t}\right)^n \left(2\pi v \sqrt{R_T(t)R_R(t)}\right)^{-n} \iint b_n(z(t), x_1, x_2) dx_1 dx_2 \quad (10.48)$$

where $n = 1$ in the 2D case and $n = 2$ in the 3D case and z in this expression and those above for R_T and R_R is now a function of time defined as:

$$z(t) = \frac{\sqrt{(v^2 t^2 - r_1^2 - r_2^2)^2 - 4r_1^2 r_2^2}}{2tv} \quad (10.49)$$

10.5.1.3 Extraction of Scattering Matrix

If the size of the scatterer is small compared to the distance between the scatterer and the array then the transmit–receive signal is given by the approximate Equation (10.15). It is assumed that the directivity function of the array elements is known. In this case the ultrasonic response of the scatterer can be fully described by the scattering matrix S. Equation (10.11) shows that the scattering matrix is related to the angular spectrum of the scatterer, which is directly connected with the parameters of the scatterer. Therefore, information about the geometry of the scatterer is encoded in the scattering matrix.

The inverse imaging concept [10] allows the transmit–receive array data, $g(t,\mathbf{r}_{(T)},\mathbf{r}_{(R)})$, for an individual scatterer to be extracted. Then the far-field relationship (10.15) can be used to calculate the scattering matrix S for the scatterer. However, the frequency spectrum of the transmitted signal, $\Phi_0(\omega)$, is usually not known and, therefore, in practice only the quantity $S' = S\Phi_0$ can be estimated:

$$S'(e_{R(T)}, e_{R(R)}, \omega) \approx \frac{g_n(\omega, \mathbf{r}_{(T)}, \mathbf{r}_{(R)})}{f_T(e_{R(T)}) f_T(e_{R(R)})} \left(\frac{R_T R_R}{4\pi^2 \lambda^2}\right)^n c^{-1} e^{ik(R_T + R_R)} \quad (10.50)$$

where $g_n(\omega,\mathbf{r}_{(T)},\mathbf{r}_{(R)})$ is the temporal frequency spectrum of the data $g_n(t,\mathbf{r}_{(T)},\mathbf{r}_{(R)})$ and coefficients $c = \exp(-i\pi/4)$, $n = \frac{1}{2}$ in the 2D case and $c = i/k$, $n = 1$ in the 3D case. The location of the scatterer can be taken as the point $\mathbf{r}_0$ where function $|b_n|$ achieves its maximum.

10.6
Defect Characterization and Sizing

10.6.1
Crack Sizing

Once the scattering matrix for a particular scatterer is extracted the next step for characterizing the scatterer is to determine its geometry from the scattering matrix. For an arbitrary scatterer this can be performed, for example, by the linear sampling method [37]. However, if there is some prior knowledge about the type of the scatterer then simpler methods can be used.

10.6.1.1 1D Array

For scatterer characterization it is more convenient to use the incident and scattering angles, φ_{in}, φ_{sc}, instead of φ_T, φ_R (Figure 10.2). If the scatterer is a crack then in a 2D media it is described by two parameters: its orientation and length. In this case the scattering matrix depends on the crack length, d, crack orientation, φ_0, and frequency, ω. From Equation (10.11) it follows that the scattering matrix has dimensions of length. The only parameter with the dimension of length is a wavelength λ. Hence:

$$S(\varphi_{in}, \varphi_{sc}, d, \varphi_0, \omega) = \lambda \tilde{S}\left(\varphi_{in} - \varphi_0, \varphi_{sc} - \varphi_0, \frac{d}{\lambda}\right) \quad (10.51)$$

where the dimensionless scattering matrix $\tilde{S}$ depends on dimensionless crack length d/λ.

The problem of scattering of elastic waves by a crack can be solved by the semi-analytical method, developed by Glushkov et al. [24], which is based on the boundary integral equation method. Figure 10.14a shows the amplitudes of the calculated scattering matrix for the longitudinal wave incident and longitudinal wave reflected from a crack 0.86λ long in aluminum. The zero angle corresponds to the normal direction to the crack surface and the reference point (x_0, z_0) is taken in the middle of the crack. It can be seen that the scattering matrix has one main lobe. Therefore, the orientation of the crack can be defined by the position of the maximum of the main lobe.

The width of the main lobe along the pulse–echo diagonal, $\varphi_{in} = \varphi_{sc}$, is related to the crack size. If the crack length increases then the main lobe becomes narrower. To estimate the crack size the full width at half maximum (FWHM) of the pulse–echo scattering amplitude can be taken, as shown in Figure 10.14b. From Equation (10.51) it follows that this parameter depends only on the ratio of crack size to the wavelength and does not depend on the orientation of the crack and frequency. The dependence of this parameter on the crack size calculated by the semi-analytical method is shown in Figure 10.14c. It can be seen that this method of crack sizing is more sensitive to smaller cracks where the gradient of the FWHM curve is steepest.

Note that in the case of the finite aperture array each scatterer can be illuminated only over a finite range of angles. Therefore, only part of the scattering matrix can

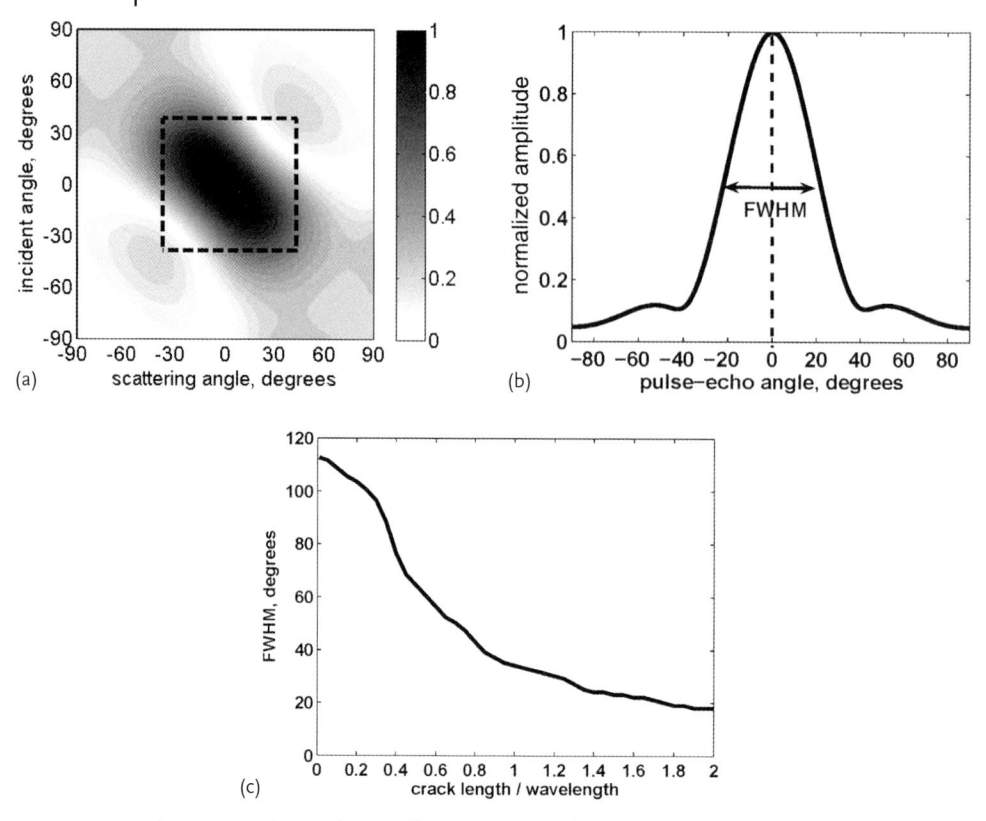

Figure 10.14 2D crack sizing from scattering matrix: (a) normalized amplitude of the scattering matrix for the longitudinal wave and 0.86λ crack inclined at 0° relative to the array with dotted area showing region probed by array; (b) normalized pulse–echo scattering amplitude showing full width at half maximum (FWHM) metric; (c) dependence of the FWHM of the pulse–echo scattering curve on the crack length.

be reconstructed using the array data. This is schematically shown by the dashed square in Figure 10.14a. This part depends on the relative position of the scatterer to the array and is different for different scatterers.

Clearly, the suggested crack characterization technique can only be applied if the part of the main lobe that is used for the measurement of the FWHM is contained in the portion of the extracted scattering matrix. However, the pulse–echo curve is symmetrical relative to the its peak position. Therefore, the half width at half maximum (HWHM) instead of the full width can be used for crack sizing. This requires that only half of the main lobe is contained in the extracted scattering matrix.

10.6.1.2 2D Array

For a 3D system the incident and scattered directions can be described by the azimuth angle, φ, and elevation angle, θ. For the given parameters the scattering

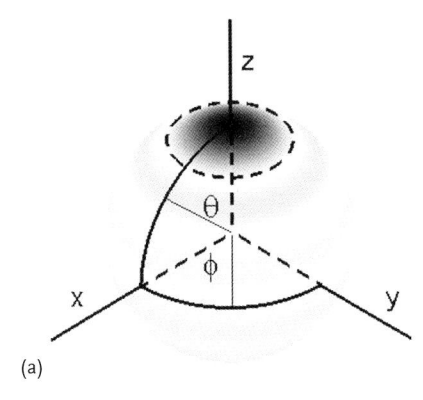

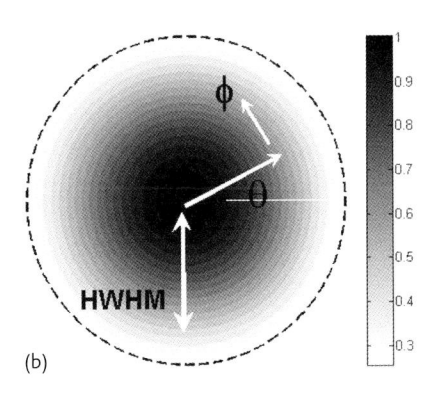

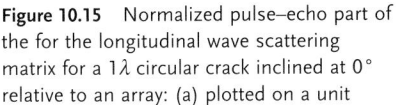

Figure 10.15 Normalized pulse–echo part of the for the longitudinal wave scattering matrix for a 1λ circular crack inclined at 0° relative to an array: (a) plotted on a unit sphere with the dashed ring indicating the region probed by array; (b) 2D projection of the probed region of the pulse–echo scattering matrix.

matrix defines the amplitude of a scattered plane wave in the (φ_{sc}, θ_{sc}) direction, which is caused by an incident plane wave of unity amplitude propagating in the (φ_{in}, θ_{in}) direction.

A 3D plane crack is described by its 2D shape and orientation. First, a circular crack is considered. The pulse–echo part of the scattering matrix ($\varphi_{in} = \varphi_{sc}$, $\theta_{in} = \theta_{sc}$) for the longitudinal wave and a 1λ diameter crack inclined at 0° relative to the array is shown in Figure 10.15a plotted onto a unit sphere. It can be seen that the pulse–echo scattering matrix has one main lobe. The position of the maximum defines the orientation of a crack. The width of the main lobe corresponds to the crack size. If the crack length increases then the main lobe becomes narrower. To estimate the crack size the HWHM of the pulse–echo scattering amplitude can be taken, which is more easily seen if the upper hemisphere of the unit sphere is projected onto a 2D circular region, as shown in Figure 10.15b. Note that the HWHM depends only on the crack size and does not depend on the orientation of the crack. The dependence of this parameter on the crack size is similar to that shown in Figure 10.14c for the 1D crack. Note that for a crack of arbitrary shape its orientation is also defined by the position of the main lobe in the pulse–echo scattering matrix and the shape of the crack is related to the shape of the main lobe.

10.6.2
Experimental Results

10.6.2.1 1D Array

To illustrate the application of inverse imaging to defect characterization an experiment is considered [10]. The 64-element Imasonic array with 5.5 MHz center frequency described previously was used on an aluminum sample with six

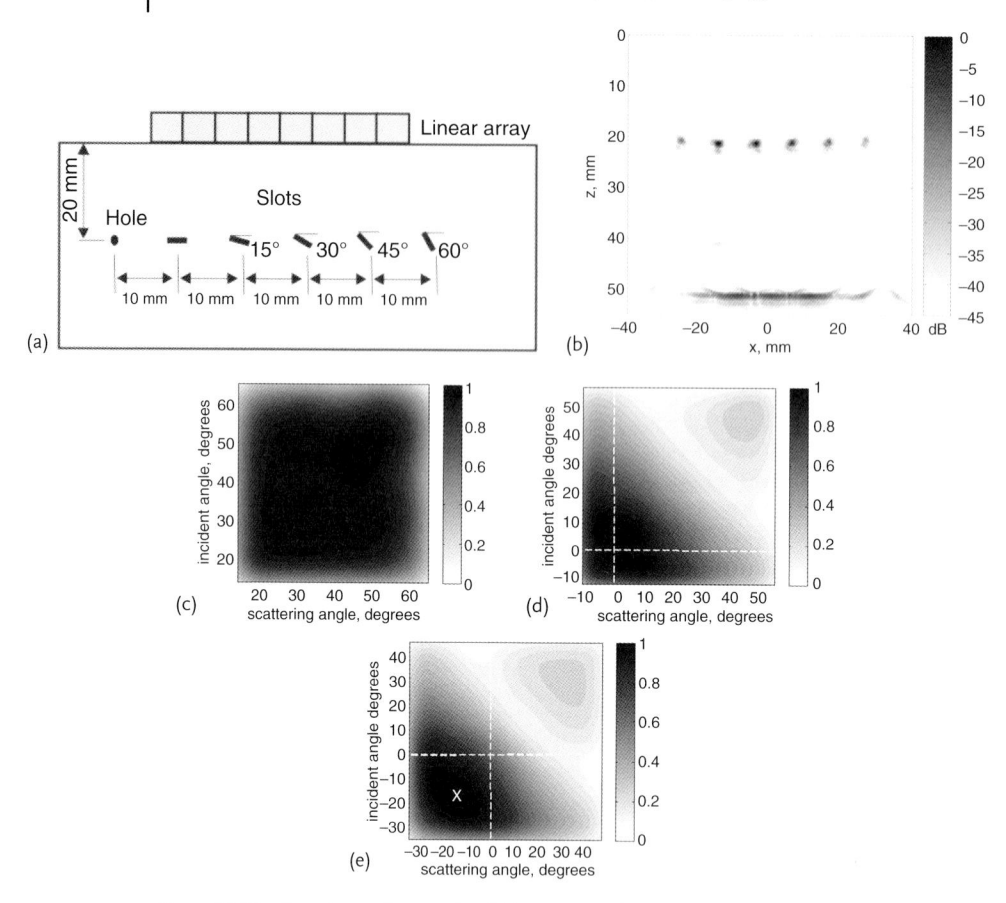

Figure 10.16 Experimental example of 2D scatterer characterization using a 1D array: (a) sample geometry; (b) back-propagation image. Normalized scattering matrices for: (c) hole, (d) 0°, and (e) 15° slots.

scatterers, shown in Figure 10.16a. The left-most scatterer is a 1 mm diameter through-thickness circular hole and the others are through-thickness slots of 1 mm length at various orientations. Note that at the frequency 5.5 MHz the wavelength of the bulk longitudinal wave is 1.2 mm, so all scatterers have a sub-wavelength size of about 0.86λ. The image obtained by the back-propagation method is shown in Figure 10.16b. The array aperture was padded by an extra 32 element positions at each end of the array to prevent aliasing. It is seen that all six scatterers are clearly visible. However, there is no visible difference between them and, therefore, it is impossible to characterize the scatterers directly from the image.

To apply inverse imaging the 3D generalized image was obtained. The image data corresponding to each scatterer was extracted from the generalized image using a spatial filter in the form of a sphere with the center at the location of the scatterer and diameter equal to ⅓ of the distance between the scatterers. Then

Table 10.1 Measured parameters of the slots (scatters) in the sample shown in Figure 10.16a at 5.5 MHz.

Scatterer	Orientation (°)	HWHM (°)	Length (mm)
1 mm slot (0°)	1.5	16.5	1.2
1 mm slot (−15°)	−16.5	18.5	1.05
1 mm slot (−30°)	−32.5	21.5	0.93
1 mm slot (−45°)	−48	18	1.08
1 mm slot (−60°)	−56	16	1.29

the inverse imaging procedure was applied to obtain the array transmit–receive data for each scatterer. Using Equation (10.50) the scattering matrices for each scatterer were calculated at the center frequency. Note that each scattering matrix can be obtained only for certain ranges of angles depending on the location of the scatterer. The extracted scattering matrix for the 1 mm diameter hole is shown in Figure 10.16c. It is seen that for the available range of incident-reflected angles the hole has approximately uniform scattering. Figure 10.16d and e shows the scattering matrices for the slots of 0° and 15° orientations. Each matrix is normalized to its maximum amplitude. It can be seen that the scattering matrices for the slots are totally different from that for the hole and have a distinguished maximum at the angle corresponding to their orientation.

The size of a slot can be calculated using the HWHM of the main lobe. Table 10.1 shows the predicted and measured results. From the table, it can be seen that the agreement between the actual scatterer length and orientation and the length and orientation measured from the experimental scattering matrices is good.

10.6.2.2 2D Array

To illustrate the application of the inverse imaging to 3D defect characterization the data from the 3 MHz 128 element sparse Poisson disk array was used. The samples with the spherical inclusion and flat bottom holes described in Sections 10.4.2.1 and 10.4.2.2 were used [18]. The transmit–receive array data corresponding to the 2.1 mm diameter spherical inclusion and 2 mm diameter flat bottom hole (FBH) was extracted from the full array data. At a frequency of 3 MHz the wavelength of the bulk longitudinal wave is 2 mm, so both scatterers are approximately wavelength size. Note that in the first approximation the FBH can be considered as a plane circular crack. The images in Figures 10.11b and 10.12b show that there is no visible difference between scatterers and, therefore, it is impossible to characterize them directly from the image.

Next, the scattering matrices for each scatterer were calculated, yielding the results shown in Figure 10.17. It can be seen that the scattering matrix for the FBH is totally different to that for the spherical inclusion. The pulse–echo scattering matrix of the spherical inclusion is approximately uniform, while the pulse–echo scattering matrix of the FBH has a distinguished maximum at the zero angle

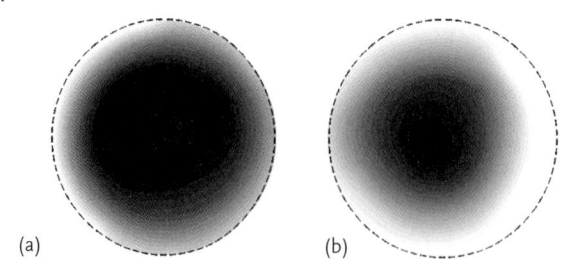

Figure 10.17 Two-dimensional projections of probed regions of normalized pulse–echo scattering matrices for (a) spherical inclusion and (b) flat bottom hole.

which corresponds to the orientation of its surface normal. The size of the FBH can be calculated using the HWHM of the main lobe in the pulse–echo scattering matrix. The measured averaged HWHM is 22°, which corresponds to a crack size of 1.2λ (=2.28 mm). The true size of the FBH is 2 mm.

10.7
Conclusions

This chapter has presented an overview of the development and application of post-processing techniques to array imaging. The starting point is full matrix capture (FMC) in which the time domain signal from every possible transmit–receive pair is recorded. This data set represents the maximum possible information a given array can obtain about its surroundings. At the time of writing this data is of a size that could be called "large but manageable." Increases in storage capability and data transfer rates over the next few years will further improve this situation. Ultimately, we envisage FMC data being stored in the majority of inspections and, for example, in safety critical applications these data will be archived for future reference.

The next challenge is to process this FMC data to produce images and, where possible, extract quantitative information about defects, such as their location, size, and orientation. A large number of imaging algorithms are available: in this chapter, total focusing method (TFM) like algorithms, in which a focus is achieved on every pixel in the image, have been described and discussed in detail. This type of imaging algorithm is particularly attractive because it (i) uses the maximum available aperture and hence achieves high-resolution throughout the image plane and (ii) is both linear and invertible. The FMC can also be post-processed to produce traditional image types such as plane and focused B-scans as well as sector-scans. Indeed, the ability to process the data in various ways is one of the attractive features of the post-processing approach.

It seems reasonable to suggest that as computational power and manufacturing technologies improve there will be a gradual shift towards 2D arrays and 3D imaging. For this reason we have shown the development of post-processing algorithms for both linear arrays (resulting in 2D images) and 2D arrays (resulting in 3D images). At present the number of channels available in commercial array controllers is limited to the low hundreds and so the optimal use of this limited number of array elements is discussed. It has been shown that a random (Poisson's disk) element distribution allows element spacing of above a wavelength to be used and hence this provides a significantly larger aperture (for a given number of elements) when compared to a traditional grid-like layout. We envisage that ever larger apertures will be facilitated by the future availability of array controllers with more channels. There are also interesting future possibilities in terms of reducing the size of the instrumentation to make it more compact, possibly of a size at which all the electronics fit within the array transducer itself.

Industry requires not only for defects to be detected but also for them to be accurately sized. Improved sizing leads to reduced conservatism, which in turn leads to extended operating cycles. High-resolution imaging, such as the TFM algorithm, provides an excellent basis for visualization (and hence quantification) of large defects. However, as the defect size tends to wavelength order and below, such imaging based sizing and characterization methods perform less well. In this chapter the extraction and quantification of scattering matrices as one possible route to the improved sizing of small defects has been described. Arrays naturally illuminate objects over a range of angles and so angular scattering information can be extracted from the FMC data at a given array location. We show how this angular scattering information (the far-field scattering matrix) can be extracted from local regions within the image domain. This spatial localization is important as it means that multiple neighboring defects can be characterized. As above, this algorithm is demonstrated in both linear and 2D arrays. This type of analysis of scattering information provides a clear basis for defect characterization and sizing. In the future, as processing power steadily increases, such analysis could be implemented in real time and fed into automated defect classification/characterization schemes.

References

1 Long, R., et al. (2010) Through-weld ultrasonic phased array inspection using full matrix capture, in *Review of Progress in Quantitative Nondestructive Evaluation, Vols 29a and 29b*, AICP Conference Proceedings vol. 1211 (eds D.O. Thompson and D.E. Chimenti), American Institute of Physics, pp. 918–925.

2 Brekow, G., et al. (2009) Phased array-based SAFT for defect sizing on power plant components, in *Review of Progress in Quantitative Nondestructive Evaluation, Vols 28a and 28b*, AICP Conference Proceedings vol. 1096 (eds D.O. Thompson and D.E. Chimenti), American Institute of Physics, pp. 872–879.

3 Lane, C.J.L., et al. (2010) The inspection of anisotropic single-crystal components using a 2-D ultrasonic array. *IEEE Trans. Ultrason. Ferroelectr. Freq. Control*, **57**, 2742–2752.

4 Johnston, P.H. (2009) Addressing the limit of detectability of residual oxide discontinuities in friction stir butt welds of aluminum using phased array ultrasound, in *Review of Progress in*

Quantitative Nondestructive Evaluation, Vols 28a and 28b, AICP Conference Proceedings vol. 1096 (eds D.O. Thompson and D.E. Chimenti), American Institute of Physics, pp. 1902–1909.

5 Satyarnarayan, L., *et al.* (2007) Phased array ultrasonic measurement of fatigue crack growth profiles in stainless steel pipes. *J. Press. Vessel Technol. – Trans. ASME*, **129**, 737–743.

6 Holmes, C., *et al.* (2005) Post-processing of the full matrix of ultrasonic transmit-receive array data for non-destructive evaluation. *NDT & E Int.*, **38**, 701–711.

7 Drinkwater, B.W. and Wilcox, P.D. (2006) Ultrasonic arrays for non-destructive evaluation: a review. *NDT & E Int.*, **39**, 525–541.

8 Wilcox, P.D., *et al.* (2007) Advanced reflector characterization with ultrasonic phased arrays in NDE applications. *IEEE Trans. Ultrason. Ferroelectr. Freq. Control*, **54**, 1541–1550.

9 Zhang, J., *et al.* (2008) Defect characterization using an ultrasonic array to measure the scattering coefficient matrix. *IEEE Trans. Ultrason. Ferroelectr. Freq. Control*, **55**, 2254–2265.

10 Velichko, A. and Wilcox, P.D. (2009) Reversible back-propagation imaging algorithm for postprocessing of ultrasonic array data. *IEEE Trans. Ultrason. Ferroelectr. Freq. Control*, **56**, 2492–2503.

11 Puel, B., *et al.* (2011) Optimization of ultrasonic arrays design and setting using a differential evolution. *NDT & E Int.*, **44**, 797–803.

12 Hunter, A.J., *et al.* (2011) Monte-Carlo inversion of travel-time data for the estimation of weld model parameters. *Rev. Prog. Quant. Nondestr. Eval.*, **30**, 604–611.

13 Freemantle, R.J., *et al.* (2005) Rapid phased array ultrasonic imaging of large area composite aerospace structures. *Insight*, **47**, 129–132.

14 Smith, R.A., *et al.* (2003) Rapid ultrasonic inspection of ageing aircraft. *Insight*, **45**, 174–177.

15 Verkooijen, J. and Boulavinov, A. (2008) Sampling phased array – a new technique for ultrasonic signal processing and imaging. *Insight*, **50**, 153–157.

16 Moreau, L., *et al.* (2009) Ultrasonic imaging algorithms with limited transmission cycles for rapid nondestructive evaluation. *IEEE Trans. Ultrason. Ferroelectr. Freq. Control*, **56**, 1932–1944.

17 Velichko, A. and Wilcox, P.D. (2010) Strategies for ultrasound imaging using two-dimensional arrays, in *Review of Progress in Quantitative Nondestructive Evaluation, Vols 29a and 29b*, AICP Conference Proceedings vol. 1211 (eds D.O. Thompson and D.E. Chimenti), American Institute of Physics, pp. 887–894.

18 Velichko, A. and Wilcox, P.D. (2011) Defect characterization using two-dimensional arrays. *Annu. Rev. Prog. QNDE*, **30**, 835–842.

19 Lerch, R. (1990) Simulation of piezoelectric devices by 2-dimensional and 3-dimensional finite-elements. *IEEE Trans. Ultrason. Ferroelectr. Freq. Control*, **37**, 233–247.

20 Mcnab, A., *et al.* (1990) The calculation of acoustic fields in solids for transient normal surface force sources of arbitrary geometry and apodization. *J. Acoust. Soc. Am.*, **87**, 1455–1465.

21 Robertson, D., *et al.* (2004) Minimisation of mechanical cross-talk in periodic piezoelectric composite arrays. *Insight*, **46**, 658–661.

22 Yaralioglu, C.G., *et al.* (2005) Finite-element analysis of capacitive micromachined ultrasonic transducers. *IEEE Trans. Ultrason. Ferroelectr. Freq. Control*, **52**, 2185–2198.

23 Wilm, M., *et al.* (2004) Cross-talk phenomena in a 1-3 connectivity piezoelectric composite. *J. Acoust. Soc. Am.*, **116**, 2948–2955.

24 Glushkov, E., *et al.* (2006) An analytically based computer model for surface measurements in ultrasonic crack detection. *Wave Motion*, **43**, 458–473.

25 Miller, G.F. and Pursey, H. (1954) The field and radiation impedance of mechanical radiators on the free surface of a semi-infinite isotropic solid. *Proc. R. Soc. London Ser. A – Math. Phys. Sci.*, **223**, 521–541.

26 Schmerr, L.W. (1998) *Fundamentals of Ultrasonic Nondestructive Evaluation: A*

Modeling Approach, Plenum Press, New York.

27 Auld, B.A. (1989) *Acoustic Fields and Waves in Solids*, 2nd edn, R.E. Krieger, Malabar, FL.

28 Bostrom, A. and Wirdelius, H. (1995) Ultrasonic probe modeling and nondestructive crack detection. *J. Acoust. Soc. Am.*, **97**, 2836–2848.

29 Velichko, A. and Wilcox, P.D. (2010) An analytical comparison of ultrasonic array imaging algorithms. *J. Acoust. Soc. Am.*, **127**, 2377–2384.

30 Hunter, A.J., *et al.* (2008) The wavenumber algorithm for full-matrix imaging using an ultrasonic array. *IEEE Trans. Ultrason. Ferroelectr. Freq. Control*, **55**, 2450–2462.

31 Chiao, R.Y. and Thomas, L.J. (1994) Analytic evaluation of sampled aperture ultrasonic-imaging techniques for NDE. *IEEE Trans. Ultrason. Ferroelectr. Freq. Control*, **41**, 484–493.

32 Simonetti, F. and Huang, L.J. (2009) Synthetic aperture diffraction tomography for three-dimensional imaging. *Proc. R. Soc. London Ser. A – Math. Phys. Eng. Sci.*, **465**, 2877–2895.

33 Schmerr, L.W. and Song, S.-J. (2007) *Ultrasonic Nondestructive Evaluation Systems: Models and Measurements*, Springer, New York; London.

34 Kerbrat, E., *et al.* (2002) Ultrasonic nondestructive testing of scattering media using the decomposition of the time-reversal operator. *IEEE Trans. Ultrason. Ferroelectr. Freq. Control*, **49**, 1103–1113.

35 Lehman, S.K. and Devaney, A.J. (2003) Transmission mode time-reversal super-resolution imaging. *J. Acoust. Soc. Am.*, **113**, 2742–2753.

36 Marengo, E.A., *et al.* (2007) Time-reversal MUSIC imaging of extended targets. *IEEE Trans. Image Process.*, **16**, 1967–1984.

37 Colton, D., *et al.* (2000) Recent developments in inverse acoustic scattering theory. *Siam Rev.*, **42**, 369–414.

38 Portzgen, N., *et al.* (2007) Inverse wave field extrapolation: a different NDI approach to imaging defects. *IEEE Trans. Ultrason. Ferroelectr. Freq. Control*, **54**, 118–127.

39 Petersen, D.P. and Middleton, D. (1962) Sampling and reconstruction of wave-number-limited functions in N-dimensional Euclidean spaces. *Inf. Control*, **5**, 279–323.

40 Cook, R.L. (1986) Stochastic sampling in computer-graphics. *ACM Trans. Graph.*, **5**, 51–72.

41 Yellott, J.I. (1983) Spectral consequences of photoreceptor sampling in the Rhesus retina. *Science*, **221**, 382–385.

42 Davies, J. and Cawley, P. (2007) The application of synthetically focused imaging techniques for high resolution guided wave pipe inspection, in *Review of Progress in Quantitative Nondestructive Evaluation, Vols 26A and 26B*, AICP Conference Proceedings vol. 894 (eds D.O. Thompson and D.E. Chimenti), American Institute of Physics, pp. 681–688.

11
Ultrasonic Force and Related Microscopies

Andrew Briggs and Oleg V. Kolosov

11.1
Introduction

The task of improving the resolution of any microscopy poses a significant challenge. In particular, the resolution of traditional acoustic microscopy (like conventional optical microscopy) is limited by the wavelength. To decrease the acoustic wavelength one must increase the ultrasonic frequency, which in turn leads to quadratically increasing attenuation of acoustic waves in the coupling fluid. Cryogenic liquids offer low attenuation but also poor impedance matching to the studied samples [1]. For awhile, it seemed that achieving spatial resolution significantly below 1 μm in routine acoustic microscopy would be impractical.

The first scheme to overcome the diffraction limit in acoustic microscopy used a sapphire pin attached to a piezoelectric transducer to detect the ultrasonic vibration of the sample surface [2, 3]. Figure 11.1 shows the original sketch of the concept. It was somewhat like an early car mechanic locating a fault in an engine using a long screwdriver, pressing its sharp end to different places, and listening to the vibration of the handle pressed against the ear. The microscope operated at 30 MHz, and was used to detect ultrasonic waves propagating through an aluminum plate. It gave a spatial resolution of 20 μm, which is the diameter of the bottom of the sapphire pin, and is about a tenth of the wavelength of the ultrasound. These experiments demonstrated the feasibility of the idea, but it was to be several years before technology would be available for acoustic imaging with nanometer-scale resolution.

The invention of scanning tunneling microscopy inaugurated a new era of scanning probe microscopy [5]. In scanning probe microscopy (SPM) a sharp tip probes the sample locally and is mechanically scanned in a raster (as in a scanning acoustic microscope) to acquire a two-dimensional image of the sample surface. The image resolution is determined by the interaction between the tip and the sample. In a scanning tunneling microscope, the interaction is the tunneling of electrons through the vacuum gap between the tip and the sample, which varies

Advances in Acoustic Microscopy and High Resolution Imaging: From Principles to Applictaions, First Edition.
Edited by Roman Gr. Maev.
© 2013 Wiley-VCH Verlag GmbH & Co. KGaA. Published 2013 by Wiley-VCH Verlag GmbH & Co. KGaA.

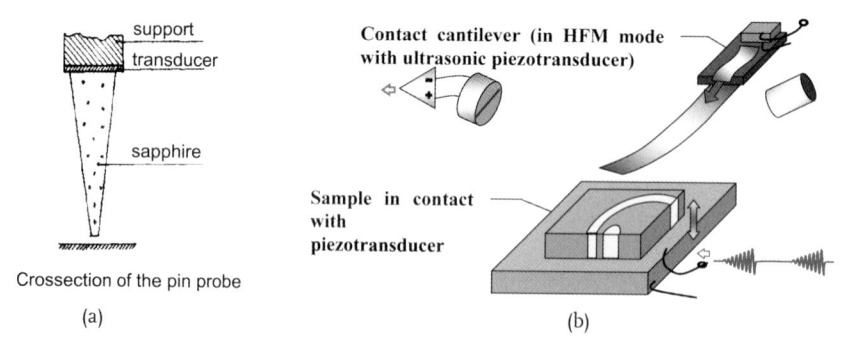

Figure 11.1 (a) Sketch of the world's first ultrasonic scanning pin microscope. The scanning pin microscope operated at 30 MHz, and gave a resolution of about 0.2 mm [2]. (b) Schematic of a UFM. Vertical and lateral deflections of this cantilever provide a measure of forces acting between the tip and the studied object. In an ultrasonic force microscope, or UFM, high frequency piezo-transducers are added to the sample stage and/or to the cantilever base. The detection is the same as in the normal SFM modes [4].

exponentially with separation. In an atomic force microscope (AFM), the interaction is the mechanical force between the tip and the sample, that variation of which with separation can also be highly nonlinear.

Scanning force microscopy can been used to image the surface elastic properties of materials, by modulating the tip–surface distance at frequencies at or below the cantilever primary resonance [6–8]. In this way the viscoelastic behavior of relatively compliant materials such as biomaterials and polymers can be imaged [8, 9]. However, for stiffer materials such as semiconductors this method does not provide good contrast. The stiffness of the contact is generally much greater than the stiffness of the cantilever, so that when relative motion occurs it is almost entirely taken up by deflection of the cantilever. The tapping mode, which is widely used for imaging in atomic force microscopy to minimize the surface in most cases, is insensitive to material properties [10, 11]. Making the cantilever stiffer can help, but reduces the deflection sensitivity and can damage the tip and sample [12]. To overcome these limitations, the cantilever can be vibrated at ultrasonic frequencies above its primary resonance [13–16]. The effective stiffness of the cantilever then increases due to inertia. This offers sensitivity to materials with much higher stiffness, such as semiconductors, ceramics, metals, and composites. Techniques using this principle include atomic force acoustic microscopy (AFAM [16]) and scanning local-acceleration microscopy (SLAM [15]). This chapter describes an approach that depends on the nonlinear nature of the interaction between tip and sample; this has become known as ultrasonic force microscopy (UFM) [13, 17–19]. The combination of acoustic excitation with scanning probe microscopy makes it possible to image and study the elastic and viscoelastic properties of materials with nanoscale spatial resolution [19].

11.2
Mechanical Diode Detection

In a scanning force microscope, a sharp needle is fabricated at the end of a flexible cantilever, whose deflection provides a measure of the force acting between the tip and the studied surface (Figure 11.1). Minute changes in the cantilever deflection can be measured by the focused laser beam bouncing off the cantilever to a four-quadrant position sensitive photodetector. When operating in the so-called contact mode, the tip–surface distance is kept in the range at which repulsive forces prevail. A feedback circuit is used to maintain a constant normal deflection, and hence a constant normal force, by vertically moving the specimen with respect to the tip. The specimen is scanned under the tip by a piezoelectric actuator while the deflection of the cantilever is measured. In some implementations, a quadrant detector is used to measure both normal and lateral deflection, thus allowing friction forces to be measured. The data can be processed to create topographical images and lateral force images [6].

To understand the UFM detection, consider the nonlinearity of the tip–surface interaction and what happens when the indentation depth is modulated. In Figure 11.2a a schematic relationship between force and displacement is plotted. The force is a highly nonlinear function of tip–sample displacement, and depends on whether approach or retraction is underway. If the tip and the surface are well separated, at the far right on the graph, there is negligible interaction. During the approach, from right to left on the graph, initially there is an attractive interaction. At closer approach the force becomes repulsive. When reversing the displacement, the tip and the surface adhere until the contact is broken at a certain pull-off distance, which is larger than the displacement at which the contact is established during the approach. The distance scale over which significant variations of force occur is only a fraction of a nanometer. Suppose the tip is in contact with the surface at an initial displacement that gives a normal force F_1. The displacement to give pull-off is h_1. An oscillatory displacement of amplitude a_0 is now introduced. If the amplitude of the displacement is small, $a_0 \ll h_1$, the average normal force does not change appreciably. As the oscillatory amplitude is increased, the nonlinearity becomes appreciable, until at an oscillatory displacement amplitude a_1 in Figure 11.2a contact can be lost at one end of the cycle.

The stiffness of the tip–sample interaction is generally much higher than the stiffness of the cantilevers, which give adequate sensitivity in SFM. Therefore, if the sample surface is vibrated at a low frequency, the tip will tend to follow this motion with little relative displacement between the two. If the vibration is applied at a frequency that is much higher than any natural resonant frequency of the cantilever, then the additional inertia of the cantilever can result in a substantial enhancement of the relative displacement, and hence greater indentation amplitude [13]. In the limit that the cantilever motion is negligible, the relative displacement is almost equal to the normal vibration amplitude of the sample surface, giving $a \approx \Delta h$. This can be considered as an inertial increase in the effective spring

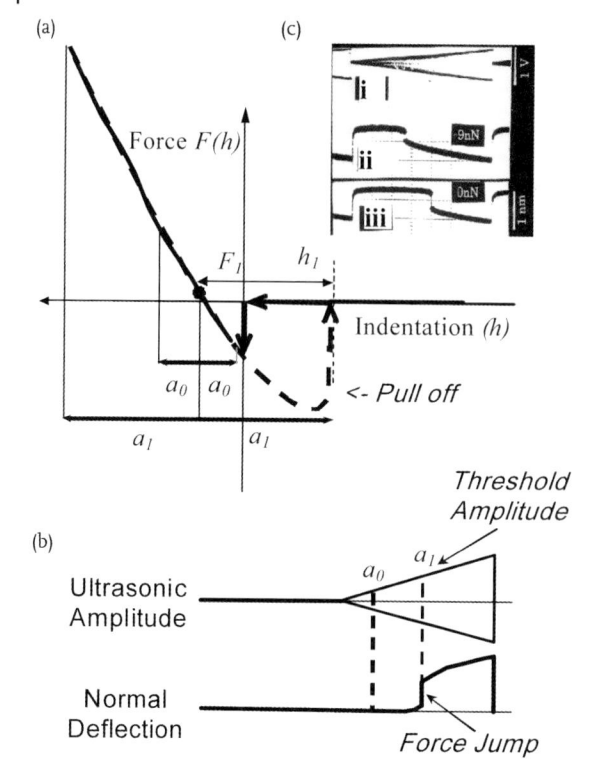

Figure 11.2 (a) Tip–surface interaction force versus indentation for approach and retraction (solid and dashed, respectively). Approach (solid) and retraction (dashed) differ at the verge of the tip–surface contact; (b) nonlinear detection of ultrasound at increasing ultrasonic vibration amplitude; (c) oscilloscope traces of ultrasound detection in a standard SFM setup [13, 18].

constant of the cantilever at high frequency, while at low frequency the cantilever retains its high compliance and hence sensitivity to normal and lateral forces.

11.3
Experimental UFM Implementation

In practical ultrasonic force microscopy, the feedback signal of the AFM is used to maintain a constant average deflection of the cantilever. This is necessary for obtaining a consistent UFM signal, and it also means that topographical information is available. But it presents a dilemma in separating the unique UFM contrast from the conventional SPM image. If the ultrasound were applied with constant amplitude, then it would introduce a small change to the SPM deflection, but it would be impossible to distinguish between this and topography. Instead, the ultrasonic vibration is modulated in amplitude. The feedback circuit of the AFM

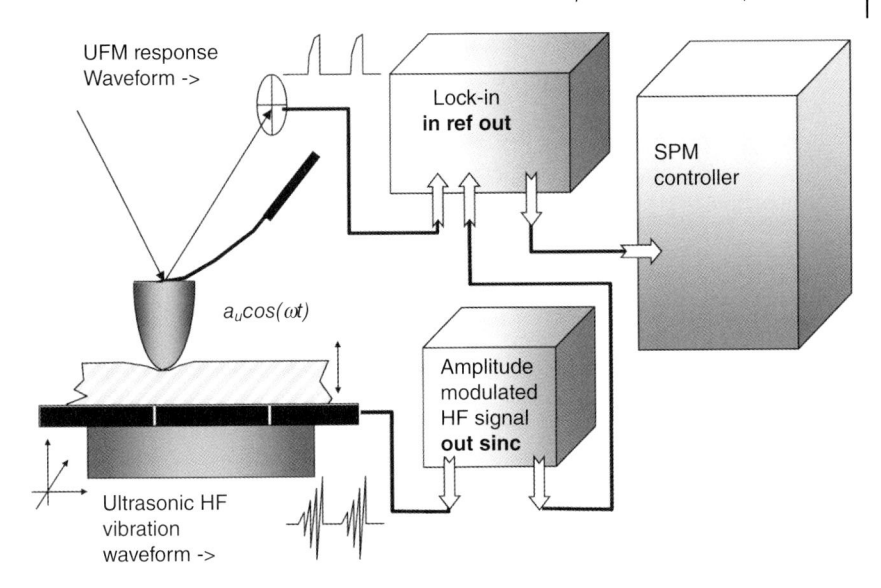

Figure 11.3 Schematic diagram of typical electronics components of a UFM setup.

will have an upper limit, or cut-off frequency, above which it does not respond. In a typical commercial AFM this may be of the order of 1 kHz, though the user may have some control over this. The error signal of the feedback loop is then used as the measure of the UFM response. Although for some analytical purposes one may wish to measure the full response illustrated in Figure 11.2c, for most practical imaging a lock-in amplifier can be used to give phase-sensitive detection of the UFM signal. The modulation signal is the reference and the error signal from the feedback circuit is the input. The output of the lock-in amplifier provides the contrast signal for the UFM image (Figure 11.3).

The lock-in phase and time constant can be selected to give the best contrast. The output can be chosen between two pairs: x (*in-phase*) and y (*quadrature*), or amplitude ($R = \sqrt{x^2 + y^2}$) and phase [$\theta = \arctan(x/y)$]. Any variation in threshold amplitude or force jump produces a variation of the shape of the ultrasonically induced normal deflection, which can be measured by the lock-in amplifier. In choosing the parameters for the ultrasonic excitation, there are three critical choices. First, the maximum applied amplitude must be higher than any threshold amplitude to be measured over the whole area to be imaged. Second, the amplitude modulation frequency must be sufficient to enable several cycles to be averaged at each pixel. Finally, the profile of the amplitude modulation can be chosen from waveforms such as sinusoidal, ramp (saw-tooth), triangular, and trapezoidal [13, 18]. Experience shows that the best contrast is generally obtained with a saw-tooth profile with a blank period of the same or larger length (as in Figure 11.2c), combined with the amplitude lock-in signal output $R = \sqrt{x^2 + y^2}$.

The ultrasonic frequency must be chosen carefully to obtain unambiguous UFM data. The frequency should be high enough to enable the inertial term ($\omega^2 m$) to give sufficient enhancement of the elastic stiffness of the cantilever to make it comparable with the tip–sample stiffness. While frequencies of up to 60 MHz or more have been used in some experiments [20] it is more common to choose a frequency in the range 1–10 MHz. The linear response of the cantilever cannot readily be measured at these frequencies, and simulations suggest that it is likely to be negligible [21]. Once the transducer has been selected and mounted, fine-tuning of the frequency then becomes pragmatic. The useful bandwidth of a piezoelectric transducer for this purpose is usually $<\pm 10\%$ of the central frequency. There may be other resonance frequencies at which a nonlinear UFM response can be excited, and many of them are not pure normal modes. If the piezoelectric transducer frequency is not high enough, cantilever higher harmonics may interfere with the UFM detection. These harmonics are usually damped, but if they are present they can reduce the effective dynamic rigidity of the cantilever, especially if the UFM is operated in a vacuum.

A practical procedure for tuning the frequency starts with acquisition of a spectrum of the nonlinear response. With amplitude modulation applied, the ultrasonic vibration frequency is swept over a wide interval centered at the nominal resonance frequency of the piezoelectric transducer. From this spectrum, one finds the frequencies at which a nonlinear response can be obtained. The second step is to analyze the ultrasonically induced normal deflection for the frequencies at which a nonlinear response has been detected. Practical guidelines for achieving reproducible and reliable UFM data are [18]:

1) The amplitude modulation frequency should generally be in the range 0.5–3 kHz. Below 0.5 kHz the feedback usually modifies the ultrasonically induced normal deflection. Above 3 kHz the characteristic time constants and delays of the cantilever response can become comparable with the amplitude modulation period.

2) The maximum ultrasonic amplitude must be chosen so that the average threshold amplitude occurs at roughly ¾ of the reference period.

3) The lateral deflection signal should also be checked. If the lateral response is unstable, it is quite likely that the normal deflection is influenced, even though it might look stable.

The experimental setup can be based on almost any commercial AFM [13, 22]. The main mechanical modification to implement UFM is made to the sample holder to allow application of a normal ultrasonic vibration. The sample holder of a commercial system is usually a thin disc of metal, often fixed to the piezoactuator (scanner) via a magnet positioned inside the actuator. A piezoplate (ultrasonic transducer) can be permanently glued to the sample holder, for example, with cyanoacrylate or epoxy. An insulating spacer can be inserted between the metallic sample holder and the piezoplate, to isolate the lower electrode of the piezoplate from the SFM unit. A piezoplate has a longitudinal resonance that depends inversely on its thickness (typically, 1 mm gives 2 MHz). A diameter from 8 to

20 mm is common (depending on the SFM system). Very soft connecting wires (especially critical if the sample is scanned, as it is in DI Multimode™ or DI/Thermal Microscopes CP systems) should be connected with a low melting point solder to avoid depolarization of the piezoplate. The top of the piezoplate can be connected to ground, to avoid electrical interaction between the tip and the sample (in particular for studying conductive samples).

For the ultrasonic vibration a programmable waveform generator can be used to generate signals in a required frequency range, typically from 1 MHz to 10 MHz. It should have amplitude modulation with various modulation shapes, particularly blanked saw-tooth (Figure 11.2b) or at least saw-tooth capability. A maximum output voltage of 10 V_{pp} is more than sufficient for the average piezoplate (for a typical experiment an amplitude of 1–4 V_{pp} is used). The ultrasonically induced normal deflection can be visualized on the oscilloscope (preferably digital for signal averaging) by using the feedback error signal (provided the ultrasonic deflection is modulated above the feedback circuit frequency cut-off). The oscilloscope and the lock-in amplifier should be synchronized to the modulation frequency.

The sample can be directly bonded to the piezoplate through an acoustic coupler such as salol (phenyl salicylate), which melts at 42 °C; a grain of crystalline salol can be used to nucleate crystallization of the supercooled liquid. Afterwards, the salol can be melted and the sample removed without damage. Epoxy and other glues are more permanent. The sample dimensions and mass must be compatible with the microscope stage and scanner, and the sample, which may be polished, cleaved, or cast, must be flat enough to avoid the kind of topographical artifacts discussed below in Section 11.7.

Figure 11.4 shows AFM and UFM images obtained in this way of a longitudinal section of a composite consisting of a silicon carbide fiber (SiC: Young modulus $E \approx 500\,GPa$) in a mullite matrix (Al_2O_3-SiO_x, Young modulus $E \approx 150\,GPa$). These are relatively stiff materials. The AFM image in Figure 11.4a, shows the topography, with a rough trench (3) between the fiber (1) and the matrix (4). The UFM image in Figure 11.4b reveals elasticity, showing that the interface contains a relatively soft layer (3) coating the silicon carbide fiber between the stiffer regions of SiC and mullite. The force modulation image (Figure 11.4d) of the area of the topographical image (Figure 11.4c) shows information that is not related to the structure of the fiber interface, highlighting less relevant surface contamination.

11.4
UFM Contrast Theory

UFM detection is obtained by measuring the cantilever deflection at low frequency (Figure 11.2). The ultrasonic vibration applied to the sample is invariably from a longitudinal wave transducer fixed to the bottom, causing normal vibration of the

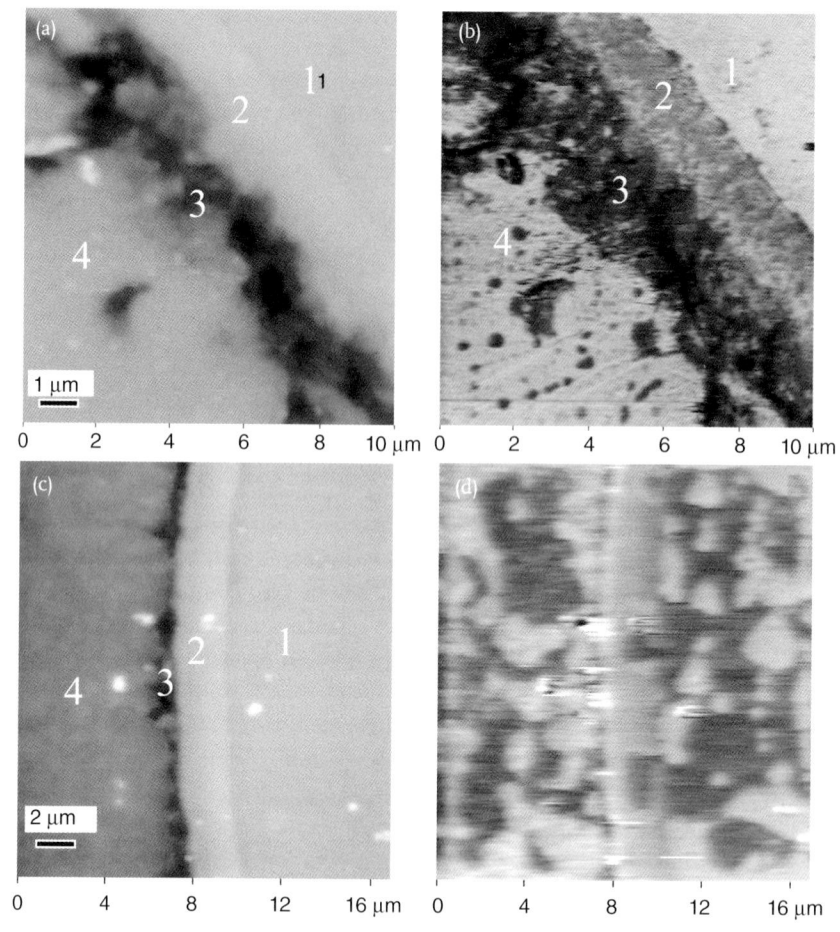

Figure 11.4 Topography AFM (a) and elasticity UFM (b) images of SiC ceramic fiber. The topography AFM image shows the interface as a jagged trench dividing a SiC fiber (1) and a mullite (Al_2O_3-SiO_x) matrix (4). The elasticity image reveals the detailed structure of this interface consisting of the relatively soft intermediate concentric carbon-rich layer (2) and a softer reaction layer (3) separating much stiffer regions of SiC and mullite. Image size is $10 \times 10\,\mu m^2$. The force modulation image (d) of the area of the topographical image (c) shows information that is not related to the structure of the fiber interface, highlighting less relevant surface contamination [4].

sample surface. As the ultrasonic amplitude is increased, contact is eventually broken at the pull-off point ($a_1 = \Delta h_1$), giving a discontinuity in the time-averaged displacement. We refer to this ultrasonic amplitude as the threshold amplitude, and the corresponding inflection in the displacement curve as the force jump. A further increase of the ultrasonic amplitude results in a steady increase of the time-averaged force and therefore of the quasi-static normal deflection [18].

The force on the cantilever under normal vibration of amplitude a from an initial indentation h_1 is found by integrating over a period [13]:

$$F_m(h_1, a) = \frac{1}{T}\int_0^T F[h_1 - a\cos(2\pi ft)]\,dt \qquad (11.1)$$

where $F(h)$ is the force dependence on the indentation depth without ultrasonic vibration, f_{ult} is the ultrasonic frequency, and the integral is taken over a period $T_{ult} = 1/f_{ult}$. Because of the nonlinearity of the force function F, the cantilever will acquire a new equilibrium deflection z_{eq} with a corresponding new mean indentation depth h_{eq}, so that now:

$$F_m(h_{eq}, a) = k_C z_{eq} \qquad (11.2)$$

This description is based on the simplified consideration that the cantilever acts as a point mass. A fuller analysis of the cantilever takes into account its distributed mass and multiple vibration modes [21]. This allows a more rigorous description of the mode of operation of the UFM and the contrast in the images, and also provides the theoretical basis for using the cantilever as a waveguide through which vibrations can be introduced [23]. For the applications described in this chapter, the key components of the UFM and the mechanical diode principle are:

1) the inertial stiffness of the cantilever at the ultrasonic vibration frequency;
2) nonlinear detection of additional forces at low frequency;
3) the compliance of the cantilever at the detection frequency.

The UFM signal depends on the local tip–surface force dependence, and therefore on everything that affects this, including elasticity, adhesion, and viscoelastic relaxation, together with the subsurface structure of the object, local topography and discontinuities such as cracks, and the shape of the particular tip. The dominant property is usually the elastic stiffness of the sample in the vicinity of the surface.

As we saw, the AFM cantilever response to the heterodyne force sample vibration of amplitude a_u can be described by the introduction of a new force-versus-separation dependence $F_m(z)$, derived from the original $F(z)$ dependence by averaging over a vibration period T [Equations (11.1) and (11.2)]. Therefore, the UFM response can be relatively easy calculated, using the well-known force balance equation in SFM, provided original $F(z)$ dependence is known.

To calculate the UFM response from Equations (11.1) and (11.2), the force-versus-separation dependence $F(z)$ must be known. The choice of a continuum mechanics description of the tip–surface force interaction $F(z)$ depends on the geometry, elastic properties, and adhesion energy. An approximation that lends itself to analytical modeling is the Johnson–Kendall–Roberts (JKR) model [24]. The force F and the displacement z may each be expressed in terms of the radius a of the contact area [25]:

$$F = \frac{4E^* a^3}{3R} + \sqrt{16\pi\gamma E^* a^3} \qquad (11.3)$$

$$z = \frac{a^2}{R} - 2\sqrt{\frac{\pi \gamma a}{E^*}} \tag{11.4}$$

The adhesion energy is γ, and the mutual plain strain modulus and radius of curvature are, respectively:

$$\frac{1}{E^*} = \frac{1-v_t^2}{E_t} + \frac{1-v_s^2}{E_s} \tag{11.5}$$

$$\frac{1}{R} = \frac{1}{R_t} + \frac{1}{R_s} \tag{11.6}$$

where E and v denote Young modulus and Poisson ratio, respectively, and subscripts t and s indicate the tip and sample. In the approximation that the material of the tip is much stiffer than the sample and the surface of the sample is flat, the terms $(1-v_t^2)/E_t$ and $1/R_s$ may be neglected. Even for perfectly elastic materials, there will be hysteresis if contact is broken in the cycle of vibration; with viscoelastic samples such as polymers there will be additional hysteretic effects [26]. The JKR model is most valid for large radius and adhesion energy and small stiffness; other models are appropriate for different regimes [27]. All the continuum mechanics models give nonlinearity, which may be simply expressed by saying that the instantaneous stiffness is determined by the contact area, and this will be greater when you push than when you pull. Atomistic modeling of nanoscale contacts suggests that the contact area and hence stiffness may be two or more times what is calculated by continuum models, but does not alter the conclusion that the force–displacement relationship is highly nonlinear [28].

The UFM response is calculated for silicon and germanium surfaces in Figure 11.6a, using the JKR model and average values of the elastic moduli for silicon and germanium [29]. The calculated UFM responses for unstrained Si and Ge (curves 1 and 2 in Figure 11.5a, b) are distinctly different, with the step in the Ge response occurring at a higher amplitude than that of Si. Careful choice of the amplitude would give a smaller UFM signal for Ge, confirming the possibility of directly mapping the elasticity of Si-Ge nanostructures. Figure 11.6 presents images showing such contrast from germanium quantum dots on silicon. Figure 11.6a was obtained by conventional AFM topography, with a line scan underneath. Figure 11.6b was obtained by UFM, with a corresponding UFM line scan. Using engineering stiffness parameters, germanium (E_{Ge} = 121 GPa) is less stiff than silicon (E_{Si} = 164 GPa), and the Ge islands give a lower signal than the surrounding wetting layer, which consists of two or three atomic layers of Ge on the Si substrate. The bright UFM signal around the dots may be due to a topographic effect, which will be discussed further in Section 11.7. At the edge of the dots the contact area between the tip and substrate may be increased, especially if the dot is surrounded by a narrow moat as some are, and this would give rise to increased contact stiffness and hence UFM signal. A model based on statistical thermodynamics has been developed to account for the size and shape distributions of these

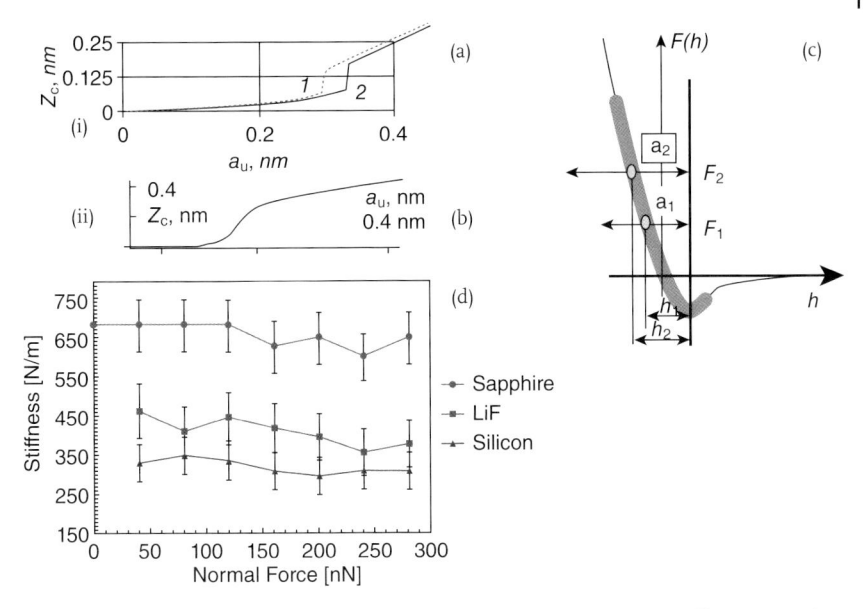

Figure 11.5 (a, b) Theoretical calculation of the UFM response on Si (1) and Ge (2) surface, using the parameters $E_{Si} = 164\,GPa$, $E_{Ge} = 121\,GPa$, surface energy in an ambient environment $W = 1\,N\,m^{-1}$, and the manufacturer's data for tip radius $R = 10\,nm$. (c) Illustration of the differential UFM approach to the measurement of contact stiffness [30]. If one measures the threshold amplitude values (a_1 and a_2) for two different normal force values (F_1 and F_2), the contact stiffness S_{eff} is given by $S_{eff} = (F_2 - F_1)/(a_2 - a_1)$. (d) Experimental stiffness measurements obtained using differential UFM for sapphire, silicon (100), and LiF (100) [31]. (Nominal cantilever stiffness was $k_c = 2.8\,N\,m^{-1}$, and radius of curvature $R = 10\,nm$).

dots [32]. UFM images have been obtained of individual antimonide particles [33]. The nanoparticles were formed by aggregation and spontaneous rapid crystallization of thermally deposited Sb onto the basal planes of highly oriented pyrolytic graphite (HOPG) and molybdenum disulfide (MoS$_2$). The UFM contrast was interpreted in terms of variations in local stiffness, which correlated with evidence from transmission electron microscopy of strained regions within the nanocrystals.

11.5
Quantitative Measurements of Contact Stiffness

The shape of the force versus indentation curve depends on surface adhesive and elastic properties. Variations in these parameters affect the ultrasonically induced deflection. Conversely, variations in the shape of the ultrasonically induced normal deflection contain information on surface adhesive and elastic properties. Figure

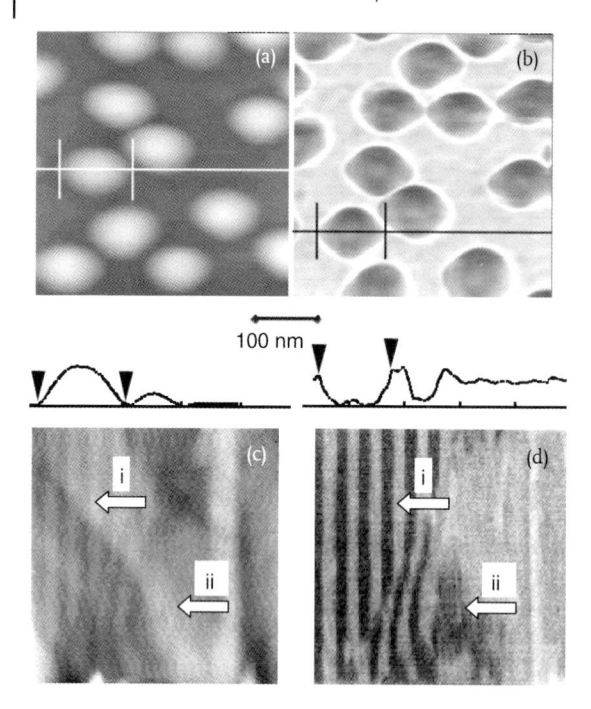

Figure 11.6 Topographical (a) and UFM (b) images of Ge quantum dots on a Si substrate (image size: 400 × 400 nm²) [18, 29]. Topography (c) and UFM (d) images of GaSb-InAs superlattice with periodicity of 40 nm (arrow i) and 8 nm (arrow ii). The superlattice was cleaved and immediately mounted on the sample stage. The topographical image shows that the surface is very flat (RMS roughness < 0.2 nm over 1 μm² area). Wider layers (arrow i) are barely visible whereas the finer ones (arrow ii) are not visible at all. The very fine superlattice (arrow ii) of only 4 nm wide layers is observable in the UFM image (d) [18].

11.2 illustrates how the threshold amplitude should depend on the normal force value. If the normal force is set at a higher value $F_2 > F_1$, then the threshold amplitude ($a_2 = h_2$) needed to reach the pull-off point should be higher than the threshold amplitude ($a_1 = h_1$) for F_1. If the threshold amplitude values (a_1 and a_2) are measured for two different normal force values (F_1 and F_2), the contact stiffness is:

$$S_{\text{eff}}(F_{\text{av}}) = \frac{F_2 - F_1}{a_2 - a_1} \tag{11.7}$$

where:

$$F_{\text{av}} = (F_2 + F_1)/2 \tag{11.8}$$

The beauty of the differential UFM approach is that the absolute value of the contact stiffness of a nanoscale contact at a known force level F is directly measured in terms of the ultrasonic vibration amplitude and the applied force (independent of the adhesion or other contact parameters). The contact geometry would need to be known to determine the elastic stiffness of the sample.

The differential UFM approach is based on three main assumptions:

1) It is possible to identify a threshold amplitude, defined as the amplitude at which the contact breaks, and pull-off occurs, for part of the ultrasonic cycle. It can be identified as the amplitude at which the inflection occurs in the normal deflection signal.

2) The threshold amplitude depends on the applied normal force.

3) The cantilever vibration at the ultrasonic working frequency is negligible. Therefore, the difference in threshold amplitude at different values of normal force is equal to the difference in indentation $\Delta h = h_2 - h_1$.

The second assumption is based on contact mechanics models in which viscoelastic effects that might influence the instability point (pull-off) and adhesion are negligible or can be allowed for. The third assumption is based on representing the cantilever with a point mass model. Simulations using a distributed mass model indicate that ultrasonic vibration of the cantilever is relatively small and in many cases less than 0.05 of the UFM normal deflection [21].

Figure 11.5d presents experimental stiffness measurements using differential UFM for three high modulus materials: sapphire, silicon (100), and LiF (100) [31]. The samples were probed with the same silicon tip on a V-shaped cantilever (nominally, the cantilever stiffness was $k_c = 2.8\,\mathrm{N\,m^{-1}}$, and the radius of curvature $R = 10\,\mathrm{nm}$). The surface RMS roughness of the surfaces was less than 0.2 nm over a few square micrometers for all three samples. The relative difference between the three sets of data reveals that the elastic properties of these three materials can be distinguished by differential UFM; the relative independence of the applied force may indicate the fact that the tip had been flattened by extended contact with such hard samples.

11.6
UFM Picture Gallery

Ultrasonic force microscopy is able to give contrast from samples over a wide range of elastic properties, from stiff crystalline materials like semiconductors or ceramics, including composites of hard materials (metals, metal oxides, carbon fiber) and polymers, to soft rubbery inclusions and proteins.

Figure 11.6c, d shows a double GaSb-InAs superlattice with periodicities of 40 and 8 nm. The superlattice was cleaved and immediately mounted on the sample stage. The topographical image shows that the surface is flat (RMS roughness < 0.2 nm over 1 µm² area). Wider layers (arrow i) are barely visible whereas the finer ones (arrow ii) are not visible at all. The UFM contrast is different from the topography, even while the difference in elastic moduli is only of about 6% ($E_{GaSb} = 88\,\mathrm{GPa}$, $E_{InAs} = 82\,\mathrm{GPa}$). UFM detects this difference, with the GaSb layers brighter than the InAs layers. The very fine superlattice (arrow ii) of only 4 nm wide layers is observable in the UFM image (Figure 11.6b).

The resolution of scanning probe techniques is full of happy surprises. The original estimates of the resolution of the scanning tunneling microscope assumed a smooth sphere of radius 100 nm. The actual resolution obtained far surpassed this, because far from being a smooth sphere the tip is always atomically rough. Though a Darwinian process of natural selection, by the time the user is ready to record a picture the tip has one atom protruding significantly further than the others, giving sub-nm resolution. For UFM, it is not obvious how the lateral resolution might compare with the corresponding AFM. During the ultrasonic vibration the contact size varies from zero to something larger than a conventional AFM working in contact mode at the same normal force. The UFM signal arises from nonlinearity in the force–displacement curve, and this nonlinearity is greatest when the diameter of the contact area is least. Hence it might be expected that the UFM lateral resolution should be at least as good as the AFM lateral resolution, and perhaps even better. This is borne out by the UFM resolution of the finer periodic lattice in Figure 11.6d.

High density interconnects on semiconductor chips link various parts of the processor with memory and other functions. In the Damascene process, the underlying silicon oxide insulating layer is patterned with open trenches, which are then overfilled with metal. Chemical-mechanical polishing is then used to remove the metal to the level of the top of the insulating layer. The term comes from metallurgical processes associated with pattern-welded swords from Damascus. Figure 11.7 shows AFM and UFM images of a Damascene interconnect test structure. They show trenches 0.32 µm wide adjacent to a chessboard contact pad. The UFM images (Figure 11.7b, d, f) display little topographical sensitivity, while UFM contrast is material specific, delineating polymer and Al regions uniformly across the scan area. The image contrast corresponding to Al is uniform across the trench region, the 10 µm wide lead, and the contact pad. The white areas in the topographic images are polymer (benzocyclobutene, BCB) spacers with low dielectric constant (low k) to increase the speed of data transfer on the chip. The dark areas in the topography image are the aluminum connects. The topography results from the different speed of removal of material during chemical-mechanical polishing. The UFM contrast arises primarily from the difference between the Al and BCB elastic moduli. The UFM images reveal elastic nonuniformity across the top of the BCB wall. Two distinct regions are apparent. The center portion of the BCB wall displays a lower contact stiffness compared to high-contrast regions near the Al/BCB interfaces that may betoken an increase in the BCB rigidity in the vicinity of the Al/BCB interface [34].

The samples shown in Figure 11.8 are of lower stiffness. Figure 11.8a, b shows a composite consisting of injection molded poly(methyl methacrylate) (PMMA) ($E_{PMMA} = 4.5$ GPa) with spherical PMMA–rubber inclusions ($E_{rubber} < 0.1$ GPa) to increase its fracture toughness [35]. The structure of the inclusions is illustrated in the inset. Each inclusion is made of a rubber core and alternate layers of PMMA

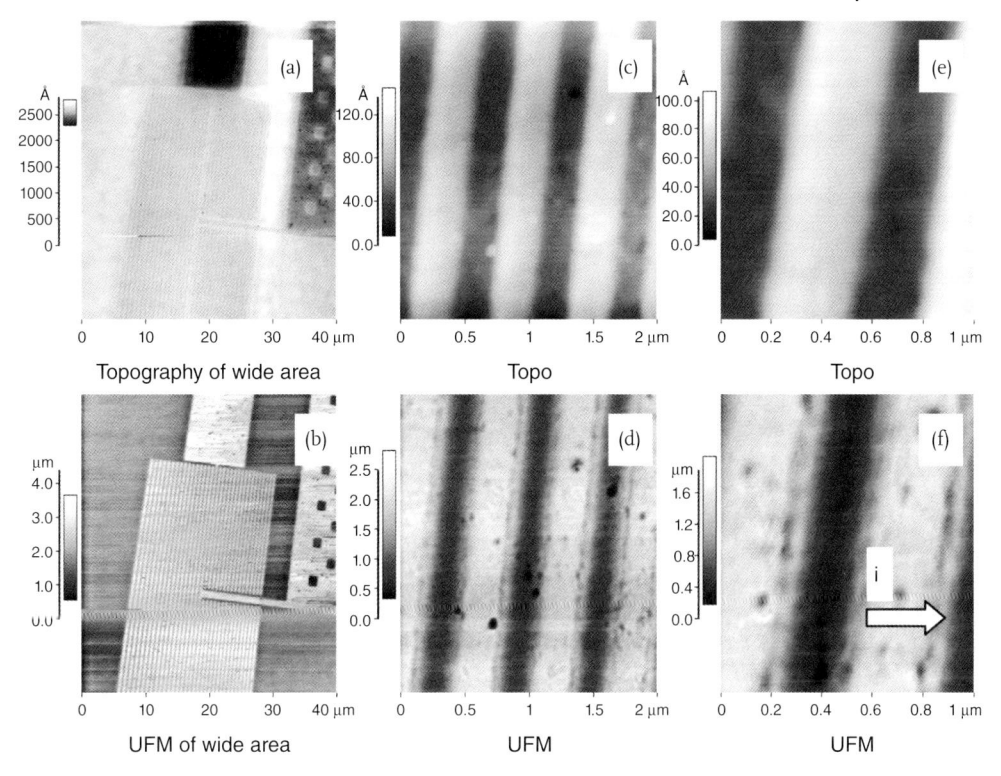

Figure 11.7 (a, c, e) AFM and (b, d, f) UFM images in the trench region of an Al/BCB (aluminum/benzocyclobutene) damascene test structure. The contrast inverts between topography and elasticity scans. There is interfacial variation of the elasticity between the Al and BCB regions. The interface between hardened and unmodified BCB is denoted by the white arrow i [34].

and rubber a few nanometers thick, with an outer layer is of PMMA. In the UFM images the inclusions have a texture that may be due to topographical deformations induced either by the manufacturing process or by the scanning itself. They exhibit different elastic behavior from the surroundings. The difference in contrast between inclusions may be due to the depth of a given inclusion relative to the surface or to the amount of the outer layer of PMMA remaining around the rubber.

Figure 11.8c, d shows images of amylin fibers deposited on mica in water solution and then dehydrated. The proteins have lower stiffness. Mica is a suitable substrate as its cleavage planes are atomically flat and it provides good bonding to the proteins. In conventional contact mode AFM, no clear topographical image of these samples could be obtained, because the scan usually sweeps soft materials away. Application of the ultrasound signal for UFM allowed a topographic image to be obtained through a mechanism described as superlubricity (Section 11.8).

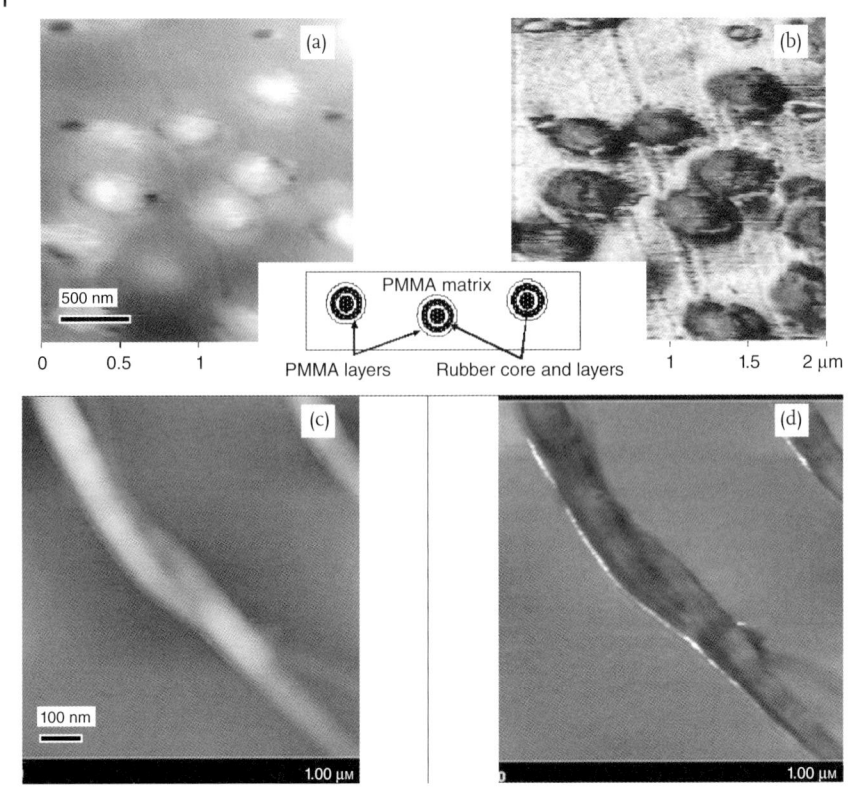

Figure 11.8 (a) Topography and (b) UFM images of a compliant sample made of injection molded poly(methyl methacrylate) PMMA ($E_{PMMA} = 4.5\,GPa$) with spherical PMMA–rubber inclusions ($E_{rubber} < 0.1\,GPa$) [35]. (c) Topography and (d) UFM images of amylin fibers on a freshly cleaved sheet of mica. The amylin fiber are less stiff than mica and UFM reveals internal structure reflecting packing of the fibers. Images courtesy of Grishin et al. [36].

The UFM image shows that the fibers of 50–100 nm in size are more compliant than the surrounding area and reveals the internal structure reflecting packing of the fibers.

This gallery of pictures illustrates how UFM can image the elastic properties of a broad range of materials from very stiff engineering ceramics to very soft polymers and biopolymers. The secret lies in the use of high frequencies to give inertial stiffness combined with low frequencies to give sensitive detection. There is an analogy with using a four-point probe to measure an electrical device, combining a low impedance current source with a high impedance voltmeter. The next section addresses some of the artifacts that may be present in UFM images, and how to recognize them, take account of them, and if possible avoid them.

11.7
Image Interpretation – Effects of Adhesion and Topography

In addition to the elastic properties that UFM is intended to image, anything else that affects the tip–surface interaction will also affect the UFM contrast. Prominent among these are surface adhesion and abrupt topographical features. The shape of the ultrasonically induced normal deflection (that can be qualitatively described using threshold amplitude and the force jump; see Figure 11.2) is affected by both elasticity and adhesion. If there is no appreciable variation in adhesive properties, a variation in threshold amplitude gives a good indication of variations in stiffness. While this is generally valid for most stiff inorganic samples, some samples, particularly polymeric samples with a plasticized surface, highly hydrophilic samples, or samples with soft surface layers, can deviate from this behavior.

Figure 11.9 shows an organic thiolipid Langmuir–Blodgett (LB) film deposited on a hydrophilic substrate of a freshly cleaved mica. The elongated molecules of the film have been compressed to such an extent that a fluid phase and a solid phase coexist at the air–water interface. In the fluid phase the molecules are not spatially organized although their mobility is inhibited. In the solid phase, the molecules are mutually aligned and tend to form stars made of six sectors, each of them with a different packing orientation [37]. The topographical image (Figure 11.9a) contains edges of such stars with liquid in between. Small dust particles on the surface can nucleate the ordering of the molecules in the fluid phase, creating islands between the main regions of solid phase. In the fluid phase the molecules are tilted at random angles to the normal giving a mean height about 2 nm lower than the solid phase. In the UFM image, a higher output from the lock-in amplifier is displayed as a brighter color. This typically results from a lower threshold amplitude and therefore a stiffer material. In the UFM image (Figure 11.9b) the fluid phase is brighter than the solid one. To avoid the naïve conclusion that the fluid phase is stiffer than the solid one, one should perform a safety check and monitor the actual shape of the ultrasonically induced normal deflection on each phase.

Figure 11.9c presents the two ultrasonically induced normal deflection signals on solid and fluid phase. The fluid threshold amplitude is higher than the solid threshold amplitude, but the force jump on reaching the pull-off point is higher for the fluid phase. The output of the lock-in amplifier is determined by the integrated area between the deflection curve and zero (indicated by the broken line; the deflection is here displayed as lower than zero). The fluid response gives a higher lock-in amplifier output because its signal area is bigger. For samples with high adhesion variation, the difference in shape in Figure 11.9c is a warning sign that the interpretation should take into account how the lock-in amplifier responds to such behavior. Theoretical modeling may be needed to determine how the threshold amplitude is affected by the variation of the adhesive properties. An alternative approach is to use the differential force method for contact stiffness measurement described in Section 11.5, to compensate for variations in

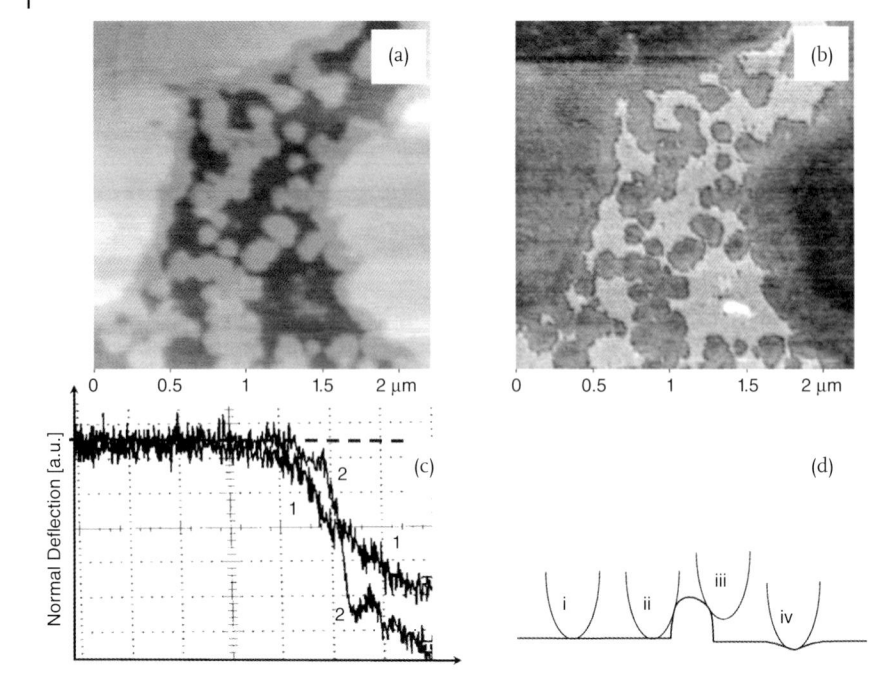

Figure 11.9 (a)–(e) Adhesive and topographic contributions to the contrast. (a) Topographical image of a Langmuir–Blodgett thiolipid film having ordered areas of solid-like (bright) and disordered liquid-like phase (dark areas). (b) UFM image of liquid areas shows inverted (brighter) contrast for liquid area linked with the variation of the shape of UFM response (c) due to strong adhesion, rather than change in the threshold amplitude alone. (d) Schematic representation of contact area increase leading to apparent increased local stiffness of the sample [18].

the adhesive contribution. Whenever adhesion may be significant, the shape of ultrasonically induced deflection and its consistency throughout the sample serves as a good check against artifacts associated with adhesion, and as a corollary can provide information on variability of adhesion properties of the sample.

Another significant influence on UFM contrast comes from its sensitivity to variations of the tip–surface geometry. The geometry is affected by both stiffness and adhesion, but it can also be affected by topography. For the contrast to arise solely from variations in stiffness, an ideal sample would be one with a surface smooth to the scale of the tip curvature. Since this is not always possible, it is important to understand how topographical variations manifest themselves to be able to identify them whenever they occur. Figure 11.9d illustrates some possible situations. An increase of the contact area, due to a negative value of the local curvature or to a high asperity, makes the contact stiffer and therefore a lower threshold amplitude is needed to break the contact. This could be wrongly interpreted as the presence of a stiffer material. On the other hand, if the contact area diminishes due to a positive value of the local curvature, the stiffness decreases and the threshold amplitude increases. As a result, the region could be incorrectly interpreted as a more compliant material. The bright halo around the Ge quantum dots in Figure 11.6b is an edge effect in areas where the rounded AFM tip simultaneously touches the protruding dot *and* the substrate, increasing the effective tip–surface contact stiffness (and UFM response). Under certain growth conditions a moat forms around Ge/Si dots, and this would introduce a further geometrical effect. It is similar to case ii in Figure 11.9d. Such a halo allows one to estimate an upper limit of the size of the contact region and hence the UFM resolution; in this case it was about 5–10 nm.

11.8
Superlubricity

A collateral benefit of using ultrasonic vibration in scanning probe microscopy is the reduction of friction between sample and cantilever. Although the ultrasound increases the additional average force acting on a tip, the friction vanishes when the tip–surface contact breaks for part of the vibration cycle [38]. The friction force is reduced even at amplitudes at which no break in contact occurs, and this reduction does not greatly depend on the normal load. Once contact is broken, the lateral force vanishes and the cantilever almost instantly slides to a new equilibrium position where there are no lateral forces – and there is therefore no friction! Figure 11.10 illustrates the effect of the ultrasonic amplitude on the friction force and cantilever deflection. Figure 11.10a is a sketch of the experiment, in which the frictional force is measured by the torsional deflection of the cantilever. Figure 11.10b shows the effect of ultrasound amplitude for a Si_3N_4 microlever on a polished Si sample, at two different values of the normal load, 0 and 2 nN. As the ultrasonic amplitude is increased from zero, the friction force decreases at first slowly, commencing at low ultrasonic amplitudes irrespective of the normal load. When the amplitude reaches a threshold value (indicated by the vertical dotted lines), which depends on the normal load, the cantilever deflects due to the strong nonlinearity of the force curve. At amplitudes above this threshold, the friction force rapidly goes to zero. This phenomenon of vanishing friction has applications to micro-electromechanical systems (MEMS) and nano-electromechanical systems (NEMS).

This reduction in friction has been exploited to image 90 nm polystyrene spheres (Figure 11.10c, d). The ordered arrangement of the spheres in the presence of ultrasonic vibration can be seen only in topography (Figure 11.10c) when ultrasonic vibrations were applied. The contact mode AFM image in the absence of ultrasound (Figure 11.10d) shows only hazy streaks. Subsequent UFM examination of the same area shows how the arrangement of the latex spheres was disrupted by the AFM scan. The ultrasonic superlubricity allows UFM to be used to study delicate samples in a way similar to the tapping mode AFM.

Magnetic recording materials have oxide particles embedded in a polymer binder. As a quaint reminder of the past, Figure 11.10e, f shows images of the

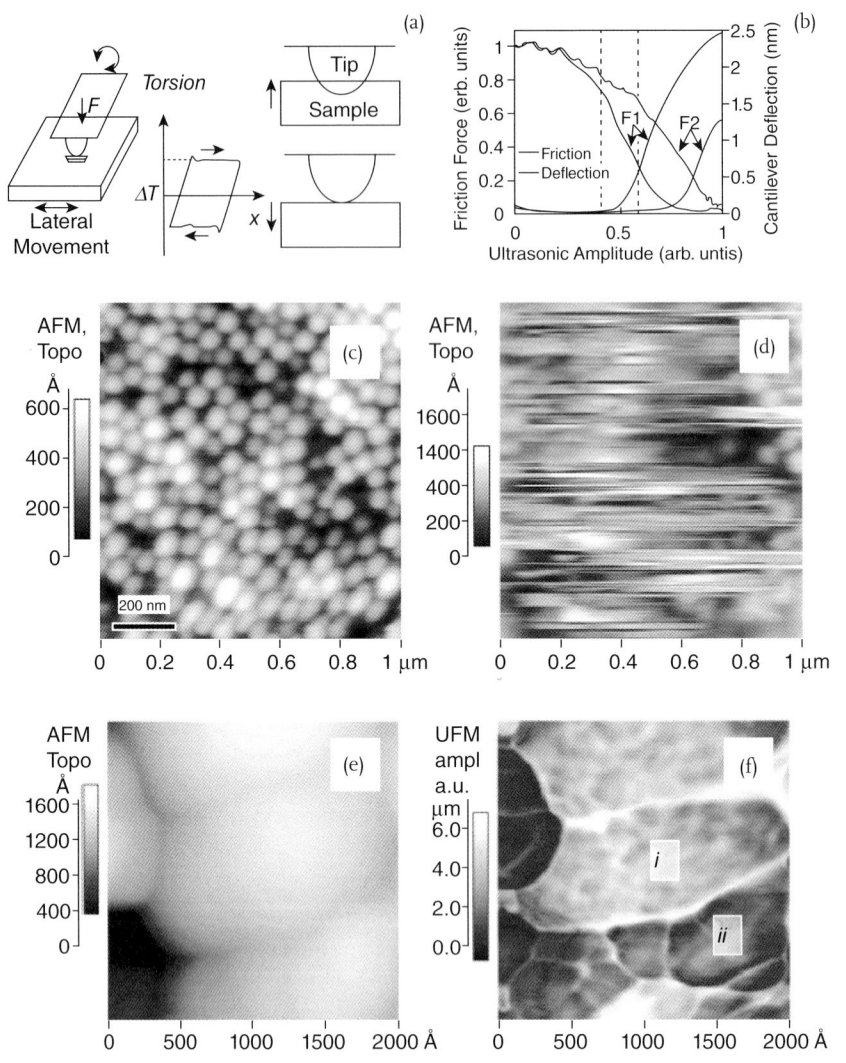

Figure 11.10 Ultrasound induced lubricity:
(a) schematic of the experiment;
(b) measurements of dynamic friction and cantilever deflection dependencies on the ultrasonic amplitude (Si sample, Si_3N_4 microlever). The loads applied are F1: 0; F2: 2 nN. The sliding speed is 50 nm s^{-1} [38].
(c, d) Topography images of 90 μm polystyrene spheres imaged in (c) with ultrasound (UFM mode) and (d) in standard contact topography mode without application of ultrasound. In (c) the ultrasound was on during the topography image; in (d) it was off. Without ultrasound the sample became damaged, as clearly seen in (d). (f) UFM image of a floppy disk and the corresponding topography image (e). The fine contrast within the magnetic metal coated oxide particles (i) observed in UFM and the contrast between different particles, for example, (i, ii), is due to the different properties of the oxide particles and the polymer binder. Image was taken underwater, which illustrates the ability of UFM (and mechanical diode) to operate in a liquid environment.

materials in a floppy disk. The images were taken underwater, and illustrate that the mechanical diode of the UFM can work well in a liquid. The topography image (Figure 11.10e) gives little contrast between the polymer and binder particles, of the kind that is readily apparent in the UFM image (Figure 11.10f), both within a region such as i and between two regions such as i and ii. In each case the contrast arises from the difference between the oxide and polymer elastic properties.

11.9
Defects Below the Surface

Since the acoustic wave responsible for the UFM excitation passes through the sample, it is to be expected that the UFM contrast will be sensitive to subsurface defects or discontinuities. An example of such subsurface defects is given in Figure 11.11, which shows a poly(ethylene terephthalate) (PET) substrate several μm thick coated with a 20 nm SiO_x layer, where $1 \leq x \leq 2$ [18]. Such nanocomposites are used in packaging, with the glass layer providing a barrier against the permeation of gas through the polymer. There is a crack running across the sample, probably caused by strain deformation, which would provide a path for gas permeation. Also visible in the UFM image (Figure 11.11b, though not in the topography image Figure 11.11a) are several discs of diameter 3 μm or so. These are interpreted as delaminations between the polymer and the SiO_x glass. The UFM frequency was 2 MHz. At this frequency the wavelength in PET is about 1 mm. Although the thickness of the delaminations is likely to be much less than this, the impedance mismatch between the air in the void and the materials on either side is so great that transmission across a delamination is negligible. The thickness of the glass layer is much less than the lateral extent of the delaminations, and the UFM signal above the delaminations is small, yielding the dark patches seen in Figure 11.11b. In this way UFM can detect subsurface defects that would not be directly revealed by other AFM techniques [39].

Nanoscale structures can be imaged in UFM even if the elastic inhomogeneity is not a strong one like a crack or delamination with a large mismatch of acoustic impedance. The surface quality must be close to atomically flat, in order not to mask the subsurface elasticity information. Figure 11.11c–e shows subsurface nanoscale resolution elastic imaging of a semiconductor quantum dot (QD) nanostructure. These are InAs quantum dots in a GaAs matrix (the difference in elastic moduli is approximately 20%) under a capping layer. The surface topography (Figure 11.11d) has some vague indication of underlying structures, whereas the UFM image (Figure 11.11e) reveals identifiable QD structures.

In cases when the depth penetration of the elastic field is not sufficient, a novel imaging principle involving UFM nanoscale resolution material sensitivity was developed recently [40]. A novel method uses Ar ion beam cross-section polishing via a beam exiting the sample (BEXP, beam exit ion cross-section polishing). In this approach, a sample is tilted at a small angle with respect to a polishing beam that enters from underneath the surface of interest and exits at a glancing angle

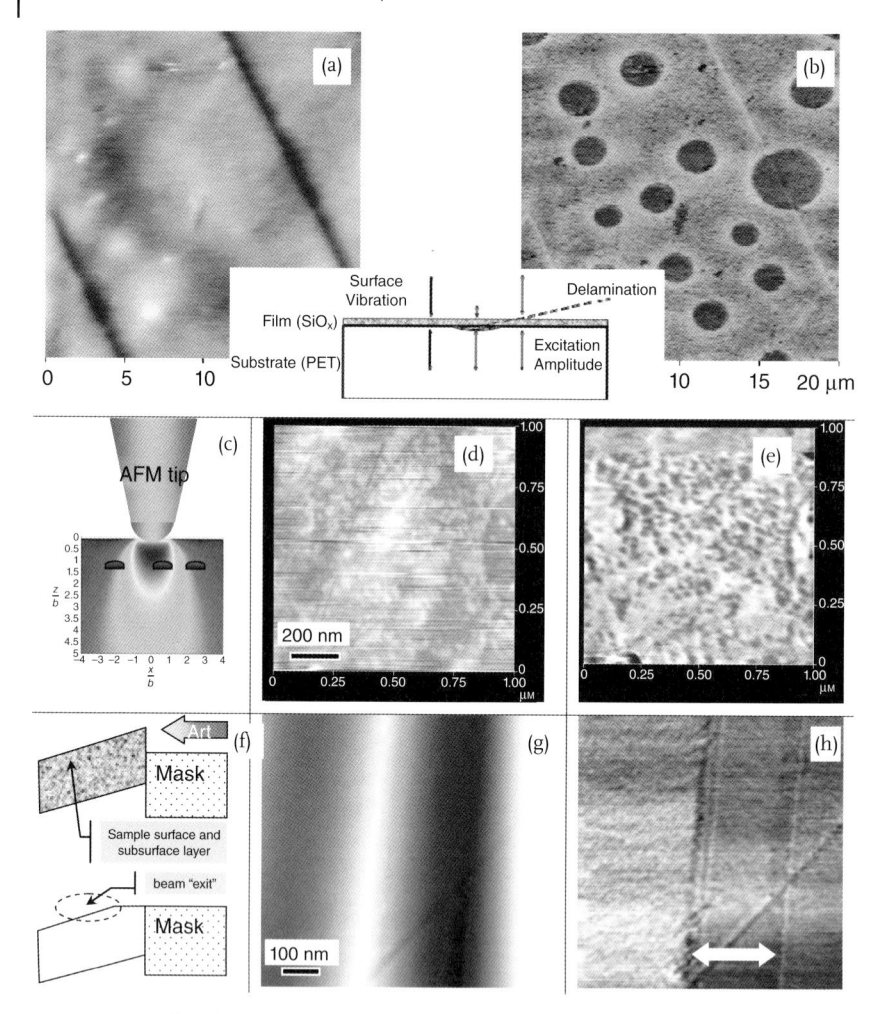

Figure 11.11 Subsurface imaging of elastic properties with UFM. (a, b) Original report on UFM observation of subsurface delaminations – the sample of a PET substrate coated with a 20 nm SiO$_x$ layer. "Bubbles" visible in the UFM image (b) are delaminations at the oxide–polymer interface – schematically illustrated by (c) – that are barely visible in the topography image (a) [39]. (c) An oscillating elastic field in UFM penetrates surface at depths on the order of several diameters of contact area. (d) Topography and UFM (e) image of InAs quantum dots (QDs) on a GaAs substrate partly covered with a GaAs capping layer. The UFM images reveal features similar to topography but provide much better discrimination and localization of QD structures (e.g., in fuzzy areas in topography indicated by arrows) (sample courtesy by Mohamed Henini). (f)–(h) Material sensitive UFM imaging of ion cross-sectional polishing. (f) Schematic illustration of the principle of beam exit Ar ion cross-section polishing (BEXP). A surface layer of interest is exposed to the beam exit only and therefore is not perturbed by the proximity of the shadow mask. (g) Topography and (h) UFM image of InSb superlattice layers in a GaAs matrix (arrow indicates superlattice region). The first two superlattice layers positioned at 5 and 25 nm depth are clearly observable (average concentration of InSb in the superlattice zone was around 1 wt%) [40].

similar to the edge mechanical polishing [41]. This creates an almost perfect nanometer scale flat cross-section with a close to open angle prismatic shape of polished and pristine sample surfaces ideal for SPM imaging. Using the new method and material sensitive UFM it was possible to map the internal structure of an InSb/InAs QD superlattice of 18 nm layer periodicity with the depth resolution on the order of 5 nm (Figure 11.11f–h). BEXP with UFM imaging was also shown to be applicable to revealing details of interfaces in very large scale integration low-k dielectric interconnects similar to the ones in Figure 11.7, porous Si, and details of mechanical nanostructure of abalone shells.

There is no universal formula for the depth of subsurface objects probed by UFM. If the ultrasonic wavelength is long compared with any relevant sample dimensions, which is usually the case, then a good approach is to consider the equivalent problem in static elasticity. A subsurface delamination, such as the ones shown in Figure 11.11a, b, will generally give edges in the UFM images whose sharpness is comparable with the depth of the delamination below the surface. If the phase of the cantilever response could be measured, then it would be possible to use depth reconstruction algorithms from optical and acoustic near-field imaging [42]. For most purposes, less abrupt variations in properties can be considered by turning the problem the other way round. A Hertzian contact gives a field with a characteristic depth comparable with the diameter of the contact area [25]. Since the resolution of the UFM is determined by the size of the contact area, the depth over which the UFM is most sensitive to small variations in elastic properties is comparable with its resolution. Bigger variations in elastic properties can give contrast from greater depths; by Saint-Venant's principle the resolution will not then depend much on the geometry of the tip–sample contact.

11.10
Time-Resolved Nanoscale Phenomena

The difference in stiffness of the AFM cantilever at low and high frequencies, which is so crucial to the operation of the UFM, can be described more rigorously in terms of a dispersive mechanical waveguide. The high compliance at low frequencies corresponds to low impedance, and the high stiffness at high frequencies corresponds to high impedance. A flexural wave can be launched down the cantilever, leading to a vibration at the tip rather similar to the effect of the ultrasonic wave propagating through the sample in the simple UFM [20]. The mechanical diode effect can still be used to detect the tip–surface interaction. This technique is known as a waveguide UFM (W-UFM). It gives contrast similar to the simple UFM, with the difference that the sample does not need to be able support ultrasonic wave propagation, and in a sample with strong bulk heterogeneities the contrast will be dominated by the properties within the Hertzian contact zone.

If a vibration is also applied through the sample, it is possible to perform heterodyne detection. Ultrasonic vibration is applied simultaneously to the cantilever (at frequency ω_c) and sample (at frequency ω_s). The AFM tip detects the oscillating force at the difference frequency $\omega_t - \omega_s$, very much like a heterodyne radio receiver. This technique is known as heterodyne force microscopy (HFM). Once again, the tip–surface force nonlinearity plays a critical role. The low frequency beating oscillation carries information on the phase of the original high frequency oscillations.

To evaluate the phase sensitivity and operation of HFM, we consider a simple model of the tip–surface force nonlinearity:

$$F(z) = k_s(z_t - z_s) + \chi_s(z_t - z_s)^2 \tag{11.9}$$

where z_t and z_s are the instantaneous displacements of the tip and the sample, and k_s, χ_s are coefficients describing the linear and quadratic force–displacement response. Any time-dependent phenomena on the time scale τ, such as viscoelastic relaxation or resonance, can be represented by a phase delay of the sample vibration $\phi = \omega_s \tau$, with:

$$z_s = a_u \cos(\omega_s t + \omega_s \tau) \tag{11.10}$$

Preserving only the low frequency terms in the AFM cantilever response (since the high frequency terms will be filtered out by the mechanical response, and possibly also the electronic circuitry), the additional force due to the vibration of the tip and the surface is:

$$F = \chi_s\{a_t^2/2 - a_t a_s \cos[(\omega_t - \omega_s)t - \omega_s t] + a_s^2/2\} \tag{11.11}$$

where a and ω are the amplitude and angular frequency of displacement, and the subscripts t and s refer to the cantilever and the sample. The first term in parentheses represents the nonlinear detection of cantilever vibration or W-UFM [43], the last term describes the nonlinear detection of the sample vibration (UFM), and the middle term describes the mixing (HFM). Figure 11.12 presents an experimental demonstration of such mixing. The HFM mechanically mixed signal has good signal-to-noise ratio and closely follows the reference signal from electronic mixing of the electrical signals driving the sample and cantilever piezoelectric transducers. Even a short relaxation time τ will cause a significant phase shift $\omega \tau$ in the resulting nonlinear LF cantilever response. For example, if the difference frequency is a few kilohertz, and the phase of the cantilever vibration is measured with a precision of 1°, a relaxation time of 300 ps can be detected by HFM.

Heterodyne force microscopy can also be performed using a combination of modulated optical excitation with ultrasonic excitation (optical HFM, OHFM). Figure 11.13a shows an experimental setup, which may be based on a commercial AFM. A transparent silicon nitride tip is vibrated while the sample is irradiated from above by light chopped at a slightly different frequency. The sample undergoes periodic temperature variations of a few degrees, causing the surface to vibrate with an amplitude of a few picometers. Because of the nonlinearity of the tip–sample force curve, a vibration of the contacting tip is induced at the difference frequency. This frequency is chosen to be lower than the fundamental cantilever resonance but higher than the response frequency of the AFM feedback loop. The cantilever deflection is measured in the standard way, generally through the

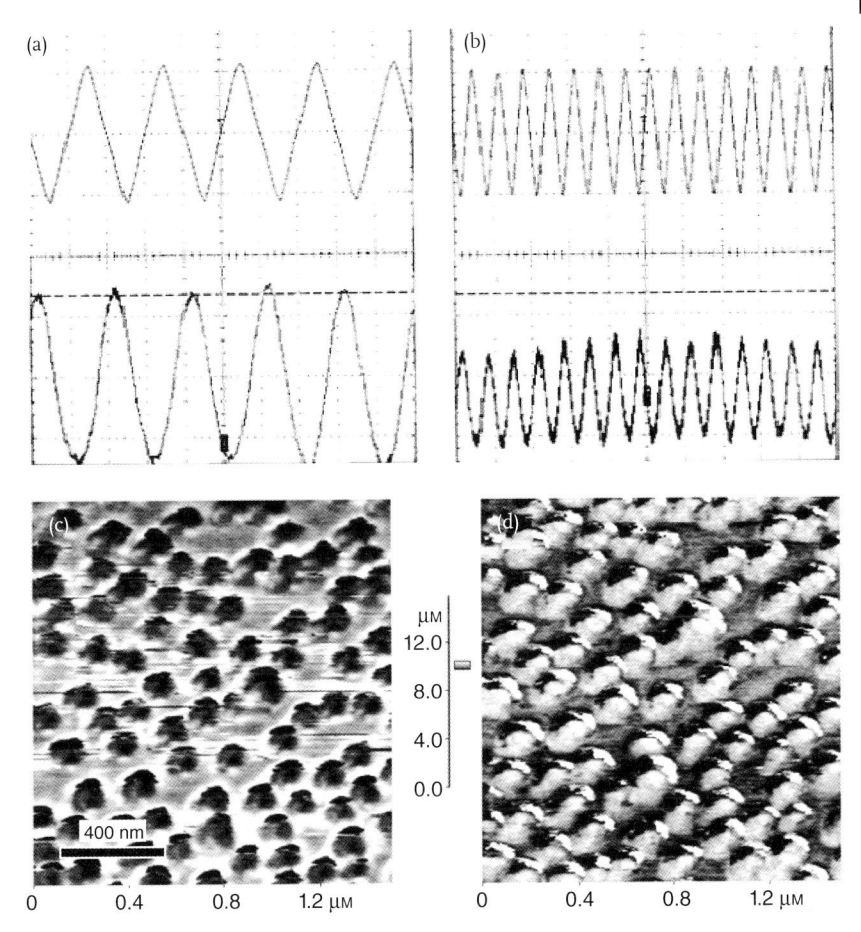

Figure 11.12 (a, b) Heterodyne mixing of two ultrasonic vibrations of sample and the cantilever at excitation frequencies around 9 MHz and difference frequencies of (a) 5 kHz and (b) 15 kHz. The HFM "mechanically" mixed signal is of good signal-to-noise ratio and closely follows oscillations of a reference signal that uses a conventional high frequency mixer [44]. (c) Amplitude heterodyne force microscopy (HFM) image of Ge quantum dots similar to those in Figure 11.6a, b, revealing the lower elastic moduli of Ge QDs. Phase HFM image (d) reveals changes in the temporal response of the QDs, linked potentially with some adhesion differences in the QDs and underlying substrate (full delay scale 10 ns).

deflection of an optical beam reflected from the cantilever. The amplitude and phase of the heterodyne signal are used to form the OHFM image through lock-in detection of the cantilever deflection. Separating the individual contributions to the signal can be difficult, but the technique can be illustrated by the thermal contrast from a sample with a flat surface covered with a homogeneous layer. Figure 11.13 shows OHFM amplitude and phase images for a copper strip

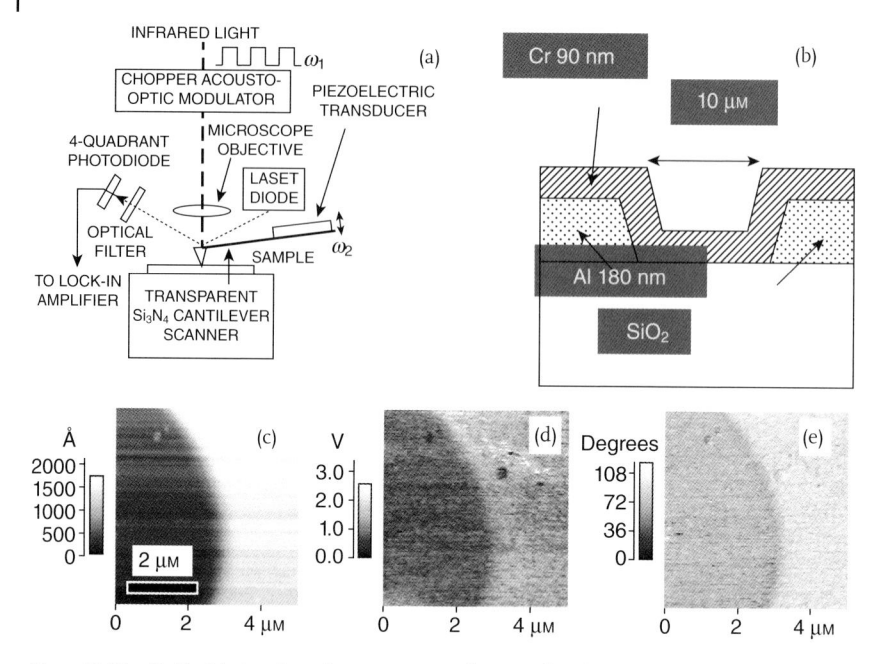

Figure 11.13 Optical heterodyne force microscopy (OHFM) – (a) principle and experimental setup. The transparent silicon nitride tip was vibrated in contact with the sample at a frequency $f_2 = 4.190$ kHz while the sample was irradiated from above by light chopped at a slightly different frequency ($f_1 = 4.193$ kHz), focused through the tip to a spot ~2 μm in diameter with incident power $P = 0.5$ mW and wavelength 830 nm. The sample undergoes periodic temperature variations of ~3 K, causing the surface to vibrate at f_1 with an amplitude of ~10 pm. Through the nonlinearity of the tip–sample force–displacement curve, a vibration of the contacting tip is induced at the difference frequency $f_1 - f_2 = 3$ kHz. The cantilever resonance frequency is 38 kHz: (c) topography, (d) amplitude, and (e) phase OHFM images for a chromium coated region of Al on a SiO_2 structure. The Al on SiO_2 layer and SiO_2-only areas are clearly visible under a 90 nm thick Cr layer with resolution on the order of 150 nm [45].

width 500 nm wide and 350 nm thick. The strip is on a silicon substrate. A silicon oxide layer was then deposited by chemical vapor deposition (CVD); this layer was then polished and finally covered with a chromium layer of uniform thickness 100 nm and flatness better than 10 nm [34]. The underlying copper strip could therefore be detected in neither topographical nor optical images, but is revealed in both amplitude and phase images of OHFM. The principle of HFM can be generalized to any combination of excitation signals, and any nonlinear detection mechanism [46], to give both nanosecond time resolution and nanometer spatial resolution [44].

Scanning probe microscopy was born when Binnig, Rohrer, and Gerber first realized the scanning tunneling microscope (STM) [5]. The invention of AFM by Binnig, Rohrer, and Quate [48] overcame one of the main STM limitations, namely that STM requires materials with some degree of electron transport [47].

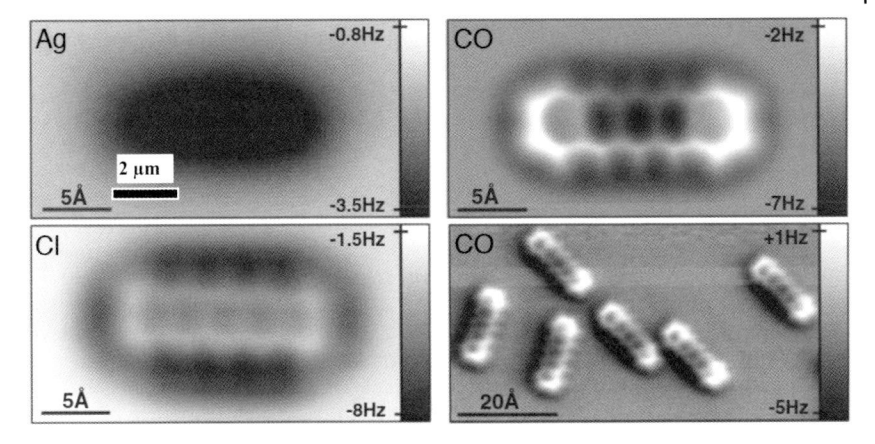

Figure 11.14 Force based SPM measurements of pentacene molecules on Cu(111). The lateral resolution of individual electron clouds rivals, if not exceeds, the resolution previously obtained only with scanning tunneling microscopy [47].

Nevertheless, the lateral resolution of SPM techniques using interatomic forces (a cornerstone of AFM, UFM, and other tip–sample interaction force based techniques) until recently had difficulty achieving a spatial resolution rivaling STM. A new development from the same laboratory that gave birth to the SPM – the IBM Zurich laboratory – uses a clever way of exploiting an oscillating probe in an ultrahigh vacuum environment when the tip of the probe is functionalized with a CO molecule. Figure 11.14 gives some images obtained in this breakthrough research, which have sufficient resolution to image individual bonds in pentacene molecules [45]. The force-versus-distance interaction is highly nonlinear, and therefore should also exhibit aspects of the response hitherto associated with UFM and HFM. And who knows? Perhaps the combination of new developments like this with the methods described in this chapter will pave the way for the new generation of scanning force microscopy tools that will image dynamic mechanical properties of molecular assemblies with angstrom resolution.

Acknowledgments

The authors would like to thank all those who have collaborated with us in the development of UFM and related techniques, in particular Walter Arnold, Nancy Burnham, Martin Castell, Teresa Cuberes, Franco Dinelli, Gerard Gremaud, Manus Hayne, Mohammed Henini, Bryan Huey, Tony Krier, Andrew Kulik, Hubert Pollock, Alex Robson, Oliver Wright, and Kazushi Yamanaka. OVK would like to thank his wife Tatiana and daughter Ksenia for tremendous and much needed support while preparing this manuscript.

Some of the material used in this chapter is based on material from *Acoustic Microscopy*, 2nd edition, by G.A.D. Briggs and O. V. Kolosov (2010), and is reproduced by © permission of Oxford University Press.

References

1. Foster, J. and Quate, C.F. (1984) Acoustic microscopy in superfluid-helium. *Phys. Today*, **37**, S4.
2. Zieniuk, J.K. and Latuszek, A. (1986) Ultrasonic pin scanning microscope: a new approach to ultrasonic microscopy, in *IEEE 1986 Ultrasonics Symposium: Proceedings* (ed. B.R. McAvoy), IEEE, pp. 1037–1039.
3. Zieniuk, J.K. and Latuszek, A. (1987) Ultrasonic pin scanning microscope: a new approach to ultrasonic microscopy. *IEEE Trans. Ultrason. Ferroelectr. Freq. Control*, **34**, 414.
4. Kolosov, O. (1998) UFM shakes out the details at the nanoscopic scale. *Mater. World*, **6**, 753–754.
5. Binnig, G., Rohrer, H., Gerber, C., and Weibel, E. (1982) Tunneling through a controllable vacuum gap. *Appl. Phys. Lett.*, **40**, 178–180.
6. Martin, Y., Williams, C.C., and Wickramasinghe, H.K. (1987) Atomic force microscope force mapping and profiling on a sub 100-A scale. *J. Appl. Phys.*, **61**, 4723–4729.
7. Weisenhorn, A.L., Maivald, P., Butt, H.J., and Hansma, P.K. (1992) Measuring adhesion, attraction, and repulsion between surfaces in liquids with an atomic-force microscope. *Phys. Rev. B*, **45**, 11226–11232.
8. Miyatani, T., Horii, M., Rosa, A., Fujihira, M., and Marti, O. (1997) Mapping of electrical double-layer force between tip and sample surfaces in water with pulsed-force-mode atomic force microscopy. *Appl. Phys. Lett.*, **71**, 2632–2634.
9. Gunther, P., Fischer, U., and Dransfeld, K. (1989) Scanning near-field acoustic microscopy. *Appl. Phys. B - Photophys. Laser Chem.*, **48**, 89–92.
10. Behrend, O.P., Oulevey, F., Gourdon, D., Dupas, E., Kulik, A.J., Gremaud, G., and Burnham, N.A. (1998) Intermittent contact: tapping or hammering? *Appl. Phys. A - Mater. Sci. Process.*, **66**, S219–SS21.
11. Burnham, N.A., Behrend, O.P., Oulevey, F., Gremaud, G., Gallo, P.J., Gourdon, D., Dupas, E., Kulik, A.J., Pollock, H.M., and Briggs, G.A.D. (1997) How does a tip tap? *Nanotechnology*, **8**, 67–75.
12. Quate, C.F., Khuri-Yakub, B.T., Akamine, S., and Hadimioglu, B.B. (1994) Near field acoustic ultrasonic microscope system and method US Patent 5,319,977.
13. Kolosov, O. and Yamanaka, K. (1993) Nonlinear detection of ultrasonic vibrations in an atomic-force microscope. *Jpn. J. Appl. Phys.*, **32**, L1095–L1098.
14. Rohrbeck, W., Chilla, E., Frohlich, H.J., and Riedel, J. (1991) Detection of surface acoustic-waves by scanning tunneling microscopy. *Appl. Phys. A - Mater. Sci. Process.*, **52**, 344–347.
15. Burnham, N.A., Kulik, A.J., Gremaud, G., Gallo, P.J., and Oulevey, F. (1996) Scanning local-acceleration microscopy. *J. Vac. Sci. Technol. B*, **14**, 794–799.
16. Rabe, U. and Arnold, W. (1994) Acoustic microscopy by atomic-force microscopy. *Appl. Phys. Lett.*, **64**, 1493–1495.
17. Yamanaka, K., Ogiso, H., and Kolosov, O. (1994) Ultrasonic force microscopy for nanometer resolution subsurface imaging. *Appl. Phys. Lett.*, **64**, 178–180.
18. Dinelli, F., Castell, M.R., Ritchie, D.A., Mason, N.J., Briggs, G.A.D., and Kolosov, O.V. (2000) Mapping surface elastic properties of stiff and compliant materials on the nanoscale using ultrasonic force microscopy. *Philos. Mag. A*, **80**, 2299–2323.
19. Huey, B.D. (2007) AFM and acoustics: fast, quantitative nanomechanical mapping. *Annu. Rev. Mater. Res.*, **37**, 351–385.
20. Inagaki, K., Kolosov, O.V., Briggs, G.A.D., and Wright, O.B. (2000) Waveguide ultrasonic force microscopy at 60MHz. *Appl. Phys. Lett.*, **76**, 1836–1838.
21. Hirsekorn, S., Rabe, U., and Arnold, W. (1997) Theoretical description of the transfer of vibrations from a sample to the cantilever of an atomic force microscope. *Nanotechnology*, **8**, 57–66.
22. Dinelli, F., Assender, H.E., Takeda, N., Briggs, G.A.D., and Kolosov, O.V. (1999) Elastic mapping of heterogeneous nanostructures with ultrasonic force

microscopy (UFM). *Surf. Interface Anal.*, **27**, 562–567.

23 Inagaki, K.K., Briggs, G.A.D., Muto, S., Horisaki, Y., and Wright, O.B. (1998) Ultrasonic force microscopy in waveguide mode up to 100 MHz, in *IEEE 1998 Ultrasonics Symposium – Proceedings*, vol. 1 (eds S.C. Schneider, M. Levy, and B.R. McAvoy), IEEE, pp. 1255–1259.

24 Johnson, K.L., Kendall, K., and Roberts, A.D. (1971) Surface energy and contact of elastic solids. *Proc. R. Soc. London, Ser. A - Math. Phys. Sci.*, **324**, 301–313.

25 Johnson, K.L. (1985) *Contact Mechanics*, Cambridge University Press, Cambridge.

26 Greenwood, J.A. and Johnson, K.L. (2006) Oscillatory loading of a viscoelastic adhesive contact. *J. Colloid Interface Sci.*, **296**, 284–291.

27 Johnson, K.L. and Greenwood, J.A. (1997) An adhesion map for the contact of elastic spheres. *J. Colloid Interface Sci.*, **192**, 326–333.

28 Luan, B.Q. and Robbins, M.O. (2005) The breakdown of continuum models for mechanical contacts. *Nature*, **435**, 929–932.

29 Kolosov, O.V., Castell, M.R., Marsh, C.D., Briggs, G.A.D., Kamins, T.I., and Williams, R.S. (1998) Imaging the elastic nanostructure of Ge islands by ultrasonic force microscopy. *Phys. Rev. Lett.*, **81**, 1046–1049.

30 Kolosov, O.V., Ogiso, H., and Yamanaka, K. (1993) Ultrasonic force microscopy a new technique for a nondestructive investigation on nanometer scale viscoelastic properties, in *Proceedings of the Third Japan International SAMPE Symposium*, vol. 2 (eds T. Kishi, N. Takeda, and Y. Kagawa), Society of the Advancement of Material and Process Engineering, pp. 2196–2201.

31 Dinelli, F., Biswas, S.K., Briggs, G.A.D., and Kolosov, O.V. (2000) Measurements of stiff-material compliance on the nanoscale using ultrasonic force microscopy. *Phys. Rev. B*, **61**, 13995–14006.

32 Rudd, R.E., Briggs, G.A.D., Sutton, A.P., Medeiros-Ribeiro, G., and Williams, R.S. (2007) Equilibrium distributions and the nanostructure diagram for epitaxial quantum dots. *J. Comput. Theor. Nanosci.*, **4**, 335–347.

33 Cuberes, M.T., Stegemann, B., Kaiser, B., and Rademann, K. (2007) Ultrasonic force microscopy on strained antimony nanoparticles. *Ultramicroscopy*, **107**, 1053–1060.

34 Geer, R.E., Kolosov, O.V., Briggs, G.A.D., and Shekhawat, G.S. (2002) Nanometer-scale mechanical imaging of aluminum damascene interconnect structures in a low-dielectric-constant polymer. *J. Appl. Phys.*, **91**, 4549–4555.

35 Porfyrakis, K., Kolosov, O.V., and Assender, H.E. (2001) AFM and UFM surface characterization of rubber-toughened poly(methyl methacrylate) samples. *J. Appl. Polym. Sci.*, **82**, 2790–2798.

36 Grishin, I., Tinker, C., Allsop, D., Robson, A., and Kolosov, O. (2012) Nanoscale SPM characterisation of nacre aragonite plates and synthetic human amyloid fibres, in *NSTI-Nanotech 2012*, vol. 1, CRC Press-Taylor & Francis Group, Santa Clara, USA, pp. 110-113.

37 Gourdon, D., Burnham, N.A., Kulik, A., Dupas, E., Oulevey, F., Gremaud G., Stamou, D., Liley, M., Dienes, Z., Vogel, H., and Duschl, C. (1997) The dependence of friction anisotropies on the molecular organisation of LB films as observed by AFM. *Tribol. Lett.*, **3**, 317-324.

38 Dinelli, F., Biswas, S.K., Briggs, G.A.D., and Kolosov, O.V. (1997) Ultrasound induced lubricity in microscopic contact. *Appl. Phys. Lett.*, **71**, 1177–1179.

39 McGuigan, A.P., Huey, B.D., Briggs, G.A.D., Kolosov, O.V., Tsukahara, Y., and Yanaka, M. (2002) Measurement of debonding in cracked nanocomposite films by ultrasonic force microscopy. *Appl. Phys. Lett.*, **80**, 1180–1182.

40 Kolosov, O.V., Grishin, I., and Jones, R. (2011) Material sensitive scanning probe microscopy of subsurface semiconductor nanostructures via beam exit Ar ion polishing. *Nanotechnology*, **22**, 8.

41 Bowden, F.P. and Tabor, D. (1973) *Friction: An Introduction to Tribology*, Anchor Press/Doubleday, New York.

42 Rosner, B.R.T. and van der Weide, D.W. (2002) High-frequency near-field

microscopy. *Rev. Sci. Instrum.*, **73**, 2505–2525.

43 Cuberes, M.T., Assender, H.E., Briggs, G.A.D., and Kolosov, O.V. (2000) Heterodyne force microscopy of PMMA/rubber nanocomposites: nanomapping of viscoelastic response at ultrasonic frequencies. *J. Phys. D - Appl. Phys.*, **33**, 2347–2355.

44 Kolosov, O.V. and Briggs, G.A.D. (1996) Atomic force microscopy and method thereof. In: UK patent application 9617380.2.

45 Tomoda, M., Shiraishi, N., Kolosov, O.V., and Wright, O.B. (2003) Local probing of thermal properties at submicron depths with megahertz photothermal vibrations. *Appl. Phys. Lett.*, **82**, 622–624.

46 Kumano, N., Inagaki, K., Kolosov, O., and Wright, O. (1998) Optical heterodyne force microscopy, in *IEEE 1998 Ultrasonics Symposium – Proceedings*, vol. 1 (eds S.C. Schneider, M. Levy, and B.R. McAvoy), IEEE, pp. 1269–1272.

47 Gross, L., Mohn, F., Moll, N., Liljeroth, P., and Meyer, G. (2009) The chemical structure of a molecule resolved by atomic force microscopy. *Science*, **325**, 1110–1114.

48 Binnig, G., Quate, C.F., and Gerber, C. (1986) Atomic force microscope. *Phys. Rev. Lett.*, **56**, 930–933.

12
Ultrasonic Atomic Force Microscopy

Kazushi Yamanaka and Toshihiro Tsuji

12.1
Introduction

Atomic force microscopy (AFM) combined with ultrasonic excitation and either or both of linear and nonlinear detection has been regarded as the most promising characterization method of surfaces and subsurfaces on the nanometer resolution. It includes ultrasonic force microscopy (UFM), ultrasonic atomic force microscopy (UAFM), atomic force acoustic microscopy (AFAM), scanning acoustic force microscopy (SAFM), and so on. Among them, UAFM uses a higher order mode cantilever vibration excited at its base. It enables precise imaging of both topography and elasticity of solids such as metals and ceramics, without the need for bonding a transducer to the sample. Thus, a range of unique analysis and techniques have been developed. This chapter includes fundamental aspects such as comparison between linear and nonlinear detection, deflection and torsion, frequency sweep and tracking. It then describes applications to subsurface defects.

12.2
Principle

12.2.1
Forced Vibration of Cantilever from the Base

An atomic force microscope [1] uses a cantilever to measure nanoscale irregularities on the surface of a sample, utilizing the deflection of the cantilever supporting a tip owing to the force acting between the sample surface and the tip. Methods have been proposed to measure the distribution of contact stiffness by detecting the vibration of the AFM cantilever when the sample is vibrated at or a higher frequency than its resonance frequency (frequency range of ultrasound), while vertical control is realized via the static cantilever deflection [2–6]. These methods can be used to measure the elasticity of stiff materials. Note, however, that it is not possible in the force modulation mode where the sample is vibrated at a frequency much lower than the resonance frequency in the vertical direction [7], not in the lateral direction [8].

However, since the sample has to be bonded to an ultrasonic vibrator:

1) selection of optimum adhesive for each sample is required;
2) adhesives contaminate the sample and cannot be used with LSI wafers and other samples requiring a high degree of cleanness;
3) a large or irregularly shaped sample is hard to vibrate uniformly;
4) unwanted resonance peaks of the sample overlap cantilever resonances, degrading the precision.

These disadvantages are overcome by ultrasonic atomic force microscopy (UAFM) [9–24] in which the cantilever rather than the sample is vibrated, without requiring the sample to be bonded to a vibrator. With suppression of both spurious vibration of the cantilever base (a chip to mount the cantilever to cantilever holder) and nonlinear jumping of the tip, a wealth of information is conveniently obtained from the clear spectra of fundamental and higher order modes of deflection, torsion, and lateral bending vibration of the cantilever.

12.2.2
Quantitative Information, Directional Control, and Resonance Frequency Tracking

The cantilever vibration spectra in contact with the sample were found to be strongly dependent on the excitation power [11]. However, if the excitation power is small enough, the resonance peak width decreases and the peak frequency increases to a certain limiting value. In this condition the tip–sample contact is kept linear, and satisfactory agreement between the measured and calculated frequency is obtained. The agreement is further improved by taking into account the lateral stiffness. More quantitative information on the elasticity of the sample is obtained from the contact load dependence of the frequency, where the contact stiffness of a non-spherical tip shape is derived from the Sneddon–Maugis formulation [11], and the tip shape index is estimated by an inverse analysis of the load–frequency relation [13]. A further advantage is the evaluation of not only the vertical but also the lateral stiffness by simultaneous measurement of deflection and torsional vibration [10]. This has been demonstrated on a ground silicon wafer [11].

The modulus can be calculated using the resonance frequency obtained from the peak frequency of a spectrum, and the loss modulus is calculated using the Q factor, which is defined as the ratio of the peak frequency to its width. However, measurement of spectra takes a long time (typically 5 s for one point for an average of ten times). Consequently, mapping the resonance frequency and the Q factor takes a very long time (~91 h for a 256×256 pixels image).

A resonance-tracking scheme was developed to reduce the time required for mapping the resonance frequency [15]. Furthermore, if we use the analytical relationship between the peak height of resonance and the Q factor obtained by the

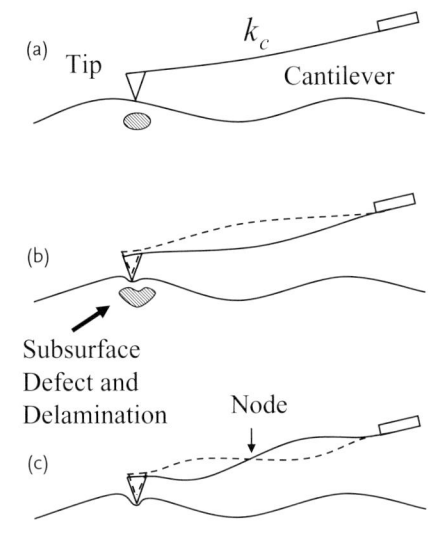

Figure 12.1 Principle of UAFM: (a) AFM; (b) UAFM in the first resonance; (c) UAFM in the second resonance.

theory of UAFM, we can measure the Q factor from the vibration amplitude at resonance. Based on this idea, advanced UAFM was developed for mapping both the resonance frequency and Q factor of the sample. Here the frequency is not fixed but automatically tuned to an instantaneous resonance frequency determined by the stiffness of the sample.

12.2.3
Effective Enhancement of Cantilever Stiffness

As an essential point of UAFM, cantilever stiffness is effectively enhanced by higher order resonances (Figure 12.1), as was pointed out in the first paper on UAFM [9]. This is the clearest demonstration of the property that a cantilever softer than the sample stiffness is effectively stiffened at higher order modes (Figure 12.1c). This property is used to measure the elasticity of stiff samples using a soft cantilever. This stiffening is explained by the inertia of the cantilever as well as the formation of nodes along the cantilever axis. The distance between the sample and the closest node is much less than the original cantilever length, resulting in the effective shortening of the cantilever.

12.2.4
Criterion to Avoid Plastic Deformation

The starting point for any reliable measurement is repeatability. Plastic deformation of the tip and/or sample is the most serious obstacle to repeatability since it

12 Ultrasonic Atomic Force Microscopy

Table 12.1 Von Mises stress for some sets of tip with radius R and contact force $F_\perp$.

	Tip radius R (nm)	Contact force $F_\perp$ (nN)	Contact diameter $2a_c$ (nm)	Max normal stress σ_0 (GPa)	Von Mises stress σ_V (GPa)
(a)	50	250	6.7	10.7	3.2
(b)	50	2400	14.2	22.7	6.8
(c)	100	250	11.4	3.6	1.1
(d)	100	2400	24.3	7.7	2.3

is an irreversible process. Luckily, we have a good criterion to judge if plastic deformation takes place.

If the von Mises stress σ_V calculated from stress components is larger than the yield stress, plastic deformation will take place. The maximum σ_V under the tip–sample contact is approximately given by:

$$\sigma_V \cong 0.3\sigma_0 \tag{12.1}$$

in which $\sigma_0 = 1.5 F_\perp/(\pi a_c^2)$ is the maximum normal stress in the contact area – where $a_c = (3RF_\perp/4K^*)^{1/3}$ is the contact radius (R is the tip radius and K^* is the effective modulus of tip/sample pair) – and $F_\perp$ is the contact force. Assuming typical parameters of AFM, a K^* of 100 GPa and contact force of 250 nN and 2400 nN, the contact diameter and von Mises stress are evaluated in Table 12.1.

In Table 12.1, it is noted that σ_V is fairly large despite a small $F_\perp$, which is due to a very small a_c. Therefore, we have the following concerns:

1) *Yield and wear of samples:*

    Cases (a) and (b) in Table 12.1 are for radius $R = 50$ nm and a slightly worn tip while and (c) and (d) are for $R = 100$ nm and a severely worn tip. In case (a), σ_V is as large as 3.2 GPa – much larger than yield stress σ_y of typical metals listed in Table 12.2, which is usually less than 1 GPa (= 1000 MPa). Since some metals such as gold (Au) are very soft, they easily undergo plastic deformation at a contact force $F_\perp$ exceeding 1000 nN. This point should be taken care of when applying not only UAFM but also related techniques in this book. One should always know the approximate tip radius R. It is useful to observe the appearance of a sample surface in the noncontact mode AFM image after spectra measurement.

2) *Wear of tip:*

    Even for a silicon tip whose σ_y is 5–9 GPa, yield may occur at $\sigma_V > 5$ GPa. Moreover, the tilt of the cantilever (typically about 11° in common AFM) further increases σ_V due to surface friction. Consequently, severe wear of a Si tip should take place at conditions above $\sigma_V = 5$ GPa. The tip will be worn and the radius increases to, for example, 100 nm (d), and the resonance frequency

Table 12.2 Yield stress σ_y of typical materials.

Material	Yield strength σ_y (MPa)
Structural steel ASTM A36 steel	250
Steel, API 5L X65 (Fikret Mert Veral)	448
High density polyethylene (HDPE)	26–33
Polypropylene	12–43
Stainless steel AISI 302–cold-rolled	520
Titanium alloy (6% Al, 4% V)	830
Aluminum alloy 2014-T6	400
Copper 99.9% Cu	70
Silkworm silk	500
Kevlar	3620

will also increase. Hence, to obtain reliable data, frequent measurement of a reference sample is required when using $F_\perp > 1000$ nN.

3) *Spatial resolution degradation:*

The contact diameter representing the resolution is more than 10 nm in Table 12.1, except for case (a). Therefore, to achieve a spatial resolution of about 5 nm, the contact force should be less than 250 nN with a tip without wear. Hence, users need to take care if images are obtained at high loads.

A worn tip also causes artifacts. The contrast of grooves and grain boundary can be inverted by the compensation of multi-asperity contact. Moreover, the simple relation:

$$k^* = 2a_c K^* \quad (12.2)$$

cannot be employed for contacts without axial symmetry. If it is used, overcompensation may easily take place, resulting in an inverted artifact. There is no established relation that can be employed to convert the contact stiffness into the local modulus value.

12.3 Theory

12.3.1 Overview

In this section we describe the theory of UAFM and its application to evaluation of subsurface delamination and cracks. Open cracks and delaminations with a finite gap can be detected by linear analysis. It is simpler than nonlinear measurements, and thus desirable when it is possible. Subsections 12.3.2 and 12.3.3 relate to linear analysis and theory. Not only in UAFM but also in any AFM measurement there is an appropriate range of contact force. This example is shown for subsurface imaging in Subsection 12.3.4. However, closed cracks and delaminations cannot be evaluated by linear analysis. Thus, Subsections 12.3.5 and 12.3.6 are concerned with nonlinear theory.

12.3.2
Linear Analysis of Stiffness and the Q Factor

The principle of UAFM for the analysis of stiffness and subsurface defects is shown in Figure 12.1 [11]. When a resonance vibration is excited to the cantilever, elastic deformation of the sample is caused by effective stiffening of the cantilever due to the inertia effect (Figure 12.1b) as well as shortening of the lever due to the formation of nodes (Figure 12.1c) as proposed previously.

In a model of a UAFM cantilever with distributed mass, the slope of the cantilever is given by:

$$V(x) = \frac{\partial z(x)}{\partial x}$$

$$= (u_0/2)\frac{\beta}{L_1}e^{i\omega t}\left[\sinh\beta\frac{x}{L_1} - \sin\beta\frac{x}{L_1} - B(\omega)\left(\sin\beta\frac{x}{L_1} + \sinh\beta\frac{x}{L_1}x\right)\right.$$

$$\left. + D(\omega)\left(\cos\beta\frac{x}{L_1}x - \cosh\beta\frac{x}{L_1}\right)\right] \quad (12.3)$$

$$B(\omega) = -\frac{SS_h + \alpha(CS_h + SC_h)}{(1+CC_h) + \alpha(CS_h - SC_h)}, \quad D(\omega) = \frac{CS_h + SC_h + 2\alpha CC_h}{(1+CC_h) + \alpha(CS_h - SC_h)}$$

where z is the deflection of cantilever, u_0 is the vibration amplitude of the cantilever base, ω is the angular frequency, L_1 is the actual distance between the tip and the base (cantilever length), and $S = \sin\beta$, $S_h = \sinh\beta$, $C = \cos\beta$, $C_h = \cosh\beta$, $\alpha = -1/\beta^3(3k^*/k_c + i\sqrt{3}\Gamma\beta^2)$, and $\beta = 3^{1/4}\sqrt{\Omega}$ [15]. The factor $\Omega = \omega/\sqrt{k_c/M}$ is the normalized frequency and $\Gamma = \gamma/\sqrt{Mk_c}$ is the normalized damping coefficient, where M is the mass of cantilever, k_c is the cantilever stiffness, and k^* and γ are, respectively, the contact stiffness and damping coefficient between tip and sample. Because the slope is proportional to the signal measured by optical-deflection AFM, Equation (12.3) is an analytical expression of the UAFM spectrum at a given location x of the laser beam spot.

Figure 12.2 shows spectra calculated using Equation (12.3) with $\Gamma \equiv \gamma/\sqrt{Mk_c}$ = 0.5, 1, 2, 5, and 10 and $k^*/k_c = 200$, for the laser beam spot at the end of cantilever ($x = L_1$) [15]. The Q factor is calculated as the ratio of peak frequency Ω_0 to the 3 dB-width $\Delta\Omega$. The inset of Figure 12.2 shows the relation between the Q factor and the peak height of resonance V_{max}, showing clear linearity between them. Though the linearity is an approximate relation, it holds over a reasonably wide range of normalized damping coefficient Γ. For example, the ratio between the Q

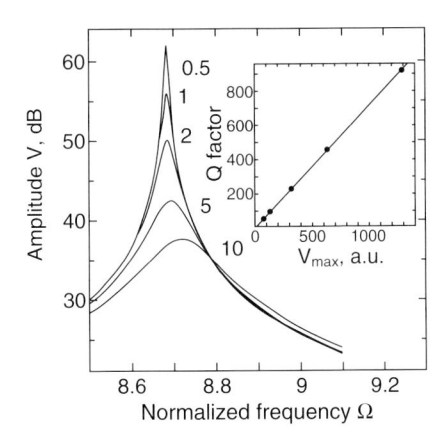

Figure 12.2 Relation between the Q factor and the maximum peak height V_{max} of a peak formed around $\Omega \approx 0.87$ when and $k^*/k_c = 200$. Five different values of $\Gamma \equiv \gamma/\sqrt{Mk_c}$ were assumed (0.5, 1, 2, 5, and 10).

factor and the maximum peak height V_{max}, Q/V_{max}, remains almost constant (0.413–0.422) over a range of Γ (0.1–10.0) for the normalized contact stiffness k^*/k_c of 200. For $k^*/k_c > 200$, the variation of Q/V_{max} is even smaller. Thus, the peak height of resonance can be employed as a measure of the Q factor. The analysis can be further improved by considering the lateral stiffness [25], tilt of the cantilever [11], and shape of the tip [13].

The Q factor is determined by the internal friction of the sample and by the water or contaminant film on the sample. Although other factors such as the air damping, clamp of the cantilever base to the ultrasonic transducer, and defects within the cantilever change the Q factor, their effect is usually small or uniform and, therefore, does not significantly affect the contrast in the image.

To verify such an effect quantitatively, we use the continuum theory to describe the vibration of the cantilever with a tip in contact with the sample. For analysis of experiments, lateral stiffness [10, 25] and the oblique sample surface [26] are considered. The frequency equation of the cantilever is:

$$C \cdot C_h(PQ - 1 + D^2) + S \cdot C_h(P + Q) + C \cdot S_h(P - Q) + 2S \cdot S_h\sqrt{PQ}D + 1 + PQ - D^2 = 0 \quad (12.4)$$

$$P = \frac{\beta^3}{3(k^*/k_c)(r\sin^2\varphi + \cos^2\varphi)} \quad Q = \frac{\beta(L_1/h)^2}{3(k^*/k_c)(r\cos^2\varphi + \sin^2\varphi)}$$

$$D = \frac{(1-r)^2 \tan^2\varphi}{(r + \tan^2\varphi)(r\tan^2\varphi + 1)} \text{ and } r = \frac{k^*_{Lat}}{k^*} \quad (12.5)$$

in which $\beta = \kappa L_1$ is the product of the wave number κ and the cantilever length L_1, k^* is the vertical (out-of-plane) contact stiffness, k^*_{Lat} is the lateral (in-plane) contact stiffness, k_c is the cantilever stiffness, and φ is the angle between the lever axis (the x-axis) and the sample surface. When the parameters (r, φ, L_1/h, k^*/k_c)

are given, Equations (12.4) and (12.5) are solved for β and the resonance frequency is given by:

$$\kappa^4 = \left(\frac{\beta}{L_1}\right)^4 = \frac{\omega^2 \rho A}{EI} = \frac{3\omega^2 \rho A}{k_c L_1^3} \tag{12.6}$$

In the limit of $r \to 0$ and $\varphi \to 0$, we obtain $Q \to \infty$ and hence:

$$\frac{k_c}{3k^*}\beta^3(1+\cos\beta\cosh\beta) = \cos\beta\sinh\beta - \sin\beta\cosh\beta \tag{12.7}$$

which is the original equation for ultrasonic AFM [5, 8].

12.3.3
Linear Theory of Subsurface Imaging

To investigate a variation of the resonance frequency due to a subsurface object, we performed a finite element method (FEM) analysis [14, 17]. For material constants we employed those of HOPG used in subsurface dislocation observation. Figure 12.3 shows a cross section of the axisymmetric models for the FEM analysis along the indentation axis. In the model shown in Figure 12.3a, we assume a 3-nm thick, low-elasticity layer with infinite width at the depth of 3 nm. It may represent a dislocation or any other defect. In the model shown in Figure 12.3b, to take into account the finite width of the defect, we limited the radius of the low-elasticity layer to 3 nm from the indentation axis. Young's moduli and Poisson's ratios of the substrate and the low-elasticity layer were assumed to be E_{sb} = 30 GPa and v_{sb} = 0.24, and E_{low} = 15 GPa and v_{low} = 0.24, respectively. We employed another model with the homogeneous elasticity, with E_{sb} and v_{sb} representing a defect-free model. In the analysis, we indented a rigid sphere of 50 nm radius to the depth of 1 nm in 0.01 nm steps.

Figure 12.4 shows the von Mises stress distribution in the finite-width-layer model under a load of (a) 25 nN and (b) 100 nN. Under the 25 nN load, the stress

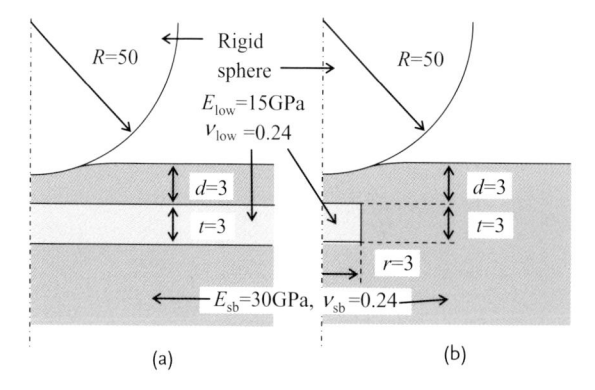

Figure 12.3 Models for the FEM analysis (unit is nm). (a) Low-elasticity layer under the surface; (b) low-elasticity layer with a radius of 3 nm under the surface.

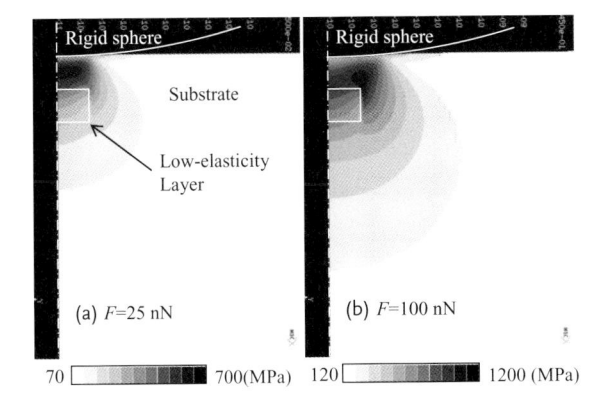

Figure 12.4 Von Mises stress distribution under different loads: $F =$ (a) 25 and (b) 100 nN.

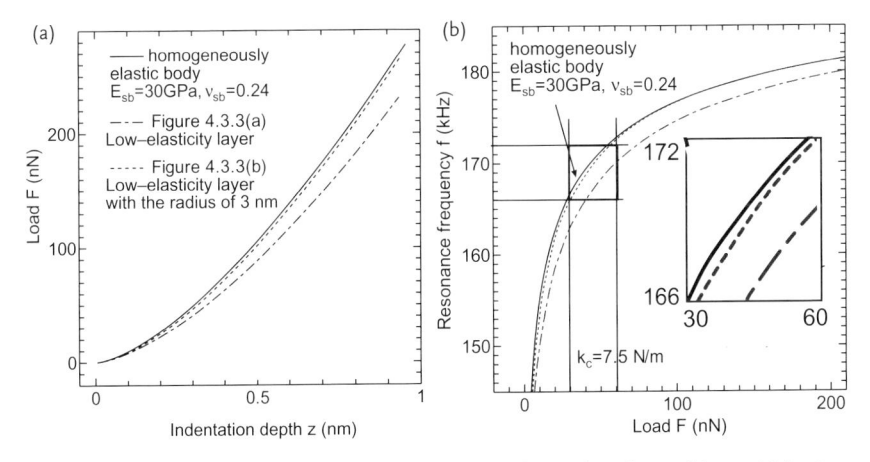

Figure 12.5 Expected difference in resonance frequency for a subsurface soft layer: (a) load displacement curve calculated by the FEM analysis; (b) the first resonance frequency calculated from (a).

distribution is affected by the presence of a subsurface low-elasticity layer. The contact radius was almost the same as the width (3 nm) of the defect. It indicates that the effect of the defect is present for the whole contact area. In contrast, under a load higher than 100 nN, the contact radius is 5 nm, which is longer than the width of defect. This indicates that the effect of defect is limited to less than 40% of the contact area.

At each indentation step, we recorded the load $F_\perp$ and displacement z as shown in Figure 12.5a. The load for the finite-width-layer model is slightly less than that for the defect-free model, whereas that for the infinite-width-layer model is significantly less than that for the defect-free model. Next, we calculated the contact stiffness $k^* = \partial F_\perp / \partial z$ using the results in Figure 12.5a. In UAFM, we can measure a contact stiffness 1–100 times higher than the cantilever stiffness when we use the first resonance frequency. Then, using the frequency equation [Equation (12.7)], we calculated the resonance frequency corresponding to the contact stiffness shown in Figure 12.5b, assuming a cantilever stiffness $k_c = 7.5\,\text{N}\,\text{m}^{-1}$.

In Figure 12.5b, the solid curve shows the resonance frequency of the defect-free model with homogeneous elasticity. The dot-dash curve shows the resonance frequency of the infinite-width-layer model in Figure 12.3a. The resonance frequency was significantly lower than that of the defect-free model. The dotted curve shows the resonance frequency of the finite-width-layer model in Figure 12.3b. Under the load higher than 100 nN where the contact radius was longer than 5 nm, the resonance frequency approached that of the defect-free model, but it was lower under the lower load. Then, the resonance frequency calculated from the model with the subsurface low-elasticity layer was lower than that calculated from the defect-free model. The frequency difference is 2 kHz for the infinite-width-layer model and 0.6 kHz for finite-width-layer model (3 nm). Even the latter is larger than the noise level for reasonably good conditions. Therefore, when there is a subsurface defect in objects, such as dislocation, delamination, or crack, we will be able to detect it by measuring a lower resonance frequency than that of the defect-free area.

12.3.4
Advantage of Appropriate Load

Not only in UAFM but also in any AFM measurement, there is an appropriate range of contact force. Now we show that this has been shown in Figures 12.4 and 12.5:

1) *Load range to avoid plastic deformation:*

   Clearly, the von Mises stress reaches 1000 MPa in the stress distribution near the contact area with a $F = 100\,\text{nN}$ in Figure 12.4. According to Table 12.2, it is more than the yield stress of most of materials. Thus plastic deformation of the sample is unavoidable and degrades the reproducibility of measurement. It is detrimental both for imaging and elasticity measurements. However, this is much less the case in the contact area with $F = 25\,\text{nN}$. Thus $F = 25\,\text{nN}$ is preferable for improved reproducibility.

2) *Load range to enhance contrast of subsurface defect layer:*

   The inset of Figure 12.5b shows that the difference between resonance frequency at a defect-free area and at defect area is small when $F > 100\,\text{nN}$ or when $F < 20\,\text{nN}$. Thus there is an appropriate range of load of $20\,\text{nN} < F < 100\,\text{nN}$.

12.3.5
Nonlinear Analysis of Spectra

Although nonlinearity is usually an obstacle for reliable measurement, it might be useful for some kinds of measurements that are employed with difficult objects

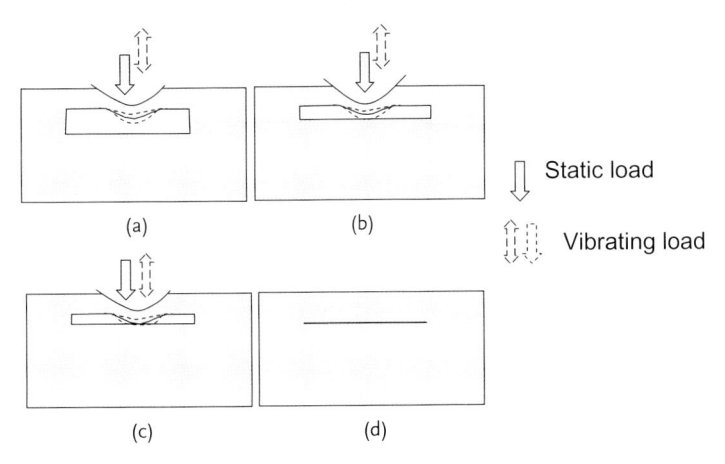

Figure 12.6 Contact acoustic nonlinearity (CAN) in closed crack. (a) A gap decreases the contact stiffness with linearity; (b) a wide gap closed during downward motion (dotted line) of the tip; (c) a closed gap more tightly closed during the downward motion and separated during the upward motion (dotted line) of the tip; (d) perfectly closed gap.

where methods using linear measurement are not available. In macroscopic ultrasonics, the typical object is closed cracks or delaminations [19]. In UAFM, dislocations in graphite are most clearly characterized not only by linear but also by nonlinear spectra [14, 17, 18]. Though the dislocation observation had been an academic subject when it was reported in the 1990s, it may have practical application in the diagnosis of graphene devices.

Figure 12.6 illustrates the origin of the linear and nonlinear spectra at a subsurface gap [18, 24]. First, note that the load acting on the contact area between the UAFM tip and sample is the sum of the static load and the vibration force. In terms of these two forces, the behavior of a subsurface gap is classified into three cases as shown in Figure 12.6a–c. Solid lines represent the tip position and the gap deformation due to the static load and dotted lines represent those due to the vibration force.

Figure 12.6 predicts the contact stiffness as a function of tip position. When the gap is sufficiently wider than the vibration amplitude of the tip, the gap is not closed during the vibration (Figure 12.6a). When the gap is slightly wider than the static displacement of the tip, the gap is not closed by static displacement but is closed by the increasing load period of the vibrations (Figure 12.6b). Since the contact stiffness increases as the gap is closed, it is called a stiffening nonlinear spring and may be called the "subsurface tapping mode." On the other hand, when the gap is narrower than the static displacement, it is closed by static displacement but opened during the decreasing load period of the vibrations (Figure 12.6c). Since the contact stiffness decreases as the gap is opened, it is called a softening nonlinear spring. It is similar to the typical behavior at the pull off, and so it may be called the "subsurface pull off." Figure 12.6d shows a permanently closed gap with compression stress. Such a closed gap may be present in micro- or nanodevices, but it is a very difficult defect to discover since it can only be opened by strong tensile stresses.

12.3.6
Duffing Model

This type of nonlinear vibration can be analyzed by several different approaches. In this work we adopt the simplest approach using the Duffing equation for a nonlinear vibration, since it gives a simple analytic solution for the stiffening or softening stiffness during vibration. The Duffing equation is given by:

$$\ddot{\zeta} + \Gamma\dot{\zeta} + \chi_1\zeta + \chi_3\zeta^3 = \zeta_E\sqrt{\Omega^2 + \Gamma^2}\cos\Omega\tau \tag{12.8}$$

where ζ is the tip displacement, τ is time, Γ is a damping coefficient, χ_1 is a linear stiffness coefficient, χ_3 is a nonlinear stiffness coefficient, ζ_E is an excitation amplitude, and Ω is an excitation frequency. All quantities are dimensionless. A positive χ_3 represents the stiffening spring and a negative χ_3 represents the softening spring. For the harmonic vibration, since the fundamental frequency component having the period $2\pi/\Omega$ predominates over the higher harmonics, the periodic solution takes the form:

$$\zeta = \zeta_1\sin\Omega\tau + \zeta_2\cos\Omega\tau \tag{12.9}$$

Substituting Equation (12.9) into Equation (12.8), and equating the coefficient of the terms containing $\sin\Omega\tau$ and $\cos\Omega\tau$ separately to zero, yields:

$$\left[(\Omega^2 - \chi_1 - \tfrac{3}{4}\chi_3\zeta_0^2)^2 + \Omega^2\Gamma^2\right]\zeta_0^2 = \zeta_E^2(\Omega^2 + \Gamma^2) \tag{12.10}$$

where $\zeta_0^2 = \zeta_1^2 + \zeta_2^2$.

Figure 12.7 shows calculated spectra for the Duffing oscillator, where the frequency is gradually increasing. The parameters were chosen so as to qualitatively

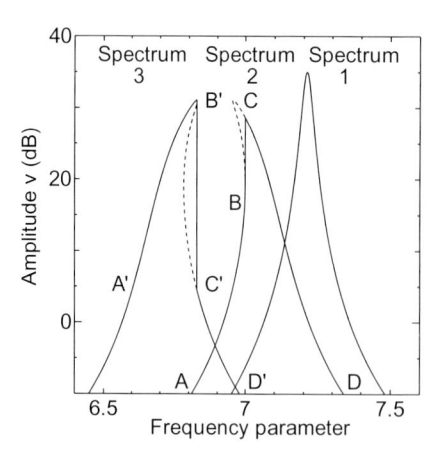

Figure 12.7 Calculated spectra for the Duffing oscillator.

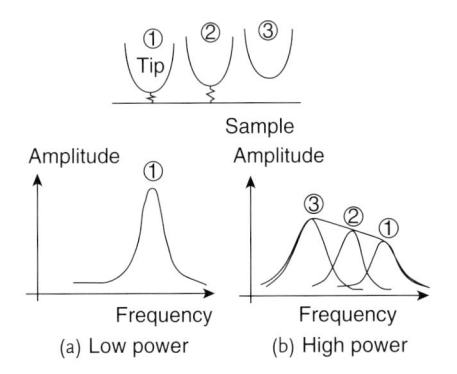

Figure 12.8 Surface tip–sample contact vibration for different excitation power levels.

reproduce the measured spectra at the subsurface dislocation in graphite shown in Figure 12.18, where $k_c = 4.6\,\text{N m}^{-1}$ and $f_0 = 38.2\,\text{kHz}$.

Spectrum 1 is a linear spectrum with $\chi_3 = 0$ ($\chi_1 = 52.0$, $\zeta_E = 0.3$, $\Gamma = 0.04$). The linear spectrum is symmetric with regard to the peak frequency. Spectrum 2 is a nonlinear spectrum with $\chi_3 = -0.9$ ($\chi_1 = 50.0$, $\zeta_E = 0.3$, $\Gamma = 0.05$), representing the softening spring since the peak is shifted to a lower frequency owing to the third-order term. Spectrum 3 is a nonlinear spectrum with $\chi_3 = 0.9$ ($\chi_1 = 45.0$, $\zeta_E = 0.3$, $\Gamma = 0.05$), representing the stiffening spring. The behavior is the opposite of that in spectrum 2 and the peak is shifted to a higher frequency. Branches represented by dashed lines are unstable solutions, which cannot be realized physically. The softening spring can be explained by the gap behavior shown in Figure 12.6c and the stiffening spring is explained by the gap behavior shown in Figure 12.6b.

To explain the shape of the observed linear and nonlinear spectra, an intuitive model is shown in Figure 12.8 [11]. If the excitation power is low, the variation of the tip–sample indentation during one cycle of vibration should be small and the contact stiffness at every moment can be regarded as constant during one vibration cycle. Then the spectra consist of a single peak corresponding to a single value of contact stiffness. However, if the excitation power is increased, an intermittent contact results and the contact stiffness changes in magnitude during the vibration cycle. The stiffness is small while the tip is detached from the sample. In this case, the vibration spectra averaged over one cycle may be approximated by the sum of component spectra for different contact stiffness. The total spectra are, therefore, broader and the peak frequency is lower than the linear spectra. If the amplitude of component vibration is not uniform, then the total spectra will be asymmetric. The qualitative features of the observed spectra are consistent with this model.

12.3.7
Numerical Model with Double Nodes

There are serious problems in closed cracks in structures. If the crack is open, it reflects ultrasound but if it is closed it transmits the ultrasound. This situation is

320 | *12 Ultrasonic Atomic Force Microscopy*

(a) Model   (b) Low amplitude wave   (c) High amplitude wave

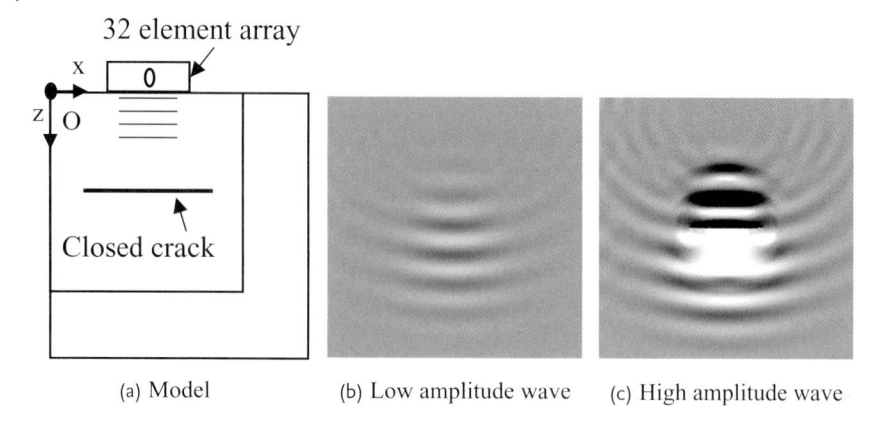

Figure 12.9 FDTD simulation of interaction with ultrasound and a closed crack: (a) model; (b) low amplitude wave; (c) high amplitude wave.

similar to the closed gap in Figure 12.6d. If it has a compression stress, it is very difficult to open it. One possibility is to apply a very high frequency vibration so that a propagating wave is generated where there is more than a wavelength between the sample surface and the gap.

To analyze this situation, a model of closed contact in two or three dimensions is needed. An example of this analysis is given in our recent simulation, shown in Figure 12.9 [27].

The simulation conditions were as follows: the number of elements in the array sensor was 32, the element pitch was four nodes in the finite difference time domain (FDTD) model, center position of crack (0, 150) nodes, angle of incident 0°, focusing at node position (−550, 1200), and compression residual stress of 100 MPs. A small amplitude incident wave is transmitted as in Figure 12.9b. Crack face vibration is generated by applying large-amplitude incident wave as in Figure 12.9c, where the tensile stress of the incident wave is comparable to the compression residual stress.

12.4
Instrumentation

Figure 12.10 shows an implementation of UAFM based on a contact mode AFM. A high frequency vibrator attached to the cantilever base is driven in the high frequency range of 50 kHz to 10 MHz. The resultant vibration of the cantilever is detected by a photodiode and processed by using a lock-in amplifier, network analyzer, or a special phase-locked loop (PLL) circuit. Linear and nonlinear detection schemes are employed. In the linear detection scheme, the high frequency signal is measured at the same frequency. In the nonlinear scheme, the high frequency signal is amplitude-modulated by a low frequency signal in the frequency range 1–10 kHz. Instead of the high frequency signal, the low frequency

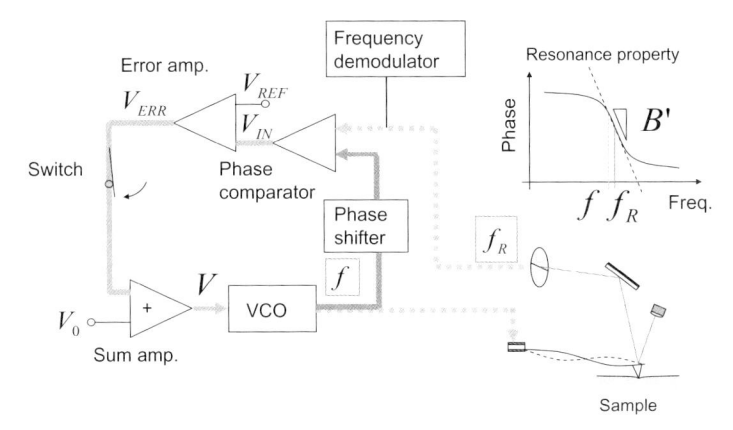

Figure 12.10 Phase-locked loop circuit for the mapping of resonance frequency and Q factor.

signal is measured. The advantage of a nonlinear scheme is that we can use commercial AFM without modification. Moreover, the use a of low frequency signal is favorable for achieving a high signal-to-noise ratio. However, the analysis is not simple and precise.

Since we directly vibrate the cantilever in UAFM, it is similar to noncontact (NC) mode and tapping mode AFM. However, the use of a higher mode is unique to UAFM. In addition, we use very low vibration amplitude of less than 1 nm, so that we can control the contact force to a very low level of less than 0.1 nN. Finally, it is emphasized that there is no problem with the inspection of large samples (e.g., silicon wafers for VLSI) and irregularly shaped samples (e.g., turbine blades or magnetic disk heads), since the tedious transducer bonding to the samples is not required.

In the present instrumentation, mapping of the resonance frequency and Q factor is realized by the PLL circuit, where the cantilever vibration is excited by a voltage-controlled oscillator (VCO) [15, 28]. The input voltage V of the VCO is adjusted to realize resonance when the tip is in contact with the sample. Then, the cantilever deflection signal detected by the photodiode (PD) is split into two parts, one of which is low-pass-filtered (LPF) to control the z position of the sample. The other part is band-pass-filtered (BPF) and its phase is compared with that of the VCO output signal. The phase difference between them is adjusted with a variable phase shifter ϕ to equate the phase comparator output V_{IN} to a reference voltage V_{REF}.

After connecting the switch, we start raster scanning of the sample. If the resonance frequency is changed, the phase signal is also changed. Then, the output V_{ERR} of the error amplifier caused by the phase change is added to the VCO input to recover the resonance. In this manner, the cantilever is always vibrated at the resonance frequency and the vibration amplitude represents the Q factor.

Whereas the resonance frequency tracking described above is similar to that of the frequency modulation mode of noncontact AFM (NC-AFM), the vibration amplitude of the cantilever is quite different. Although it is very large (>10 nm) in NC-AFM, it should be small (<1 nm) in UAFM, to keep the tip always in linear contact with the sample, that is, the contact stiffness k^* should not deviate from its static value. Thus, we should control carefully the driving power of the cantilever in UAFM. To determine the optimum driving power, we monitor the vibration spectrum and find the largest possible power where the spectrum still remains symmetrical and sharp, immediately before becoming asymmetrical and broad. Thus, both a good signal-to-noise ratio and linear contact are realized.

12.5
Experiments

12.5.1
Effort to Avoid Nonlinearity at Tip–Sample Contact

An obstacle to achieve reliable measurement is the nonlinearity at tip–sample contact. To show this, three silicon cantilevers with a silicon tip were used. The nominal length, width, thickness, and stiffness of the cantilever were, respectively, $L_1 = 444\,\mu m$, $a = 73\,\mu m$, $b = 3.5\,\mu m$ and $k_c = 1.5\,Nm^{-1}$, respectively. The spectra of tip 1 around the second resonance at the static load of 30 nN on a soda-lime glass are shown in Figure 12.11. Different levels of excitation power were applied to the piezoelectric transducer attached to the cantilever holder. For clear comparison, each spectrum is shifted by 5 dB from the preceding spectra [11].

Although the first step for quantitative evaluation is the precise measurement of resonance frequency, it turned out that this is not an easy task. It was found

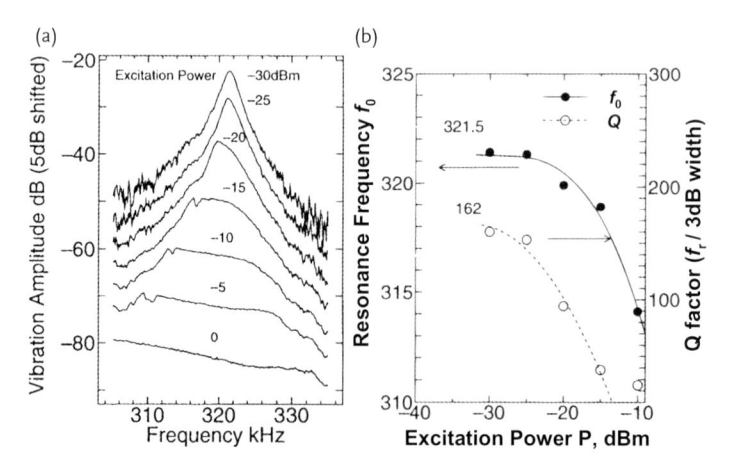

Figure 12.11 Pursuit for linearity of spectra in UAFM and related techniques: (a) deflection vibration spectra with different excitation power of cantilever; (b) variation of resonance frequency and Q factor (3 dB) with excitation power of cantilever.

that the width decreases and the peak frequency increases as the excitation power is reduced from −10 to −25 dBm. However, if the resonance frequency f_r and Q factor are plotted as functions of excitation power (Figure 12.11b) it is possible to estimate the limiting values of f_r and Q as 321.5 kHz and 162, respectively, as the power was extrapolated to zero. Moreover, approximately constant values were obtained by reducing the excitation power to less than −25 dBm. We call the spectra measured under such conditions a "linear spectra."

It has been suggested that the width of the spectra is related to the viscosity or energy dissipation at the tip–sample contact [6, 25, 26] and it is possible to evaluate the viscosity from the peak width. However, because the width is generally dependent on the excitation power, it is not possible to evaluate correctly the viscosity from the nonlinear spectra. It is essential to suppress the nonlinearity for reliable viscosity measurement.

12.5.2
Relation between UAFM and UFM

A similar but other type of nonlinearity is described in the model of UFM [2–4]. When an amplitude-modulated (AM) high-frequency vibration is excited on the sample or on the cantilever support, a vibration at the modulation frequency is generated by a demodulation or "mechanical diode" effect due to the nonlinear tip–sample contact. To understand the relation between these two nonlinear effects, it is useful to compare the threshold power for linear spectra P_L with that for nonlinear demodulation of the AM vibration P_D.

According to investigations under the same experimental conditions as in Figure 12.11, the threshold power P_D was ~0 dBm, which is significantly higher than the threshold for linear spectra P_L (−25 dBm, as shown previously). Correspondingly, the carrier frequency providing the maximum demodulation was 300 kHz at just above P_D, which is lower than the linear resonance frequency of 321 kHz. This result shows that the tip–sample contact is still nonlinear even at a power of $< P_D$. The reason why P_D is higher than P_L is because a stronger nonlinearity is required for the demodulation effect than for the resonance shift.

Although both effects have the same physical origin, that is, the nonlinear force–distance (or indentation) relation, they can be distinguished because the influence of the excitation power on the visibility of these effects is opposite. As shown in Figure 12.11, the resonance frequency can be measured only in the low power range, where well-defined spectra are obtained. It becomes difficult to identify the peak at powers above −5 dBm, where the spectra are severely broadened and distorted. On the other hand, in our theoretical study on the nonlinear imaging method for UFM, we showed that the demodulation effect becomes significant only at powers higher than P_D. Moreover, it was found that above the threshold power P_D the tip is pulled off from the sample surface, overcoming the adhesion force F_c. Therefore, the nonlinear effect responsible for the demodulation depends not only on the elasticity but also on the adhesion force, bringing more complexity.

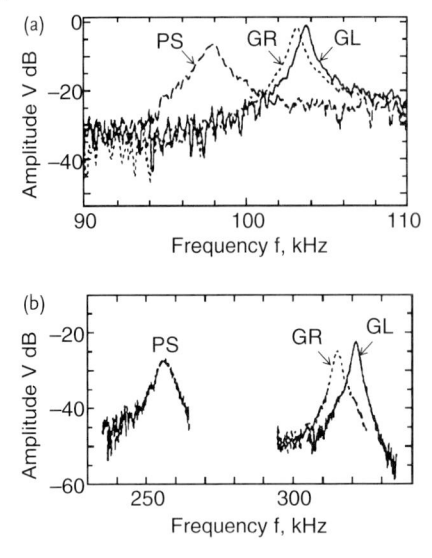

Figure 12.12 Spectra of three different materials: glass (GL), graphite (HOPG) (GR), and polystyrene (PS).

The images obtained in the demodulation or the mechanical diode mode are sometimes very clear and sensitive to variation of material properties or to the existence of subsurface defects. However, it is difficult to extract quantitative information from these nonlinear modes because the nature of the tip–sample contact is quite complex at large amplitude vibration under the adhesion force, as discussed above. Further study is needed on this subject.

12.5.3
Quantitative Evaluation of Elasticity

Notably, quantitative evaluation of elasticity becomes feasible only when the linear spectra are obtained [11]. The linear spectra of three different materials (soda-lime glass, GL; graphite, GR; and polystyrene, PS) are presented in Figure 12.12. Figure 12.12a represents the first resonance (such as in Figure 12.1b) and Figure 12.12b the second resonance (Figure 12.1c). Both spectra of different materials are separated according to the difference in contact stiffness, but the second resonance has better separation due to the effective stiffening by inertia and the formation of nodes as predicted in Reference [9]. The resonance frequency f_R of each spectra together with the free resonance frequency are plotted in Figure 12.13 against the contact stiffness k^* normalized by the lever stiffness k_c. The contact stiffness was estimated using the approximated expression Equation (12.7). To estimate the contact radius, the tip was imaged using a porous silicon sample with remaining sharp silicon crystals within it. The contact radius a_c was estimated to be 20 nm. More precise estimation of the contact radius is presented below.

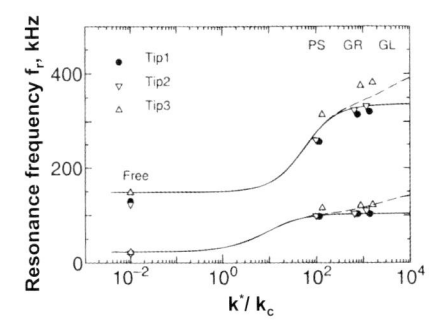

Figure 12.13 Calculated and measured resonance frequency as a function of relative contact stiffness divided by the cantilever stiffness.

The calculated resonance frequency, assuming $E = 169\,\text{GPa}$, $\rho = 2.3 \times 10^3\,\text{kg m}^{-3}$ and using Equation (12.7), is plotted as the solid curve in Figure 12.13. The agreement between measured and calculated frequencies is acceptable. Because the agreement for tip 3 at a large value of k^*/k_c was rather poor, we examined the effect of lateral stiffness by Equation (12.5) and the ratio $r = 0.8$. The angle φ was assumed to be zero. The agreement was then improved as shown by dashed curve and it is shown that quantitative evaluation of the sample contact stiffness is thus possible using the linear theory of UAFM and linear spectra. We may conclude that the "linear" spectra really reflect the linear tip–sample contact described by linear theory. In contrast, under conditions where the resonance frequency and Q factor vary as the excitation power varies, we have a certain kind of nonlinearity. Such nonlinear spectra should not be used in quantitative analysis, as shown before.

12.6
Observation of Defects in Layered Materials

12.6.1
Defects in Graphene Sheets

Graphite is a well-known material with a layered structure of graphene sheets in which hexagonal carbon (C) planes are stacked periodically along the c-axis due to weak interlayer interaction forces, as schematically shown in Figure 12.14. Each C plane can easily slide against another, making graphite a good solid lubricant. In addition, the interlayer space accepts a large amount of ions or molecules so that graphite forms graphite intercalation compounds (GICs) [29], which are applied in useful devices such as Li-ion rechargeable batteries. The excellent electron mobility in graphene sheets is expected to revolutionize high-speed electronic devices.

Despite its significant usefulness, however, the atomic nature of the interlayer interaction in graphite is still not fully understood. Although shear-induced motion

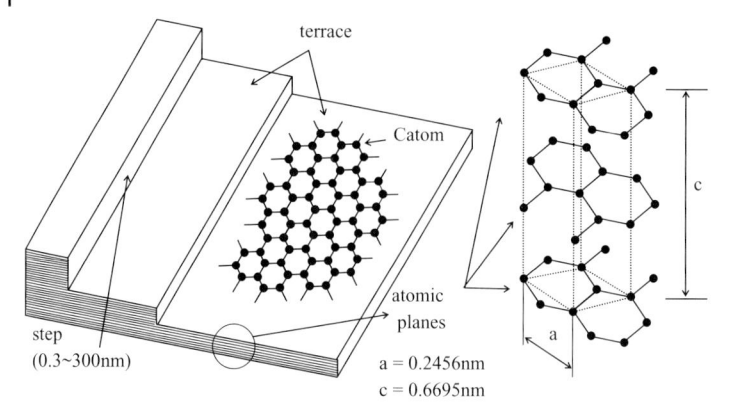

Figure 12.14 Crystal structure of HOPG and graphene sheets.

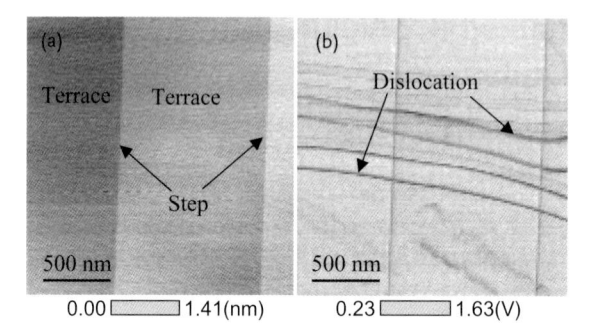

Figure 12.15 Images of the cleaved HOPG surface: (a) contact AFM topography; (b) amplitude distribution of the cantilever deflection vibration at the resonance frequency measured at position A (180 kHz). Image area was 2000 nm × 2000 nm and contact load was 97 nN, including adhesion force of 27 nN.

of partial dislocations in highly oriented pyrolytic graphite (HOPG) has been observed using scanning tunneling microscopy (STM), the depth of observation is only one or two atomic layers. We have reported that subsurface dislocations in HOPG can be clearly observed using UFM or UAFM with an observation depth of more than 3 nm. Although previous observations were limited in their static conditions, we found that the dislocations are extremely mobile. In this section, we describe detailed dynamic observations of the subsurface dislocations in HOPG, using UAFM.

The sample was freshly cleaved HOPG (Union Carbide grade B). To obtain the images, experiments were performed in ambient air at 23 °C without humidity control.

Figure 12.15 shows a typical cleaved HOPG surface. Figure 12.15a is a contact-mode AFM topography showing monolayer steps and atomically flat terraces. The UAFM amplitude image in Figure 12.15b shows the amplitude distribution of the cantilever deflection vibration at the first resonance frequency measured at

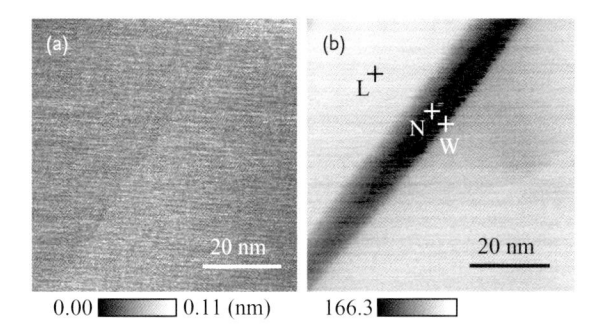

Figure 12.16 UAFM image of a dislocation: (a) topography and (b) resonance frequency image.

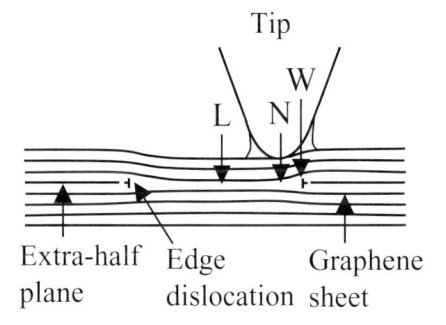

Figure 12.17 Schematic illustration of a section of HOPG and a tip.

position A (180 kHz). The dark shades indicate low contact stiffness, which is due to subsurface dislocations [3, 4, 14, 16]. We observed many dislocations that were not visible in the topography (Figure 12.15a).

Figure 12.16 shows the observation of the subsurface edge dislocation in HOPG crystal where $k_c = 2.4\,\text{N}\,\text{m}^{-1}$ and $f_0 = 28\,\text{kHz}$. Figure 12.16a is the topography, showing a small depression due to subsurface stacking fault. Figure 12.16b is an UAFM image, showing a dislocations. Figure 12.17 is a schematic illustration showing narrow and wide gaps with respect to the tip vibration amplitude (~0.1 nm).

Though the dislocation was observed in the shape of a string or a stripe, ranging in width from 10 to 20 nm, it was not clear whether the width represents the resolution limit or it is due to a detailed structure of dislocation. Thus, we examined cantilever vibration spectra at positions labeled in Figure 12.16b and found a significant difference between positions L, N, and W. The notation "L" was adopted to mean a Linear spectrum, a Narrow subsurface space apart from the dislocation core, and "W" a Wider space close to the core.

Figure 12.18 shows the spectra measured with the frequency increasing from 157.2 to 177.2 kHz. The excitation power of the cantilever vibration was −40 dBm and the contact load was 40 nN. The spectrum measured at position L on the

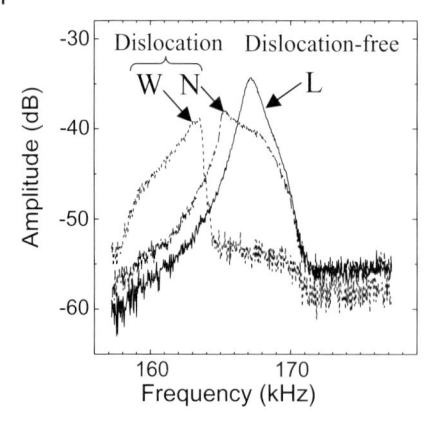

Figure 12.18 Spectra obtained at positions shown in Figure 12.17.

normal area was symmetric with regard to the peak frequency. This result confirms that the excitation power was sufficiently low to avoid the partial pulling off of the tip from the sample, which results in asymmetry in the spectra.

However, even in this case, the spectrum measured at position N in the center of the low frequency area was asymmetric. As clearly seen, it had a sharp increase of the amplitude (upward amplitude jump) at the low-frequency side of the peak and a gradual decrease at the high-frequency side.

The spectrum measured at position W close to the edge of the low-frequency area was also asymmetric but almost inverted in the frequency axis from that at position N. The spectrum had a gradual increase of the amplitude at the low frequency side of the peak and a sudden decrease (downward amplitude jump) at the high-frequency side. Moreover, the center frequency of the spectrum at position W was significantly lower (5 kHz) than that at position N.

These results verify that two positions W and N have distinctly different subsurface atomic structures, and the spatial resolution of UAFM is high enough to resolve them. Since the distance between W and N is 5 nm, the resolution for subsurface object is also 5 nm.

12.6.2
Dislocation in Molybdenum Disulfide

Analogous behavior was observed on a cleaved MoS_2 surface (Figure 12.19). Figure 12.20 shows images of MoS_2 as (a) topography and (b) UAFM resonance frequency image on an atomically flat terrace. The image area was 500 nm × 500 nm and the contact load was about 100 nN. We observed two dislocations.

Figure 12.21a shows another UAFM image of a dislocation in MoS_2. Figure 12.21b shows the spectra measured at positions 1, 2, and 3 in Figure 12.21a. The spectrum measured at position 1 on the normal area was symmetric. At position 2 in the center of the low-frequency area, the spectrum was asymmetric. At

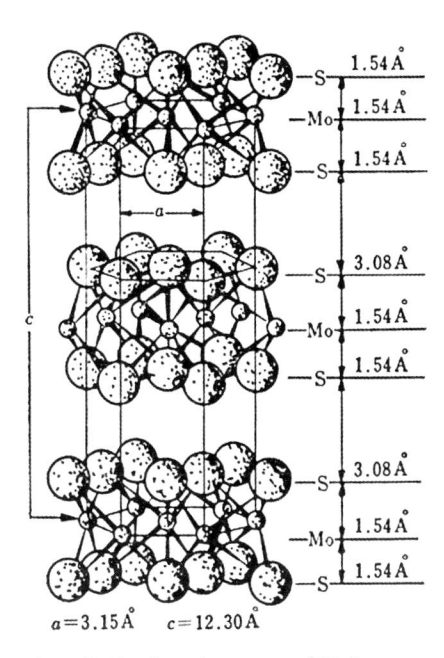

Figure 12.19 Crystal structure of MoS$_2$.

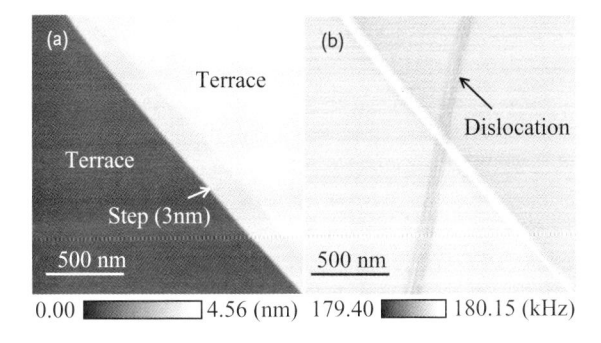

Figure 12.20 (a) Topography and (b) UAFM image of MoS$_2$.

position 3 close to the edge of the low-frequency area it was asymmetric but with a steep slope at the opposite side. The behavior of the spectra at positions 1 and 2 was almost the same as at positions N and W in HOPG.

12.6.3
Observation of Dislocation Behavior under Different Loads

Figure 12.22a–d shows magnified UAFM amplitude images of two dislocations in Figure 12.15 under different loads F including an adhesion force of 27 nN.

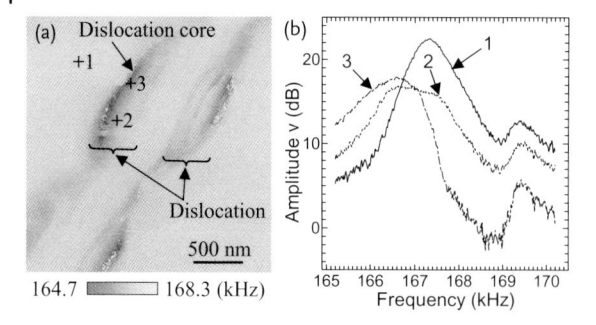

Figure 12.21 Analysis of dislocation in MoS$_2$: (a) UAFM image of a dislocation in MoS$_2$ and (b) spectra at positions shown in (a).

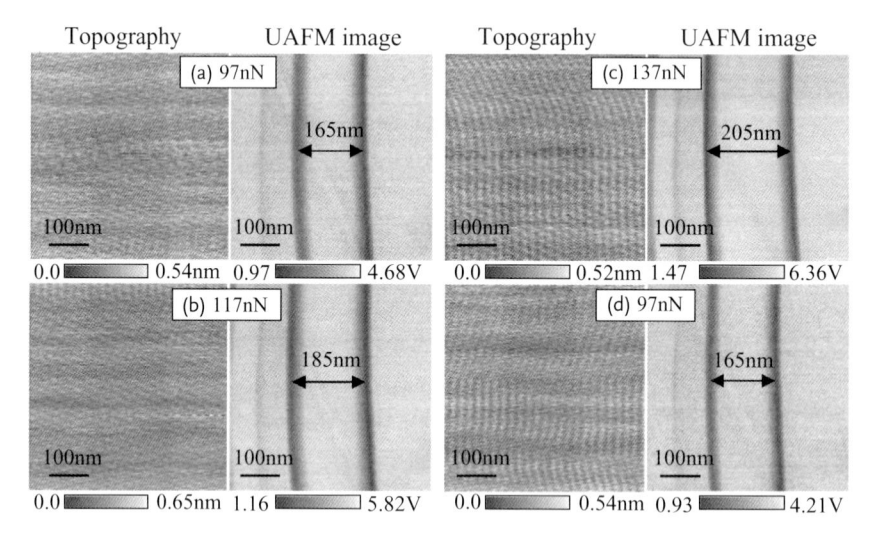

Figure 12.22 Reversible lateral motion of dislocations under applied load. Amplitude distribution of the cantilever deflection vibration at the resonance frequency measured at position A (186.6 kHz). The load F including an adhesion force of 27 nN is shown in each image. Image area: 500 nm × 500 nm.

As the load was increased, the distance between the two dislocations increased, as shown in Figure 12.22a–c. The distance under 137 nN was 40 nm longer than that under 97 nN. However, when the load was decreased to 97 nN, the distance returned to the initial value of 165 nm (Figure 12.22d). Since the motion was almost symmetrical, these dislocations moved 20 nm laterally as the load increased and returned to the original position as the load decreased. We confirmed that this motion is not an artifact due to distortion by scanning, since other objects such as the steps did not move even when the load was increased by the same amount.

12.6.4
Analysis of Dislocation Motion under Varying Applied Load

To show the reproducibility of the motion and to analyze it more quantitatively, we present a resonance frequency mapping of dislocations in other areas of the same sample. Figure 12.23 shows the resonance frequency image of the first deflection mode with an area of 1000 nm × 1000 nm where the contact load was 115 nN, including an adhesion force of 40 nN. The grayscale indicates the resonance frequency in the range 170.9–171.4 kHz. We observed dislocations C and D, with a resonance frequency 0.1–0.2 kHz less than that in the normal area. Upon careful investigation of the topography (not shown here), the area between the dislocations was found to be slightly depressed from the overall level by 0.03–0.05 nm. This depression is represented by the cross-sectional model shown in Figure 12.17, where two edge dislocations (Frank partial dislocations) with different polarity face each other and the lack of an extra-half plane between the dislocations depresses the surface.

We measured the resonance frequency profile along the line AB as the applied load was varied from 115 to 400 nN, including an adhesion force of 40 nN, with the scanning direction being from A to B (rightward scanning). Then, we investigated the motion of dislocations C and D. A horizontal line in Figure 12.24 shows the resonance frequency profile along the line AB shown in the grayscale. The brighter color shows the higher frequency. As the load is changed, the profile is shifted to lower position, so that the whole sequence is observed as an image.

From Figure 12.24a we note that dislocation C moved to the left and dislocation D moved to the right as the load increased; and both of them returned to their original positions as the load decreased. The direction of motion was such that the

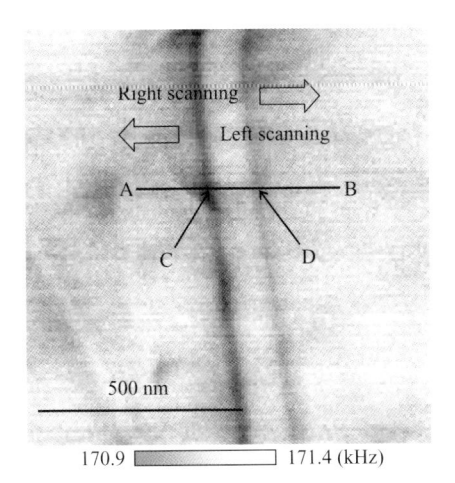

Figure 12.23 Resonance frequency mapping of the cleaved HOPG surface at the first deflection mode. Image area was 1000 nm × 1000 nm and the contact load was 115 nN, including adhesion force of 40 nN.

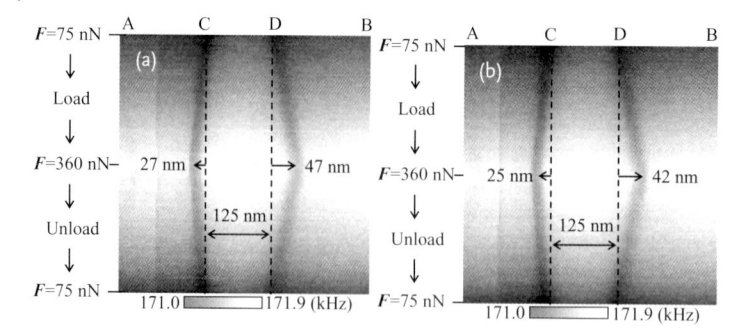

Figure 12.24 Reversible lateral motion of dislocations under varying applied load. Resonance frequency profile along the line AB. (a) Rightward scanning; (b) leftward scanning. The load F, including an adhesion force of 40 nN, is shown on the left-hand side. Length: 500 nm.

length of the extra-half plane was shortened. Under the maximum load, dislocation D moved by 47 nm, corresponding to 330 times the C–C bond length of 0.142 nm in graphite. Dislocation C moved by 27 nm, showing that the mobility differs for each dislocation.

If the effect of lateral force on the dislocation motion is great, the direction or magnitude of dislocation motion would be changed by inverting the scanning direction. Thus, to evaluate the effect of lateral force, we inverted the scanning direction. Figure 12.24b shows the variation of the resonance frequency profile in leftward scanning. It is noted that the direction and the magnitude of dislocation motion were almost the same as those in rightward scanning within experimental error. Therefore, we found that the dislocation motion is mainly affected not by the lateral force due to scanning but by the applied normal load. Note here that we do not exclude the effect of shear stress produced by the indented tip, which is large in the peripheral region of indentation even without lateral force.

At another position of the sample, a strange bending of the dislocation was observed as shown in Figure 12.25. At a low load ($F_0 = 10$ nN), a dislocation of finite length was observed in UAFM image, whereas a small depression was observed on its left side in the topography in (Figure 12.25a). This feature is consistent with the schematic illustration in Figure 12.17. However, on increasing the load by 15 nN the dislocation split into bent and straight dislocations (Figure 12.25b). The bent angle was enlarged by increasing the load up to 30 nN (Figure 12.25c). At higher load the straight dislocation became unclear while the bent one was still clear. As the load was decreased, the same feature was observed reversibly.

This observation suggests that there were two dislocations at different subsurface depths and, probably, the shallow one was the straight one since its image resolution was higher. The contrast of the shallow dislocation was eliminated by increasing the load. However, the reason for the eliminated contrast is unknown at present.

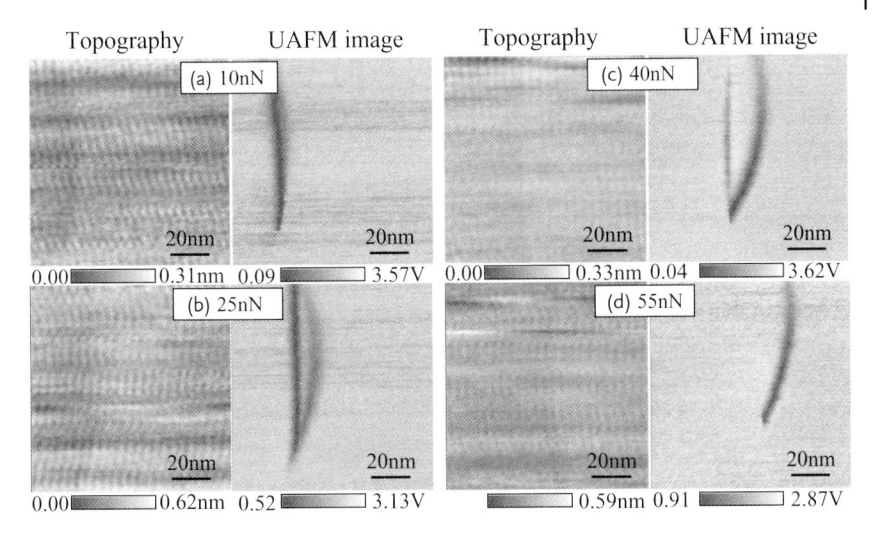

Figure 12.25 Bending motion of dislocations under applied load. Image area: 500 nm × 500 nm.

12.6.5
Model for the Reversible Long-Range Motion of Dislocation

If the dislocations observed here are of the edge-type, images in the previous sections indicate that the dislocations move in the direction of climb, which is known as dislocation motion normal to the slip plane and requires the creation of vacancies and interstitial atoms and their diffusion. However, since the experiment was performed at room temperature and the formation energy for a vacancy or an interstitial atom in graphite is very high (7.0 eV), the creation of vacancies or interstitial atoms should have been negligible. Therefore, the observed motion cannot be considered as a normal climb.

A possible explanation of the motion can be given using the model shown in Figure 12.26, where we assume a pinning point P and an elastic spring S_D due to C–C bonds in the extra-half plane (EHP). Figure 12.26a shows the initial position of the tip and dislocation. As the load increases, the stress and strain energy at the deformed area under the tip increases (Figure 12.26b). Then, the EHP is removed from that area while storing the elastic energy in S_D due to the in-plane elastic deformation of the EHP; thus, the dislocation moves to the right. Conversely, as the load decreases, the EHP returns to the original configuration while releasing the elastic energy (Figure 12.26c); thus, the dislocation returns to the original position.

Since this model implies the long-range travel of C atoms at the edge of the EHP over many lattice constants, it may appear unacceptable from the viewpoint of dislocation theories [30]. However, we do not consider it to be impossible because of the exceptionally weak interlayer interaction in layered crystals such as

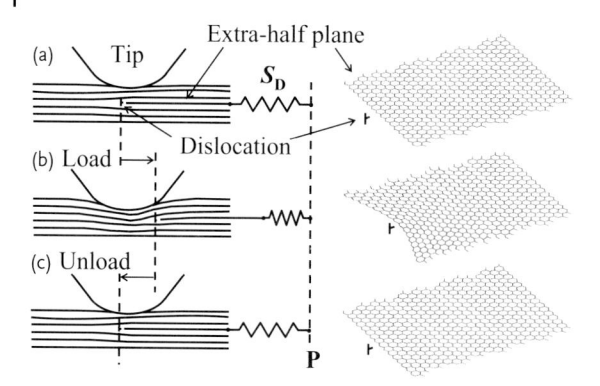

Figure 12.26 Model for the reversible long-range motion of edge dislocation under applied load.

HOPG and the large local stress around the indented tip in UAFM. If this model proves to be correct, it would provide a new concept in dislocation theories. Moreover, since the dislocation motion is well controlled, it suggests the possibility of manipulating subsurface atoms by scanning probe microscopy.

Finally, we note that the surprisingly long range reversible motion of dislocation in graphite during load/unload process may provide a useful insight in the design and diagnosis of graphene devices, being developed as future electronic and MEMS devices.

12.6.6
Delamination in Microelectronic and Mechanical Devices

Though cracks and dislocations are different objects, there is a similarity between a particular type of crack and an edge dislocation. Since the width of the gap near the core of an edge dislocation decreases as the distance from the core increases, the gap has a shape of wedge. On the other hand, a crack with one side partially open and the other side completely closed accompanies a wedge shaped vacancy. Therefore, the edge dislocation and the crack have a similar shape, although their dimension is very different. Consequently, the edge dislocation may be regarded as a model of a crack on the nanoscale.

In this context, we note that the softening spring presented in this study is similar to a closed crack that is opened by the tensile stress operating in a half period of a longitudinal acoustic wave. In contrast, the stiffening spring model is similar to an open crack that is closed by the compressive stress operating in the other half period of a longitudinal acoustic wave. In addition, evaluation of the mechanical properties of subsurface dislocation by using UAFM has some analogy with the evaluation of cracks by using nonlinear vibration of an object that contains micro-cracks.

Although the nonlinear vibration has been observed on macro- and micro-cracks, this is the first time that analogous nonlinear behavior was observed on an atomic-scale discontinuity. This finding will provide some insight into the

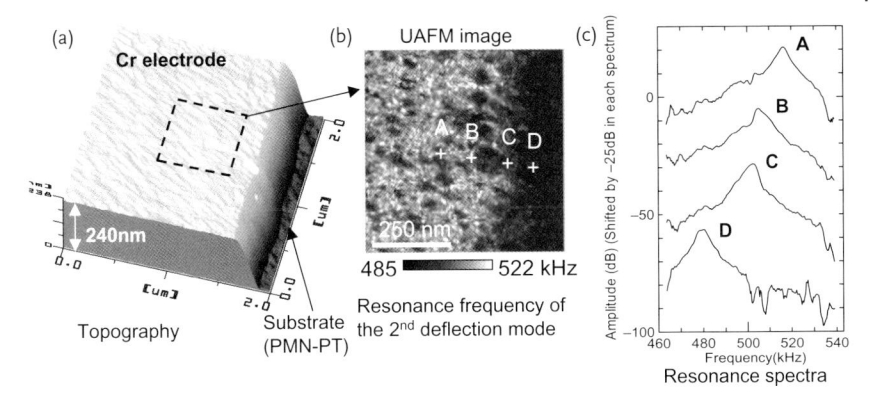

Figure 12.27 Observation of subsurface delamination of Cr electrode: (a) Cr electrode deposited on a substrate 240 nm thick; (b) UAFM resonance frequency image in second deflection mode; (c) spectra measured at positions A to D.

physics and mechanics of nanoscale structures that will be extensively developed in the near future.

A more practical example of nonlinear spectra is found in microelectronic devices [22, 24]. A chromium (Cr) electrode was fabricated on a lead magnesium niobate–lead titanate [0.65 Pb(Mg$_{1/3}$Nb$_{2/3}$)O$_3$–0.35 PbTiO$_3$, PMN-PT] substrate using the lift-off process. Figure 12.27 was obtained using a cantilever with $k_c = 5.0$ N m^{-1} and $f_0 = 36.1$ kHz. The topography of an area on the edge of an electrode shows a thickness of 240 nm (Figure 12.27a). In the UAFM resonance frequency image shown in Figure 12.27b, the darker area had lower resonance frequency, indicating lower contact stiffness. The low frequency region was probably due to delamination. To confirm this, spectra were measured at positions A, B, C, and D in Figure 12.27b. The peak frequency decreased from A to D (Figure 12.27c), indicating a decrease in contact stiffness. Moreover, the asymmetric shapes of spectra B and C indicate the contact vibration of the gap, typically predicted by the calculated spectra of Figure 12.7.

12.7 Conclusion

The concept of UAFM is as follows: the principle is effective enhancement of cantilever stiffness using higher deflection mode. The forced vibration of the cantilever from the base overcomes the disadvantage due to bonding of the vibrator to the sample. A small excitation power is important for quantitative measurement. Not only vertical but also lateral stiffness can be evaluated by simultaneous measurement of the deflection and torsional vibrations. The resonance frequency tracking scheme significantly reduces the time for mapping the resonance frequency. For reliable measurement, an appropriate load should be set, taking plastic deformation of the sample and tip into account.

Linear analysis of the stiffness and Q factor has been shown theoretically using the continuum theory of the cantilever. The peak height of the spectrum is approximately linear with Q factor. The frequency equation has been derived, considering lateral stiffness and the oblique sample surface. The linear theory of subsurface imaging was then developed by an FEM analysis. A subsurface defect in objects, such as dislocation, delamination, or crack, is detected by a resonance frequency lower than that of defect-free area. Moreover, it is indicated that there is an appropriate load range for imaging a subsurface defect layer. However, the nonlinearity may be useful for the evaluation of subsurface closed cracks or delaminations. The origin of the stiffening and softening springs due to the subsurface gap has been illustrated by "subsurface tapping mode" and "subsurface pull off," respectively.

Experimentally, it was shown that the vibration spectra of UAFM and related techniques are strongly dependent on the excitation power of the cantilever vibration. The resonance peak width decreases and the peak frequency increases as the excitation power is reduced while the power is above a certain threshold level. Controlling the excitation power, we obtained linear spectra independent of the excitation power. Using the linear spectra, satisfactory agreement between the measured and calculated peak frequency was obtained by assuming a consistent tip–sample contact stiffness. Further improvement by taking into account the non-spherical tip shape can be attained using the Sneddon–Maugis formulation of contact stiffness [11] and the tip shape index estimation by inverse analyses of load-frequency relation [13].

We observed subsurface dislocations in layered crystals using the amplitude image and the resonance frequency image. Upon close observation of the dislocation behavior, we found that the dislocation in HOPG moved laterally by more than 40 nm as the load increased by 285 nN, and that it returned to the original position as the load decreased. This behavior was caused not by lateral force due to scanning but by the applied normal load. We then proposed a possible model to explain this motion. Though the dislocation was observed with a width of several tens nm, it was caused by the subsurface structure accompanied by edge dislocation due to the shape variation of the resonance spectrum. Finally, this measurement was applied to the evaluation of an electrode of a microelectronic device and its validity was shown.

References

1 Binnig, G., Quate, C.F., and Gerber, C. (1986) *Phys. Rev. Lett.*, **12**, 930.
2 Kolosov, O. and Yamanaka, K. (1993) *Jpn. J. Appl. Phys.*, **32**, L1095.
3 Yamanaka, K., Ogiso, H., and Kolosov, O. (1994) *Appl. Phys. Lett.*, **64**, 178.
4 Yamanaka, K. (1996) *Thin Solid Films*, **273**, 116.
5 Rabe, U. and Arnold, W. (1994) *Ann. Phys.*, **3**, 589.
6 Rabe, U., Janser, K., and Arnold, W. (1996) *Rev. Sci. Instrum.*, **67**, 3281.
7 Radmacher, M., Tillmann, R.W., and Gaub, H.E. (1993) *Biophys. J.*, **64**, 735.
8 Yamanaka, K. and Tomita, E. (1995) *Jpn. J. Appl. Phys.*, **34**, 2879.

9 Yamanaka, K. and Nakano, S. (1996) *Jpn. J. Appl. Phys.*, **35**, 3787.
10 Yamanaka, K. and Nakano, S. (1998) *Appl. Phys. A*, **66**, 313.
11 Yamanaka, K., Noguchi, A., Tsuji, T., Koike, T., and Goto, T. (1999) *Surf. Interface Anal.*, **27**, 600.
12 Yamanaka, K. (1999) U.S. Patent 6,006,593.
13 Yamanaka, K., Tsuji, T., Noguchi, A., Koike, T., and Mihara, T. (2000) *Rev. Sci. Instrum.*, **71**, 2403.
14 Tsuji, T. and Yamanaka, K. (2001) *Nanotechnology*, **12**, 301.
15 Yamanaka, K., Maruyama, Y., Tsuji, T., and Nakamoto, K. (2001) *Appl. Phys. Lett.*, **78**, 1939.
16 Tsuji, T., Irihama, H., and Yamanaka, K. (2002) *Jpn. J. Appl. Phys.*, **41**, 832.
17 Tsuji, T., Irihama, H., and Yamanaka, K. (2002) *JSME Int. J. Ser. A*, **45**, 561.
18 Yamanaka, K., Tsuji, T., Irihama, H., and Mihara, T. (2003) *Proc. SPIE*, **5045**, 104.
19 Yamanaka, K., Mihara, T., and Tsuji, T. (2004) *Jpn. J. Appl. Phys.*, **43**, 3082.
20 Tsuji, T., Ogiso, H., Akedo, J., Saito, S., Fukuda, K., and Yamanaka, K. (2004) *Jpn. J. Appl. Phys.*, **43**, 2907.
21 Tsuji, T., Saito, S., Fukuda, K., Yamanaka, K., Ogiso, H., Akedo, J., and Kawakami, K. (2005) *Appl. Phys. Lett.*, **87**, 071909.
22 Tsuji, T., Kobari, K., Ide, S., and Yamanaka, K. (2007) *Rev. Sci. Instrum.*, **78**, 103703.185.
23 Ide, S., Kobari, K., Tsuji, T., and Yamanaka, K. (2007) *Jpn. J. Appl. Phys.*, **46**, 4446.
24 Yamanaka, K., Kobari, K., and Tsuji, T. (2008) *Jpn. J. Appl. Phys.*, **47**, 6070.
25 Wright, O. and Nishiguchi, N. (1997) *Appl. Phys. Lett.*, **71**, 626.
26 Rabe, U., Turner, J., and Arnold, W. (1998) *Appl. Phys. A*, **66**, S277.
27 Yamanaka, K., Ohara, Y., Oguma, M., and Shintaku, Y. (2011) *Appl. Phys. Express*, **4**, 076601.
28 Kobayashi, K., Yamada, H., and Matsushige, K. (2002) *Surf. Interface Anal.*, **33**, 89.
29 Zabel, H. and Solin, S.A. (1990) *Graphite Intercalation Compounds I: Structure and Vibrations* (eds H. Zabel and S.A. Solin), Springer, Berlin, p. 1.
30 Read, W.T. (1953) *Dislocations in Crystals*, McGraw-Hill, New York.

13
Acoustical Near-Field Imaging

Walter Arnold

13.1
Principle of Near-Field Imaging

13.1.1
Early Systems of Acoustical Near-Field Imaging

Scanning acoustic microscopy (SAM) is discussed in detail by authors of the articles in Chapter 3 of this book. Because in an SAM a focusing lens is used for imaging, its resolution is directly related to the used wavelength. The practical limit for the shortest wavelength employable ($\lambda \approx 0.75\,\mu m$) at room temperature corresponds to a frequency of 2 GHz. The main limitation comes from the excessive attenuation in the coupling water. Furthermore, scattering and attenuation in the material of the component under test rarely allow one to obtain images from a depth of more than a few wavelengths in the GHz frequency range. Therefore, both resolution and penetration depth is limited to the lower μm range.

The resolution limit may be overcome by using the concept of "super resolution" or near-field imaging. As the name indicates, the contrast originates in the near-field of the antenna employed. Near-field imaging is actually quite an old technique, also in acoustical systems. One of the first experiments using the concept of near-field imaging was performed by Ash and Nichols [1] who used the electromagnetic leakage-field emanating from a diaphragm with a small aperture in a 10 GHz microwave cavity. In the near-field of the aperture of 1.5 mm diameter, which served as an antenna, the spatial resolution was $\lambda/60$ when scanning the resonator over the surface of the material or component to be tested.

One of the first near-field acoustical systems for non-destructive testing in the aeronautical industry is the so-called Fokker bond tester or mechanical impedance spectrometer [2–4]. Such a system possesses an exciting oscillator and a receiving oscillator. When contacting a component (Figure 13.1), one measures with a Fokker bond tester the shift in resonance frequencies of the oscillator that depend on changes of the adhesive properties of layers within the stress-field of the contactor, usually a sphere attached to the exciter. The response can be calibrated in

Advances in Acoustic Microscopy and High Resolution Imaging: From Principles to Applictaions, First Edition.
Edited by Roman Gr. Maev.
© 2013 Wiley-VCH Verlag GmbH & Co. KGaA. Published 2013 by Wiley-VCH Verlag GmbH & Co. KGaA.

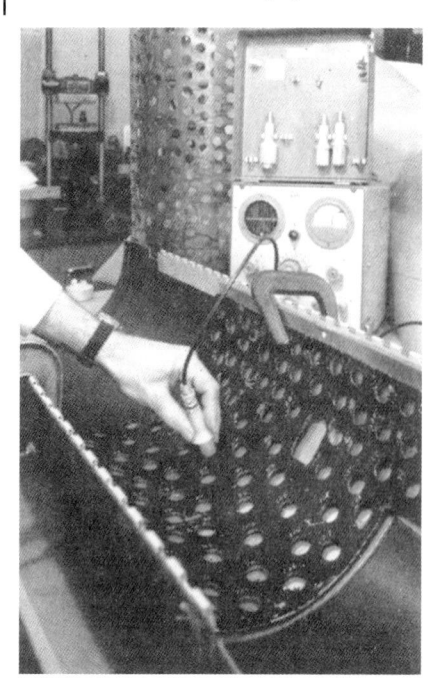

Figure 13.1 Fokker bond test applied to test the adhesion of a missile structure by measuring contact resonances. A contact-resonance near-field image was obtained manually at the holes in the template. Taken from Reference [2], with permission from the American Society of Non-Destructive Testing.

terms of adhesive properties, which has been discussed in detail in Reference [5]. The quantity that is measured is the local flexibility [3]. The inverse of this quantity is the contact stiffness, which is discussed later in the context of AFM cantilever contact resonances. Combined with a scanning system, the Fokker bond tester becomes a near-field acoustic imaging system whose spatial resolution is given by the contact radius of the contacting sphere. In the Fokker bond tester, the exciter oscillates in the kHz range and the corresponding wavelengths are in the decimeter range. Contact radii are in the millimeter range, hence we have a resolution of $\lambda/100$, which is much smaller than the wavelength employed [3, 4], that is, we have "super-resolution."

Combinations of an aperture with an acoustic lens were designed by Dürr et al. [6] to obtain super-resolution (here $\lambda/4$). Khuri-Yakub et al. [7] used a similar concept. A scanning near-field acoustic microscope using a horn or a pin was built by Zieniuk and Latuszek [8] with a resolution of 10 μm at 35 MHz. In their set-up, the sample under test was insonified with plane waves from below and the receiving antenna was brought in close proximity to the sample. The resolution is again determined by the diameter of the contacting horn; see schematic design in Figure 13.2a in comparison to a focusing SAM (Figure 13.2b).

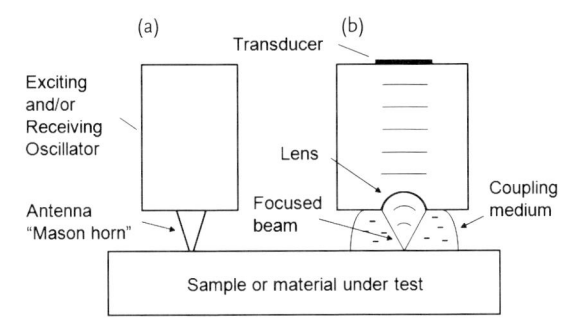

Figure 13.2 Principle of near-field imaging in comparison to scanning acoustic microscopy. In near-field imaging the size of the antenna in contact with, or in proximity to, the sample determines the spatial resolution (a) whereas in SAM the resolution is given by the size of the focused beam, which is determined by the wavelength employed divided by the numerical aperture of the lens (b).

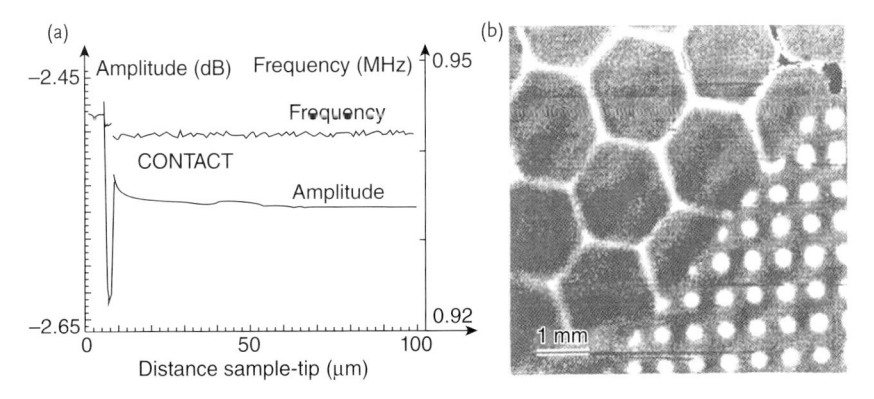

Figure 13.3 (a) Contact-resonance s_{11} signal obtained with a Mason horn (see Figure 13.2a, left-hand side) 100 µm in diameter at its end. There is a change of resonance as well of the amplitude when approaching the examined structure. (b) Image obtained with a Mason horn operated at 1 MHz. The test sample shows the structures etched in a copper layer adhering to the glass-fiber reinforced epoxy material. In addition to a clear contrast due to the structures, some of the subsurface fibers become visible as well (dark diagonal stripes). Taken from Reference [9], with permission.

Kulik *et al.* [9] employed a continuous wave system where the reflection amplitude of a CW signal at a frequency of 1 MHz propagating in a Mason horn [10] and its contact resonances were used as a contrast quantity in the imaging system (Figure 13.3a). The horn had a tapered diameter of 100 µm. Acoustical images were obtained with the resolution of the horn diameter (Figure 13.3b), that is, super-resolution with $\lambda/d \approx 50$.

A different set-up was built by Güthner *et al.* [11]. They used a vibrating tuning-fork positioned near the sample surface. The sharp corner of the tuning fork acted as a tip. The frictional properties of the moving air between tip and surface determined the contrast. The spatial resolution was better than 3 µm.

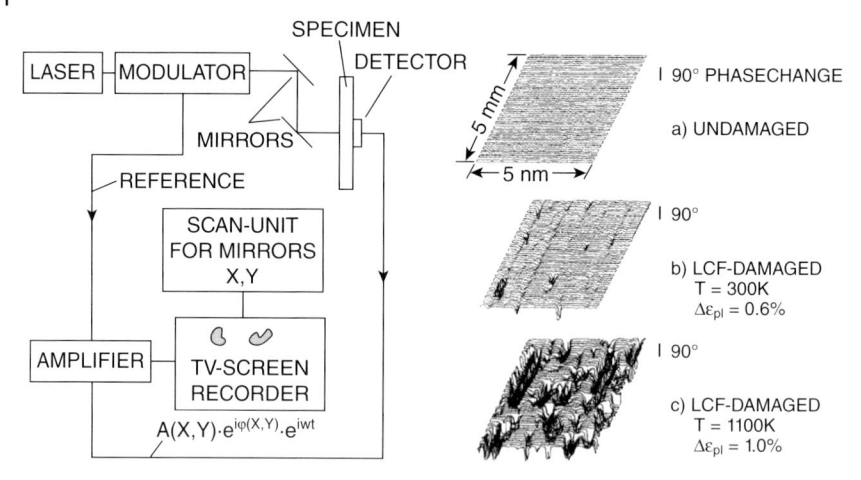

Figure 13.4 An amplitude-modulated laser beam generates elastic waves by the thermoelastic effect in the sample under test, which was in this case an iron-based superalloy (Incoloy 800 H) with various degrees of damage, that is, cracks caused by low-cycle fatigue. The thermal penetration of the heat source was 16 μm, corresponding to the modulation frequency of $\omega/2\pi = 7.4$ kHz and an elastic wavelength in the material ≈ 0.67 m. Because the cracks were in the extreme near-field of the thermoelastic source, elastic scattering at the crack flanks changes the signal picked by an accelerometer which served as a detector. The width w of the indication is about twice the crack depth d. Taken from Reference [12], with permission from Springer.

In a photoacoustic imaging system near-field effects have also been observed. The contrast was due to the scattering of a long-wavelength ultrasonic field in the kHz frequency range at cracks, allowing one to measure the crack depth (Figure 13.4). Both the generation of the ultrasonic field by a modulated laser beam and its detection using an accelerometer was spot-like in comparison to the wavelength employed, thus providing the near-field resolution [12].

Finally, Khuri-Yakub et al. [13] and Takata [14] combined scanning tunneling microscopy (STM) and acoustical technology by using the STM tips as an acoustical antenna and in this way achieved a resolution below 0.1 μm. Here, due to an extremely short distance between the tip and the surface, short-range interactions were sufficiently strong to excite a detectable acoustical signal in transmission. Uozumi and Yamamuro [15] propagated pulsed ultrasound of 1.4 MHz in an STM tip of 100 nm radius and of 1 mm diameter. Again the tunneling current was used as an imaging quantity.

13.2
Near-Field Acoustical Imaging and Atomic Force Microscopy

In atomic force microscopy (AFM), a microfabricated elastic beam with a sensor tip at its end is scanned over the sample surfaces [16]. Dynamic modes, where the cantilever or the sample surfaces are vibrated, belong to the standard equipment of most commercial instruments. With a variety of these techniques, such as force modulation microscopy, scanning local acceleration microscopy, scanning microdeformation microscopy, pulsed-force microscopy, tapping mode, or intermittent contact modes, images can be obtained whose contrasts depend on the elasticity, friction, and adhesion of the tip–sample contact.

13.2.1
Force Modulation

As soon as the atomic force microscope was invented, attempts were made to use it for elasticity measurements. Force modulation in atomic force microscopy is a technique by which to measure tip–surface interactions, which in turn are determined by local elastic restoring forces, local frictional forces, and local adhesion between a tip and the surface under inspection. The tip or sample is oscillated at a given frequency and pushed into the repulsive regime. Data on local forces can be acquired along with data on the topography, which allows a comparison of both height and material properties. In the force modulation technique [17], the z-position of the sample is modulated with amplitudes Δz_m ranging from sub nm to some 10 nm (Figure 13.5).

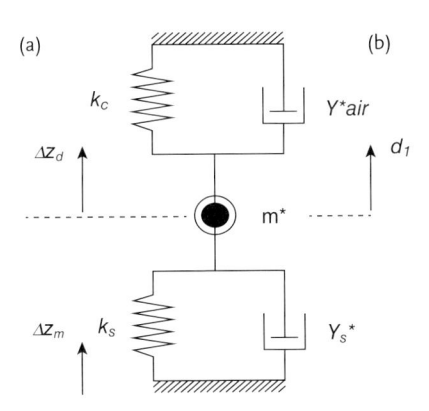

Figure 13.5 Principle of force modulation (a) and scanning local acceleration microscopy (b). In force modulation, the z-piezo of the AFM instrument is modulated by an amplitude Δz_m. The cantilever follows with an amplitude Δz_d that depends on the local stiffness k_s of the material and on the stiffness of the cantilever, k_c (Equation 13.1). The dashpot represents viscoelasticity and adhesion in the contact, as well as damping of the cantilever in air; (b) scanning local acceleration microscopy (SLAM) [18] is identical with force modulation for frequencies below the cantilever contact-resonance frequency. For frequencies above the resonance frequency, there is a still an amplitude response of the cantilever, which is given by the accelerated mass (Equation 13.4), that corresponds to the effective mass m^* of the cantilever, which is $m^* \approx m/4$ where m is the mass of the cantilever. Burnham et al. [18] also discuss the cases when either the suspension of the cantilever or the cantilever itself is excited by a dynamic force inducing a corresponding cantilever amplitude.

Therefore, at a constant static load applied to the cantilever, the total force exerted on the surface of the sample is modulated as well and the ensuing cantilever amplitude Δz_d depends on the local contact stiffness k_s:

$$\frac{\Delta z_d}{\Delta z_m} = \frac{k_s}{k_s + k_c} \tag{13.1}$$

where k_c is the stiffness of the cantilever. One measures the corresponding vertical photodiode output signal ΔV_m as well as the signal ΔV_d at a given static load without modulation. Using Hertzian contact mechanics for the tip–sample contact and a calibration sample of given indentation modulus M_s, it is possible to calibrate the photodiode output signal to obtain the factor R [17]:

$$M_s = \frac{\sqrt{2}k_c}{a_c[R(\Delta V_m/\Delta V_d) - 1]} \tag{13.2}$$

Here, a_c is the contact radius, also called sample depression radius. If the experimental parameters are set such that larger depressions occur, the modulation force develops a signal at the double frequency due to the nonlinearity of a Hertzian contact [19]. Furthermore, the adhesion in the contact, as well as the viscoelasticity of the examined material, lead to a phase shift between the force excitation signal and the output signal of the AFM photodiode.

13.2.2
Local Acceleration Microscopy

From a mechanical engineering point of view, the heart of an AFM consists of the cantilever, and a tip in contact with the sample. Provided that the suspension of the cantilever as well as the scanning unit is infinitely stiff, the oscillations of these AFM components may be described in a simplified way. If one assigns to the cantilever an effective mass m^*, one can describe the oscillatory behavior of the whole system with the replacement model shown in Figure 13.5. Neglecting the damping effects, at frequencies small compared to the resonance frequency of the system in contact with a sample, the amplitude d_1 of the cantilever oscillation is given by [18]:

$$\frac{d_1}{\Delta z_m} = \frac{k_s}{k_s + k_c} \tag{13.3}$$

where Δz_m is again the excitation amplitude of the sample. Equation (13.3) is equivalent to Equation (13.1) and it leads after calibration, using the Hertzian contact theory again, to Equation (13.2), that is, it encompasses the force modulation. Regarding the high-frequency limit above, the contact resonance of the system leads to:

$$\frac{d_1}{z_m} = \frac{k_s}{m^*\omega^2} \tag{13.4}$$

This means that the amplitude of the cantilever is determined by the accelerated local mass; hence the name scanning local acceleration microscopy (SLAM) was coined. This method has no relation to the so-called scanning laser acoustic microscope whose acronym is also SLAM [20]. There are several related AFM techniques based on force modulation where instead of a cantilever with its indenting tip, a capacitive load-displacement indenter is used, allowing one to perform, besides force modulation, nanoindentation at higher loads than the loads possible with an AFM [21].

13.2.3
Pulsed-Force Microscopy

In the pulsed-force microscopy the force–distance curve is rapidly scanned through point by point at the surface and parameters are determined that depend on elasticity, adhesion, and other surface properties [22]. Peak forces are detected in special commercial systems [23]. In principle, this can be done with any force acting between tip and sample, for example, piezoelectricity or magnetic forces.

In all these techniques, quantitative determination of the Young's modulus of a sample surface with an AFM is a challenge. Especially when stiff materials such as metals or ceramics are encountered, the image contrast due to elasticity becomes very low, because the spring constants of common AFM cantilevers, ranging from 0.01 to $70\,\mathrm{N\,m^{-1}}$, are then much lower than the tip–sample contact stiffness. The exploitation of cantilevers' higher modes for imaging can be used to image stiff materials as discussed in the following section.

13.2.4
Atomic Force Acoustic Microscopy or AFM Contact-Resonance Imaging

13.2.4.1 Principle of Operation
In atomic force acoustic microscopy (AFAM) or in ultrasonic atomic force microscopy (UAFM) one measures the resonances of atomic force cantilevers with the tip contacting the specimen surface [24–26]. From such measurements one can derive the local indentation modulus M, using a suitable mechanical model relating the resonance frequencies of the oscillating cantilever to the tip–sample contact stiffness k^*, which in turn is related to the local indentation modulus M. The indentation modulus is an elastic constant that accounts for the compressive and the shear deformations in the contact zone between isotropic or anisotropic materials. Though originally used for nanoindentation [27], it has been shown that M is the elastic quantity that can be derived from contact-resonance techniques such as AFAM, UAFM, and scanning microdeformation microscopy [25, 26, 28] resonance ultrasound microscopy (RUM) [29] and related techniques. All these techniques emerged as tools for the measurement of elastic properties of the near-surface zone of materials with micrometer (RUM) or nanometer resolution (AFAM and UAFM).

13.2.4.2 Flexural Cantilever Resonances

The cantilever resonances can be excited either by transmitting ultrasound through the sample (AFAM technique [30]) or the component under test or by exciting the cantilever base (UAFM technique [31]), or by exciting the cantilever itself [32]. All these techniques have advantages and disadvantages and are discussed in detail elsewhere. In AFAM, the cantilever with its tip plays the role of the horn in the impedance spectroscopy technique, as discussed above, and the tip–sample contact serves to probe the local mechanical impedance [33].

Vibrations of beams are treated in many textbooks of acoustics and engineering mechanics. Some of them date back to the 19th century (Tyndall, 1869 [34]). With regard to applications in AFAM and ultrasonic force microscopy, I refer to Rabe et al. [35], Yamanaka et al. [36], Dupas et al. [37], Turner and Hurley [38], Rabe [39], Huey [40], and Song and Bhushan [41]. There are quite a few others as well.

For a beam of uniform cross-section, the equation of motion for flexural vibrations is a fourth-order differential equation. For a forced rectangular beam the following equation of motion holds:

$$EI\frac{\partial^4 y}{\partial x^4} + \eta_{air}\frac{\partial y}{\partial t} + \rho A \frac{\partial^2 y}{\partial t^2} = F\delta(x - x_0)e^{i\omega t} \tag{13.5}$$

where E is the Young's modulus of the beam material, I is the cantilever area moment of inertia, η_{air} is the damping constant for frictional losses in air, ρ is the beam mass-density, A is its cross section, and $y(x,t)$ is the vertical displacement of the beam. The position along the beam is x and t is the time. F is the force that acts on the tip at the position x_0.

The interaction forces normal to the sample surface are represented as a linear spring with a characteristic spring constant k^*, which is the negative derivative of the tip–sample force in the equilibrium position:

$$k^* = -\frac{\partial F(z)}{\partial z}\bigg|_{z=z_e} \tag{13.6}$$

where z is the tip–sample distance, $F(z)$ is the tip–sample interaction force, and z_e is the equilibrium position. The lateral contact stiffness k^*_{Lat} is defined analogously. The ensuing mechanical model of a vibrating cantilever in AFAM is presented in Figure 13.6. The superposition of vertical and lateral tip–sample interaction forces may be represented by a set of springs and dashpots.

Taking into account the boundary conditions for a free cantilever, one obtains the characteristic equation for which the solutions $k_n L$ (n = 1, 2, 3. . . .) yield the wave numbers k_n of an infinite set of flexural modes. The dispersion relations are used to calculate the resonance frequency f_n of the n-th mode [35]:

$$f_n = \left(\frac{k_n L}{c_c}\right)^2 \text{ with } c_c = L\sqrt{2\pi}\sqrt[4]{\frac{\rho A}{EI}} \tag{13.7}$$

where c_c is a constant that depends on the geometry of the cantilever and on the elastic modulus of the beam in length direction. The boundary conditions

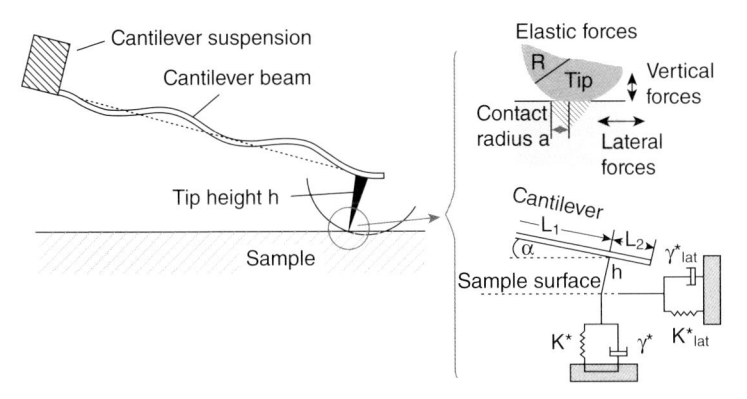

13.2 Near-Field Acoustical Imaging and Atomic Force Microscopy

Figure 13.6 Mechanical model of the AFM cantilever as a rectangular elastic beam. The contact itself is described as two parallel linear springs with spring constants or contact stiffnesses k^* and k^*_{lat}, and two damping dashpots γ^* and γ^*_{lat} for both vertical and lateral contact forces. In the figure $L = L_1 + L_2$ is the cantilever length, L_2 is the distance between the tip position and the free end of the cantilever, h is the length of the tip, and k^* and k^*_{lat} and γ and γ^*_{lat} are the contact stiffnesses and the damping constants for vertical and lateral forces, respectively. Owing to technical reasons the cantilever is inclined by an angle α.

change when the tip is in contact with the sample surface. When approaching the sample surface, the tip first senses long-range attractive forces before contact and when in contact repulsive tip–sample forces. If the static load $F_o = k_c \times d_c$, applied to the tip by a cantilever deflection d_c, is large enough, attractive forces can be neglected.

The contribution of the imaginary part to the total contact stiffness is usually smaller than 1%. Therefore, in many applications of AFAM, the dashpots γ^* and γ^*_{lat} were neglected in the calculation of the contact stiffness [24, 42]. For a spring-coupled cantilever with the ratio $k^*/k^*_{lat} = c_k$, one obtains the characteristic equation:

$$A_0 c_P \left(\frac{h}{L_1}\right)^2 \left(3\frac{k^*}{k_c}\right)^2 + 3A_1 \frac{k^*}{k_c}(\cos^2\alpha + c_k \sin^2\alpha) + 3A_2 \frac{h}{L_1}\frac{k^*}{k_c}\sin\alpha \times \cos\alpha(c_k - 1)$$
$$+ 3A_3 \left(\frac{h}{L_1}\right)^2 \frac{k^*}{k_c}(\sin^2\alpha + c_k \cos^2\alpha) + A_4 = 0 \qquad (13.8)$$

In this equation the parameters A_0, A_1, A_2, A_3, and A_4 are defined as follows:

$$A_0 = (1 - \cos k_n L_1 \cosh k_n L_1)(1 + \cos k_n L_2 \cosh k_n L_2) \qquad (13.9a)$$

$$A_1 = k_n L_1 [-(1 - \cos k_n L_1 \cosh k_n L_1)(\sin k_n L_2 \cosh k_n L_2 - \sinh k_n L_2 \cos k_n L_2)$$
$$+ (1 + \cos k_n L_2 \cosh k_n L_2)(\sin k_n L_1 \cosh k_n L_1 - \sinh k_n L_1 \cos k_n L_1)] \qquad (13.9b)$$

$$A_2 = 2(k_n L_1)^2 [\sin k_n L_1 \sinh k_n L_1 (1 + \cos k_n L_2 \cosh k_n L_2)$$
$$+ \sin k_n L_2 \sinh k_n L_2 (1 - \cos k_n L_1 \cosh k_n L_1)] \qquad (13.9c)$$

$$A_3 = (k_n L_1)^3 [(\sin k_n L_1 \cosh k_n L_1 + \sinh k_n L_1 \cos k_n L_1)(1 + \cos k_n L_2 \cosh k_n L_2) \\ - (\sin k_n L_2 \cosh k_n L_2 + \sinh k_n L_2 \cos k_n L_2)(1 - \cos k_n L_1 \cosh k_n L_1)] \quad (13.9d)$$

$$A_4 = 2(k_n L_1)^4 (1 + \cos k_n L \cosh k_n L) \quad (13.9e)$$

The spring constant k_c of the cantilever is given by:

$$k_c = \frac{Ewt^3}{4L_1^3} \quad (13.10)$$

assuming a cantilever with rectangular cross-section of width w and thickness t.

The normalized wave number $k_n L$ of each contact resonance is related to its experimentally measured frequency f_{cont} by the dispersion relation of Equation (13.7). One obtains for k^* [24]:

$$k^* = \frac{k_c}{3}\left(z \pm \sqrt{z^2 - \frac{A_4}{A_0 c_p}\left(\frac{L_1}{h}\right)^2}\right) \quad (13.11)$$

where z is given by:

$$z = -\frac{A_1(\cos^2\alpha + c_p \sin^2\alpha) + A_2 \dfrac{h}{L_1}(c_p - 1)\sin\alpha\cos\alpha + A_3\left(\dfrac{h}{L_1}\right)^2(\sin^2\alpha + c_p \cos^2\alpha)}{2A_0 c_p \left(\dfrac{h}{L_1}\right)^2} \quad (13.12)$$

Let us briefly discuss how a quantitative data evaluation is performed. The above set of equations can be solved in several ways. One may use a MATLAB program to obtain the dispersion curves as shown in Figure 13.7 [43] or a Labview program [39]. First it is important to estimate the expected contact stiffness of the unknown material with the resolution wanted (i.e., the contact radius), using the equations below. Then, one measures the free resonance of the cantilever and the contact resonances on the material to be measured. Taking the ratio of these two values, one determines which point on the dispersion curve is used to obtain the ratio of the contact stiffness k^* to the cantilever stiffness k_c. One should select the parameters k_c, static force F_0, and tip radius R, so as to exploit the region of highest slope of the cantilever dispersion curve of a given mode. Stiffer materials are measured more easily when using a higher mode.

Recently, AFAM has been used to measure the stiffness distribution on the metallic glass PdCuSi caused by its energy landscape. The contact resonances showed a certain distribution of their frequencies absent in the crystallized PdCuSi material or in a crystalline material (Figure 13.7). To convert the contact resonances into a contact stiffness distribution, one needs the dispersion curves for a given cantilever. The parameters used for the calculation of the dispersion curve, shown in Figure 13.8, were $L_1 + L_2 = 137\,\mu m$, $L_2 = 15.1\,\mu m$, $h = 17\,\mu m$, and $\alpha = 11°$. The lateral contact stiffness is determined by the stiffness parallel to the surface of the contact, and hence by the shear modulus, and has been calculated by

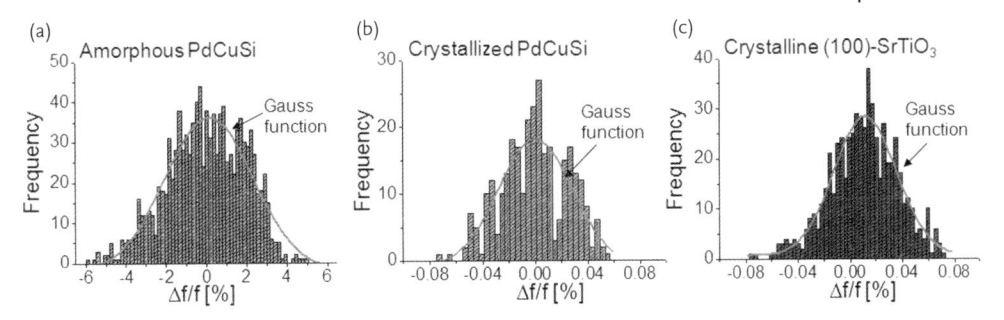

Figure 13.7 Statistics of AFAM data: (a) statistical variation of $\Delta f/f$ for amorphous PdCuSi measured at 1089 different positions. The averaged contact-resonance frequency was 1.313 MHz with a free cantilever resonance at 308 kHz. Fitting with a Gauss function yields a relative width of 4.4% at half-maximum; (b) for crystalline PdCuSi, 300 measurements were made, which were analyzed with a resulting full width at half maximum of 0.06%; (c) for crystalline SrTiO$_3$, where 600 measuring points were taken into account, yielding a relative width of 0.05% at half maximum. Taken from Reference [43], with permission.

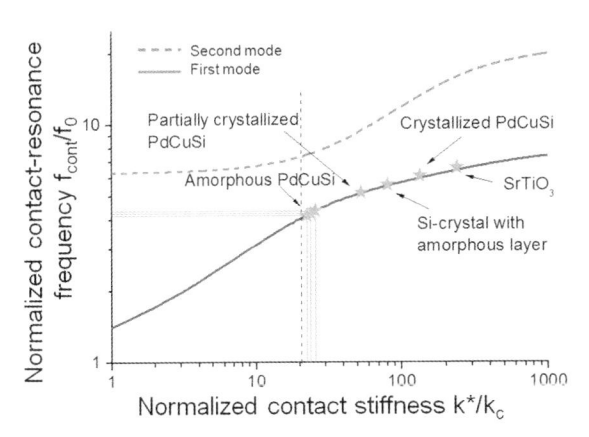

Figure 13.8 Normalized contact-stiffness k^*/k_c as a function of normalized contact-resonance frequency f_{cont}/f_0 for the first two different cantilever modes on amorphous PdCuSi and other materials for comparison. Owing to the change of slope of the dispersion curve, variations in contact resonances may entail a large variation in contact stiffness and consequently in indentation modulus. For frequencies ratios f_{cont}/f_0, which yield k^*/k_c values beyond the dashed line, the second mode becomes more sensitive to changes in contact stiffness than the first mode.

Mazeran and Loubet [44]. The ratio of the lateral contact stiffness to the vertical contact-stiffness was taken as $k^*/k^*_{lat} = c_k = 4G^*/E^*$, where E^* and G^* are the reduced Young modulus and shear modulus of the contact zone, respectively. With a Poisson ratio of 0.4, which is typical for metallic glasses, one obtains $c_k = 0.85$. Knowing then k^* and the contact area or contact radius of the tip–sample contact, one can determine the local indentation modulus as discussed in the next section. An analog distribution of anelastic properties in a ZrCuNiAl metallic glass film was observed by tapping mode AFM [45].

Previously, AFAM was used for the measurement of elastic properties of thin-film nanocrystalline ferrites [46], piezoelectric ceramic materials [47], clay [48], precipitates in polycrystalline metals [42], and elastic properties of carbon nanotubes [49], and size dependent effects in tellurium nanowires [50]. AFAM has also been applied to study the contribution of grain boundaries to the overall elasticity of nanocrystalline materials [24].

13.2.4.3 Relationship of Contact Stiffness to Indentation Modulus

The concepts of contact mechanics is presented in detail in the book of K.L. Johnson which was published in 1985 [51]. As an authoritative reference, it discusses many aspects of contact mechanics, for example, the role of the shape of the contacting bodies, the extent of the stress-fields, viscoelastic, frictional effects, and dynamic loads.

Real Part of the Contact Stiffness Oliver et al. [52] calculated the contact stiffness k^* for the case of an elastic contact between a symmetric indenter of revolution and a flat surface as:

$$k^* = \frac{2}{\sqrt{\pi}} E^* \sqrt{A} \tag{13.13}$$

where A is the contact area and E^* is the reduced modulus of elasticity. For isotropic solids E^* is given by:

$$\frac{1}{E^*} = \frac{1-v_s^2}{E_s} + \frac{1-v_{tip}^2}{E_{tip}} \tag{13.14}$$

Here, E_s and E_{tip} are the Young's modulus of the sample and of the tip, respectively, and v_s and v_{tip} are the corresponding Poisson ratios. In the case of anisotropic solids an indentation modulus M is introduced, which can be calculated from the elastic single-crystal constants in a given direction [25]. If there exists a three- or fourfold rotational symmetry axis perpendicular to the boundary, the contact area is circular. The required symmetry holds for silicon sensor-tips, which are oriented in the (001) crystallographic direction. In this case Equation (13.14) can be replaced by:

$$\frac{1}{E^*} = \frac{1}{M_S} + \frac{1}{M_{tip}} \tag{13.15}$$

where M_S and M_{tip} are, respectively, the indentation moduli of the sample and the tip, respectively [25]. For isotropic bodies the indentation modulus is $M = E/(1-v^2)$, and Equation (13.15) becomes Equation (13.14). Finally, the elastic properties of the sample can be extracted by comparing measurements:

$$E_s^* = E_r^* \left(k_s^* / k_r^* \right)^m \tag{13.16}$$

Here, r and s refer to the reference sample and the unknown sample, respectively, and m describes the tip geometry; for a flat punch $m = 1$ and for a spherical tip $m = 3/2$. For a flat punch $m = 1$ and for a spherical tip $m = 3/2$. When making comparative measurements, for example, on nanocrystalline materials, one may use elastic data from ultrasonic velocity measurements to calculate the mean macroscopic value of M_s as well [24].

As explained above, to obtain absolute AFAM data one has to go through several steps to invert the resonance frequency data into indentation modulus data, which entails corresponding inaccuracies as in all inverse problems. AFAM experiments have been carried out to check the reproducibility of contact-resonance frequencies [25]. In these experiments two different silicon surfaces with (111) and (100) orientations and surface roughness $R_q < 2$ nm were repeatedly contacted. In addition, the static tip force was varied from 420 nN to 1.68 μN consecutively in four steps, and finally different spots on the surface were measured. The reproducibility of the contact-resonance frequencies was better than 0.5% when returning to the same surface and using the same static force. This entailed an inaccuracy of the contact stiffness of at most 3%. Better repeatability of AFAM data was obtained by another group with an inaccuracy of 0.1% in contact-resonance frequencies [53]. Furthermore, it has been shown that by employing one calibration material, assuming that the tip elasticity is known, the absolute accuracy in determining M is about 20%–30% [24, 46] whereas by calibration using two calibration materials [53] one can achieve an inaccuracy less of than 2%. Calibration by global ultrasonic velocity data yielded an absolute error of maximal 10% [42]. AFAM measurements using several cantilever modes increase the accuracy of measurement. Because of redundant information, as they allow one to judge whether the cantilever oscillation behavior can be described by Equations (13.8)–(13.12) [46].

The spatial resolution is given by the contact radius a_c, which depends on the tip radius R_t, the applied static force F_0 acting on the tip, including the forces of adhesion F_0, and the reduced elastic modulus E^* of the contact. In the case of a Hertzian contact with a spherical tip, this value is given by:

$$a_c = \sqrt[3]{3F_0 R_t / 4E^*} \qquad (13.17)$$

This resolution has been tested and exploited in several applications of contact-resonance AFM [24, 42, 47–50, 53–56].

The influence of surface roughness on the contact stiffness is largest if the roughness is periodically, conforming to the tip shape, that is, when the surface waviness is twice the tip radius with *large* amplitudes. Notably, even multi-asperity contacts between tip and surface of granular gold films with grains of 5 nm diameter and having height differences of 10 nm show contact-resonance variations of less than 1.4% [55].

An important aspect and difficult to control parameter in contact-resonance measurements is tip stability during a measurement cycle or when imaging. When inverting contact-resonance data into indentation moduli data, one often notices that the contact stiffness is no longer of the same value at the end of a measurement in comparison to its value in the beginning, although one measures still on the same material. This is usually caused by an increase of contact area because the tip has worn and then a nominally different tip radius R_t has evolved or the tip shape became a punch and then the contact radius is equal the tip radius. Tip wear may result in large elliptical contact area as well. Such a behavior can easily be judged when measuring contact stiffness k^* versus load F_0 in order to see whether Equation (13.17) is still fulfilled [24, 36]. There are various efforts to control this effect but in the long run, only measurements with small tip loads and using higher flexural modes avoid this phenomenon or at least makes it controllable [57]. This is feasible as it has been shown in a number of recent applications of contact-resonances with contact stiffnesses as low as 10 N/m [58, 59].

An AFAM image is shown in Figure 13.9. The first and the second contact-resonance spectra were obtained at every point of the area examined, using a Solver P47H (manufactured by NT-MDT Co., Russia) scanning probe microscope with AFAM capability. The distribution of k^*/k_c values were derived using the contact-resonance spectra and the mechanical model shown in Figure 13.6 with the appropriate dispersion curve. An average value M_{Iso} of the matrix was used to determine M of the $M_{23}C_6$ precipitates in two alloys. As can be seen, the AFAM technique is a tool for determining the modulus of submicron-sized precipitates with a spatial resolution of less than 10 nm in polycrystalline materials.

There are a number of cantilever designs for special applications, for example, cantilevers with an additional concentrated mass on the top of the cantilever position [60] of cantilevers with a simplified dispersion behavior [61]. Furthermore, the matching of free and contact resonance frequencies can be improved by taking into account modified boundary conditions at the suspension of the cantilever [37, 62], however, this leads to a more complex modeling which may only be handled with FEM techniques [63].

Imaginary Part of the Contact Stiffness   It is of much interest to know not only the local contact stiffness and indentation modulus but also the local damping that might be caused by various physical mechanisms. Ultrasonic attenuation studies and internal friction studies have a long tradition and were used to gain information on elementary excitations in solids such as phonons, electrons and their interaction with a time varying stress-field, on dislocation–phonon interactions, on the diffusion of precipitates, on the gap function in superconductors, on tunneling sites in disordered materials, and many other phenomena [64–68]. But until now it was not possible to probe these interactions locally on the scale of the microstructure or the nanostructure of a material. AFAM, UAFM, or in general the AFM contact-resonance techniques would allow study of the local damping in the tip–sample interaction volume if it were possible to determine the local material damping from the cantilever damping values.

There are now studies that relate the Q-value or width of the contact-resonance curve to the local absorption or internal friction values more generally to the local loss modulus E''. Yua et al. [69, 70] worked out in detail the relation between the imaginary part of the contact stiffness and the Q-value of the cantilever of the

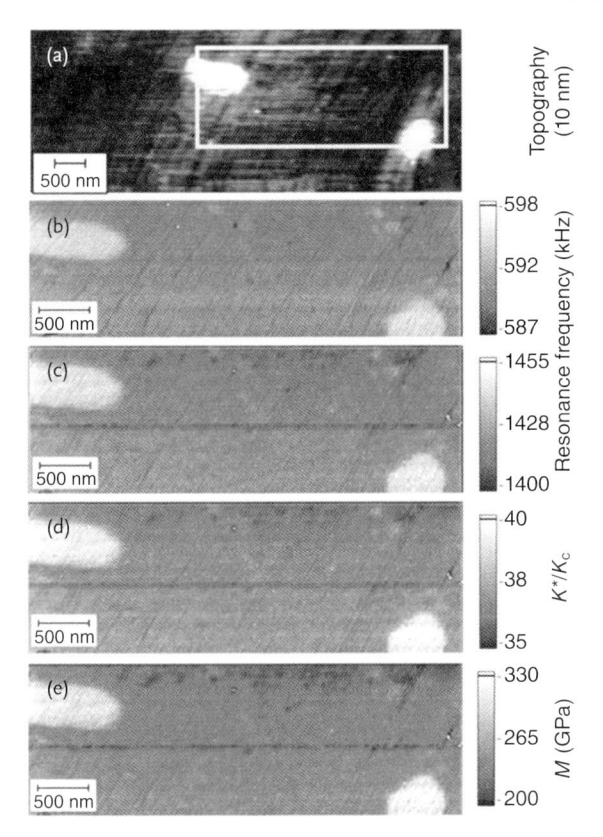

Figure 13.9 Sequence of AFAM or contact-resonance measurement, here on a 9Cr-1Mo ferritic steel specimen containing $M_{23}C_6$ type carbides: (a) the topography image allows one to judge the suitability of the surface area for AFAM measurement. If the topography is too rough, the contact area A of the tip–sample contact will influence the results; (b, c) cantilever contact-resonance frequencies measured for the first (b) and (c) second contact resonance; (d) from the contact resonances, k^*/k_c is determined using the appropriate dispersion curve for the cantilever employed; (e) the indentation modulus M is determined by calibration with a reference sample according to Equation (13.16) to determine the contact area A or contact radius a_c (see Equation (13.17)). The reference sample may be the host material itself and the global values for the elastic moduli for calibration may be determined by ultrasonic time-of-flight data. As can be seen in (e), the distribution of the indentation modulus of the $M_{23}C_6$ type carbide precipitate becomes visible as bright islands having a value of $M = 310 \pm 10$ GPa. The contact radius given in Equation (13.17) determines the spatial resolution. Typical values range from some nm to some 10 nm (data modified from Reference [42], along with additional data, by kind permission of American Institute of Physics).

mode employed. This relation is not straightforward because the viscous force transmitted to the cantilever via the dashpot representing the damping of the material (Figure 13.6) leads to a stiffening of the contact with increasing contact damping [33]. As a result, there is an interdependence between the Q-value of the cantilever measured, the real part of the complex contact stiffness k^*, and the local damping γ^* that also depends on the mode employed [70]. The damping may then be related to a physical mechanism operating in the contact [71]. Furthermore, it is possible using the contact resonances to image the local damping distribution or loss modulus E'' on a surface, which has been shown for a polymer blend [72].

As stated above, the local frequency dependent reduced complex modulus is defined as [69, 70]:

$$E^*(\omega) = E'^* + E''^* \tag{13.18}$$

E'^* and E''^* can be related to the contact stiffness and the contact damping in the following way:

$$k^* = 2E'^*/(\sqrt{\pi}/A_c) \tag{13.19}$$

(equivalent to Equation 13.13) where A_c is the contact area. The imaginary part of the contact stiffness is related to the loss modulus by:

$$\omega\gamma^* = 2E''^*/(\sqrt{\pi}/A_c) \tag{13.20}$$

It is tempting to use the relation for the damping factor in internal friction measurements given by:

$$Q^{-1} = E''/E' \tag{13.21}$$

in the same manner here [71]. Using Equations (13.18) and (13.19) one obtains:

$$Q^{-1} = \omega\gamma^*/k^* \tag{13.22}$$

where:

$$k^*_{complex} = k^* + i\omega\gamma^* \tag{13.23}$$

Using the relations given for the damping β of the cantilever motion [69, 70] one gets:

$$Q^{-1} \approx 1.76\omega\gamma^*/k^* \tag{13.24}$$

very close to Equation (13.22). As stated above, Equations (13.22) and (13.24) are not necessarily valid when measuring the cantilever damping β, and relating it to the contact damping factor because Q^{-1} depends on the mode employed.

It is important to note that the cantilever resonance-curves are Lorentzians both for the free and the contact resonances [69, 70]. Thus, as long as the point-mass model is valid, for $k^*/kc \leq 5$ [35], the individual Q^{-1} can be obtained by employing $Q^{-1}_{total} = \Sigma_i Q_i^{-1}$ and $Q_i^{-1} = \Delta\omega_i/\omega_i$ should hold. For the free cantilever:

$$Q^{-1} = \eta'_{air}/\omega_{res} \tag{13.25}$$

and for the cantilever in contact:

$$Q_{total}^{-1} = (\eta'_{air} + \eta'_{contact})/\omega_{res} \tag{13.26}$$

should apply as well. The damping constants η_{air} and γ^* of the beam model (Equations 13.5 and 13.20) (Figure 13.6) are related to the damping constants of Equations (13.5) and (13.20) and to the damping constants in the point-mass or FMA model in the following way:

$$\eta_{air} = \rho A \times \eta'_{air} \text{ and } \eta_{contact} = \gamma^*/m^* \tag{13.27}$$

with $m^* \approx m/4$, where m is the cantilever mass and m^* its effective mass. The resonance frequency ω_{res} is given by $\omega_{res} = \sqrt{(k+k^*)/m^*}$ [35]. Similarly to the beam model there is a mechanical block diagram for the point mass or FMA model as shown in Figure 13.5.

The above relations were applied to measure local contact damping in nanocrystalline Ni as a function of grain size [71] in order to monitor the influence of homogeneous dislocation generation on the damping factor Q^{-1}. In these experiments, there was a large static load F_0 applied to the cantilever with relatively blunt tips, in order to generate dislocation in the stress-field of the tip, which indeed were observed. There were sudden changes of Q^{-1} observable with increasing load at certain F_0 values caused by homogeneous dislocation generation. To explain the contact damping peaks and the simultaneous reduction of the contact stiffness that were observed on three nc-Ni samples out of five, it has been proposed that plasticity effects play a role in the contact zone because of the high mechanical stresses exerted by the tip [71]. Also of interest is the background of the measured Q^{-1} values; typical data are shown in Table 13.1.

The contact damping values obtained by AFAM may be compared to data obtained by internal friction, respectively ultrasonic measurements carried out in nanocrystalline Ni or other nanocrystalline materials. It is well known that Q^{-1} of nc-materials is almost independent of frequency and is similar for different materials; however, Q^{-1} does depend on the grain size [73]. Using such ultrasonic absorption data at a few MHz and with $Q^{-1} = \alpha\lambda/\pi$, where α is the ultrasonic absorption coefficient and λ the ultrasonic wavelength, one obtains $Q \approx 100$ for nc-materials of comparable grain size. Compared with other data obtained on nc-Ni at room temperatures, however, at much lower frequencies of a few Hz or kHz, values of $Q^{-1} \approx 3.5 \times 10^{-3}$ ($Q \approx 300$) [74, 75] are found. These data are indeed comparable to those in Table 13.1, in particular if one keeps in mind that in the contact damping there is also a contribution to the adhesion tip–sample, which may be neglected in the contact stiffness measurement but not in the damping evaluation.

There are few other quantitative measurements dealing with the origin of the contact-damping values of AFAM or related techniques. Besides the work of Yua et al. [69, 70] and Killgore et al. [72], it was suggested that the Q-value of AFM contact resonances be exploited to perform local internal friction measurements, for example, in polymers, and relate them to the various relaxation mechanisms, both locally and globally [76]. Similarly, Q-mapping of the contact resonances of AFM cantilevers was suggested to image areas of high absorption coefficients [77]

Table 13.1 Q-values of contact resonances at a frequency of ≈ 0.85 MHz and damping values for cantilevers of $232 \times 38 \times 6.8\,\mu\text{m}^3$ $[(L_1 + L_2) \times b \times a]$ with mass $m = 1.4 \times 10^{-10}$ kg; the static force was $F_o = 4510$ nN. The number attached to nc-Ni designates the time tempered in min at 200 °C in order to obtain a certain grain size from the starting material with a grain size of 14 nm. The Q-values for the free cantilever were measured separately. For the free cantilever $Q \approx 600$. Table in modified form taken from Reference [71], with permission from Elsevier.

	Cantilever 1 contacting quartz glass	Cantilever 2 contacting a Ni(111) crystal	Cantilever 3 contacting nc-Ni90, grain size 67 nm	Cantilever 4 contacting nc-Ni15, grain size 35 nm
Angular contact-resonance frequency ω_{res} [s^{-1}]	5.26	5.5	5.22	5.26
Q_{total} from contact-resonance curve; $Q_{total} = \omega_{res}/\Delta\omega_{res}$	50	165	102	104
$Q'_{contact}$	54	225	120	128
Damping constant [s^{-1}] $\eta'_{air} + \eta'_{contact}$ and	1.05×10^5	3.3×10^4	5.1×10^4	5.05×10^4
$\eta'_{contact}$ according to Equations (13.26) and (13.25)	9.7×10^4	2.4×10^4	4.4×10^4	4.1×10^4
Damping constant γ^* [Ns/m] and	3.6×10^{-6}	1.4×10^{-6}	2.5×10^{-6}	5.65×10^{-6}
$\eta_{contact}$ [s^{-1}] from $Q^{-1}_{contact}$ using Equation (13.24) and the measured	5.2×10^5	1.7×10^5	3.7×10^5	1.7×10^6
k^* [N/m]	1795	3108	2835	6697

in materials. As stated above, similarly to AFAM, the so-called resonant ultrasound microscopy (RUM) was used to map local Q^{-1} values in dual-phase steels and in CuNbTi composites by Ogi et al. [78]. They suggested that dislocations are at the origin of the local internal friction and that variations in their Peierls stress in the α- and γ-phases of the microstructure cause the contrast variations in the damping scans.

A decrease in contact stiffness and an increase in contact damping have been observed when generating an edge dislocation in graphite at a particular load [54]. Finally, the radiation loss of the cantilever vibration in the contact zone was calculated. It is minimal, corresponding to a damping of $Q^{-1} = 10^{-6}$ [79].

13.2.4.4 Torsional Resonances

Using the torsional cantilever contact resonances, one can obtain data on the shear elasticity [80] as well as on frictional properties, particularly when lubrication layers are present in the contact [81–83]. For torsional vibrations, the cantilever equation of motion is a second-order differential equation not showing the large dispersion effects as the bending cantilever motions. The procedure used to obtain the in-plane or reduced shear modulus from the torsional-contact resonances is very similar as with the flexural mode; however, there is an additional factor that is not easy to circumvent. The tip is relatively compliant for shear motion and this is difficult to take into account – see Scherer *et al.* [80]. The combined use of torsional resonances and flexural resonances allows one to measure the local Poisson ratio [84].

The onset of micro-slip friction was observed in the contact-zone of a tip on bare and lubricated Si surfaces by measuring the increase in the contact damping Q^{-1} of torsional resonances [85]. Atomic stip-slick phenomena have been observed as well [59].

13.2.4.5 Piezo-mode Imaging

The imaging of piezoelectric materials on a nano-scale using the technique called piezo-mode AFM was first presented by Güthner and Dransfeld [86]. An AC voltage at a frequency higher than the cut-off frequency of the feedback loop of the AFM but lower than the first resonance frequency of the cantilever was applied between the tip and an electrode below the sample. This voltage excites a local sample vibration that is detected by the AFM sensor tip in contact with the sample surface (Figure 13.10). The local piezo-activity of piezoelectric polymer films [86] and of PZT films [87] can be imaged simultaneously with the topography based on this technique. The applied voltage generates an additional signal at the second harmonic of the excitation frequency that is related to electrostriction and the permittivity of the piezoelectric material [87]. The second harmonic signal in conjunction with the first harmonic signal can be used to extract the spontaneous polarization from the measured signals [88]. On $BaTiO_3$ ceramics the domain

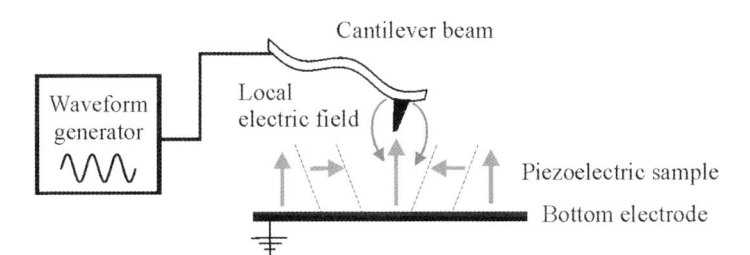

Figure 13.10 Principle of the piezo-mode technique. An AC voltage is applied to a cantilever coated with a conductive layer. This creates an electric field around the tip that interacts with the surface of a piezoelectric sample via the inverse piezoelectric effect. The movement of the surface is translated into vibrations of the cantilever. The amplitude of the cantilever vibrations is detected by the lock-in technique. Resonance amplification is obtained when one of the cantilever-contact resonances is used. Taken from M. Kopycinska-Müller, PhD thesis, Faculty 8, Saarland University, Germany, 2005, unpublished, with permission.

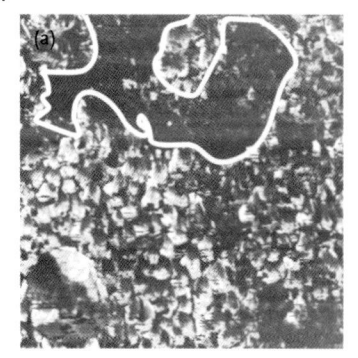

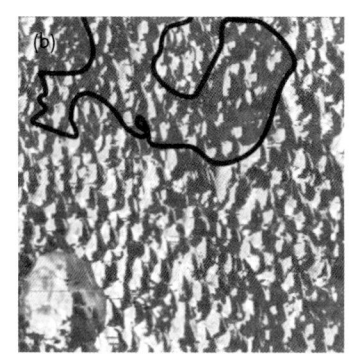

Figure 13.11 Ultrasonic piezo-mode images obtained on a fully crystallized PTC sample annealed at 650 °C. Again, the bright islands separated by the dark areas can be distinguished in the image obtained at the first bending contact mode (a). When the same area was imaged with a torsional mode (b), an additional structure appeared whereas the bending mode showed no contrast. This proves the presence of in-plane-oriented domains in the PTC film annealed at 650 °C. The cantilever used in this experiment was PT-Ir coated with $k_c = 1.8\,\text{N}\,\text{m}^{-1}$. The size of the images is $2.5 \times 2.5\,\mu\text{m}^2$. Taken from M. Kopycinska-Müller, PhD thesis, Faculty 8, Saarland University, Germany, 2005, unpublished, with permission.

orientation or the grain orientation can be reconstructed by comparison of the amplitudes of the components of the surface vibration in different directions relative to the surface [89, 90].

Labardi *et al.* [91] found various resonances on triglycine sulfate at frequencies above the first free resonance, identified them as cantilever contact resonances, and used them for imaging. In this case the piezo-mode signal is enhanced through resonance amplification. By quantitative evaluation of the contact-resonance spectra in the AFAM mode, both for flexural and torsional cantilever modes and in the ultrasonic piezo-mode, material properties such as elasticity, domain orientation, and nonlinear behavior as a function of amplitude or pressure can be simultaneously studied (Figure 13.11) [47, 92]. In several recent publications on piezo-mode imaging, the effects of local elasticity and piezoelectric properties on the image contrast determination are discussed [93–97].

13.2.4.6 Nonlinear Contact Resonances and Related Phenomena

When in an AFAM set-up the amplitude of surface vibration is increased, a DC offset of the cantilever develops, that is, the sensor tip lifts off the sample surface. This rectifying effect has been known since the early experiments combining ultrasonic technology with AFM and was exploited in ultrasonic force microscopy (UFM) [98]. When the threshold amplitude for the DC lift-off at two different static cantilever loads is compared, a qualitative acoustical image of elastic sample properties can be obtained [99]. The advantage of this technique as compared to AFAM or UAFM is that not only a low electronic bandwidth of the cantilever optical detection system of the AFM is required. Nevertheless, as both elasticity and adhesion contribute to the image contrast in a nonlinear fashion, their separation in a quantitative manner remains difficult. In addition, the UFM technique can be employed to study frictional properties [100]. Further details of this technique can be found in Chapter 11 ("Acoustically Excited Probe Microscopy" by Briggs and Kolosov) and Chapter 12 ("Ultrasonic Atomic Force Microscopy" by Yamanaka and Tsuji).

There are also nonlinear effects that lead to a deviation of the contact-resonance Lorentzian line shapes. For example, the effect of adhesion counteracts the elastic restoring force and therefore leads to a distortion of the contact resonances and to hysteretic behavior [62, 101, 102]. Piezoelectric forces may lead to the opposite behavior of the contact resonance [103]. In several publications nonlinear effects in cantilever contact resonances, such as mixing [104, 105], subharmonic generation [106], modal interactions [107], and the reconstruction of tip force–distance curves [108, 109] is discussed. All these effects give a deeper insight into forces acting in the contact; however, to exploit these means to control a mechanical nonlinear oscillator is not easy and complicates the experimental setup considerably. The tip–sample contact is inherently nonlinear because, already, the elastic Hertzian contact forces are nonlinear in their distance dependence and this leads to nonlinear contact resonances [102], as has been known for a long time in mechanical engineering [110–112].

A somewhat related phenomenon is mode coupling. Let us assume that the tip position is not in the center line of the cantilever beam. When the tip is then in contact with the surface, torsional forces and bending forces act on the cantilever leading to an excitation of both flexural and torsional modes. Such effects were indeed observed experimentally [113] and the underlying theory of modal interaction was discussed by Turner and Hurley [38] and by Arafat et al. [114]. When a tapping cycle operated at its lowest, flexural frequency mode covers a full force-distance curve; this mode coupling can be deliberately exploited to transfer the harmonic frequencies generated to the torsional mode below its lowest resonance frequency, thus linearizing the experimental set-up for imaging [115] and obtaining force-distance curves.

13.2.4.7 Subsurface Imaging Using Contact Resonances

Quite a few efforts have been made to achieve subsurface imaging using various schemes of AFAM or UAFM. Subsurface imaging is possible if the physical quantity measured exhibits a gradient into the material within the penetration of the stress-field emanating from the tip–sample contact. For example, the stiffness of the material changes at the surface when there are defects within a certain depth. This has been exploited for detecting dislocations generated below the surface [54], to probe the thickness of thin-solid films [116], to monitor subsurface adhesion on buried interfaces [117], to detect subsurface defects [36, 118, 119], or to detect the effects of crack generation and propagation [120]. Cavities can be seen at depths of about 200–500 nm and have diameters of ≈300 nm. The depth range is roughly given by the depth range of the stress field that the tip generates at the surface, the stress field of the defect, or their joint action. Therefore, the depth range

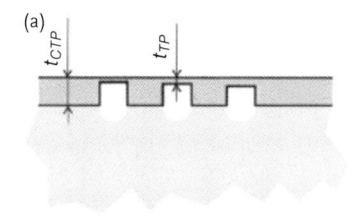

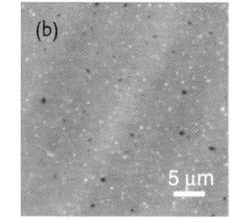

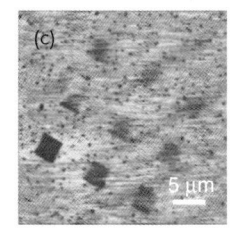

Figure 13.12 (a) Schematic drawing of square holes machined into SiN$_3$ of thickness t_{CTP} at different depths t_{TP} ranging from 30 to 270 nm. Their size was $3.7 \times 3.7\,\mu m^2$. (b) Topography image and (c) AFAM amplitude image obtained close to a contact-resonance frequency of 760 kHz at a static load of $F_0 = 0.7\,\mu N$ and with a contact radius of $a_c \approx 10$ nm. The hole with the thinnest cover of 30 nm is visible in the bottom left-hand corner. The thickest cover of 270 nm is visible in the upper right-hand corner. The z-scale in the topography images is 100 nm. In (c), the detection depth is about 30 times the contact radius! Taken from Reference [123], with permission from Elsevier.

depends on the static load, the tip radius, and the tip shape as well as the cantilever mode employed [121, 122]. Elliptical shapes should have a larger depth range [51] but no one seems to have exploited this fact yet for depth sensing. So far the most detailed study of the depth range of AFAM for defect detection has been reported by Striegel et al. [123]. This group examined various defect shapes such as voids below silicon membranes, rectangular flat bottoms voids of various depths machined into silicon nitride, and wedges with increasing thickness. They could detect defects as deep as 900 nm depending on their size, which of course changes the local contact stiffness k^*. As is the case for all interior acoustical imaging techniques, the contrast determines whether an image of the defect is obtained with an acceptable signal-to-noise ratio (SNR). Here, the ratio $\Delta k^* = (k_{def} - k^*)/k^*$ of the stiffness change of the defect area relative to the host material determines the contrast. As a rule of thumb, one can assume that the depth range z is about at least three times the contact radius if the Hertzian contact mechanics between tip and surface prevails, depending on the parameters of the tip–sample contact as discussed above. This rule of thumb was corroborated by Strieger et al. [123] (Figure 13.12).

Recently, a subsurface imaging scheme combining ultrasound and AFM [124] called scanning near-field ultrasonic holography (SNFUH) has been reported. In SNFUH the device to be tested is insonified from below at a frequency ω_1 and simultaneously the cantilever is excited via its holder at the slightly different frequency ω_2 by a piezoelectric transducer. The phase of the ω_1 wave is influenced by the stiffness difference of the defect relative to its surrounding and by its geometrical shape. This difference is transferred to the phase of the surface displacement and hence to the cantilever vibration, picked up by the AFM photodetector and monitored by a phase-lock amplifier for imaging. By using heterodyning, the phase contrast is enhanced because the phase difference is enlarged by the factor

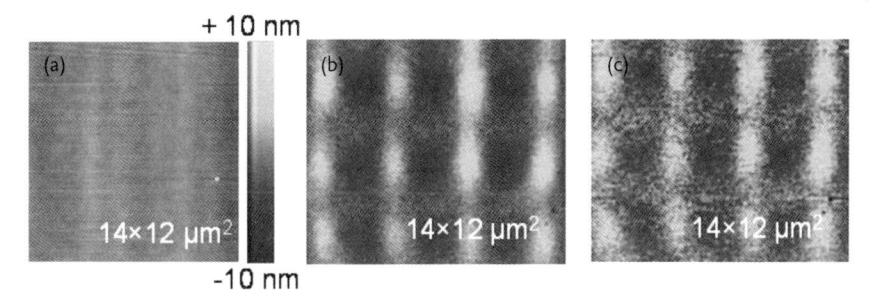

Figure 13.13 (a) Topography image of gold-line structures covered with 7 μm PMMA photo-resist layer. If there is any topography left, it is less than 1 nm peak-to-peak. (b) Amplitude image and (c) phase image of the gold lines (height 50 nm, pitch 3 μm), buried 7 μm under the photo-resist layer, which is much larger than the depth range due to contact stiffness changes. The signal-to-noise ratio in the amplitude image was 6–8 dB and the phase contrast was 3° ± 0.5°. Taken from Reference [131], with permission from the American Institute of Physics.

$\omega_1/(\omega_2 - \omega_1)$, which originates from the nonlinear force–distance curve along which the tip is operated [125–128]. The enhancement of phase changes by heterodyning is a technique known from the time when one could monitor analog signal processing in the GHz range only indirectly [129, 130].

Furthermore, a scheme has been reported to detect GHz ultrasonic waves in an AFAM to render possible the detection of small defects in materials much below the Hertzian stress-field. When the tip of this cantilever contacts the sample, an RF signal is applied to a transducer attached to the sample at a frequency of $\omega_c/2\pi = 1.04$ GHz. This signal is amplitude modulated at $\omega_m/2\pi$, for example, at the expected contact-resonance frequency $\omega_{cr}/2\pi$ of the cantilever. As the 1 GHz wave traverses the sample it gets scattered and/or attenuated when a buried defect is present, producing diffraction or a shadow image at the surface of the device under test (Figure 13.13). Either the amplitude or the phase of the cantilever oscillation signal is picked up by the optical detection scheme of the AFM. The transmitted signal at the carrier frequency of 1 GHz is detected indirectly by demodulating the signal at the frequency $\omega_m/2\pi$, which is only possible if the interaction between the tip and the sample surface is nonlinear. This is provided if the set-point in the force–distance curve of the AFM is chosen appropriately [131]. A similar scheme was developed to image the amplitude distribution of a 1.6 GHz bulk resonator by amplitude modulating the exciting GHz signal at the second contact flexural cantilever-mode [132].

In this subchapter, various techniques are discussed to allow elasticity, anelasticity, and plastic deformation measurements, or in more general terms force measurements, and to image these quantities in the near-field of a sensor. Several efforts have been undertaken to combine acoustical imaging with atomic force microscopies. These are near-field imaging systems exploiting the contrast from modulation of the forces between tip and surfaces of the component under test.

Manifold factors determine the obtainable spatial resolution and the contrast when working in the near-field of an antenna.

Acknowledgment

I thank S. Amelio, H. Bentaher, A. Boub, A. Caron, A. Ganea, R. Ghokale, K. Janser, M. Kopycinska-Müller, E. Kullenberg, A. Kumar, K. Schwarz, M. Reinstädtler, D. Rupp, and V. Scherer for their contributions in the field of atomic force acoustic microscopy, and S. Hirsekorn and U. Rabe for long-term collaboration in that field during my employment from 1980 until retirement at Fraunhofer IZFP end of 2007. Furthermore, it is a pleasure to thank Chanmin Su and Shuiqing Hu, Bruker-Nano, S. Barbara, and D. Bedorf, M. Büchsenschütz-Göbeler, S. Küchemann, H. Wagner, B. Zhang, and last but not least K. Samwer from the University of Göttingen for fruitful collaboration.

Financial support by the Germany Science Foundation, by the Federal Ministry of Research and Technology, by the State of the Saarland, by the Volkswagen Foundation, by the Alexander von Humboldt Foundation, and by the industrial partners in the various projects carried out at Fraunhofer IZFP is thankfully acknowledged.

References

1. Ash, E.A. and Nichols, G. (1972) Super-resolution aperture scanning microscope. *Nature*, **237**, 510–512.
2. Smith, D.F. and Cagle, C.V. (1966) Ultrasonic testing of adhesive bonds using Fokker bond tester. *Mater. Eval.*, **24**, 362–370.
3. Lange, V.V. (1994) The mechanical impedance analysis method of non-destructive testing: a review. *Non-Destructive Testing Eval.*, **11**, 177–193.
4. Ermolov, I.N. and Lange, V.V. (2009) Ultrasonic testing, in *Handbook on Nondestructive Testing*, vol. 3 (ed. V.V. Kluev), SPEKTR, Moscow, pp. 282–302.
5. Guyott, C.C.H., Cawley, P., and Adams, R.D. (1986) The non-destructive testing of adhesively bonded structure: a review. *J. Adhesion*, **20**, 129–159.
6. Dürr, W., Sinclair, D.A., and Ash, E.A. (1980) A high resolution acoustic probe, in *Proceedings IEEE 1980 Ultrasonic Symposium* (ed. B.R. McAvoy), IEEE, New York, pp. 594–597.
7. Khuri-Yakub, B.T., Cinbis, C., Chou, C.H., and Reinholdtsen, P.A. (1989) Near-field scanning acoustic microscope, in *Proceedings 1989 Ultrasonics Symposium* (ed. B.R. McAvoy), IEEE, New York, pp. 805–807.
8. Zienuk, J.K. and Latuszek, A. (1989) Non-conventional pin scanning ultrasonic microscopy, in *Proceedings 17th International Symposium Acoustical Imaging* (eds H. Shimizu, N. Chubachi, and J. Kushibiki), Plenum Press, New York, pp. 219–224.
9. Kulik, A., Attal, J., and Gremaud, G. (1993) Near-field scanning acoustic microscopy, in *Proceedings 20th International Symposium on Acoustical Imaging* (eds Y. Wei and B. Gu), Plenum Press, New York, pp. 241–244.
10. Mason, W.P. (1958) *Physical Acoustics and the Properties of Solids*, Academic Press, New York.
11. Güthner, P., Fischer, U.C., and Dransfeld, K. (1989) Scanning near-field acoustic microscopy. *Appl. Phys. B*, **48**, 89–92.

12 Arnold, W., Hoffmann, B., and Willems, H. (1986) Crack-depth estimation by photoacoustic microscopy. *Z. Phys B.*, **64**, 31–34.

13 Khuri-Yakub, B.T., Akamine, S., Hadimioglu, B., Yamada, Y., and Quate, C.F. (1991) Near-field acoustic microscopy, in *Scanning Microscopy Instrumentation* (ed. G.S. Kino), SPIE, vol. 1556, pp. 30–38.

14 Takata, K. (1992) Tunneling acoustic microscopy. *Jpn. J. Appl. Phys. Suppl.*, **31-1**, 3–8.

15 Uozumi, K. and Yamamuro, K. (1989) A possible novel scanning ultrasonic tip microscope. *Jpn. J. Appl. Phys.*, **28**, L1297–L1299.

16 Binnig, G., Quate, C.F., and Gerber, C. (1986) Atomic force microscope. *Phys. Rev. Lett.*, **56**, 930–933.

17 Maivald, P., Butt, H.-J., Gould, S.A.C., Prater, C.B., Drake, B., Gurley, J.A., Elings, V.B., and Hansma, P.K. (1991) Using force modulation to image surface elasticities with the atomic force microscope. *Nanotechnology*, **2**, 103–106.

18 Burnham, N.A., Kulik, A., Gremaud, G., Gallo, P.J., and Oulevey, P.J.F. (1996) Scanning local acceleration microscopy. *J. Vac. Sci. Technol. B*, **14**, 794–799.

19 Radmacher, M., Tilmann, R.W., and Gaub, H.E. (1993) Imaging viscoelasticity by force modulation with the atomic force microscope. *Biophys. J.*, **64**, 735–742.

20 Kessler, L.W. and Yuhas, D.E. (1979) Acoustic microscopy–1979. *Proc. IEEE*, **67**, 526–536.

21 Syed Asif, S.A., Wahl, K.J., Colton, R.J., and Warren, O.L. (2001) Quantitative imaging of nanoscale mechanical properties using hybrid nanoindentation and force modulation. *J. Appl. Phys.*, **90**, 1192–1200, and references contained therein.

22 Krotil, H.-U., Stifter, T., and Marti, O. (2000) Concurrent measurement of adhesive and elastic surface properties with a new modulation technique for scanning force microscopy. *Rev. Sci. Instrum.*, **71**, 2765–2771.

23 Su, C., Hu, S., Hu, Y., Erina, N., and Slade, A. (2010) Quantitative mechanical mapping of biomolecules and cells in fluid. *Mater. Res. Soc. Symp. Proceedings*, **1261**, 1261-U01-05.

24 Kopycinska-Müller, M., Caron, A., Hirsekorn, S., Rabe, U., Natter, H., Hempelmann, R., Birringer, R., and Arnold, W. (2008) Quantitative evaluation of elastic properties of nano-crystalline nickel using atomic force acoustic microscopy. *Z. Phys. Chem.*, **222**, 471–498.

25 Rabe, U., Amelio, S., Kopycinska, M., Hirsekorn, S., Kempf, M., Göken, M., and Arnold, W. (2002) Imaging and measurement of local mechanical material properties by atomic force acoustic microscopy. *Surf. Interface Anal.*, **33**, 65–70.

26 Yamanaka, K., Tsuji, T., Noguchi, A., Koike, T., and Mihara, T. (2000) Nanoscale elasticity measurement with in situ tip shape estimation in atomic force microscopy. *Rev. Sci. Instrum.*, **71**, 2403–2408.

27 Vlassak, J.J. and Nix, W.D. (1993) Indentation modulus of elastically anisotropic half-spaces. *Philos. Mag. A*, **67**, 1045–1056.

28 Robert, L. and Cretin, B. (1999) Determination of the observation depth in scanning microdeformation microscopy. *Surf. Interf. Analys.*, **27**, 568–571.

29 Ogi, H., Niho, H., and Hirao, M. (2006) Elastic-stiffness distribution on dual phase stainless steel studied by resonance ultrasound microscopy. *Acta Mater.*, **54**, 4143–4148.

30 Rabe, U. and Arnold, W. (1994) Acoustic microscopy by atomic force microscopy. *Appl. Phys. Lett.*, **64**, 1493–1495.

31 Yamanaka, K. and Nakano, S. (1996) Ultrasonic atomic force microscope with overtone excitation of cantilever. *Jpn. J. Appl. Phys.*, **35**, 3787–3792. doi: 10.1143/JJAP.35.3787.

32 Schwarz, K., Rabe, U., Hirsekorn, S., and Arnold, W. (2008) Excitation of atomic force microscope cantilever vibrations by Schottky barriers. *Appl. Phys. Lett.*, **92**, 183105.

33 Turner, J.A., Hirsekorn, S., Rabe, U., and Arnold, W. (1997) High-frequency response of atomic-force microscope

cantilevers. *J. Appl. Phys.*, **82**, 966–979.

34 Tyndall, J. (1869) *Der Schall* (eds. H. Helmholtz and G. Wiedeman), Vieweg and Son, Braunschweig, Germany.

35 Rabe, U., Janser, K., and Arnold, W. (1996) Vibrations of free and surface-coupled atomic-force microscope cantilevers: theory and experiment. *Rev. Sci. Instrum.*, **67**, 3281–3293.

36 Yamanaka, K., Noguchi, A., Tsuji, T., Koike, T., and Goto, T. (1999) Quantitative material characterization by ultrasonic AFM. *Surf. Interface Anal.*, **27**, 600–606.

37 Dupas, E., Gremaud, G., Kulik, A., and Loubet, J.L. (2001) High-frequency mechanical spectroscopy with an atomic force microscope. *Rev. Sci. Instrum.*, **72**, 3891–3897.

38 Turner, J.A. and Hurley, D.C. (2003) Ultrasonic methods in contact atomic force microscopy, in *Ultrasonic Methods for Material Characterization (Instrumentation, Mesure, Métrologie III)* (eds D. Placko and T. Kundu), Lavoisier, Cachan, France, pp. 117–148.

39 Rabe, U. (2006) Atomic force acoustic microscopy, in *Applied Scanning Probe Methods*, vol. II (eds B. Bushan and H. Fuchs), Springer, Berlin, pp. 37–90.

40 Huey, B.D. (2007) AFM and acoustics: fast, quantitative nanomechanical mapping. *Annual Rev. Mat. Research*, **37**, 351–385. doi: 10.1146/annurev.matsci.37.052506.084331.

41 Song, Y. and Bhushan, B. (2008) Atomic force microscopy dynamic modes: modeling and applications. *J. Phys. Condens. Matter*, **20**, 225012.

42 Kumar, A., Rabe, U., Hirsekorn, S., and Arnold, W. (2008) Elasticity mapping of precipitates in polycrystalline materials using atomic force acoustic microscopy. *Appl. Phys. Lett.*, **92**, 183106.

43 Wagner, H., Bedorf, D., Küchemann, S., Schwabe, M., Zhang, B., Arnold, W., and Samwer, K. (2011) Local elastic properties of a metallic glass. *Nat. Mater.*, **10**, 1–4. doi: 10.1038

44 Mazeran, P.E. and Loubet, J.L. (1999) Normal and lateral modulation with a scanning force microscope, an analysis: implication in quantitative elastic and friction imaging. *Tribol. Lett.*, **7**, 199–212.

45 Liu, Y.H., Wang, D., Nakajima, K., Zhang, W., Hirata, A., Nishi, T., Inoue, A., and Chen, M.W. (2011) Characterization of Nanoscale Mechanical Heterogeneity in a Metallic Glass by Dynamic Force Microscopy. *Phys. Rev. Lett.*, **106**, 125504.

46 Kester, E., Rabe, U., Presmanes, L., Tailhades, P., and Arnold, W. (2000) Measurement of Young's modulus of nanocrystalline ferrites with spinel structures by atomic force acoustic microscopy. *J. Phys. Chem. Solids*, **61**, 1275–1284.

47 Rabe, U., Kopycinska, M., Hirsekorn, S., Munoz-Saldana, J., Schneider, G.A., and Arnold, W. (2002) High-resolution characterisation of piezoelectric ceramics by ultrasonic scanning force microscopy techniques. *J. Phys. D*, **35**, 2621–2635.

48 Prasad, M., Kopycinska, M., Rabe, U., and Arnold, W. (2002) Measurement of Young's modulus of clay minerals using atomic force acoustic microscopy. *Geophys. Res. Lett.*, **29**, 13–16.

49 Passeri, D., Rossi, M., Alippi, A., Bettucci, A., Terranova, M.L., Tamburri, E., and Toschi, T. (2008) Characterization of epoxy/single-walled carbon nanotubes composite samples via atomic force acoustic microscopy. *Phys. E-Low-Dim. Syst. Nanostruct.*, **40**, 2419–2424.

50 Stan, G., Krylyuk, S., Davydov, A.V., Vaudin, M.D., Bendersky, L.A., and Cook, R.F. (2009) Contact-resonance microscopy for nanoscale elastic property measurements: spectroscopy and imaging. *Ultramicroscopy*, **109**, 929–936.

51 Johnson, K.L. (1985) *Contact Mechanics*, Cambridge University Press, Cambridge, UK, pp. 420–421.

52 Oliver, W.C., Pharr, G.M., and Brotzen, F.R. (1992) On the generality of the relationship among contact stiffness, contact area, and elastic modulus during indentation. *J. Mater. Res.*, **7**, 613–617.

53 Stan, G. and Price, W. (2006) Quantitative measurements of indentation moduli by atomic force

acoustic microscopy using a dual reference method. *Rev. Sci. Instrum.*, **77**, 103707.

54 Yamanaka, K., Kobari, K., and Tsuji, T. (2008) Evaluation of functional materials and devices using atomic force microscopy with ultrasonic measurements. *Jpn. J. Appl. Phys.*, **47**, 6070–6076.

55 Stan, G. and Cook, R.F. (2008) Mapping of elastic properties of granular Au films by contact resonance atomic force microscopy. *Nanotechnology*, **19**, 2235701. doi: 10.1088/0957-4484/19/23/235701.

56 Stan, G., Ciobanu, C.V., Thayer, T.P., Wang, G.T., Creighton, J.R., Purushotham, K.P., Bendersky, L.A., and Cook, R.F. (2009) Elastic moduli of faceted aluminum nitride nanotube measured by contact resonance atomic force microscopy. *Nanotechnology*, **20**, 035706. doi: 10.1088./0957-4484/20/3/035706

57 Killgore, J.P., Geiss, R.H., and Hurley, D.C. (2011) Continuous measurement of atomic force microscope tip wear by contact resonance force microscopy. *Small*, **7**, 1018–1022.

58 Killgore, J.P. and Hurley, D.C. (2012) Low-force AFM nanomechanics with higher-eigenmode contact resonance spectroscopy. *Nanotechnology*, **23**, 055702. doi: 10.1088/0957-4484/23/5/055702.

59 Steiner, P., Roth, R., Gnecco, E., Glatzel, T., Baratoff, A., and Meyer, E. (2009) Modulation of contact-resonance frequency accompanying atomic-scale stick-slip in friction force microscopy. *Nanotechnology*, **20**, 495701. doi: 1088/0957-4484/20/49/495701.

60 Muraoka, M. (2007) Vibrational dynamics of concentrated-mass cantilevers in atomic force acoustic microscopy: presence of modes with selective enhancement of vertical or lateral tip motion. *J. Phys. Conf. Series*, **61**, 836–840. doi: 10.1888/1742-6596/6171/167.

61 Le Rouzic, J., Cretin, B., Vairac, P., and Cavalier, B. (2009) Specific geometries of resonant cantilevers for scanning force microscopy, in *Proceedings IEEE Frequency Control Symposium 2009 and 22nd European Frequency and Time Forum*, IEEE Archives, NY, pp. 822–825. doi: "http://dx.doi.org/10.1109/FREQ.2009.5168301" \t "blank" 10.1109/FREQ.2009.5168301.

62 Vairac, P., Boucenna, R., Le Rouzic, J., and Cretin, B. (2008) Scanning microdeformation microscopy: experimental investigations on non-linear contact spectroscopy. *J. Phys. D: Appl. Phys.*, **41**, 155503.

63 Rabe, U., Hirsekorn, S., Reinstädtler, M., Sulzbach, T., Lehrer, Ch., and Arnold, W. (2007) Influence of the cantilever holder on the vibrations of AFM cantilevers. *Nanotechnology*, **18**, 044008. doi: 10.1088/0957-4484/18/4/044008.

64 Truell, R., Elbaum, C., and Chick, B.B. (1969) *Ultrasonic Methods in Solid State Physics*, Academic Press, New York and London.

65 Nowick, A.S. and Berry, B.S. (1972) *Anelastic Relaxation in Crystalline Solids*, Academic Press, New York and London.

66 Tucker, J.W. and Rampton, V.W. (1972) *Microwave Ultrasonics in Solid State Physics*, North-Holland Publishing Company.

67 Hunklinger, S. and Arnold, W. (1976) Ultrasonic properties of glasses at low temperatures, in *Physical Acoustics*, vol. 12 (eds W.P. Mason and R.N. Thurston), Academic Press, pp. 155–121.

68 Schaller, R., Fantozzi, G., and Gremaud, G. (eds) (2001) *Mechanical Spectroscopy Q−1 2001*, Trans Tech Publications Ltd, Uetikon, Switzerland.

69 Yua, P.A., Hurley, D.C., and Turner, J.A. (2008) Contact-resonance atomic force microscopy for viscoelasticity. *J. Appl. Phys.*, **104**, 0474916.

70 Yua, P.A., Hurley, D.C., and Turner, J.A. (2011) Relationship between Q-factor and sample damping for contact resonance atomic force microscopy measurement of viscoelastic properties. *J. Appl. Phys.*, **109**, 113528.

71 Caron, A. and Arnold, W. (2009) Observation of local internal friction and plasticity onset in nanocrystalline nickel by atomic force acoustic microscopy. *Acta Mater.*, **57**, 4353–4363.

72 Killgore, J.P., Yablon, D.G., Tsou, A.H., Gannepalli, A., Yua, P.A., Turner, J.A., Proksch, R., and Hurley, D.C. (2011) Viscoelastic property mapping with contact resonance force microscopy. *Langmuir*, **27**, 13983–13987.

73 Lang, M.J., Duarte-Dominguez, M., Birringer, R., Hempelmann, R., Natter, H., and Arnold, W. (1999) Measurement of elastic and anelastic properties of nanocrystalline metals. *NanoStructured Mater.*, **12**, 811–816.

74 Lohmiller, J., Eberl, C., Schwaiger, R., Kraft, O., and Balk, T.J. (2008) Mechanical spectroscopy of nanocrystalline nickel near room temperature. *Scr. Mater.*, **59**, 467–470.

75 Li, P.-Y., Zhang, X.-Y., Wu, X.-L., Huang, Y.-N., and Meng, X.-K. (2008) Internal friction of bend-deformed nanocrystalline nickel by mechanical spectroscopy. *Chin. Phys Lett*, **25**, 4339–4341.

76 Oulevey, F., Gremaud, G., Sémoroz, A., Kulik, A.J., Burnham, N.A., Dupas, E., and Gourdon, D. (2001) Local mechanical spectroscopy with nanometer-scale lateral resolution. *Rev. Sci. Instrum.*, **72**, 2085–2094.

77 Yamanaka, K., Maruyama, Y., Tsuji, T., and Nakomoto, K. (2001) Resonance frequency and Q factor mapping by ultrasonic atomic force microscopy. *Appl. Phys. Lett.*, **78**, 1939–1941.

78 Ogi, H., Niho, H., and Hirao, M. (2006) Internal-friction mapping on solids by resonance ultrasound microscopy. *Appl. Phys. Lett.*, **88**, 141110.

79 Hirsekorn, S., Rabe, U., and Arnold, W. (2001) Ultrasonic radiation in dynamic force microscopy. *Appl. Phys. A*, **72**, S87–S92.

80 Scherer, V., Reinstädtler, M., and Arnold, W. (2003) Atomic force microscopy with lateral modulation, in *Applied Scanning Probe Methods* (eds H. Fuchs, B. Bhushan, and S. Hosaka), Springer, Berlin, pp. 75–116, ISBN 3-540-00527-7, and references contained therein.

81 Scherer, V., Bhushan, B., Rabe, U., and Arnold, W. (1997) Local elasticity and lubrication measurements using atomic force and friction force microscopy at ultrasonic frequencies. *IEEE Trans. Magn.*, **33**, 4077–4079.

82 Scherer, V., Arnold, W., and Bhushan, B. (1998) Active friction control using ultrasonic vibration, in *NSF/ASME Workshop on Tribology Issues and Opportunities in MEMS* (ed. B. Bhushan), Kluwer Academic Publishers, Dordrecht, pp. 463–469.

83 Reinstädtler, M., Kasai, T., Rabe, U., Bhushan, B., and Arnold, W. (2005) Imaging and measurement of elasticity and friction using the TRmode. *J. Phys. D Appl. Phys.*, **38**, R269–R282.

84 Hurley, D.C. and Turner, J.A. (2007) Measurement of Poisson's ratio with contact-resonance atomic force microscopy. *J. Appl. Phys.*, **102**, 033509.

85 Reinstädtler, M., Rabe, U., Scherer, V., Hartmann, U., Goldade, A., Bhushan, B., and Arnold, W. (2003) On the nanoscale measurement of friction using atomic-force microscope cantilever torsional resonances. *Appl. Phys. Lett.*, **82**, 2604–2606.

86 Güthner, P. and Dransfeld, K. (1992) Local poling of ferroelectric polymers by scanning force microscopy. *Appl. Phys. Lett.*, **61**, 1137–1139.

87 Franke, K., Besold, J., Haessler, W., and Seegebarth, C. (1994) Modification and detection of domains on ferroelectric PZT films by scanning force microscopy. *Surf. Sci.*, **302**, L283–L288.

88 Franke, K., Huelz, H., and Weihnacht, M. (1998) How to extract spontaneous polarization information from experimental data in electric force microscopy. *Surf. Sci.*, **41**, 178–182.

89 Eng, L.M., Güntherodt, H.J., Schneider, G.A., Köpke, U., and Munoz-Saldana, J. (1999) Nanoscale reconstruction of surface crystallography from three-dimensional polarization distribution in ferroelectric barium-titanate ceramics. *Appl. Phys. Lett.*, **74**, 233–235.

90 Muñoz-Saldaña, J., Schneider, G.A., and Eng, L.M. (2001) Stress induced movement of ferroelastic domain walls in $BaTiO_3$ single crystals evaluated by scanning force microscopy. *Surf. Sci. Lett.*, **480**, L402–L410.

91 Labardi, M., Likodimos, V., and Allegrini, M. (2000) Force microscopy contrast mechanisms in ferroelectric domain imaging. *Phys. Rev. B*, **61**, 14390–14398.
92 Kopycinska-Müller, M., Reinstädtler, M., Rabe, U., Caron, A., Hirsekorn, S., and Arnold, W. (2004) Ultrasonic modes in atomic force microscopy, in *Proceedings 27th International Symposium Acoustical Imaging* (eds W. Arnold and S. Hirsekorn), Kluwer Plenum Press, pp. 699–704.
93 Harnaega, C., Pignolet, A., Alexe, M., and Hesse, D. (2002) Piezoresponse scanning force microscopy: what quantitative information can we really get out of piezoresponse measurements on thin films. *Integr. Ferroelectr.*, **44**, 113–124.
94 Liu, X.X., Heiderhoff, R., Abicht, H.P., and Balk, L.J. (2002) Scanning near-field acoustic study of ferroelectric BaTiO3 ceramics. *J. Phys. D.: Appl. Phys.*, **35**, 74–87.
95 Tsuji, T., Ogiso, H., Akedo, J., Satto, S., Fukuda, K., and Yamanaka, K. (2004) Evaluation of domain boundary of piezo/ferroelectric material by ultrasonic atomic force microscopy. *Jpn. J. Appl. Phys.*, **43**, 2907–2913.
96 Kalinin, S.V., Jesse, S., Rodriguez, B.J., Shin, J., Baddorf, A.P., Lee, H.N., Borisevich, A., and Pennycook, S.J. (2006) Spatial resolution, information limit, and contract transfer in piezoresponse force microscopy. *Nanotechnology*, **17**, 3400–3411.
97 Jesse, S., Guo, S., Kumar, A., Rodriguez, B.J., Proksch, R., and Kalinin, S.V. (2010) Resolution theory, and static and frequency-dependent cross-talk in piezoresponse force microscopy. *Nanotechnology*, **21**, 405703.
98 Yamanaka, K., Ogiso, H., and Kolosov, O. (1994) Ultrasonic force microscopy for nanometer resolution subsurface imaging. *Appl. Phys. Lett.*, **64**, 178–180.
99 Dinelli, F., Biswas, S.K., Briggs, G.A.D., and Kolosov, O.V. (2000) Measurement of stiff-material compliance on the nanoscale using ultrasonic force microscopy. *Phys. Rev. B*, **61**, 13995–14006.
100 Cuberes, M.T. (2007) Nanoscale friction and ultrasound, in *Fundamentals of Friction and Wear* (eds E. Gnecco and E. Meyer), Springer, Berlin, pp. 49–71.
101 Muraoka, M. and Arnold, W. (2001) A method to evaluate local elasticity and adhesion energy based on nonlinear response of AFM cantilever vibration. *JSME Int. J., A*, **44**, 396–405.
102 Turner, J.A. (2004) Non-linear behavior of a beam with cantilever-Hertzian contact boundary conditions. *J. Sound Vibr.*, **275**, 177–191.
103 Rabe, U., Kopycinska-Müller, M., Reinstädtler, M., Hirsekorn, S., and Arnold, W. (2002) Nonlinear effects in ultrasonic transmission in atomic force microscope contacts, in *Proceedings 16th International Symposium Nonlinear Acoustics, Moscow, August 19–23, 2002*, vol. 2 (eds O.V. Rudenko and O.A. Sapozhnikov), Faculty of Physics, Moscow State University, pp. 711–718, ISBN 5-879 0034-6.
104 Dupas, E., Kulik, A., Gourdon, D., Oulevey, F., Burnham, N.A., Gremaud, G., and Arnold, W. (1997) Mixing of ultrasonic signals with an AFM, Paper Th1.3P13, Session Novel Instrumentation/Sensors, STM 97, Hamburg, Germany, July 20–25, 1997.
105 Tetard, L., Passian, A., Eslami, S., Jalili, N., Farahi, R.H., and Thundat, T. (2011) Virtual resonance and frequency difference generation by Van der Waals interaction. *Phys. Rev. Lett.*, **106**, 180801.
106 Burnham, N.A., Kulik, A., Gremaud, G., and Briggs, G.A.D. (1996) Nanosubharmornics – the dynamics of small contacts. *Phys. Rev. Lett.*, **74**, 5092–5095. doi: 10.1103
107 Abdel-Rahmann, E.M. and Nayfeh, A.H. (2005) Contact force identification using the subharmonic resonance of a contact-mode atomic force microscopy. *Nanotechnology*, **16**, 199–207.
108 Rupp, D., Rabe, U., Hirsekorn, S., and Arnold, W. (2007) Nonlinear contact resonance spectroscopy in atomic force acoustic microscopy. *J. Phys. D Appl. Phys.*, **40**, 7136–7145.
109 Hutter, C., Platz, D., Tholén, E.A., Hansson, T.H., and Haviland, D.B. (2010) Reconstruction nonlinearities

with intermodulation spectroscopy. *Phys. Rev. Lett.*, **104**, 050801.
110 Nayak, P.R. (1972) Contact vibrations. *J. Sound Vibr.*, **22**, 297–322.
111 Sabot, J., Krempf, P., and Janolin, C. (1998) Non-linear vibrations of a sphere-plane contact excited by a normal load. *J. Sound Vibr.*, **214**, 359–375.
112 Bichri, A., Belhaq, M., and Perret-Liaudet, J. (2001) Control of vibroimpact dynamics of a single-sided Hertzian contact force oscillator. *Nonlinear Dynamics*, **63**, 51–60.
113 Reinstädtler, M., Rabe U., Scherer V., Turner, J.A, and Arnold, W. (2003) Imaging of flexural and torsional resonance modes of atomic force microscopy cantilevers using optical interferometry. *Surface Science*, **532–535**, 1152–1158.
114 Arafat, H.N., Nayfeh, A.H., and Abdel-Rahmann, E.M. (2008) Modal interactions in contact-mode atomic force microscopy. *Nonlinear Dynamics*, **54**, 151–166.
115 Sahin, J., Maganov, S., Su, C., Quate, C.F., and Solgaard, O. (2007) An atomic force microscope tip designed to measure time-varying nanomechanical forces. *Nature Nanotechnology*, **2**, 507-514. doi: 10.1038/nnano.2007.226.
116 Yaralioglu, G.G., Degertekin, F.L., Crozier, K.B., and Quate, C.F. (2000) Contact stiffness of layered materials for ultrasonic force microscopy. *J. Appl. Phys.*, **87**, 7491–7496.
117 Hurley, D.C., Kopycinska-Müller, M., Langlois, E.D., Kos, A.B., and Barbosa, N., III (2006) Mapping substrate/film adhesion with contact-resonance-frequency atomic force microscopy. *Appl. Phys. Lett.*, **89**, 021911.
118 Crozier, K.B, Yaralioglu, G.G., Degertekin, F.L., Adams, J.D., Minne, S.C., and Quate, C.F. (2000) Thin film characterization by atomic force microscopy at ultrasonic frequencies. *Appl. Phys. Lett.*, **76**, 1950–1952.
119 Sarioglu, A.F., Atalar, A., and Degertekin, F.L. (2004) Modeling the effect of subsurface interface defects on contact stiffness for ultrasonic atomic force microscopy. *Appl. Phys. Lett.*, **84**, 5368–5370.
120 Caron, A., Rabe, U., Rödel, J., and Arnold, W. (2007) Near-field acoustical imaging using lateral bending mode of atomic force microscope cantilevers – Applications to fracture mechanics of nc-zirconia, in *Proceedings 28th International Symposium Acoustical Imaging* (ed. M. André), Springer, Berlin, pp. 31–41.
121 Parlak, Z. and Degertekin, F.L. (2008) Contact stiffness of finite size subsurface defects for atomic force microscopy: three-dimensional finite element modeling and experimental verification. *J. Appl. Phys.*, **103**, 114910.
122 Killgore, J.P., Kelly, J.Y., Stafford, C.M., Fasolka, M.J., and Hurley, D.C. (2011) Quantitative subsurface contact-resonance force microscopy of model polymer nanocomposites. *Nanotechnology*, **22**, 175706.
123 Striegler, A., Köhler, B., Bendjus, B., Roellig, M., Kopycinska-Mueller, M., and Meyendorf, N. (2011) Detection of buried reference structures by use of atomic force acoustic microscopy. *Ultramicroscopy*, **111**, 104–116.
124 Shekhawat, G.S. and Dravid, V.P. (2005) Nanoscale imaging of buried structures via scanning near-field ultrasound holography. *Science*, **310**, 89–92.
125 Rohrbeck, W. and Chilla, E. (1992) Detection of surface acoustic waves by scanning force microscopy. *phys. stat. sol.*, **131**, 69–71.
126 Chilla, E., Hesjedal, T., and Fröhlich, H.J. (1997) Nanoscale determination of phase velocity by scanning acoustic force microscopy. *Phys. Rev. B*, **55**, 15852.
127 Cuberes, M.T., Assender, H.E., Briggs, G.A.D., and Kolosov, O.V. (2000) Heterodyne force microscopy of PMMA/rubber nanocomposites: nanomapping of viscoelastic response at ultrasonic frequencies. *J. Phys. D Appl. Phys.*, **33**, 2347–2355.
128 Cantrell, S.A., Cantrell, J.H., and Lillei, P. (2007) Nanoscale subsurface imaging via resonant-difference atomic force ultrasonic microscopy. *J. Appl. Phys.*, **101**, 114324.
129 Dooley, J.W. (1970) Measurement of small changes in sound velocity in the

UHF range. *J. Acoust. Soc. Am.*, **47**, 1232–1235.

130 Edgerton, R.F. (1970) Simple measurement technique for small relative velocity changes of gigahertz acoustic waves. *J. Acoust. Soc. Am.*, **47**, 1229–1231.

131 Hu, S., Su, C., and Arnold, W. (2011) Imaging of subsurface structures using atomic force acoustic microscopy at GHz frequencies. *J. Appl. Phys.*, **109**, 084324.

132 Paulo, A.S., Black, J.P., White, R.M., and Bokor, J. (2007) Detection of nanomechanical vibrations by dynamic microscopy in higher cantilever eigenmodes. *Appl. Phys. Lett.*, **91**, 053116.

Index

a

A-scan techniques
– acoustic scattering parameters   102–104
– definition   234
– inverse echo signal filtering   101, 102
– plane wave propagation   102–104
– pulse transfer properties   99, 100
– spectral range/resolution properties   97–99
– transmitter/receiver components   95–97
absolute difference/deviation   36
abstract objects   29–31
acoustic field mathematical modeling   160
acoustic impedance microscopy
– see also quantitative ultrasonic microscopy
– acquired signal   63, 64
– basics   60, 61
– calibration for CAI   63–65
– experimental set-up   61, 62
– rat cerebellar cortex   65–69
– specimen   62, 63
acoustic impedance (Z)   162, 210, 212–222
acoustic lenses   164–167, 178, 181, 340, 341
acoustic radiation force imaging (ARFI)   14
acoustic scattering   102–104
acoustic velocity
– bulk wave velocity   141–150
– compressional wave velocity   210, 211
– high frequency acoustic microscopy   175–177
– leaky surface acoustic waves   126–141
– $V(x)$ scheme   125
– $V(z)$ method   125, 177, 178, 192, 212
acoustical near-field imaging see near-field imaging
acousto–optical tomography   5
acute myocardial infarction (AMI)   196
adenocarcinoma   194, 195, 201

adhesive properties   291–293, 297–299, 339, 340
adhesive sample bonding   308, 321
aerospace industry   71, 339, 340
aluminum   260, 261, 267, 268, 294, 295
amplitude   217–219, 285, 286, 288
anatomy   155, 156, 197, 207–209
angular dependent reflectance function ($R(\theta)$)   213
animal models   65–69, 117, 118, 190, 225, 226
anisotropy, elastic   173, 222, 223
annular array transducers   97
aperture function B 246
apertures, super-resolution   340, 341
apparent elastic coefficient ($c(\varphi)$)   219–222
Archimedes principle   219
array performance indicator (API)   252, 253
arrays
– see also post-processing approaches; quantitative acoustic microscopy; two-dimensional (2D) arrays
– imaging with   38, 39, 96–98
– – 1D arrays   245–255
– portable devices   75–91
– speed of array scanning   235, 236
articular tissues   192, 202, 203
artifacts   9, 311
astrocytes, rat   65–69
astronomy   23–29
atherosclerosis   197–200
atmospheric scattering   25–31, 33, 34
atomic force acoustic microscopy (AFAM)   282, 345–361
– contact stiffness/indentation modulus relationship   349–361
– nonlinear contact resonances   358, 359
– principle of operation   345–349

a

atomic force microscopy (AFM)   282
– *see also* ultrasonic atomic force microscopy (UAFM)
– forced vibration of samples   307, 308
– near-field imaging   342–361
– types   307
– ultrasonic force microscopy   290, 304
attenuation   126–141, 194, 196, 197
Auld's electromechanical reciprocity argument   244
averaging methods   23
axial resolution (δ_{axial})   98, 99

b

B-mode (brightness) imaging
– B/D scan technique   106
– high-resolution ranging/imaging   105–113, 116
– myocardial infarction   195–197
– portable devices   73, 80, 81, 89
– post-processing methods   234
– thickly sectioned skin   170, 171
– ultrasound echo systems, single-element transducers   104, 105
back-propagation method   7, 249, 250
backscattering   113, 114, 190
beam exit ion cross-section polishing (BEXP)   301, 302
beam-forming image post-processing   245, 246
beam-spreading term   242
benzocyclobutene (BCB) spacers   294, 295
biomechanics
– articular cartilage   202, 203
– scanning acoustic microscopy   187–203
– – articular tissues   202, 203
– – atherosclerosis   197–200
– – gastric cancer   193, 195
– – hard tissues   191, 192
– – kidney   197
– – myocardial infarction   195–197
biomedical applications
– bone quantitative acoustic scanning microscopy   207–227
– high-resolution ranging/imaging   115–118
– melanoma skin tissues, contrast mechanism studies   155–183
– multiwave imaging   3–20
– portable scanning devices   71, 72
– scanning acoustic microscopy, biomechanics   187–203
biopsy, melanoma   156, 157
blurring function (H)   58–60

bone
– degree of mineralization   193, 220, 221, 224
– elastic anisotropy   222–225
– healing   225–227
– hierarchical structure   207–209
– mechanical properties   191, 192
– multiscale elastic properties   209, 210
– quantitative acoustic scanning microscopy   207–227
– research   225, 226
– scanning acoustic microscopy   192, 193
– time-gated amplitude detection   217–219
– time-resolved measurements   216, 217
boundary conditions   346
boundary of flat reflectors   78–86
brain   6, 65–69, 71, 72
breast cancer   16, 60, 61
Brownian motion   8
bulk (K)   9
bulk modulus   49, 50, 188, 195, 196
bulk wave velocity   141–150

c

C-mode imaging   74, 155–183
calcification, bone   193, 220, 221, 224
calibration   63–65, 218, 219
cancer
– breast   16, 60, 61
– kidney   197
– leukemia   190
– melanoma skin tissues   155–183
– small animal studies   117
– stomach   193–195, 201
cantilever motion
– *see also* contact stiffness
– atomic force acoustic microscopy   345–347
– atomic force microscopy   307, 308
– – as near-field imaging system   342–344
– torsional resonance   356
– ultrasonic atomic force microscopy   312–314
– ultrasonic force microscopy   282–284, 289
carotid artery atherosclerosis   199
cartilage tissue   192
cells
– *see also* pathological specimens
– mechanical properties   189–191
– motility   189, 190
– scanning acoustic microscopy   187–203
– size   68, 69
characteristic acoustic impedance (CAI)   50
charge-coupled device (CCD) cameras   7, 8

chicken heart muscle cells   190
chromium electrodes   335
classical beam-forming method   245, 246
closed cracks   317, 319, 320
coherent regime   3–5
collagen fibrils   207, 208
collagen type I molecules   207, 208
colon adenocarcinoma   201
combined imaging method costs   4
compressional wave velocity   210, 211
computational burden   251, 252
computer modeling methods
– acoustic field mathematical modeling   160
– finite difference time domain modeling   227, 228, 320
– finite element modeling   198, 242, 314, 315
– five layer wave propagation   158–162
– high frequency acoustic microscopy   177–181
– melanoma skin tissues   158–162
– time-of-flight mathematical modeling   159, 171–173, 177, 217
confocal reflection amplitude   210, 212, 216, 218, 219, 227
contact force   316
contact inspection geometry   237
contact radius a_c 324, 325
contact stiffness
– atomic force acoustic microscopy   348–362
– near-field imaging   340
– ultrasonic atomic force microscopy   309
– ultrasonic force microscopy   282, 291–293
continuous wave systems   341, 342
continuum theory   313, 314, 336
contrast mechanism
– melanoma skin tissues   155–183
– – aims of study   157, 158
– – clinical aspects   155–157
– – digital imaging   163–173
– – high frequency acoustic microscopy   174–183
– – mathematical modeling   158–162
– – sample preparation   162, 163
– near-field imaging   339
– ultrasonic atomic force microscopy   316
– ultrasonic force microscopy   287–291
coronary arteries   197–199
cortex–callus interface   225, 226
cortical bone   209, 211
cost of combined imaging methods   4
coupling media   200
– see also water

cracks
– closed   317, 319, 320
– photoacoustic imaging system near-field effects   342
– sizing, post-processing method   267–269
– ultrasonic atomic force microscopy   307–336
– – subsurface imaging   314–316
critical angle (θ_R)   128
cystic kidney   197

d

D waves   144–146
Damascene process   294, 295
damping   346, 347, 352, 354–356
decalcification of bone   193
decomposition of the time reversal operator (DORT)   250
defect characterization
– see also cracks; subsurface defects
– array imaging   233–273
– delamination   303, 311–320, 334, 335
– dislocations   325–335
– post-processing methods   234
– subsurface defects   301–303, 311–320, 359–362
– ultrasonic atomic force microscopy   307–336
degree of mineralization of bone (DMB)   193, 220, 221, 224
delamination   303, 311–320, 334, 335
delay and sum (DAS) reconstruction   107
delta function (δ)   77
demodulation effect   323
dental caries   192
destructive measurement principle   209
deviation/absolute difference   36
dialyzed kidneys   197
differential UFM approach   292, 293
diffuse backscattering   113, 114
diffusive regimes   3, 4
digital imaging   163–173
– acoustic images   169–171
– optical   163, 164
– pulse-wave model   164–167
– resolution   168
– waveform analysis   171–173
digital storage oscilloscope (DSO)   100
Dirac's delta function (δ)   77, 128
directional control   308, 309
directivity function, water-fused   133
dislocations   325–335
dispersion curves   348, 349
Doppler imaging   24

double nodes   319, 320
Duffing model   318, 319
duplexers   95, 96

e
Earth's atmosphere   24–31, 33, 34
edge dislocations   327, 333, 334
eigenvalue approaches   250
elastic anisotropy   173, 222–225
elastic bulk modulus (K)   49, 50, 188, 195–197
elastic coefficient (C)   210
elastic properties
– see also Young's modulus (E)
– atomic force acoustic microscopy   350
– bone   209, 210, 219–222
– focal point pushing force   11, 12
– mapping   8–14
– multiwave imaging   8–14
– principle   11–13
– quantitative ultrasonic microscopy   49–70
– skin, strain imaging   117
– speckle noise   11
– ultrasonic force microscopy   291–293
elastodynamic wave equation   13
electromechanical reciprocity argument of Auld   244
endothelial lung arterial cells   191
equation of motion   346
errors calculation   129–135
esophagus   60, 61
excisional biopsy   156, 157
excitation power   319
external granular layers (EGL)   66, 67
extra-half planes (EHPs)   333
– see also Young's modulus (E)
– atomic force acoustic microscopy   350
– bone   209, 210, 219–222
– focal point pushing force   11, 12
– mapping   8–14
– multiwave imaging   8–14
– principle   11–13
– quantitative ultrasonic microscopy   49–70
– skin, strain imaging   117
– speckle noise   11
– ultrasonic force microscopy   291–293
elastodynamic wave equation   13
electromechanical reciprocity argument of Auld   244
endothelial lung arterial cells   191
equation of motion   346
errors calculation   129–135
esophagus   60, 61
excisional biopsy   156, 157
excitation power   319
external granular layers (EGL)   66, 67
extra-half planes (EHPs)   333

f
fast Fourier transform   159, 171–173, 177, 217
fast time windows   86, 87
finite difference time domain (FDTD) modeling   228, 320
finite element modeling (FEM)   198, 199, 242, 314, 315
finite-width-layer model   314–316
five layer wave propagation   158–162
flowing media
– Doppler imaging   24
– general system   35
– nonlinear imaging methods   24, 34–44
– static regions   35
– three-dimensional imaging   43, 44
focused ultrasound beams   11, 12, 215
focusing coefficients   250
focusing performance   252, 253
Fokker bond tester   339, 340
force balance equation   289
force modulation technique   343, 344
force-versus-separation dependence ($F(z)$)   289
forced vibration   307, 308
forward modeling problem   237
Fourier space/transform
– high-resolution ranging/imaging   98
– image improvement   23, 25, 26, 29, 36–38
– – Hilbert transform based methodology   39–41, 42, 43
– – Knox–Thompson method   31–33
– quantitative acoustic microscopy   128
– sound speed profile   53, 54, 59
– time-of-flight mathematical modeling   159, 171–173, 177, 217
– two-dimensional arrays layout optimization   256
– wavenumber 1D array method   252
fractional errors   133–135
fracture healing   225–227
frequency domain analysis   52–54, 56, 57
full matrix capture (FMC)   235
full width at half maximum (FWHM)   260

g
gallium antimonide-indium arsenide superlattices   292–294
gallium arsenide matrix   302

gastric cancer   193–195
Gauss's theorem   247
germanium/silicon dots   299
glass   51, 52, 101, 188, 189, 191, 324
gold   310, 361
granule cells   66
graphene   317
graphite   324, 325
graphite intercalation compounds (GICs)   325
Green's function   77, 244

h
hard tissues   191–193
– *see also* bone
Haversian canals   209
healing, bone   225–227
heart   117, 190, 195–197
Helmholtz's theorem   247
Hertzian contact zone   303, 360, 361
heterodyne force microscopy (HFM)   304–306
high-frequency acoustic microscopy
– contrast mechanism
– – abnormal skin   175–183
– – acoustic velocity   175–177
– – computer simulation   177–181
– – normal skin   174, 175
high-frequency ultrasound (HFUS)
– biomedical applications   115–118
– engineering concepts   104–115
– system components   94–104
– – acoustic scattering parameter   102–104
– – echo systems   94, 95
– – inverse echo signal filtering   101, 102
– – plane wave propagation   102–104
– – pulse transfer properties   99, 100
– – spectral/range resolution   97–99
– – transmitters/receivers   95–97
highly ordered pyrolytic graphite (HOPG)   324, 326–328, 329–333
Hilbert transform   39–44, 78
history   25, 210–213, 339–342
hole measurement   74, 75, 80–91, 253–255
Huygens' principle   242

i
image quality
– nonlinear methods   24, 34–44
– quantitative ultrasonic microscope   58–60
– speckle interferometry   24–34
– wave penetration   158
imaginary part of AFAM contact stiffness   352, 354–356

implant materials   192
in vivo intravascular ultrasound (IVUS)   200
incisional melanoma biopsy   156, 157
indentation depth   283, 284
indentation modulus M 350–362
– contact stiffness   352, 354–356
– nonlinear contact resonances   358, 359
– piezo-mode imaging   357, 358
– relationship to contact stiffness   350–362
– subsurface imaging using contact-resonances   359–362
– torsional resonances   356
indium antimonide/indium arsenide quantum dot superlattice   303
indium arsenide   292–294, 303
inertia effect   309, 312
infinite-width-layer models   314–316
integral transform method   242–245
internal granular layer (IGL)   66, 67
inverse echo signal filtering   101, 102
inverse modeling   235, 237
inverse wave field extrapolation (IWEX) method   251
inversion algorithms   5, 13, 15
iron-based superalloys   342
isotropic substrates
– sapphire   162, 165–167, 172
– water-fused quartz   131–134, 138–140, 146, 149

j
Johnson–Kendall–Roberts (JKR) model   289, 290

k
kidney   197

l
L (longitudinal) waves   144, 145, 146
Labeyrie's method   25–29, 30, 31, 33–34
Lame elastic constant K_s 243
Langmuir–Blodgett films   297–298
lateral resolution ($\delta_{lateral}$)   98, 99, 105–110, 168
lateral stiffness   308
layered materials   102, 103, 104, 314–316, 325–335
leaky surface acoustic waves (LSAWs)   126–141
least-squares method   130, 145
length scales   207, 208
leukemia cells   190
limited angle spatial compounding (LASC)   110–112

linear analysis   312–316, 325
linear detection schemes   320
linear regression equation   129, 130, 131, 145
linear ultrasonic arrays   125–150
liquid-like media   188
liver cancer   17, 18
loads   316, 329–333
local acceleration microscopy   344–345
local damping   352, 354–356
lock-in phase   285
longitudinal mode ray-paths   239, 240, 243, 245
longitudinal waves   178, 179
lower frequency ultrasound   187
lung   191

m

Mach cone   12, 13, 14
magnetic induction magneto-acoustic tomography (MAT-MI)   6
magnetic resonance (MR)   8
magnetoacoustic tomography (MAT)   6
Mason horn   341
mass density (ρ)   221
mathematical modeling *see* computer modeling methods
matrix arrays   75, 76, 77, 78, 79
matrix notations   214
mechanical diodes   283–284, 289, 323
mechanical properties
– pathological specimens   187–203
– – articular tissues   202, 203
– – cellular imaging   189–191
– – hard tissues   191–193
– – principle of acoustic microscopy   188–189
– – soft tissues   193–200
– – ultrasound speed microscopy   200, 201
melanoma skin tissues
– anatomy   155–156
– biopsy diagnosis   156–157
– contrast mechanism   155–183
– digital optical/ultrasonic imaging   163–173
– mathematical modeling   158–162
metallic glasses   349, 350
Michelson stellar interferometry   25
micro-mechanical systems (MEMS)   299
microtubules   191
microwaves, thermoacoustic imaging   6
moles (melanocytic nevi)   155, 156
molybdenum disulfide   328–330
motion theory   8, 329–334, 346

movies, multiwave imaging   5, 8–14, 15
mucosal tissue, gastric   193–195
mullite matrix   288
multi-eyed acoustic microscopes   72–75
multidirectional imaging   110–115
multimodal imaging methods, costs   4
multiscale elastic properties   209–210
multiwave imaging   3–20
– clinical applications   16–19
– concept   4–5
– inversion algorithms   5, 15
– spatial resolution regimes   3–4
– super-resolution in supersonic shear wave imaging   14–15
– three different wave interactions   5
– wave to wave generation   5–7
– wave to wave imaging, elasticity mapping   8–14
– wave to wave tagging   5, 7–8
muscle pathologies   19, 20
MUSIC eigenvalue approach   250
myocardial infarction   195–197

n

nano-electromechanical systems (NEMS)   299
nanocomposites   301–303
nanoindentation   209
nanoscale time-resolved phenomena   303–307
natural body waves   19
near-field imaging   339–362
– atomic force microscopy   342–362
– – acoustical imaging   345–362
– – force modulation   343, 344
– – local acceleration microscopy   344, 345
– – piezo-mode imaging   357, 358
– – pulsed-force microscopy   345
– – torsional resonance   356, 357
– early systems   339–342
– spatial resolution   3, 4
network analyzer (NWA)   100
non-uniform sampling   257, 358
noncontact AFM (NC-AFM)   321, 322
nonlinear contact resonances   358, 359
nonlinear detection schemes   320, 321
nonlinear imaging (NLI) methods   24, 34–44
– deviation/absolute difference   36
– Fourier transform   36–39
– Hilbert transform   39–41, 42–44
– post-processing approaches   250
nonlinear response frequency tuning   286, 289

nonlinear spectral analysis 316–319, 325
nonlinear tip–sample contact 323, 324
normal skin tissues 174–175, 180, 182, 183
normalized step response ($u(x)$) 78, 79
numerical models 319, 320
Nyquist's criteria/limit 77, 218, 254

o

one-dimensional (1D) arrays 233, 234, 245–256
one-dimensional (1D) spatial coordinates 102–103
optical heterodyne force microscopy (OHFM) 304–306
optical imaging
– acute myocardial infarction 196
– carotid artery 199
– chicken heart muscle cells 190
– coronary artery 197–200
– embryonic chicken heart muscle cells 190
– normal/abnormal skin 163–164, 174, 175, 176
– papillary adenocarcinoma 193
optical waves, thermoacoustic imaging 6
organelles 191
osteonal lamellae 209, 219, 222–223
output voltage $V(z)$ technique 125, 192, 212

p

papillary adenocarcinoma 193
parallel fibers 66
pathological specimens
– mechanical properties 187–203
– – articular tissues 201–203
– – cellular imaging 189–191
– – hard tissues 191–193
– – principle of acoustic microscopy 188, 189
– – soft tissues 193–200
– – ultrasound speed microscopy 200, 201
periodic sampling 256
phantom measurements 111, 112, 114, 116
phase information 27, 29, 32–34, 39–41, 53, 54
phase-locked loop (PLL) circuit 320, 321
phased array systems 75–91
photoacoustic imaging 5, 6, 342, 343
photodiodes 320, 321, 323
photons, tagged 7, 8
piezo-mode imaging 357, 358
piston transducers 96, 97
plane wave propagation 102–104

plantar flexors 19, 20
plastic deformation 309–311, 316
point spread function (PSF) 98, 101, 108, 109
Poisson disk sampling 257, 258
Poisson's ratio (σ)
– atomic force acoustic microscopy 348, 350
– bone elastic coefficient 225
– solid materials 188
– UAFM subsurface imaging linear theory 314
– ultrasonic force microscopy 290
poly(ethylene terephthalate) (PET) substrate 301
polymer films 357, 361
polymer substrates 62, 64
poly(methyl methacrylate) (PMMA) 294–296
polysilicone covers 162, 165–167, 172
polystyrene 138–141, 299, 300
poly(vinylidene fluoride) film (PVDF) 135
portable scanning devices 71–91
– aerospace industry 71
– B-scans 73, 80, 81, 89
– biomedical applications 71, 72
– fast/slow time windows 86, 87
– flat reflector boundaries 78–86
– hole measurement 74, 75, 80–91
– matrix arrays 75–79
– multi-eyed acoustic microscopes 72–75
– phased array systems 75–91
– slit widths 82, 84, 86
– spot welds 74–76
– step response 78–80
post-processing approaches 233–273
– advantages 235
– array data modeling 237–245
– defect characterization/sizing 267–272
– imaging with 1D arrays 245–255
– imaging with 2D arrays 255–260
– scattering matrices experimental extraction 260–266
power industry 234
printed circuit boards (PCBs) 110
propagation distances 145, 148
pulse transfer properties 99, 100
pulse-wave models 164–167
pulsed-force microscopy 345
pulser/receiver apparatus 99, 100
pulsers 96
Purkinje layer (PL) 66, 67
pushing forces 11, 12

q

Q-value/factor
- atomic force acoustic microscopy 351, 354–356
- ultrasonic atomic force microscopy 308, 309, 312–314, 321, 323, 336

quantitative acoustic microscopy
- applications 126
- array approach 125–150
- bulk wave velocity 141–150
- errors calculation 129–135
- leaky wave velocity 126–141
- output voltage 125, 212
- specimen thickness 141–150
- spherical-planar-pair lenses 125
- velocity/attenuation of leaky waves 126–141

quantitative acoustic scanning microscopy
- bone 207–228
- – elastic anisotropy
- – – lamellar level 222, 223
- – – tissue level 223–225
- – hierarchical structure 207–209
- – history of measurement principles 210–213
- – multiscale elastic properties 209, 210
- – quantitative SAM-based impedance 213–220
- – research 225, 226
- – time-gated amplitude detection 217–219
- – time-resolved measurements 216, 217
- – tissue mineralization/acoustic impedance/stiffness 219–222

quantitative elasticity evaluation 324, 325
quantitative ultrasonic force microscopy 291–293
quantitative ultrasonic microscopy 49–70
- acoustic impedance profile 60–70
- – acquired signal 63, 64
- – calibration for CAI 63–65
- – call size observation 68, 69
- – commercialized equipment 69, 70
- – rat cerebellar cortex 65–69
- – specimen 62, 63
- basics 50
- biomechanics 200–202
- sound speed profile 50–60, 61
- – calibration 52–56
- – experimental setup/acquired signal 51, 52
- – frequency domain analysis 52–58
- – spatial resolution improvement 58–61
- – specimen 50, 51
- – time–frequency domain analysis 53–57
- tissue mechanical properties 200–202

quantum dots 292, 299, 301–303
quartz 131–134, 138–140, 146, 149

r

radius bone 227
range resolution 97–99, 101, 102
ranging, high-resolution, see high-frequency ultrasound (HFUS)
rat cerebellar cortex 65–69
ray-paths, post-processing 237–245
Rayleigh waves 125–141, 212
receiver components 95–97
reference materials 62–65
reflection, post-processing 239
reflection ratio 49
renal cell carcinoma 197, 198
resolution 168, 281, 339
resonance frequency tracking
- ultrasonic atomic force microscopy
- – closed cracks 320, 321
- – dislocation motion analysis 331
- – elasticity quantitative evaluation 324, 325
- – finite/infinite-width-layer models 315, 316
- – principles 308, 309

resonance ultrasound microscopy (RUM) 345, 355
reversible long-range dislocation behavior 332–334

s

Saint-Venant's principle 303
saline coupling medium 200, 201
samples
- see also tip–sample interaction
- biopsy tissue deterioration 157
- holders for ultrasonic force microscopy 286
- normal/abnormal skin 162, 163
- slicing damage 61
- thickness 125–150, 193, 194
- ultrasonic sound speed microscopy 50, 51, 62, 63

sapphire 162, 165–167, 172
saw-tooth waveforms 285, 287
scanning acoustic microscopy (SAM) 109
- advantages for histopathology 187, 188
- biomechanics 187–203
- bone 210, 213–219, 220

– contrast mechanism 155–183
– near-field imaging comparison 341
– resolution 168
scanning force microscopy (SFM) 282–284
scanning local-acceleration microscopy (SLAM) 282, 343–345
scanning near-field ultrasonic holography (SNFUH) 360
scanning pin microscope 281, 282
scanning tunneling microscopy (STM) 281, 294, 306, 342
scattered signals
– Labeyrie's method 25–29, 30, 31, 33, 34
– scattering matrices 236, 262–266
– small/large scatterers 239, 241, 243, 244, 247, 248
sensitivity varying method 195
shear Lame elastic constant K_s 243
shear modulus (μ) 9, 11
side looking airborne radar (SLAR) systems 109
side-drilled holes 246, 247
signal prediction 240
signal-to-noise ratio 133, 149, 150, 358
silicon carbide ceramic fiber 288
silicon trinitride 359
silicon UAFM tip wear 310, 311
single-element transducers 96–98, 105, 106
– arrays comparison 233
– B-scan techniques 104, 105
– limited angle spatial compounding 110, 111
sizing of defects 267–272
skin
– acoustic imaging 169–171, 175–183
– – biopsy problems 157
– – carcinoma 175, 176
– – contrast mechanism 155–183
abnormal skin 175–183
– – melanoma 155–157
– – normal skin 174, 175, 180, 182, 183
– – sample preparation 162, 163
– – waveform analysis 171–173
– *in vivo* high-frequency ultrasound imaging 112, 114–117
– optical imaging 163, 164
slit widths 82, 84, 86
slope of UAFM cantilever 312
slow time window 86, 87
slowness parameter 142–144
small animal studies 65–69, 117, 118
Sneddon-Maugis formulation 308, 336
Snell's law 142

soft tissues 193–200
softening nonlinear spring 317–319, 336
solid materials acoustical relation 188
sonic shear waves 5, 8–15
sono-elastography concept 10
sound speed
– biomedical images 194–196, 198, 199
– calibration 52–56
– profile 50–61
– ultrasound speed microscopy 201–203
spatial mapping 216
spatial resolution 3, 4, 78–85, 311
specific gravity (ρ) 49
speckle interferometry 23–34
speckle noise 11
spectral resolution 97–99
specular reflections 113, 114
speed of array scanning 235, 236
spherical inclusions 259, 260
spherical-planar-pair lenses 125
spherically-focused sound fields 215
spherically-focused transducers 96–98, 105–107, 109
spot welds 74–76
spring constant k^* 346, 347
square pulse generation 136
static regions of flowing media 35
stationary phase method 243, 245
statistical calculations 23, 113, 114, 149, 348, 349
– *see also* computer modeling methods
steel
– array imaging 258, 259, 260, 262
– water–steel interfaces 131–134, 138–140
– welds 234
step response 78–80
stiffening nonlinear spring 317–319, 336
stiffness *see* Young's modulus (*E*)
stiffness coefficients (GPa) 224
stomach 60, 61, 194, 195, 200, 201
strain imaging 117
substrates
– glass 188, 189, 191
– poly(ethylene terephthalate) 301
– polymer substrates 62, 64
– sapphire substrates 162, 165–167, 172
– water-fused quartz 131–134, 138–140, 146, 149
subsurface defects 301–303, 311–320, 359–361
subsurface pull-off 317–336
subsurface tapping mode 317–336
subtraction of SAM (SubSAM) 189, 190

super-resolution   14, 15
– *see also* near-field imaging
superlubricity   299–301
superposition principle   35
supersonic shear wave imaging (SSI)   5, 8–15
surface acoustic waves (SAW)   177, 178, 210–217
surface adhesive properties   291–293, 297–299, 339, 340
surface roughness   351, 352
surface-mounted devices (SMD)   110
symmetry centers   27
synchrotron radiation micro-computed tomography (SRμCT)   220
synthetic aperture focusing techniques (SAFT)   106–110, 236
synthetic aperture radar (SAR)   106

t
tagging   5, 7, 8
tendons   193
thermoacoustic imaging   5, 6
thickness of tissue sections   155–183, 200–211
three-dimensional (3D) images
– *see also* two-dimensional (2D) arrays
– Hilbert methodology   43, 44
threshold power P_D 323
thyroid nodules   17–19
time constant   285
time delays   143, 144, 147–149
time domain reflectometry (TDR)   100
time summation equation   42
time-frequency domain analysis   53–57
time-gated amplitude detection   217–219
time-of-flight (TOF)
– mathematical modeling   158–162
– – fast Fourier transform   217, 222
– ranging/imaging   93–95
– sound velocity   210
time-resolution   190, 216–217, 303–307
tip–sample interaction   283, 284, 299–301, 322–324
tissue characterization   112–115
tissue mimicking phantom   14, 15
tissue thickness (*d*)   200
tooth enamel   192
topography atomic force microscopy   287, 288
torsional resonance   356
total focusing method (TFM)   235, 238, 246–247, 260
trabecular bone   209

transducers   96–98, 159
transient elastography   5, 10
transmittance function   161
transmitter components   95–97
transverse isotropic model   222
triglycine sulfate   358
triple-correction techniques   34
tuberculosis   202, 201
tubular adenocarcinoma   193–194, 201
tuning fork as tip   341
two-dimensional (2D) arrays   233, 234
– aluminum block example   260, 261, 267–268
– experimental comparison   258–260
– non-uniform sampling   257–258
– optimization of array layout   255–258
– periodic sampling   256
– spherical inclusions   259–260
two-dimensional (2D) images
– *see also* one-dimensional (1D) arrays
– confocal reflection amplitude   227
– high frequency ultrasonic   104–115
– sound speed profiles   56–58
– speckle interferometry   33–34
– three-dimensional image creation   42
two-dimensional (2D) matrix array transducers   75–91

u
ultrafast scanners   10, 12
ultrasonic atomic force microscopy (UAFM)   307–336
– cantilever stiffness enhancement   309
– defects observation in layered materials   326–335
– – delamination in microelectronic/mechanical devices   334–335
– – dislocation behavior versus loads   329–330
– – dislocation motion under varying loads   331–333
– – graphene sheets   325–328
– – molybdenum disulfide   328–330
– – reversible long-range motion of dislocation   333–334
– directional control   308–309
– forced vibration of cantilever from base   307–308
– instrumentation   320–322
– nonlinearity avoidance at tip–sample contact   322–323
– plastic deformation avoidance   309–311
– principle of action   309
– Q factor   308–309, 312–314, 321, 323, 336

- quantitative information   308–309, 324–325
- resonance frequency tapping   308–309
- sample bonding non-requirement   308
- theory   311–320
- – appropriate load advantage   316
- – Duffing model   318–319
- – numerical model with double nodes   319–320
- – Q factor   312–314
- – spectral nonlinear analysis   316–317
- – stiffness linear analysis   312–314
- – subsurface imaging linear theory   314–316
- UAFM/UFM relation   323–324

ultrasonic force microscopy (UFM)   281–307
- adhesion effects on image   297–299
- contact stiffness quantitative measurement   291–293
- contrast theory   287–291
- experimental implementation   284–288
- hole measurement example   253–255
- image interpretation   297–299
- key components   289
- mechanical diode detection   283–284, 289
- picture gallery   292–296
- schematic diagram   285
- subsurface defects   301–303
- superlubricity   299–301
- time-resolved nanoscale phenomena   303–307
- topography effects on image   297–299

ultrasound speed microscopy *see* quantitative ultrasonic microscopy

v

vector field f 247–249
velocity *see* acoustic velocity

vertical resolution (Δr)   168
viscosity, acoustic derivation   188
Volkmann canals   209
von Mises stress σ_v 310, 311
$V(x)$ scheme   125
$V(z)$ method   125, 177–178, 192, 212

w

water
- skin high frequency ultrasound imaging   116
- sound propagation   105
- specimen immersion   125–150, 165, 166, 171

water-fused quartz   131–134, 138–140, 146, 149
wave interactions   5
wave penetration   158
wave propagation   160
wave to wave generation   5–7
wave to wave imaging   8–14
waveform analysis   171–173
waveguide UFM (W-UFM)   303
wavelength   3, 4
wavenumber   1D array method   247–249

x

X-scans   200

y

yield stress σ_v 310, 311
Young's modulus (E)
- atomic force acoustic microscopy   348, 351
- melanoma skin tissues   155, 181, 183
- solid materials   188
- ultrasonic atomic force microscopy   312–314, 324–325
- ultrasonic force microscopy   291–293
- wave to wave imaging   8–14